Introduction To
Stochastic Calculus
With Applications

SECOND EDITION

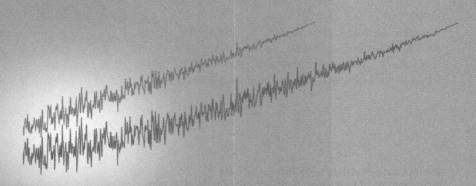

Introduction To

Stochastic Calculus

With Applications

Second Edition

Fima C Klebaner
Monash University, Australia

Imperial College Press

Published by

Imperial College Press
57 Shelton Street
Covent Garden
London WC2H 9HE

Distributed by

World Scientific Publishing Co. Pte. Ltd.
5 Toh Tuck Link, Singapore 596224
USA office: 27 Warren Street, Suite 401-402, Hackensack, NJ 07601
UK office: 57 Shelton Street, Covent Garden, London WC2H 9HE

British Library Cataloguing-in-Publication Data
A catalogue record for this book is available from the British Library.

ISBN-13 978-1-86094-555-7
ISBN-10 1-86094-555-4
ISBN-13 978-1-86094-566-3 (pbk)
ISBN-10 1-86094-566-X (pbk)

Printed in Singapore

Preface

Preface to the Second Edition

The second edition is revised, expanded and enhanced. This is now a more complete text in Stochastic Calculus, from both a theoretical and an applications point of view. Changes came about, as a result of using this book for teaching courses in Stochastic Calculus and Financial Mathematics over a number of years. Many topics are expanded with more worked out examples and exercises. Solutions to selected exercises are included. A new chapter on bonds and interest rates contains derivations of the main pricing models, including currently used market models (BGM). The change of numeraire technique is demonstrated on interest rate, currency and exotic options. The presentation of Applications in Finance is now more comprehensive and self-contained. The models in Biology introduced in the new edition include the age-dependent branching process and a stochastic model for competition of species. These Markov processes are treated by Stochastic Calculus techniques using some new representations, such as a relation between Poisson and Birth-Death processes. The mathematical theory of filtering is based on the methods of Stochastic Calculus. In the new edition, we derive stochastic equations for a non-linear filter first and obtain the Kalman-Bucy filter as a corollary. Models arising in applications are treated rigorously demonstrating how to apply theoretical results to particular models. This approach might not make certain places easy reading, however, by using this book, the reader will accomplish a working knowledge of Stochastic Calculus.

Preface to the First Edition

This book aims at providing a concise presentation of Stochastic Calculus with some of its applications in Finance, Engineering and Science.

During the past twenty years, there has been an increasing demand for tools and methods of Stochastic Calculus in various disciplines. One of the greatest demands has come from the growing area of Mathematical Finance, where Stochastic Calculus is used for pricing and hedging of financial derivatives, such as options. In Engineering, Stochastic Calculus is used in filtering and control theory. In Physics, Stochastic Calculus is used to study the effects of random excitations on various physical phenomena. In Biology, Stochastic Calculus is used to model the effects of stochastic variability in reproduction and environment on populations.

From an applied perspective, Stochastic Calculus can be loosely described as a field of Mathematics, that is concerned with infinitesimal calculus on non-differentiable functions. The need for this calculus comes from the necessity to include unpredictable factors into modelling. This is where probability comes in and the result is a calculus for random functions or stochastic processes.

This is a mathematical text, that builds on theory of functions and probability and develops the martingale theory, which is highly technical. This text is aimed at gradually taking the reader from a fairly low technical level to a sophisticated one. This is achieved by making use of many solved examples. Every effort has been made to keep presentation as simple as possible, while mathematically rigorous. Simple proofs are presented, but more technical proofs are left out and replaced by heuristic arguments with references to other more complete texts. This allows the reader to arrive at advanced results sooner. These results are required in applications. For example, the change of measure technique is needed in options pricing; calculations of conditional expectations with respect to a new filtration is needed in filtering. It turns out that completely unrelated applied problems have their solutions rooted in the same mathematical result. For example, the problem of pricing an option and the problem of optimal filtering of a noisy signal, both rely on the martingale representation property of Brownian motion.

This text presumes less initial knowledge than most texts on the subject (Métivier (1982), Dellacherie and Meyer (1982), Protter (1992), Liptser and Shiryayev (1989), Jacod and Shiryayev (1987), Karatzas and Shreve (1988), Stroock and Varadhan (1979), Revuz and Yor (1991), Rogers and Williams (1990)), however it still presents a fairly complete and mathematically rigorous treatment of Stochastic Calculus for both continuous processes and processes with jumps.

A brief description of the contents follows (for more details see the Table of Contents). The first two chapters describe the basic results in Calculus and Probability needed for further development. These chapters have examples but no exercises. Some more technical results in these chapters may be skipped and referred to later when needed.

In Chapter 3, the two main stochastic processes used in Stochastic Calculus are given: Brownian motion (for calculus of continuous processes) and Poisson process (for calculus of processes with jumps). Integration with respect to Brownian motion and closely related processes (Itô processes) is introduced in Chapter 4. It allows one to define a stochastic differential equation. Such equations arise in applications when random noise is introduced into ordinary differential equations. Stochastic differential equations are treated in Chapter 5. Diffusion processes arise as solutions to stochastic differential equations, they are presented in Chapter 6. As the name suggests, diffusions describe a real physical phenomenon, and are met in many real life applications. Chapter 7 contains information about martingales, examples of which are provided by Itô processes and compensated Poisson processes, introduced in earlier chapters. The martingale theory provides the main tools of stochastic calculus. These include optional stopping, localization and martingale representations. These are abstract concepts, but they arise in applied problems, where their use is demonstrated. Chapter 8 gives a brief account of calculus for most general processes, called semimartingales. Basic results include Itô's formula and stochastic exponential. The reader has already met these concepts in Brownian motion calculus given in Chapter 4. Chapter 9 treats Pure Jump processes, where they are analyzed by using compensators. The change of measure is given in Chapter 10. This topic is important in options pricing, and for inference for stochastic processes. Chapters 11-14 are devoted to applications of Stochastic Calculus. Applications in Finance are given in Chapters 11 and 12, stocks and currency options (Chapter 11); bonds, interest rates and their options (Chapter 12). Applications in Biology are given in Chapter 13. They include diffusion models, Birth-Death processes, age-dependent (Bellman-Harris) branching processes, and a stochastic version of the Lotka-Volterra model for competition of species. Chapter 14 gives applications in Engineering and Physics. Equations for a non-linear filter are derived, and applied to obtain the Kalman-Bucy filter. Random perturbations to two-dimensional differential equations are given as an application in Physics. Exercises are placed at the end of each chapter.

This text can be used for a variety of courses in Stochastic Calculus and Financial Mathematics. The application to Finance is extensive enough to use it for a course in Mathematical Finance and for self study. This text is suitable for advanced undergraduate students, graduate students as well as research workers and practitioners.

Acknowledgments

Thanks to Robert Liptser and Kais Hamza who provided most valuable comments. Thanks to the Editor Lenore Betts for proofreading the 2nd edition. The remaining errors are my own. Thanks to my colleagues and students from universities and banks. Thanks to my family for being supportive and understanding.

Fima C. Klebaner
Monash University
Melbourne, 2004.

Contents

Chapter 1

Preliminaries From Calculus

Stochastic calculus deals with functions of time t, $0 \leq t \leq T$. In this chapter some concepts of the infinitesimal calculus used in the sequel are given.

1.1 Functions in Calculus

Continuous and Differentiable Functions

A function g is called continuous at the point $t = t_0$ if the increment of g over small intervals is small,

$$\Delta g(t) = g(t) - g(t_0) \to 0 \text{ as } \Delta t = t - t_0 \to 0.$$

If g is continuous at every point of its domain of definition, it is simply called continuous.

g is called differentiable at the point $t = t_0$ if at that point

$$\Delta g \sim C \Delta t \quad \text{or} \quad \lim_{\Delta t \to 0} \frac{\Delta g(t)}{\Delta t} = C,$$

this constant C is denoted by $g'(t_0)$. If g is differentiable at every point of its domain, it is called differentiable.

An important application of the derivative is a theorem on finite increments.

Theorem 1.1 (Mean Value Theorem) *If f is continuous on $[a, b]$ and has a derivative on (a, b), then there is c, $a < c < b$, such that*

$$f(b) - f(a) = f'(c)(b - a). \tag{1.1}$$

1

Clearly, differentiability implies continuity, but not the other way around, as continuity states that the increment Δg converges to zero together with Δt, whereas differentiability states that this convergence is at the same rate or faster.

Example 1.1: The function $g(t) = \sqrt{t}$ is not differentiable at 0, as at this point

$$\frac{\Delta g}{\Delta t} = \frac{\sqrt{\Delta t}}{\Delta t} = \frac{1}{\sqrt{\Delta t}} \to \infty$$

as $t \to 0$.

It is surprisingly difficult to construct an example of a continuous function which is not differentiable at *any* point.

Example 1.2:   An example of a continuous, nowhere differentiable function was given by the Weierstrass in 1872: for $0 \le t \le 2\pi$

$$f(t) = \sum_{n=1}^{\infty} \frac{\cos(3^n t)}{2^n}. \tag{1.2}$$

We don't give a proof of these properties, a justification for continuity is given by the fact that if a sequence of continuous functions converges uniformly, then the limit is continuous; and a justification for non-differentiability can be provided in some sense by differentiating term by term, which results in a divergent series.

To save repetition the following notations are used: *a continuous function f is said to be a C function; a differentiable function f with continuous derivative is said to be a C^1 function; a twice differentiable function f with continuous second derivative is said to be a C^2 function; etc.*

Right and Left-Continuous Functions

We can rephrase the definition of a continuous function: a function g is called continuous at the point $t = t_0$ if

$$\lim_{t \to t_0} g(t) = g(t_0), \tag{1.3}$$

it is called right-continuous (left-continuous) at t_0 if the values of the function $g(t)$ approach $g(t_0)$ when t approaches t_0 from the right (left)

$$\lim_{t \downarrow t_0} g(t) = g(t_0), \quad (\lim_{t \uparrow t_0} g(t) = g(t_0).) \tag{1.4}$$

If g is continuous it is, clearly, both right and left-continuous.

The left-continuous version of g, denoted by $g(t-)$, is defined by taking left limit at each point,

$$g(t-) = \lim_{s \uparrow t} g(s). \tag{1.5}$$

From the definitions we have: g is left-continuous if $g(t) = g(t-)$.
The concept of $g(t+)$ is defined similarly,

$$g(t+) = \lim_{s\downarrow t} g(s). \qquad (1.6)$$

If g is a right-continuous function then $g(t+) = g(t)$ for any t, so that $g_+ = g$.

Definition 1.2 *A point t is called a discontinuity of the first kind or a jump point if both limits $g(t+)$ and $g(t-)$ exist and are not equal. The jump at t is defined as $\Delta g(t) = g(t+) - g(t-)$. Any other discontinuity is said to be of the second kind.*

Example 1.3: The function $\sin(1/t)$ for $t \neq 0$ and 0 for $t = 0$ has discontinuity of the second kind at zero, because the limits from the right or the left don't exist.

An important result is that a function can have at most countably many jump discontinuities (see for example Hobson (1921), p.286).

Theorem 1.3 *A function defined on an interval $[a, b]$ can have no more than countably many jumps.*

A function, of course, can have more than countably many discontinuities, but then they are not all jumps, i.e. would not have limits from right or left.
Another useful result is that a derivative cannot have jump discontinuities at all.

Theorem 1.4 *If f is differentiable with a finite derivative $f'(t)$ in an interval, then at all points $f'(t)$ is either continuous or has a discontinuity of the second kind.*

PROOF: If t is such that $f'(t+) = \lim_{s\downarrow t} f'(s)$ exists (finite or infinite), then by the Mean Value Theorem the same value is taken by the derivative from the right

$$f'(t) = \lim_{\Delta t\downarrow 0} \frac{f(t + \Delta) - f(t)}{\Delta} = \lim_{\Delta\downarrow 0, 0<c<\Delta} f'(c) = f'(t+).$$

Similarly for the derivative from the left, $f'(t) = f'(t-)$. Hence $f'(t)$ is continuous at t. The result follows.

$\square$

This result explains why functions with continuous derivatives are sought as solutions to ordinary differential equations.

Functions considered in Stochastic Calculus

Functions considered in stochastic calculus are functions without discontinuities of the second kind, that is functions that have both right and left limits at any point of the domain and have one-sided limits at the boundary. These functions are called *regular* functions. It is often agreed to identify functions if they have the same right and left limits at any point.

The class $D = D[0, T]$ of right-continuous functions on $[0, T]$ with left limits has a special name, *càdlàg* functions (which is the abbreviation of "right continuous with left limits" in French). Sometimes these processes are called R.R.C. for regular right continuous. Notice that this class of processes includes C, the class of continuous functions.

Let $g \in D$ be a càdlàg function, then by definition, all the discontinuities of g are jumps. By Theorem 1.3 such functions have no more than countably many discontinuities.

Remark 1.1: In stochastic calculus $\Delta g(t)$ usually stands for the size of the jump at t. In standard calculus $\Delta g(t)$ usually stands for the increment of g over $[t, t + \Delta]$, $\Delta g(t) = g(t + \Delta) - g(t)$. The meaning of $\Delta g(t)$ will be clear from the context.

1.2   Variation of a Function

If g is a function of real variable, its variation over the interval $[a, b]$ is defined as

$$V_g([a, b]) = \sup \sum_{i=1}^{n} |g(t_i^n) - g(t_{i-1}^n)|, \tag{1.7}$$

where the supremum is taken over partitions:

$$a = t_0^n < t_1^n < \ldots < t_n^n = b. \tag{1.8}$$

Clearly, (by the triangle inequality) the sums in (1.7) increase as new points are added to the partitions. Therefore variation of g is

$$V_g([a, b]) = \lim_{\delta_n \to 0} \sum_{i=1}^{n} |g(t_i^n) - g(t_{i-1}^n)|, \tag{1.9}$$

where $\delta_n = \max_{1 \leq i \leq n} (t_i - t_{i-1})$. If $V_g([a, b])$ is finite then g is said to be a function of finite variation on $[a, b]$. If g is a function of $t \geq 0$, then the variation function of g as a function of t is defined by

$$V_g(t) = V_g([0, t]).$$

Clearly, $V_g(t)$ is a non-decreasing function of t.

Definition 1.5 *g is of finite variation if $V_g(t) < \infty$ for all t. g is of bounded variation if $\sup_t V_g(t) < \infty$, in other words, if for all t, $V_g(t) < C$, a constant independent of t.*

Example 1.4:

1. If $g(t)$ is increasing then for any i, $g(t_i) > g(t_{i-1})$ resulting in a telescoping sum, where all the terms excluding the first and the last cancel out, leaving

$$V_g(t) = g(t) - g(0).$$

2. If $g(t)$ is decreasing then, similarly,

$$V_g(t) = g(0) - g(t).$$

Example 1.5: If $g(t)$ is differentiable with continuous derivative $g'(t)$, $g(t) = \int_0^t g'(s)ds$, and $\int_0^t |g'(s)|ds < \infty$, then

$$V_g(t) = \int_0^t |g'(s)|ds.$$

This can be seen by using the definition and the mean value theorem. $\int_{t_{i-1}}^{t_i} g'(s)ds = g'(\xi_i)(t_i - t_{i-1})$, for some $\xi_i \in (t_{i-1}, t_i)$. Thus $|\int_{t_{i-1}}^{t_i} g'(s)ds| = |g'(\xi_i)|(t_i - t_{i-1})$, and

$$V_g(t) = \lim \sum_{i=1}^n |g(t_i) - g(t_{i-1})| = \lim \sum_{i=1}^n |\int_{t_{i-1}}^{t_i} g'(s)ds|$$

$$= \sup \sum_{i=1}^n |g'(\xi_i)|(t_i - t_{i-1}) = \int_0^t |g'(s)|ds.$$

The last equality is due to the last sum being a Riemann sum for the final integral.

Alternatively, the result can be seen from the decomposition of the derivative into the positive and negative parts,

$$g(t) = \int_0^t g'(s)ds = \int_0^t [g'(s)]^+ ds - \int_0^t [g'(s)]^- ds.$$

Notice that $[g'(s)]^-$ is zero when $[g'(s)]^+$ is positive, and the other way around. Using this one can see that the total variation of g is given by the sum of the variation of the above integrals. But these integrals are monotone functions with the value zero at zero. Hence

$$V_g(t) = \int_0^t [g'(s)]^+ ds + \int_0^t [g'(s)]^- ds$$

$$= \int_0^t ([g'(s)]^+ + [g'(s)]^-)ds = \int_0^t |g'(s)|ds.$$

Example 1.6: (Variation of a pure jump function).
If g is a regular right-continuous (càdlàg) function or regular left-continuous (càglàd), and changes only by jumps,

$$g(t) = \sum_{0 \leq s \leq t} \Delta g(s),$$

then it is easy to see from the definition that

$$V_g(t) = \sum_{0 \leq s \leq t} |\Delta g(s)|.$$

Example 1.7: The function $g(t) = t \sin(1/t)$ for $t > 0$, and $g(0) = 0$ is continuous on $[0, 1]$, differentiable at all points except zero, but has infinite variation on any interval that includes zero. Take the partition $1/(2\pi k + \pi/2), 1/(2\pi k - \pi/2)$, $k = 1, 2, \ldots$.

The following theorem gives necessary and sufficient conditions for a function to have finite variation.

Theorem 1.6 (Jordan Decomposition) *Any function* $g : [0, \infty) \to \mathbb{R}$ *of finite variation can be expressed as the difference of two increasing functions*

$$g(t) = a(t) - b(t).$$

One such decomposition is given by

$$a(t) = V_g(t) \quad b(t) = V_g(t) - g(t). \tag{1.10}$$

It is easy to check that $b(t)$ is increasing, and $a(t)$ is obviously increasing. The representation of a function of finite variation as difference of two increasing functions is not unique. Another decomposition is

$$g(t) = \frac{1}{2}(V_g(t) + g(t)) - \frac{1}{2}(V_g(t) - g(t)).$$

The sum, the difference and the product of functions of finite variation are also functions of finite variation. This is also true for the ratio of two functions of finite variation provided the modulus of the denominator is larger than a positive constant.

The following result follows by Theorem 1.3, and its proof is easy.

Theorem 1.7 *A finite variation function can have no more than countably many discontinuities. Moreover, all discontinuities are jumps.*

PROOF: It is enough to establish the result for monotone functions, since a function of finite variation is a difference of two monotone functions.

A monotone function has left and right limits at any point, therefore any discontinuity is a jump. The number of jumps of size greater or equal to $\frac{1}{n}$ is no more than $(g(b) - g(a))n$. The set of all jump points is a union of the sets of jump points with the size of the jumps greater or equal to $\frac{1}{n}$. Since each such set is finite, the total number of jumps is at most countable.

□

A sufficient condition for a *continuous* function to be of finite variation is given by the following theorem, the proof of which is outlined in Example 1.5.

Theorem 1.8 *If g is continuous, g' exists and $\int |g'(t)|dt < \infty$ then g is of finite variation.*

Theorem 1.9 (Banach) *Let $g(t)$ be a continuous function on $[0, 1]$, and denote by $s(a)$ the number of t's with $g(t) = a$. Then the variation of g is $\int_{-\infty}^{\infty} s(a)da$.*

Continuous and Discrete Parts of a Function

Let $g(t)$, $t \geq 0$, be a right-continuous increasing function. Then it can have at most countably many jumps, moreover the sum of the jumps is finite over finite time intervals. Define the discontinuous part g^d of g by

$$g^d(t) = \sum_{s \leq t} (g(s) - g(s-)) = \sum_{0 < s \leq t} \Delta g(s), \qquad (1.11)$$

and the continuous part g^c of g by

$$g^c(t) = g(t) - g^d(t). \qquad (1.12)$$

Clearly, g^d changes only by jumps, g^c is continuous and $g(t) = g^c(t) + g^d(t)$. Since a finite variation function is the difference of two increasing functions, the decomposition (1.12) holds for functions of finite variation. Although representation as the difference of increasing functions is not unique, decomposition (1.12) is essentially unique, in a sense that any two such decomposition differ by a constant. Indeed, if there were another such decomposition $g(t) = h^c(t) + h^d(t)$, then $h^c(t) - g^c(t) = g^d(t) - h^d(t)$, implying that $h^d - g^d$ is continuous. Hence h^d and g^d have the same set of jump points, and it follows that $h^d(t) - g^d(t) = c$ for some constant c.

Quadratic Variation

If g is a function of real variable, define its quadratic variation over the interval $[0, t]$ as the limit (when it exists)

$$[g](t) = \lim_{\delta_n \to 0} \sum_{i=1}^{n} (g(t_i^n) - g(t_{i-1}^n))^2, \qquad (1.13)$$

where the limit is taken over partitions: $0 = t_0^n < t_1^n < \ldots < t_n^n = t$, with $\delta_n = \max_{1 \leq i \leq n} (t_i^n - t_{i-1}^n)$.

Remark 1.2: Similarly to the concept of variation, there is a concept of Φ-variation of a function. If $\Phi(u)$ is a positive function, increasing monotonically with u then the Φ-variation of g on $[0, t]$ is

$$V_\Phi[g] = \sup \sum_{i=1}^{n} \Phi(|g(t_i^n) - g(t_{i-1}^n)|), \qquad (1.14)$$

where supremum is taken over *all* partitions. Functions with finite Φ-variation on $[0, t]$ form a class V_Φ. With $\Phi(u) = u$ one obtains the class VF of functions of finite variation, with $\Phi(u) = u^p$ one obtains the class of functions of p-th finite variation, VF_p. If $1 \leq p < q < \infty$, then finite p-variation implies finite q-variation.

The stochastic calculus definition of quadratic variation is different to the classical one with $p = 2$ (unlike for the first variation $p = 1$, when they are the same). In stochastic calculus the limit in (1.13) is taken over *shrinking* partitions with $\delta_n = \max_{1 \leq i \leq n} (t_i^n - t_{i-1}^n) \to 0$, and not over all possible partitions. We shall use only the stochastic calculus definition.

Quadratic variation plays a major role in stochastic calculus, but is hardly ever met in standard calculus due to the fact that smooth functions have zero quadratic variation.

Theorem 1.10 *If g is continuous and of finite variation then its quadratic variation is zero.*

PROOF:

$$
\begin{aligned}
[g](t) &= \lim_{\delta_n \to 0} \sum_{i=0}^{n-1} (g(t_{i+1}^n) - g(t_i^n))^2 \\
&\leq \lim_{\delta_n \to 0} \max_i |g(t_{i+1}^n) - g(t_i^n)| \sum_{i=0}^{n-1} |g(t_{i+1}^n) - g(t_i^n)| \\
&\leq \lim_{\delta_n \to 0} \max_i |g(t_{i+1}^n) - g(t_i^n)| V_g(t).
\end{aligned}
$$

Since g is continuous, it is uniformly continuous on $[0, t]$, hence $\lim_{\delta_n \to 0} \max_i |g(t_{i+1}^n) - g(t_i^n)| = 0$, and the result follows.

$\square$

Remark that there are functions with zero quadratic variation and infinite variation (called functions of *zero energy*).

Define the quadratic covariation (or simply covariation) of f and g on $[0, t]$ by the following limit (when it exists)

$$[f, g](t) = \lim_{\delta_n \to 0} \sum_{i=0}^{n-1} \left(f(t_{i+1}^n) - f(t_i^n) \right) \left(g(t_{i+1}^n) - g(t_i^n) \right), \qquad (1.15)$$

when the limit is taken over partitions $\{t_i^n\}$ of $[0, t]$ with $\delta_n = \max_i (t_{i+1}^n - t_i^n)$.

The same proof as for Theorem 1.10 works for the following result

Theorem 1.11 *If f is continuous and g is of finite variation, then their co-variation is zero $[f, g](t) = 0$.*

Let f and g be such that their quadratic variation is defined. By using simple algebra, one can see that covariation satisfies

Theorem 1.12 (Polarization identity)

$$[f, g](t) = \frac{1}{2} \left([f + g, f + g](t) - [f, f](t) - [g, g](t) \right), \qquad (1.16)$$

It is obvious that covariation is symmetric, $[f, g](t) = [g, f](t)$, it follows from (1.16) that it is linear, that is, for any constants α and β

$$[\alpha f + \beta g, h](t) = \alpha [f, h](t) + \beta [g, h](t). \qquad (1.17)$$

Due to symmetry it is bilinear, that is, linear in both arguments. Thus the quadratic variation of the sum can be opened similarly to multiplication of sums $(\alpha_1 f + \beta_1 g)(\alpha_2 h + \beta_2 k)$. It follows from the definition of quadratic variation, that it is a non-decreasing function in t, and consequently it is of finite variation. By the polarization identity, covariation is also of finite variation. More about quadratic variation is given in the Stochastic Calculus chapter.

1.3   Riemann Integral and Stieltjes Integral

Riemann Integral

The Riemann Integral of f over interval $[a, b]$ is defined as the limit of Riemann sums

$$\int_a^b f(t)dt = \lim_{\delta \to 0} \sum_{i=1}^{n} f(\xi_i^n)(t_i^n - t_{i-1}^n), \qquad (1.18)$$

where t_i^n's represent partitions of the interval,

$$a = t_0^n < t_1^n < \ldots < t_n^n = b, \delta = \max_{1 \leq i \leq n} (t_i^n - t_{i-1}^n), \text{ and } t_{i-1}^n \leq \xi_i^n \leq t_i^n.$$

It is possible to show that Riemann Integral is well defined for continuous functions, and by splitting up the interval, it can be extended to functions which are discontinuous at finitely many points.

Calculation of integrals is often done by using the antiderivative, and is based on the the following result.

Theorem 1.13 (The fundamental theorem of calculus) *If f is differentiable on $[a, b]$ and f' is Riemann integrable on $[a, b]$ then*

$$f(b) - f(a) = \int_a^b f'(s)ds.$$

In general, this result cannot be applied to discontinuous functions, see example below. For such functions a jump term must be added, see (1.20).

Example 1.8: Let $f(t) = 2$ for $1 \leq t \leq 2$, $f(t) = 1$ for $0 \leq t < 1$. Then $f'(t) = 0$ at all $t \neq 1$. $\int_0^t f'(s)ds = 0 \neq f(t)$. f is continuous and is differentiable at all points but one, the derivative is integrable, but the function does not equal the integral of its derivative.

Main tools for calculations of Riemann integrals are change of variables and integration by parts. These are reviewed below in a more general framework of the Stieltjes integral.

Stieltjes Integral

The Stieltjes Integral is an integral of the form $\int_a^b f(t)dg(t)$, where g is a function of finite variation. Since a function of finite variation is a difference of two increasing functions, it is sufficient to define the integral with respect to monotone functions.

Stieltjes Integral with respect to Monotone Functions

The Stieltjes Integral of f with respect to a monotone function g over an interval $(a, b]$ is defined as

$$\int_a^b fdg = \int_a^b f(t)dg(t) = \lim_{\delta \to 0} \sum_{i=1}^n f(\xi_i^n)\big(g(t_i^n) - g(t_{i-1}^n)\big), \qquad (1.19)$$

with the quantities appearing in the definition being the same as above for the Riemann Integral. This integral is a generalization of the Riemann Integral, which is recovered when we take $g(t) = t$. This integral is also known as the Riemann-Stieltjes integral.

Particular Cases

If $g'(t)$ exists, and $g(t) = g(0) + \int_0^t g'(s)ds$, then it is possible to show that

$$\int_a^b f(t)dg(t) = \int_a^b f(t)g'(t)dt.$$

If $g(t) = \sum_{k=a}^{[t]} h(k)$ (a integer, and $[t]$ stands for the integer part of t) then

$$\int_a^b f(t)dg(t) = \sum_a^b f(k)h(k).$$

This property allows us to represent sums as integrals.

Example 1.9:

1. $g(t) = 2t^2 \qquad \int_a^b f(t)dg(t) = 4\int_a^b tf(t)dt$

2. $g(t) = \begin{cases} 0 & t < 0 \\ 2 & 0 \le t < 1 \\ 3 & 1 \le t < 2 \\ 5 & 2 \le t \end{cases}$

$$\int_{-\infty}^{\infty} f(t)dg(t) = 2f(0) + f(1) + 2f(2)$$

If, for example, $f(t) = t$ then $\int_{-\infty}^{\infty} tdg(t) = 5$. If $f(t) = (t+1)^2$ then $\int_{-\infty}^{\infty}(t+1)^2 dg(t) = 2 + 4 + 18 = 24$.

Let g be a function of finite variation and

$$g(t) = a(t) - b(t)$$

with $a(t) = V_g(t)$, $b(t) = V_g(t) - g(t)$, which are non-decreasing functions. If

$$\int_0^t |f(s)|da(s) = \int_0^t |f(s)|dV_g(s) := \int_0^t |f(s)||dg(s)| < \infty$$

then f is Stieltjes-integrable with respect to g and its integral is defined by

$$\int_{(0,t]} f(s)dg(s) = \int_{(0,t]} f(s)da(s) - \int_{(0,t]} f(s)db(s).$$

Notation: $\int_a^b f(s)dg(s) = \int_{(a,b]} f(s)dg(s)$.

Note: $\int_{(0,t]} dg(s) = g(t) - g(0)$ and $\int_{(0,t)} dg(s) = g(t-) - g(0)$.

If f is Stieltjes-integrable with respect to a function g of finite variation, then the variation of the integral is

$$V(t) = \int_0^t |f(s)||dg(s)| = \int_0^t |f(s)|dV_g(s).$$

Impossibility of a direct definition of an integral with respect to functions of infinite variation

In stochastic calculus we need to consider integrals with respect to functions of infinite variation. Such functions arise, for example, as models of stock prices. Integrals with respect to a function of infinite variation, cannot be defined as a usual limit of approximating sums. The following result explains, see for example, Protter (1992), p.40.

Theorem 1.14 *Let* $\delta_n = \max_i(t_i^n - t_{i-1}^n)$ *denote the largest interval in the partition of* $[a, b]$. *If*

$$\lim_{\delta_n \to 0} \sum_{i=1}^{n} f(t_{i-1}^n)[g(t_i^n) - g(t_{i-1}^n)]$$

exists for any continuous function f *then* g *must be of finite variation on* $[a, b]$.

This shows that if g has infinite variation then the limit of the approximating sums does not exist for some functions f.

Integration by Parts

Let f and g be functions of finite variation. Denote here $\Delta g(s) = g(s) - g(s-)$, then (with integrals on $(a, b]$)

$$f(b)g(b) - f(a)g(a) = \int_a^b f(s-)dg(s) + \int_a^b g(s-)df(s) + \sum_{a < s \leq b} \Delta f(s) \Delta g(s)$$

$$= \int_a^b f(s-)dg(s) + \int_a^b g(s)df(s). \qquad (1.20)$$

The last equation is obtained by putting together the sum of jumps with one of the integrals.

Note that although the sum in (1.20) is written over uncountably many values $a < s \leq b$, it has at most countably many non-zero terms. This is because a finite variation function can have at most a countable number of jumps.

If g is continuous so that $g(s-) = g(s)$ for all s then the formula simplifies and in this case we have the familiar integration by parts formula

$$f(b)g(b) - f(a)g(a) = \int_a^b f(s)dg(s) + \int_a^b g(s)df(s).$$

Example 1.10: Let $g(s)$ be of finite variation, $g(0) = 0$, and consider $g^2(s)$. Using the integration by parts with $f = g$, we have

$$g^2(t) = 2 \int_0^t g(s-)dg(s) + \sum_{s \leq t} (\Delta g(s))^2.$$

In other words,

$$\int_0^t g(s-)dg(s) = \frac{g^2(t)}{2} - \frac{1}{2} \sum_{s \leq t} (\Delta g(s))^2.$$

Now using the formula (1.20) above we also have

$$\int_0^t g(s)dg(s) = g^2(t) - \int_0^t g(s-)dg(s) = \frac{g^2(t)}{2} + \frac{1}{2} \sum_{s \leq t} (\Delta g(s))^2.$$

Thus it follows that

$$\int_0^t g(s-)dg(s) \leq \frac{g^2(t)}{2} \leq \int_0^t g(s)dg(s)$$

Change of Variables

Let f have a continuous derivative ($f \in C^1$) and g be of finite variation and continuous, then

$$f(g(t)) - f(g(0)) = \int_0^t f'(g(s))dg(s) = \int_{g(0)}^{g(t)} f'(u)du.$$

If g is of finite variation has jumps, and is right-continuous then

$$
\begin{aligned}
f(g(t)) - f(g(0)) &= \int_0^t f'(g(s-))dg(s) \\
&\quad + \sum_{0 < s \leq t} \Big(f(g(s)) - f(g(s-)) - f'(g(s-))\Delta g(s) \Big),
\end{aligned}
$$

where $\Delta g(s) = g(s) - g(s-)$ denotes the jump of g at s. This is known in stochastic calculus as Itô's formula.

Example 1.11: Take $f(x) = x^2$, then we obtain

$$g^2(t) - g^2(0) = 2 \int_0^t g(s-)dg(s) + \sum_{s \leq t} (\Delta g(s))^2.$$

Remark 1.3: Note that for a continuous f and finite variation g on $[0, t]$ the approximating sums converge as $\delta = \max_i (t_{i+1}^n - t_i^n) \to 0$,

$$\sum_i f(g(t_i^n))(g(t_{i+1}^n) - g(t_i^n)) \to \int_0^t f(g(s-))dg(s).$$

Remark 1.4: One of the shortcomings of Riemann or Stieltjes integrals is that they don't preserve the monotone convergence property, that is, for a sequence of functions $f_n \uparrow f$ does not necessarily follow that their integrals converge. The Lebesgue (or Lebesgue-Stieltjes) Integral preserves this property.

1.4   Lebesgue's Method of Integration

While Riemann sums are constructed by dividing the domain of integration on the x-axis, the interval $[a, b]$, into smaller subintervals, Lebesgue sums are constructed by dividing the range of the function on the y-axis, the interval $[c, d]$, into smaller subintervals $c = y_0 < y_1 < \ldots < y_k < \ldots y_n = d$ and forming sums

$$\sum_{k=0}^{n-1} y_k \text{length}(\{t : y_k \leq f(t) < y_{k+1}\}).$$

The Lebesgue Integral is the limit of the above sums as the number of points in the partition increases. It turns out that the Lebesgue Integral is more general than the Riemann Integral, and preserves convergence. This approach also allows integration of functions in abstract probability spaces more general than $\mathbb{R}$ or $\mathbb{R}^n$; it requires additional concepts and is made more precise in the next chapter (see Section 2.3).

Remark 1.5: In folklore the following analogy is used. Imagine that money is spread out on a floor. In the Riemann method of integration, you collect the money as you progress in the room. In the Lebesgue method, first you collect \$100 bills everywhere you can find them, then \$50, etc.

1.5   Differentials and Integrals

The differential $df(t)$ of a differentiable function f at t is defined as the *linear in Δt part* of the increment at t, $f(t + \Delta) - f(t)$. If the differential of the independent variable is denoted $dt = \Delta t$, then $f(t + dt) - f(t) = df(t) +$ smaller order terms, and it follows from the existence of the derivative at t, that

$$df(t) = f'(t)dt. \tag{1.21}$$

If g is also a differentiable function of t, then $f(g(t))$ is differentiable, and

$$df(g(t)) = f'(g(t))g'(t)dt = f'(g(t))dg(t), \tag{1.22}$$

which is known as the *chain rule*.

Differential calculus is important in applications because many physical problems can be formulated in terms of differential equations. The main relation between the integral and the differential (or derivative) is given by the fundamental theorem of calculus, Theorem 1.13.

For differentiable functions, differential equations of the form

$$df(t) = \varphi(t)dw(t)$$

can be written in the integral form

$$f(t) = f(0) + \int_0^t \varphi(s)dw(s).$$

In Stochastic Calculus stochastic differentials do not formally exist and the random functions $w(t)$ are not differentiable at any point. By introducing a new (stochastic) integral, stochastic differential equations can be defined, and, by definition, solutions to these equations are given by the solutions to the corresponding stochastic integral equations.

1.6 Taylor's Formula and Other Results

This section contains Taylor's Formula and conditions on functions used in results on differential equations. It may be treated as an appendix, and referred to only when needed.

Taylor's Formula for Functions of One Variable

If we consider the increment of a function $f(x) - f(x_0)$ over the interval $[x_0, x]$, then provided $f'(x_0)$ exists, the differential at x_0 is the linear part in $(x - x_0)$ of this increment and it provides the first approximation to the increment. Taylor's formula gives a better approximation by taking higher order terms of powers of $(x - x_0)$ provided higher derivatives of f at x_0 exist. If f is a function of x with derivatives up to order $n + 1$, then

$$
\begin{aligned}
f(x) - f(x_0) &= f'(x_0)(x - x_0) + \frac{1}{2}f''(x_0)(x - x_0)^2 + \frac{1}{3!}f^{(3)}(x_0)(x - x_0)^3 \\
&+ \ldots + \frac{1}{n!}f^{(n)}(x_0)(x - x_0)^n + R_n(x, x_0),
\end{aligned}
$$

where R_n is the remainder, and $f^{(n)}$ is the derivative of $f^{(n-1)}$. The remainder can be written in the form

$$R_n(x, x_0) = \frac{1}{(n+1)!}f^{(n+1)}(\theta_n)(x - x_0)^{n+1}$$

for some point $\theta_n \in (x_0, x)$.

In our applications we shall use this formula with two terms.

$$f(x) - f(x_0) = f'(x)(x - x_0) + \frac{1}{2}f''(\theta)(x - x_0)^2, \qquad (1.23)$$

for some point $\theta \in (x_0, x)$.

Taylor's Formula for Functions of Several Variables

Similarly to the one-dimensional case, Taylor's formula gives successive approximations to the increment of a function. A function of n real variables $f(x_1, x_2, \ldots, x_n)$ is differentiable at point $\boldsymbol{x} = (x_1, x_2, \ldots, x_n)$ if the increment at this point can be approximated by a linear part, which is the differential of f at $\boldsymbol{x}$.

$$\Delta f(\boldsymbol{x}) = \sum_{i=1}^{n} C_i \Delta x_i + o(\rho), \text{ when } \rho = \sqrt{\sum_{i=1}^{n}(\Delta x_i)^2} \text{ and } \lim_{\rho \to 0} \frac{o(\rho)}{\rho} = 0. \quad (1.24)$$

If f is differentiable at $\boldsymbol{x} = (x_1, x_2, \ldots, x_n)$, then in particular it is differentiable as a function of any one variable x_i at that point, when all the other coordinates are kept fixed. The derivative with respect to x_i is the called the partial derivative $\partial f / \partial x_i$. Unlike in the one-dimensional case, the existence of all partial derivatives $\partial f / \partial x_i$ at $\boldsymbol{x}$, is necessary but not sufficient for differentiability of f at $\boldsymbol{x}$. But if all partial derivatives exist and are continuous at that point, then f is differentiable at that point, moreover, C_i in (1.24) is given by the value of $\partial f / \partial x_i$ at $\boldsymbol{x}$. If we define the differential of the independent variable as its increment $dx_i = \Delta x_i$, then we have

Theorem 1.15 *For f to be differentiable at a point, it is necessary that f has partial derivatives at that point, and it is sufficient that it has continuous partial derivatives at that point. If f is differentiable at $\boldsymbol{x}$, then its differential at $\boldsymbol{x}$ is given by*

$$df(x_1, x_2, \ldots, x_n) = \sum_{i=1}^{n} \frac{\partial f}{\partial x_i}(x_1, x_2, \ldots, x_n) dx_i. \qquad (1.25)$$

The first approximation of the increment of a differentiable function is the differential,

$$\Delta f(\boldsymbol{x}) \approx df(\boldsymbol{x}).$$

If f possesses higher order partial derivatives, then further approximation is possible and it is given by Taylor's formula. In Stochastic Calculus the second order approximation plays an important role.

Let $f : \mathbb{R}^n \to \mathbb{R}$ be C^2, ($f(x_1, x_2, \ldots, x_n)$ has continuous partial derivatives up to order two), $\mathbf{x} = (x_1, x_2, \ldots, x_n)$, $\mathbf{x} + \Delta\mathbf{x} = (x_1 + \Delta x_1, x_2 + \Delta x_2, \ldots, x_n + \Delta x_n)$ then by considering the function of one variable $g(t) = f(\mathbf{x} + t\Delta\mathbf{x})$ for $0 \leq t \leq 1$, the following result is obtained.

$$\Delta f(x_1, x_2, \ldots, x_n) = f(\mathbf{x} + \Delta\mathbf{x}) - f(\mathbf{x}) \approx \sum_{i=1}^{n} \frac{\partial f}{\partial x_i}(x_1, x_2, \ldots, x_n) dx_i$$

$$+ \frac{1}{2} \sum_{i=1}^{n} \sum_{j=1}^{n} \frac{\partial^2 f}{\partial x_i \partial x_j}(x_1 + \theta\Delta x_1, \ldots, x_n + \theta\Delta x_n) dx_i dx_j, \quad (1.26)$$

where just like in the case of one variable the second derivatives are evaluated at some "middle" point, $(x_1 + \theta\Delta x_1, \ldots, x_n + \theta\Delta x_n)$ for some $\theta \in (0, 1)$, and $dx_i = \Delta x_i$.

Lipschitz and Hölder Conditions

Lipschitz and Hölder conditions describe subclasses of continuous functions. They appear as conditions on the coefficients in the results on the existence and uniqueness of solutions of ordinary and stochastic differential equations.

Definition 1.16 f *satisfies a Hölder condition (Hölder continuous) of order* α, $0 < \alpha \leq 1$, *on* $[a, b]$ *(* $\mathbb{R}$ *) if there is a constant* $K > 0$, *so that for all* $x, y \in [a, b]$ *(* $\mathbb{R}$ *)*

$$|f(x) - f(y)| \leq K|x - y|^{\alpha}. \quad (1.27)$$

A Lipschitz condition is a Hölder condition with $\alpha = 1$,

$$|f(x) - f(y)| \leq K|x - y|. \quad (1.28)$$

It is easy to see that a Hölder continuous of order α function on $[a, b]$ is also Hölder continuous of any lesser order.

Example 1.12: The function $f(x) = \sqrt{x}$ on $[0, \infty)$ is Hölder continuous with $\alpha = 1/2$ but is not Lipschitz, since its derivative is unbounded near zero. To see that it is Hölder, it is enough to show that for all $x, y \geq 0$ the following ratio is bounded,

$$\frac{|\sqrt{x} - \sqrt{y}|}{\sqrt{|x - y|}} \leq K. \quad (1.29)$$

It is an elementary exercise to establish that the left hand side is bounded by dividing through by $\sqrt{y}$ (if $y = 0$, then the bound is obviously one), and applying l'Hôpital's rule. Similarly $|x|^r$, $0 < r < 1$ is Hölder of order r.

A simple sufficient condition for a function to be Lipschitz is to be continuous and piecewise smooth, precise definitions follow.

Definition 1.17 f *is smooth on* $[a, b]$ *if it possesses a continuous derivative* f' *on* (a, b) *such that the limits* $f'(a+)$ *and* $f'(b-)$ *exist.*

Definition 1.18 f *is piecewise continuous on* $[a, b]$ *if it is continuous on* $[a, b]$ *except possibly a finite number of points at which right-hand and left-hand limits exist.*

Definition 1.19 f *is piecewise smooth on* $[a, b]$ *if it is piecewise continuous on* $[a, b]$ *and* f' *exists and is also piecewise continuous on* $[a, b]$.

Growth Conditions

Linear growth condition also appears in the results on existence and uniqueness of solutions of differential equations. $f(x)$ satisfies the linear growth condition if

$$|f(x)| \le K(1 + |x|). \tag{1.30}$$

This condition describes the growth of a function for large values of x, and states that f is bounded for small values of x.

Example 1.13: It can be shown that if $f(0, t)$ is a bounded function of t, $|f(0, t)| \le C$ for all t, and $f(x, t)$ satisfies the Lipschitz condition in x uniformly in t, $|f(x, t) - f(y, t)| \le K|x - y|$, then $f(x, t)$ satisfies the linear growth condition in x, $|f(x, t)| \le K_1(1 + |x|)$.

The polynomial growth condition on f is the condition of the form

$$|f(x)| \le K(1 + |x|^m), \quad \text{for some } K, m > 0. \tag{1.31}$$

Theorem 1.20 (Gronwall's inequality) *Let* $f(t)$, $g(t)$ *and* $h(t)$ *be non-negative on* $[0, T]$, *and for all* $0 \le t \le T$

$$f(t) \le g(t) + \int_0^t h(s)f(s)ds. \tag{1.32}$$

Then for $0 \le t \le T$

$$f(t) \le g(t) + \int_0^t h(s)g(s) \exp\left(\int_s^t h(u)du\right)ds. \tag{1.33}$$

This form can be found for example, in Dieudonné (1960).

Solution of First Order Linear Differential Equations

Linear differential equations, by definition, are linear in the unknown function and its derivatives. A first order linear equation, in which the coefficient of $\frac{dx(t)}{dt}$ does not vanish, can be written in the form

$$\frac{dx(t)}{dt} + g(t)x(t) = k(t). \tag{1.34}$$

These equations are solved by using the Integrating Factor Method. The integrating factor is the function $e^{G(t)}$, where $G(t)$ is chosen by $G'(t) = g(t)$. After multiplying both sides of the equation by $e^{G(t)}$, integrating, and solving for $x(t)$, we have

$$x(t) = e^{-G(t)} \int_0^t \left(e^{G(s)}k(s)\right) ds + x(0)e^{G(0)-G(t)}. \tag{1.35}$$

The integrating factor $G(t)$ is determined up to a constant, but it is clear from (1.35), that the solution $x(t)$ remains the same.

Further Results on Functions and Integration

Results given here are not required to understand subsequent material. Some of these involve the concepts of a set of zero Lebesgue measure. This is given in the next chapter (see Section 2.2); any countable set has Lebesgue measure zero, but there are also uncountable sets of zero Lebesgue measure. A partial converse to Theorem 1.8 also holds (see, for example, Saks (1964), Freedman (1983) p.209, for the following results).

Theorem 1.21 (Lebesgue) *A finite variation function g on $[a, b]$ is differentiable almost everywhere on $[a, b]$.*

In what follows sufficient conditions for a function to be Lipschitz and not to be Lipschitz are given.

1. If f is continuously differentiable on a finite interval $[a, b]$, then it is Lipschitz. Indeed, since f' is continuous on $[a, b]$, it is bounded, $|f'| \le K$. Therefore

$$|f(x) - f(y)| = |\int_x^y f'(t)dt| \le \int_x^y |f'(t)|dt \le K|x - y|. \tag{1.36}$$

2. If f is continuous and piecewise smooth then it is Lipschitz, the proof is similar to the above.

3. A Lipschitz function does not have to be differentiable, for example $f(x) = |x|$ is Lipschitz but it is not differentiable at zero.

4. It follows from the definition of a Lipschitz function (1.28), that if it is differentiable, then its derivative is bounded by K.

5. A Lipschitz function has finite variation on finite intervals, since for any partition $\{x_i\}$ of a finite interval $[a, b]$,

$$\sum |f(x_{i+1}) - f(x_i)| \leq K \sum (x_{i+1} - x_i) = K(b - a). \qquad (1.37)$$

6. As functions of finite variation have derivatives almost everywhere (with respect to Lebesgue measure), a Lipschitz function is differentiable almost everywhere.

   (Note that functions of finite variation have derivatives which are integrable with respect to Lebesgue measure, but the function does not have to be equal to the integral of the derivative.)

7. A Lipschitz function multiplied by a constant, and a sum of two Lipschitz functions are Lipschitz functions. The product of two bounded Lipschitz functions is again a Lipschitz function.

8. If f is Lipschitz on $[0, N]$ for any $N > 0$ but with the constant K depending on N, then it is called locally Lipschitz. For example, x^2 is Lipschitz on $[0, N]$ for any finite N, but it is not Lipschitz on $[0, +\infty)$, since its derivative is unbounded.

9. If f is a function of two variables $f(x, t)$ and it satisfies Lipschitz condition in x for all t, $0 \leq t \leq T$, with same constant K independent of t, it is said that f satisfies Lipschitz condition in x uniformly in t, $0 \leq t \leq T$.

A necessary and sufficient condition for a function f to be Riemann integrable was given by Lebesgue (see, for example, Saks (1964), Freedman (1983) p.208).

Theorem 1.22 (Lebesgue) *A necessary and sufficient condition for a function f to be Riemann integrable on a finite closed interval $[a, b]$ is that f is bounded on $[a, b]$ and almost everywhere continuous on $[a, b]$, that is, continuous at all points except possibly on a set of Lebesgue measure zero.*

Remark 1.6: (This is not used anywhere in the book, and directed only to readers with knowledge of Functional Analysis)
Continuous functions on $[a, b]$ with the supremum norm $\|h\| = \sup_{x \in [a,b]} |h(x)|$ is a Banach space, denoted $C([a, b])$. By a result in Functional Analysis, any linear functional on this space can be represented as $\int_{[a,b]} h(x) dg(x)$ for some function g of finite variation. In this way, the Banach space of functions of finite variation on $[a, b]$ with the norm $\|g\| = V_g([a, b])$ can be identified with the space of linear functionals on the space of continuous functions, in other words, the dual space of $C([a, b])$.

Chapter 2

Concepts of Probability Theory

In this chapter we give fundamental definitions of probabilistic concepts. Since the theory is more transparent in the discrete case, it is presented first. The most important concepts not met in elementary courses are the models for information, its flow and conditional expectation. This is only a brief description, and a more detailed treatment can be found in many books on Probability Theory, for example, Breiman (1968), Loeve (1978), Durret (1991). Conditional expectation with its properties is central for further development, but some of the material in this chapter may be treated as an appendix.

2.1 Discrete Probability Model

A probability model consists of a filtered probability space on which variables of interest are defined. In this section we introduce a discrete probability model by using an example of discrete trading in stock.

Filtered Probability Space

A filtered probability space consists of: a sample space of elementary events, a field of events, a probability defined on that field, and a filtration of increasing subfields.

Sample Space

Consider a single stock with price S_t at time $t = 1, 2, \ldots T$. Denote by Ω the set of all possible values of stock during these times.

$$\Omega = \{\omega : \ \omega = (S_1, S_2, \ldots, S_T)\}.$$

If we assume that the stock price can go up by a factor u and down by a factor d, then the relevant information reduces to the knowledge of the movements at each time.

$$\Omega = \{\omega : \ \omega = (a_1, a_2, \ldots, a_T)\} \quad a_t = u \text{ or } d.$$

To model uncertainty about the price in the future, we "list" all possible future prices, and call it possible states of the world. The unknown future is just one of many possible outcomes, called the true state of the world. As time passes more and more information is revealed about the true state of the world. At time $t = 1$ we know prices S_0 and S_1. Thus the true state of the world lies in a smaller set, subset of Ω, $A \subset \Omega$. After observing S_1 we know which prices did not happen at time 1. Therefore we know that the true state of the world is in A and not in $\Omega \setminus A = \bar{A}$.

Fields of Events

Define by $\mathcal{F}_t$ the information available to investors at time t, which consists of stock prices before and at time t.

For example when $T = 2$, at $t = 0$ we have no information about S_1 and S_2, and $\mathcal{F}_0 = \{\emptyset, \Omega\}$, all we know is that a true state of the world is in Ω. Consider the situation at $t = 1$. Suppose at $t = 1$ stock went up by u. Then we know that the true state of the world is in A, and not in its complement $\bar{A}$, where

$$A = \{(u, S_2), \ S_2 = u \text{ or } d\} = \{(u, u), (u, d)\}.$$

Thus our information at time $t = 1$ is

$$\mathcal{F}_1 = \{\emptyset, \Omega, A, \bar{A}\}.$$

Note that $\mathcal{F}_0 \subset \mathcal{F}_1$, since we don't forget the previous information.

At time t investors know which part of Ω contains the true state of the world. $\mathcal{F}_t$ is called a *field* or *algebra* of sets.
$\mathcal{F}$ is a field if

1. $\emptyset, \Omega \in \mathcal{F}$

2. If $A \in \mathcal{F}$, and $B \in \mathcal{F}$ then $A \cup B \in \mathcal{F}$, $A \cap B \in \mathcal{F}$, $A \setminus B \in \mathcal{F}$.

Example 2.1: (Examples of fields.)
It is easy to verify that any of the following is a field of sets.
1. $\{\Omega, \emptyset\}$ is called the trivial field $\mathcal{F}_0$.
2. $\{\emptyset, \Omega, A, \bar{A}\}$ is called the field generated by set A, and denoted by $\mathcal{F}_A$.
3. $\{A : A \subseteq \Omega\}$ the field of all the subsets of Ω. It is denoted by 2^{Ω}.

A partition of Ω is a collection of exhaustive and mutually exclusive subsets,

$$\{D_1, \ldots, D_k\}, \text{ such that } D_i \cap D_j = \emptyset, \text{ and } \bigcup_i D_i = \Omega.$$

The field generated by the partition is the collection of all finite unions of D_j's and their complements. These sets are like the basic building blocks for the field. If Ω is finite then any field is generated by a partition.

If one field is included in the other, $\mathcal{F}_1 \subset \mathcal{F}_2$, then any set from $\mathcal{F}_1$ is also in $\mathcal{F}_2$. In other words, a set from $\mathcal{F}_1$ is either a set or a union of sets from the partition generating $\mathcal{F}_2$. This means that the partition that generates $\mathcal{F}_2$ has "finer" sets than the ones that generate $\mathcal{F}_1$.

Filtration

A filtration $\mathbb{F}$ is the collection of fields,

$$\mathbb{F} = \{\mathcal{F}_0, \mathcal{F}_1, \ldots, \mathcal{F}_t, \ldots, \mathcal{F}_T\} \quad \mathcal{F}_t \subset \mathcal{F}_{t+1}.$$

$\mathbb{F}$ is used to model a flow of information. As time passes, an observer knows more and more detailed information, that is, finer and finer partitions of Ω. In the example of the price of stock, $\mathbb{F}$ describes how the information about prices is revealed to investors.

Example 2.2: $\mathbb{F} = \{\mathcal{F}_0, \mathcal{F}_A, 2^{\Omega}\}$, is an example of filtration.

Stochastic Processes

If Ω is a finite sample space, then a function X defined on Ω attaches numerical values to each $\omega \in \Omega$. Since Ω is finite, X takes only finitely many values x_i, $i = 1, \ldots, k$.

If a field of events $\mathcal{F}$ is specified, then any set in it is called a *measurable* set. If $\mathcal{F} = 2^{\Omega}$, then any subset of Ω is measurable.

A function X on Ω is called $\mathcal{F}$-*measurable* or a *random variable* on $(\Omega, \mathcal{F})$ if all the sets $\{X = x_i\}$, $i = 1, \ldots, k$, are members of $\mathcal{F}$. This means that if we have the information described by $\mathcal{F}$, that is, we know which event in $\mathcal{F}$ has occurred, then we know which value of X has occurred. Note that if $\mathcal{F} = 2^{\Omega}$, then any function on Ω is a random variable.

Example 2.3:   Consider the model for trading in stock, $t = 1, 2$, where at each time the stock can go up by the factor u or down by the factor d.
$\Omega = \{\omega_1 = (u, u), \omega_2 = (u, d), \omega_3 = (d, u), \omega_4 = (d, d)\}$. Take $A = \{\omega_1, \omega_2\}$, which is the event that at $t = 1$ the stock goes up. $\mathcal{F}_1 = \{\emptyset, \Omega, A, \bar{A}\}$, and $\mathcal{F}_2 = 2^\Omega$ contains all 16 subsets of Ω. Consider the following functions on Ω. $X(\omega_1) = X(\omega_2) = 1.5$, $X(\omega_3) = X(\omega_4) = 0.5$. X is a random variable on $\mathcal{F}_1$. Indeed, the set $\{\omega : X(\omega) = 1.5\} = \{\omega_1, \omega_2\} = A \in \mathcal{F}_1$. Also $\{\omega : X(\omega) = 0.5\} = \bar{A} \in \mathcal{F}_1$. If $Y(\omega_1) = (1.5)^2$, $Y(\omega_2) = 0.75$, $Y(\omega_3) = 0.75$, and $Y(\omega_4) = 0.5^2$, then Y is not a random variable on $\mathcal{F}_1$, it is not $\mathcal{F}_1$-measurable. Indeed, $\{\omega : Y(\omega) = 0.75\} = \{\omega_2, \omega_3\} \notin \mathcal{F}_1$. Y is $\mathcal{F}_2$-measurable.

Definition 2.1 *A stochastic process is a collection of random variables $\{X(t)\}$. For any fixed t, $t = 0, 1, \ldots, T$, $X(t)$ is a random variable on $(\Omega, \mathcal{F}_T)$.*
    A stochastic process is called adapted to filtration $\mathbb{F}$ if for all $t = 0, 1, \ldots, T$, $X(t)$ is a random variable on $\mathcal{F}_t$, that is, if $X(t)$ is $\mathcal{F}_t$-measurable.

Example 2.4: (Example 2.3 continued.)
$X_1 = X$, $X_2 = Y$ is a stochastic process adapted to $\mathbb{F} = \{\mathcal{F}_1, \mathcal{F}_2\}$. This process represents stock prices at time t under the assumption that the stock can appreciate or depreciate by 50% in a unit of time.

Field Generated by a Random Variable

Let $(\Omega, 2^\Omega)$ be a sample space with the field of all events, and X be a random variable with values x_i, $i = 1, 2, \ldots k$. Consider sets

$$A_i = \{\omega : \ X(\omega) = x_i\} \subseteq \Omega.$$

These sets form a partition of Ω, and the field generated by this partition is called the field generated by X. It is the smallest field that contains all the sets of the form $A_i = \{X = x_i\}$ and it is denoted by $\mathcal{F}_X$ or $\sigma(X)$. The field generated by X represents the information we can extract about the true state ω by observing X.

Example 2.5: (Example 2.3 continued.)
$\{\omega : \ X(\omega) = 1.5\} = \{\omega_1, \omega_2\} = A$, $\{\omega : \ X(\omega) = 0.5\} = \{\omega_3, \omega_4\} = \bar{A}$.

$$\mathcal{F}_X = \mathcal{F}_1 = \{\emptyset, \Omega, A, \bar{A}\}.$$

Filtration Generated by a Stochastic Process

Given $(\Omega, \mathcal{F})$ and a stochastic process $\{X(t)\}$ let $\mathcal{F}_t = \sigma(\{X_s, \ 0 \leq s \leq t\})$ be the field generated by random variables X_s, $s = 0, \ldots, t$. It is all the information available from the observation of the process up to time t. Clearly, $\mathcal{F}_t \subseteq \mathcal{F}_{t+1}$, so that these fields form a filtration. This filtration is called the natural filtration of the process $\{X(t)\}$.

If $A \in \mathcal{F}_t$ then by observing the process from 0 to t we know at time t whether the true state of the world is in A or not. We illustrate this on our financial example.

Example 2.6: Take $T = 3$, and assume that at each trading time the stock can go up by the factor u or down by d.

$$
\Omega = \begin{array}{cc}
\begin{array}{|ccc|}
\hline
u & u & u \\
u & u & d \\
\hline
\end{array} &
\begin{array}{|ccc|}
\hline
d & u & u \\
d & u & d \\
\hline
\end{array} \; \bar{B} \\[2mm]
\begin{array}{|ccc|}
\hline
u & d & u \\
u & d & d \\
\hline
\end{array} &
\begin{array}{|ccc|}
\hline
d & d & u \\
d & d & d \\
\hline
\end{array} \; B \\[2mm]
\quad A & \quad \bar{A}
\end{array}
$$

Look at the sets generated by information about S_1. This is a partition of Ω, $\{A, \bar{A}\}$. Together with the empty set and the whole set, this is the field $\mathcal{F}_1$. Sets generated by information about S_2 are B and $\bar{B}$. Thus the sets formed by knowledge of S_1 and S_2 is the partition of Ω, consisting of all intersections of the above sets. Together with the empty set and the whole set this is the field $\mathcal{F}_2$. Clearly any set in $\mathcal{F}_1$ is also in $\mathcal{F}_2$, for example $A = (A \cap B) \cup (A \cap \bar{B})$. Similarly if we add information about S_3 we obtain all the elementary sets, ω's and hence all subsets of Ω, $\mathcal{F}_3 = 2^\Omega$. In particular we will know the true state of the world when $T = 3$.
$\mathcal{F}_0 \subset \mathcal{F}_1 \subset \mathcal{F}_2 \subset \mathcal{F}_3$ is the filtration generated by the price process $\{S_t, \; t = 1, 2, 3\}$.

Predictable Processes

Suppose that a filtration $\mathbb{F} = (\mathcal{F}_0, \mathcal{F}_1, \ldots, \mathcal{F}_t, \ldots, \mathcal{F}_T)$ is given. A process H_t is called *predictable* (with respect to this filtration) if for each t, H_t is $\mathcal{F}_{t-1}$-measurable, that is, the value of the process H at time t is determined by the information up to and including time $t - 1$. For example, the number of shares held at time t is determined on the basis of information up to and including time $t - 1$. Thus this process is predictable with respect to the filtration generated by the stock prices.

Stopping Times

τ is called a random time if it is a non-negative random variable, which can also take value ∞ on $(\Omega, \mathcal{F}_T)$. Suppose that a filtration $\mathbb{F} = (\mathcal{F}_0, \mathcal{F}_1, \ldots, \mathcal{F}_t, \ldots, \mathcal{F}_T)$ is given. τ is called a *stopping time* with respect to this filtration if for each $t = 0, 1, \ldots, T$ the event

$$\{\tau \le t\} \in \mathcal{F}_t. \tag{2.1}$$

This means that by observing the information contained in $\mathcal{F}_t$ we can decide whether the event $\{\tau \leq t\}$ has or has not occurred. If the filtration $\mathbb{F}$ is generated by $\{S_t\}$, then by observing the process up to time t, $S_0, S_1, \ldots, S_t$, we can decide whether the event $\{\tau \leq t\}$ has or has not occurred.

Probability

If Ω is a finite sample space, then we can assign to each outcome ω a probability, $P(\omega)$, that is, the likelihood of it occurring. This assignment can be arbitrary. The only requirement is that $P(\omega) \geq 0$ and $\Sigma P(\omega) = P(\Omega) = 1$.

Example 2.7: Take $T = 2$ in our basic example 2.3. If the stock goes up or down independently of its value and if, say, the probability to go up is 0.4 then

$$
\begin{array}{ccccc}
\Omega = & \{(u, u); & (u, d); & (d, u) & (d, d)\} \\
P(\omega) & 0.16 & 0.24 & 0.24 & 0.36
\end{array}
$$

Distribution of a Random Variable

Since a random variable X is a function from Ω to $\mathbb{R}$, and Ω is finite, X can take only finitely many values, as the set $X(\Omega)$ is finite. Denote these values by x_i, $i = 1, 2, \ldots k$. The collection of probabilities p_i of sets $\{X = x_i\} = \{\omega : X(\omega) = x_i\}$ is called the probability distribution of X; for $i = 1, 2, \ldots k$

$$
p_i = P(X = x_i) = \sum_{\omega : X(\omega) = x_i} P(\omega).
$$

Expectation

If X is a random variable on $(\Omega, \mathcal{F})$ and P is a probability, then the expectation of X with respect to P is

$$
E_P X = \sum X(\omega) P(\omega),
$$

where the sum is taken over all elementary outcomes ω. It can be shown that the expectation can be calculated by using the probability distribution of X,

$$
E_P X = \sum_{i=1}^{k} x_i P(X = x_i).
$$

Of course if the probability P is changed to another probability Q, then the same random variable X may have a different probability distribution $q_i = Q(X = x_i)$, and a different expected value, $E_Q X = \sum_{i=1}^{k} x_i q_i$. When the probability P is fixed, or it is clear from the context with respect to which probability P the expectation is taken, then the reference to P is dropped from the notation, and the expectation is denoted simply by $E(X)$ or EX.

Conditional Probabilities and Expectations

Let $(\Omega, 2^\Omega, P)$ be a finite probability space, and $\mathcal{G}$ be a field generated by a partition of Ω, $\{D_1, \ldots, D_k\}$, such that $D_i \cap D_j = \emptyset$, and $\cup_i D_i = \Omega$. Recall that if D is an event of positive probability, $P(D) > 0$, then the conditional probability of an event A given the event D is defined by

$$P(A|D) = \frac{P(A \cap D)}{P(D)}.$$

Suppose that all the sets D_i in the partition have a positive probability. The conditional probability of the event A given the field $\mathcal{G}$ is the random variable that takes values $P(A|D_i)$ on D_i, $i = 1, \ldots k$. Let I_D denote the indicator of the set D, that is, $I_D(\omega) = 1$ if $\omega \in D$ and $I_D(\omega) = 1$ if $\omega \in \bar{D}$. Using this notation, the conditional probability can be written as

$$P(A|\mathcal{G})(\omega) = \sum_{i=1}^{k} P(A|D_i) I_{D_i}(\omega). \qquad (2.2)$$

For example, if $\mathcal{G} = \{\emptyset, \Omega\}$ is the trivial field, then

$$P(A|\mathcal{G}) = P(A|\Omega) = P(A).$$

Let now Y be a r.v. that takes values $y_1, \ldots, y_k$, then the sets $D_i = \{\omega : Y(\omega) = y_i\}$, $i = 1, \ldots k$, form a partition of Ω. If $\mathcal{F}_Y$ denotes the field generated by Y, then the conditional probability given $\mathcal{F}_Y$ is denoted by

$$P(A|\mathcal{F}_Y) = P(A|Y).$$

It was assumed so far that all the sets in the partition have a positive probability. If the partition contains a set of zero probability, call it N, then the conditional probability is not defined on N by the above formula (2.2). It can be defined for an $\omega \in N$ arbitrarily. Consequently any random variable which is defined by (2.2) and is defined arbitrarily on N is a version of the conditional probability. Any two versions only differ on N, which is a set of probability zero.

Conditional Expectation

In this section let X take values $x_1, \ldots, x_p$ and $A_1 = \{X = x_1\}, \ldots, A_p = \{X = x_p\}$. Let the field $\mathcal{G}$ be generated by a partition $\{D_1, D_2, \ldots, D_k\}$ of Ω. Then the conditional expectation of X given $\mathcal{G}$ is defined by

$$E(X|\mathcal{G}) = \sum_{i=1}^{p} x_i P(A_i|\mathcal{G}).$$

Note that $E(X|\mathcal{G})$ is a linear combination of random variables, so that it is a random variable. It is clear that $P(A|\mathcal{G}) = E(I_A|\mathcal{G})$, and $E(X|\mathcal{F}_0) = EX$, where $\mathcal{F}_0 = \{\emptyset, \Omega\}$ is the trivial field.

By the definition of measurability, X is $\mathcal{G}$-measurable if and only if for any i, $\{X = x_i\} = A_i$ is a member of $\mathcal{G}$, which means that it is either one of the D_j's or a union of some of the D_j's. Since $X(\omega) = \sum_{i=1}^p x_i I_{A_i}(\omega)$, a $\mathcal{G}$-measurable X can be written as $X(\omega) = \sum_{j=1}^k x_j I_{D_j}(\omega)$, where some x_j's may be equal. It is easy to see now that

$$\text{if } X \text{ is } \mathcal{G}\text{-measurable, then } E(X|\mathcal{G}) = X.$$

Note that since the conditional probabilities are defined up to a null set, so is the conditional expectation.

If X and Y are random variables both taking a finite number of values, then $E(X|Y)$ is defined as $E(X|\mathcal{G})$, where $\mathcal{G} = \mathcal{F}_Y$ is the field generated by the random variable Y. In other words if X takes values $x_1, \ldots, x_p$ and Y takes values $y_1, \ldots, y_k$, and $P(Y = y_i) > 0$ for all $i = 1, \ldots k$, then $E(X|Y)$ is a random variable, which takes values $\sum_{j=1}^p x_j P(X = x_j | Y = y_i)$ on the set $\{Y = y_i\}$ $i = 1, \ldots k$. These values are denoted by $E(X|Y = y_i)$. It is clear from the definition that $E(X|Y)$ is a function of Y,

$$E(X|Y)(\omega) = E(X|\mathcal{F}_Y)(\omega) = \sum_{i=1}^k \left(\sum_{j=1}^p x_j P(X = x_j | Y = y_i) \right) I_{\{Y = y_i\}}(\omega).$$

2.2   Continuous Probability Model

In this section we define similar probabilistic concepts for a continuous sample space. We start with general definitions.

σ-Fields

A σ-field is a field, which is closed with respect to countable unions and countable intersections of its members, that is a collection of subsets of Ω that satisfies

1. $\emptyset, \Omega \in \mathcal{F}$.

2. $A \in \mathcal{F} \Rightarrow \bar{A} \in \mathcal{F}$.

3. $A_1, A_2, \ldots, A_n, \ldots \in \mathcal{F}$ then $\bigcup_{n=1}^\infty A_n \in \mathcal{F}$ (and then also $\bigcap_{n=1}^\infty A_n \in \mathcal{F}$).

Any subset B of Ω that belongs to $\mathcal{F}$ is called a measurable set.

Borel σ-field

Borel σ-field is the most important example of a σ-field that is used in the theory of functions, Lebesgue integration, and probability. Consider the σ-field $\mathcal{B}$ on $\mathbb{R}$ ($\Omega = \mathbb{R}$) generated by the intervals. It is obtained by taking all the intervals first and then all the sets obtained from the intervals by forming countable unions, countable intersections and their complements are included into collection, and countable unions and intersections of these sets are included, etc. It can be shown that we end up with the smallest σ-field which contains all the intervals. A rigorous definition follows. One can show that the intersection of σ-fields is again a σ-field. Take the intersection of all σ-fields containing the collection of intervals. It is the smallest σ-field containing the intervals, the Borel σ-field on $\mathbb{R}$. In this model a measurable set is a set from $\mathcal{B}$, a Borel set.

Probability

A probability P on $(\Omega, \mathcal{F})$ is a set function on $\mathcal{F}$, such that

1. $P(\Omega) = 1$,

2. If $A \in \mathcal{F}$, then $P(\bar{A}) = 1 - P(A)$,

3. Countable additivity (σ-additivity): if $A_1, A_2, \ldots, A_n, \ldots \in \mathcal{F}$ are mutually exclusive, then $P\left(\bigcup_{n=1}^{\infty} A_n\right) = \sum_{n=1}^{\infty} P(A_n)$.

The σ-additivity property is equivalent to finite additivity plus the continuity property of probability, which states: if $A_1 \supseteq A_2 \supseteq \cdots \supseteq A_n \supseteq \cdots \supseteq A = \cap_{n=1}^{\infty} A_n \in \mathcal{F}$, then

$$\lim_{n \to \infty} P(A_n) = P(A).$$

A similar property holds for an increasing sequence of events.

How can one define a probability on a σ-field? It is not hard to see that it is impossible to assign probabilities to all individual ω's since there are too many of them and $P(\{\omega\}) = 0$. On the other hand it is difficult to assign probabilities to sets in $\mathcal{F}$ directly, since in general we don't even know what a set from $\mathcal{F}$ looks like. The standard way is to define the probability on a field which generates the σ-field, and then extend it to the σ-field.

Theorem 2.2 (Caratheodory Extension Theorem) *If a set function* P *is defined on a field $\mathcal{F}$, satisfies $P(\Omega) = 1$, $P(\bar{A}) = 1 - P(A)$, and is countably additive, then there is a unique extension of* P *to the σ-field generated by $\mathcal{F}$.*

Lebesgue Measure

As an application of the above Theorem 2.2 we define the Lebesgue measure on $[0, 1]$. Let $\Omega = [0, 1]$, and take for $\mathcal{F}$ the class of all finite unions of disjoint intervals contained in $[0, 1]$. It is clearly a field. Define the probability $P(A)$ on $\mathcal{F}$ by the length of A. It is not hard to show that P is σ-additive on $\mathcal{F}$. Thus there is a unique extension of P to $\mathcal{B}$, the Borel σ-field generated by $\mathcal{F}$. This extension is the Lebesgue measure on $\mathcal{B}$. It is also a probability on $\mathcal{B}$, since the length of $[0, 1]$ is one.

Any point has Lebesgue measure zero. Indeed, $\{x\} = \cap_n(x - 1/n, x + 1/n)$. Therefore $P(\{x\}) = \lim_{n \to \infty} 2/n = 0$. By countable additivity it follows that any countable set has Lebesgue measure zero. In particular the set of rationals on $[0, 1]$ is of zero Lebesgue measure. The set of irrationals on $[0, 1]$ has Lebesgue measure 1.

The term "almost everywhere" (for "almost all x") means everywhere (for all x) except, perhaps, a set of Lebesgue measure zero.

Random Variables

A random variable X on $(\Omega, \mathcal{F})$ is a measurable function from $(\Omega, \mathcal{F})$ to $(\mathbb{R}, \mathcal{B})$, where $\mathcal{B}$ is the Borel σ-field on the line. This means that for any Borel set $B \in \mathcal{B}$ the set $\{\omega : X(\omega) \in B\}$ is a member of $\mathcal{F}$. Instead of verifying the definition for all Borel sets, it is enough to have that for all real x the set $\{\omega : X(\omega) \leq x\} \in \mathcal{F}$. In simple words, for a random variable we can assign probabilities to sets of the form $\{X \leq x\}$, and $\{a < X \leq b\}$.

Example 2.8: Take $\Omega = \mathbb{R}$ with the Borel σ-field $\mathcal{B}$. By a measurable function on $\mathbb{R}$ is usually understood to be a $\mathcal{B}$-measurable function, that is, a random variable on $(\mathbb{R}, \mathcal{B})$. To define a probability on $\mathcal{B}$, take $f(x) = \frac{1}{\sqrt{2\pi}}e^{-\frac{x^2}{2}}$ and define $P(A) = \int_A f(x)dx$ for any interval A. It is easy to show that P so defined is a probability on the algebra containing the intervals, and it is continuous at the $\emptyset$. Thus it extends uniquely to $\mathcal{B}$. The function $X(x) = x$ on this probability space is called the standard Normal random variable.

An important question is how to describe measurable functions, and how to decide whether a given function is measurable. It is easy to see that indicators of measurable sets are measurable. (An indicator of a set A is defined as $I_A(\omega) = 1$ if and only if $\omega \in A$.) Conversely, if an indicator of a set is measurable then the set is measurable, $A \in \mathcal{F}$. A simple function (simple random variable) is a finite linear combination of indicators of measurable sets. By definition, a simple function takes finitely many values and is measurable.

Theorem 2.3 *X is a random variable on $(\Omega, \mathcal{F})$ if and only if it is a simple function or a limit of simple functions.*

The sufficiency part follows from the example below. The necessity is not hard to prove, by establishing that the supremum and infimum of random variables is a random variable.

Example 2.9: The following sequence of simple functions approximates a random variable X.

$$X_n(\omega) = \sum_{k=-n2^n}^{n2^n-1} \frac{k}{2^n} I_{[\frac{k}{2^n}, \frac{k+1}{2^n})}(X(\omega)).$$

These variables are constructed by taking the interval $[-n, n]$ and dividing it into $n2^{n+1}$ equal parts. X_n is zero on the set where $X \geq n$ or $X < -n$. On the set where the values of X belong to the interval $[\frac{k}{2^n}, \frac{k+1}{2^n})$, X is replaced by $\frac{k}{2^n}$ its smallest value on that set. Note that all the sets $\{\omega : \frac{k}{2^n} \leq X(\omega) < \frac{k+1}{2^n}\}$ are measurable, by definition of a random variable. It is easy to see that the X_n's are increasing, therefore converge to a limit, and for all ω, $X(\omega) = \lim_{n \to \infty} X_n(\omega)$.

This example is due to Lebesgue, who gave it for non-negative functions, demonstrating that a measurable function X is a limit of a monotone sequence of simple functions X_n, $X_{n+1} \geq X_n$.

The next result states that a limit of random variables and a composition of random variables is again a random variable.

Theorem 2.4

1. *If X_n are random variables on $(\Omega, \mathcal{F})$ and $X(\omega) = \lim_{n \to \infty} X_n(\omega)$, then X is also a random variable.*

2. *If X is a random variable on $(\Omega, \mathcal{F})$ and g is a $\mathcal{B}$-measurable function, then $g(X)$ is also a random variable.*

Remark 2.1: In the above theorem the requirement that the limit $X(\omega) = \lim_{n \to \infty} X_n(\omega)$ exists for all ω can be replaced by its existence for all ω outside a set of probability zero, and on a subsequence. Such a limit is in probability, it is introduced later.

σ-field Generated by a Random Variable

The σ-field generated by a random variable X is the smallest σ-field containing sets of the form $\{\omega : a \leq X(\omega) \leq b\}$, for any $a, b \in \mathbb{R}$.

Distribution of a Random Variable

The probability distribution function of X is defined as

$$F(x) = F_X(x) = P(X \leq x).$$

It follows from the properties of probability that F is non-decreasing and right-continuous. Due to monotonicity, it is of finite variation.

Joint Distribution

If X and Y are random variables defined on the same space, then their joint distribution function is defined as

$$F(x, y) = P(X \leq x, Y \leq y),$$

for any choice of real numbers x and y.

The distribution of X is recovered from the joint distribution of X and Y by $F_X(x) = F(x, \infty)$, and similarly the distribution of Y is given by $F(\infty, y)$, they are called the marginal distributions.

The joint distribution of n random variables $X_1, X_2, \ldots, X_n$ is defined by

$$P(X_1 \leq x_1, X_2 \leq x_2, \ldots, X_n \leq x_n).$$

The collection of random variables $\boldsymbol{X} = (X_1, X_2, \ldots, X_n)$ is referred to as a random vector $\boldsymbol{X}$, and the joint distribution as a multivariate probability distribution of $\boldsymbol{X}$. One can consider $\boldsymbol{X}$ as $\mathbb{R}^n$-valued random variable, and it is possible to prove that $\boldsymbol{X}$ is an $\mathbb{R}^n$-valued random variable if and only if all of its components are random variables.

Transformation of Densities

A random vector $\boldsymbol{X}$ has a density $f(\boldsymbol{x}) = f(x_1, x_2, \ldots, x_n)$ if for any set B (a Borel subset of $\mathbb{R}^n$),

$$P(\boldsymbol{X} \in B) = \int_{\boldsymbol{x} \in B} f(\boldsymbol{x}) dx_1 dx_2 \ldots dx_n. \tag{2.3}$$

If $\boldsymbol{X}$ is transformed into $\boldsymbol{Y}$ by a transformation $\boldsymbol{y} = \boldsymbol{y}(\boldsymbol{x})$, i.e.

$$
\begin{aligned}
y_1 &= y_1(x_1, x_2, \ldots, x_n) \\
y_2 &= y_2(x_1, x_2, \ldots, x_n) \\
\cdots &\quad \cdots\cdots \\
y_n &= y_n(x_1, x_2, \ldots, x_n),
\end{aligned}
$$

then, provided this transformation is one-to-one, and the inverse transformation has a non-vanishing Jacobian

$$
J = det
\begin{bmatrix}
\frac{\partial x_1}{\partial y_1} & \frac{\partial x_1}{\partial y_2} & \cdots & \frac{\partial x_1}{\partial y_n} \\
\frac{\partial x_2}{\partial y_1} & \frac{\partial x_2}{\partial y_2} & \cdots & \frac{\partial x_2}{\partial y_n} \\
\cdots & \cdots & \cdots & \\
\frac{\partial x_n}{\partial y_1} & \frac{\partial x_n}{\partial y_2} & \cdots & \frac{\partial x_n}{\partial y_n}
\end{bmatrix},
$$

Y has a density given by

$$f_{\mathbf{Y}}(\mathbf{y}) = f(x_1(\mathbf{y}), x_2(\mathbf{y}), \dots, x_n(\mathbf{y}))|J(\mathbf{y})|. \qquad (2.4)$$

This is easily established by using the change of variables in multiple integrals (see Example 2.15 for calculation of the bivariate density).

2.3 Expectation and Lebesgue Integral

Let X be a random variable on $(\Omega, \mathcal{F})$, and P be a probability on $\mathcal{F}$. Recall that in the discrete case the expectation of X is defined as $\sum_\omega X(\omega)\mathrm{P}(\omega)$. The expectation in the continuous model is defined by the means of an integral

$$\mathrm{E}X = \int_\Omega X(\omega)d\mathrm{P}(\omega).$$

The expectation is defined for positive random variables first. The general case is obtained by using the decomposition $X = X^+ - X^-$, where $X^+ = \max(X, 0)$ and $X^- = \max(-X, 0)$, and letting $\mathrm{E}X = \mathrm{E}X^+ - \mathrm{E}X^-$, provided both $\mathrm{E}X^+$ and $\mathrm{E}X^-$ are finite.

If $X \geq 0$ is a random variable, then it can be approximated by simple random variables (see Theorem 2.3 and Example 2.9). The expectation of a simple random variable is defined as a sum, that is if

$$X = \sum_{k=1}^n c_k I_{A_k}, \quad \text{then} \quad \mathrm{E}X = \sum_{k=1}^n c_k \mathrm{P}(A_k).$$

Note that for a simple random variable, $X \geq 0$ implies $\mathrm{E}X \geq 0$. This in turn implies that if $X \geq Y$, where X and Y are simple random variables, then $\mathrm{E}X \geq \mathrm{E}Y$.

Any positive random variable X can be approximated by an increasing sequence X_n of simple random variables, such approximation is given in Example 2.9. It now follows that since X_n is an increasing sequence, $\mathrm{E}X_n$ is also an increasing sequence, hence has a limit. The limit of $\mathrm{E}X_n$ is taken to be $\mathrm{E}X$. It can be shown that this limit does not depend on the choice of the approximating sequence of simple random variables, so that the expectation is defined unambiguously.

Definition 2.5 *A random variable X is called integrable if both $\mathrm{E}X^+$ and $\mathrm{E}X^-$ are finite. In this case $\mathrm{E}X = \mathrm{E}X^+ - \mathrm{E}X^-$.*

Note that for X to be integrable both $\mathrm{E}X^+$ and $\mathrm{E}X^-$ must be finite, which is the same as $\mathrm{E}|X| = \mathrm{E}X^+ + \mathrm{E}X^- < \infty$.

Lebesgue-Stieltjes Integral

A distribution function on $\mathbb{R}$ is a non-decreasing right-continuous function which approaches 0 at $-\infty$ and 1 at $+\infty$. Such a distribution function, say F, defines uniquely a probability on the Borel σ-field $\mathcal{B}$ by setting $P((a, b]) = F(b) - F(a)$.

Take now $(\Omega, \mathcal{F}) = (\mathbb{R}, \mathcal{B})$ and a probability on $\mathcal{B}$ given by a distribution function $F(x)$. A random variable on this space is a measurable function $f(x)$. Its expected value is written as $\int_{\mathbb{R}} f(x)F(dx)$ and is called the Lebesgue-Stieltjes integral of f with respect to F.

The distribution function F can be replaced by any function of finite variation, giving rise to the general Lebesgue-Stieltjes integral.

The probability distribution of a random variable X on $(\Omega, \mathcal{F})$ is the probability on $\mathcal{B}$ carried from $\mathcal{F}$ by X: for any $B \in \mathcal{B}$,

$$P_X(B) = P(X \in B). \tag{2.5}$$

The distribution function is related to this probability by $F(x) = P_X((-\infty, x])$. Equation (2.5) gives the relation between the expectations of indicator functions,

$$\int_{\Omega} I(X(\omega) \in B)dP(\omega) = \int_{-\infty}^{\infty} I(x \in B)P_X(dx).$$

This can be extended from indicators to measurable functions, using an approximation by simple functions, and we have the following result.

Theorem 2.6 *If X is a random variable with distribution function $F(x)$, and h is a measurable function on $\mathbb{R}$, such that $h(X)$ is integrable, then*

$$Eh(X) := \int_{\Omega} h(X(\omega))dP(\omega) = \int_{-\infty}^{\infty} h(x)P_X(dx) := \int_{-\infty}^{\infty} h(x)F(dx). \tag{2.6}$$

Lebesgue Integral on the Line

The Lebesgue-Stieltjes integral with respect to $F(x) = x$ is known as the Lebesgue integral.

Example 2.10: Let $\Omega = [0, 1]$, its elements ω are real numbers x, and take for probability the Lebesgue measure. Take $X(\omega) = X(x) = x^2$. Then $EX = \int_0^1 x^2 dx = 1/3$. Construct an approximating sequence of simple functions and verify the value of the above integral.

Similarly, for any continuous function $f(x)$ on $[0, 1]$, $X(\omega) = X(x) = f(x)$ is a random variable (using that a continuous function is measurable) with expectation $EX = Ef = \int_0^1 f(x)dx$.

Theorem 2.7 *If f is Riemann integrable on $[a, b]$, then it is Lebesgue integrable on $[a, b]$ and the integrals coincide.*

On the other hand there are functions which are Lebesgue integrable but not Riemann integrable. Recall that for a function to be Riemann integrable, it must be continuous at all points except for a set of Lebesgue measure zero. Some everywhere discontinuous functions are Lebesgue integrable.

Example 2.11: $\Omega = [0, 1]$, and probability is given by the Lebesgue measure. Take $X(x) = I_Q(x)$ be the indicator function of the set Q of all rationals. Q has Lebesgue measure zero. As the expectation of an indicator is the probability of its set, $EX = \int_0^1 I_Q(x)dx = 0$. However, $I_Q(x)$ is discontinuous at every point, so that the set of discontinuities of $I_Q(x)$ is $[0, 1]$ which has Lebesgue measure 1, therefore $I_Q(x)$ is not Riemann integrable.

The next result is the fundamental theorem for the Lebesgue integral on the line.

Theorem 2.8 *If f is Lebesgue integrable on $[a, b]$ then the derivative of the integral exists for almost all $x \in (a, b)$, and*

$$\frac{d}{dx} \int_a^x f(t)dt = f(x). \tag{2.7}$$

Properties of Expectation (Lebesgue Integral)

It is not hard to show that the expectation (Lebesgue Integral) satisfies the following properties

1. Linearity. If X and Y are integrable and α and β are constants, then $E(\alpha X + \beta Y) = \alpha EX + \beta EY$.

2. If random variables X and Y satisfy, $|X| \leq Y$ and Y is integrable, then X is also integrable and $E|X| \leq EY$.

3. If a random variable $X \geq 0$ then $EX = 0$ if and only if $P(X = 0) = 1$.

Jumps and Probability Densities

The jump of F at x gives the probability $P(X = x)$, $F(x) - F(x-) = P(X = x)$. Since F is right-continuous it has at most countably many jumps.

Definition 2.9 *F is called discrete if it changes only by jumps.*

If $F(x)$ is continuous at x then $P(X = x) = 0$.

Definition 2.10 *$F(x)$ is called absolutely continuous if there is a function $f(x) \geq 0$, such that for all x, $F(x)$ is given by the Lebesgue integral $F(x) = \int_{-\infty}^x f(t)dt$. In this case $F'(x) = f(x)$ for almost all x (Theorem 2.8).*

f is called the probability density function of X. It follows from the definition that for any $a < b$

$$P(a \leq X \leq b) = \int_a^b f(x)dx.$$

There are plenty of examples in any introductory book on probability or statistics of continuous random variables with densities: Normal, Exponential, Uniform, Gamma, Cauchy, etc.

The random variables X and Y with the joint distribution $F(x,y)$ possess a density $f(x,y)$ (with respect to the Lebesgue measure) if for any x, y

$$F(x,y) = \int_{-\infty}^x \int_{-\infty}^y f(u,v)dudv,$$

and then for almost all (with respect to the Lebesgue measure on the plane) x, y,

$$f(x,y) = \frac{\partial^2 F}{\partial x \partial y}(x,y).$$

A density for an n-dimensional random vector is defined similarly.

Decomposition of Distributions and FV Functions

Any distribution function can be decomposed into a continuous part and a jump part. Continuous distribution functions can be decomposed further (Lebesgue decomposition).

If F is a continuous distribution function, then it can be decomposed into sum of two continuous distribution functions, the absolutely continuous part and the singular part, i.e. for some $0 \leq a \leq 1$

$$F = aF_{\text{ac}} + (1-a)F_{\text{sing}}. \tag{2.8}$$

F_{ac} is characterized by its density that exists at almost all points. For the singular part, $F'_{\text{sing}}(x)$ exists for almost all x and is zero. An example of such function is the Cantor function, see Example 2.13, where the distribution function is a constant between the points of the Cantor set. In most applications in statistics continuous distributions are absolutely continuous with zero singular part.

Example 2.12: (Cantor set.) Consider the set $\{x : x = \sum_{n=1}^\infty \alpha_n/3^n, \alpha_n \in \{0,2\}\}$. It is possible to show that this set does not have isolated points (a perfect set), that is, any point of the set is a limit of other points. Indeed, for a given sequence of α_n's that contains infinitely many 2s, consider a new sequence which is the same up to the $m-th$ term with all the rest being zeros. The distance between these two numbers is given by $\sum_{n=m+1}^\infty \alpha_n/3^n < 3^{-m}$, which can be made arbitrarily small as

m increases. For a number with finitely many 2s, say k 2s, the numbers with the first k same places, and the rest zeros except the m-th place which is 2, approximate it. Indeed, the distance between these two numbers is $2/3^m$. It is also not hard to see that this set is uncountable (by the diagonal argument) and that it has Lebesgue measure zero. Although the Cantor set seems to be artificially constructed, Cantor type sets arise naturally in study of Brownian motion; for example, the set of zeros of Brownian motion is a Cantor type set.

Example 2.13: (Cantor distribution.)
The distribution function F of the random variable $X = \sum_{n=1}^{\infty} \alpha_n/3^n$, where α_n are independent identically distributed random variables taking values 0 and 2 with probability $1/2$, is continuous and its derivative is zero almost everywhere.

It takes a rather pathological example to construct a continuous singular distribution in one dimension. In dimension two such examples can be simple.

Example 2.14: (Continuous singular distributions on the plane.)
Take F such that $\frac{\partial^2 F}{\partial x \partial y} = 0$ almost everywhere on the plane. If F is a linear function in x and y, or a distribution function that does not depend on one of the variables x or y, then it is singular. For example, $0 \leq X, Y \leq 1$ such that their joint distribution is determined by $F(x,y) = \frac{1}{2}(x+y)$, for x,y satisfying $0 \leq x,y \leq 1$. In this case only sets that have non-empty intersection with the axis have positive probability.

Functions of finite variations have a similar structure to distribution functions. They can be decomposed into a continuous part and a jump part, and the continuous part can be decomposed further into an absolutely continuous part and a singular part.

2.4 Transforms and Convergence

If X^k is integrable, $\mathrm{E}|X|^k < \infty$, then the k-th moment of X is defined as $\mathrm{E}(X^k)$. The moment generating function of X is defined as

$$m(t) = \mathrm{E}(e^{tX}),$$

provided e^{tX} is integrable, for t in a neighbourhood of 0.

Using the series expansion for the exponential function

$$e^x = \sum_{n=0}^{\infty} \frac{x^n}{n!},$$

we can formally write, by interchanging summation and the expectation,

$$m(t) = \mathrm{E}e^{tX} = \mathrm{E}\sum_{n=0}^{\infty} \frac{t^n X^n}{n!} = \sum_{n=0}^{\infty} \frac{t^n}{n!}\mathrm{E}(X^n). \tag{2.9}$$

Thus $E(X^n)$ can be obtained from the power series expansion of the moment generating function.

The characteristic function of X is defined as

$$\phi(t) = E(e^{itX}) = E(\cos(tX)) + iE(\sin(tX)),$$

where $i = \sqrt{-1}$. The characteristic function determines the distribution uniquely, so does the moment generating function when it exists on an interval containing 0. The advantage of the characteristic function over the moment generating function is that it exists for any random variable X, since the functions $\cos(tx)$ and $\sin(tx)$ are bounded for any t on the whole line, whereas e^{tx} is unbounded and the moment generating function need not exist. Existence of the moment generating function around zero implies existence of all the moments. If X does not have all the moments, then its moment generating function does not exist.

Convergence of Random Variables

There are four main concepts of convergence of a sequence of random variables. We give the definitions of progressively stronger concepts and some results on their relations.

Definition 2.11 $\{X_n\}$ *converge in distribution to X, if their distribution functions $F_n(x)$ converge to the distribution function $F(x)$ at any point of continuity of F.*

It can be shown that $\{X_n\}$ converge in distribution to X if and only if their characteristic functions (or moment generating functions) converge to that of X. Convergence in distribution is also equivalent to the requirement that $Eg(X_n) \to Eg(X)$ as $n \to \infty$ for all bounded continuous functions g on $\mathbb{R}$.

Definition 2.12 $\{X_n\}$ *converge in probability to X if for any $\epsilon > 0$* $P(|X_n - X| > \epsilon) \to 0$ *as $n \to \infty$.*

Definition 2.13 $\{X_n\}$ *converge almost surely (a.s.) to X if for any ω outside a set of zero probability $X_n(\omega) \to X(\omega)$ as $n \to \infty$.*

Almost sure convergence implies convergence in probability, which in turn implies convergence in distribution. It is also not hard to see that convergence in distribution to a constant is the same as the convergence in probability to the same constant. Convergence in probability implies the almost sure convergence on a subsequence, namely, if $\{X_n\}$ converge in probability to X then there is a subsequence n_k that converges almost surely to the same limit.

L^r-convergence (convergence in the r-th mean), $r \geq 1$, is defined as follows.

Definition 2.14 $\{X_n\}$ *converge to* X *in* L^r *if for any* n $E(|X_n|^r) < \infty$, *and* $E(|X_n - X|^r) \to 0$ *as* $n \to \infty$.

Using the concept of uniform integrability, given later in Chapter 7, convergence in L^r is equivalent to convergence in probability and uniform integrability of $|X_n|^r$ (see for example, Loeve (1978) p.164).

The following result, which is known as Slutskii theorem, is frequently used in applications.

Theorem 2.15 *If* X_n *converges to* X *and* Y_n *converges to* Y, *then* $X_n + Y_n$ *converges to* $X + Y$, *for any type of stochastic convergence, except for convergence in distribution. However, if* $Y = 0$ *or* X_n *and* Y_n *are independent, then the result is also true for convergence in distribution.*

Convergence of Expectations

Theorem 2.16 (Monotone convergence) *If* $X_n \geq 0$, *and* X_n *are increasing to a limit* X, *which may be infinite, then* $\lim_{n \to \infty} EX_n = EX$.

Theorem 2.17 (Fatou's lemma) *If* $X_n \geq 0$ *(or* $X_n \geq c > -\infty$), *then* $E(\liminf_n X_n) \leq \liminf_n EX_n$.

Theorem 2.18 (Dominated Convergence) *If* $\lim_{n \to \infty} X_n = X$ *in probability and for all* n $|X_n| \leq Y$ *with* $EY < \infty$, *then* $\lim_{n \to \infty} EX_n = EX$.

2.5 Independence and Covariance

Independence

Two events A and B are called independent if $P(A \cap B) = P(A)P(B)$.

A collection of events A_i, $i = 1, 2, \ldots$ is called independent if for any finite n and any choice of indices i_k, $k = 1, 2, \ldots n$

$$P\left(\bigcap_{k=1}^{n} A_{i_k}\right) = \prod_{k=1}^{n} P(A_{i_k}).$$

Two σ-fields are called independent if for any choice of sets from each of them, these sets are independent.

Two random variables X and Y are independent if the σ-fields they generate are independent. It follows that their joint distribution is given by the product of their marginal distributions (since the sets $\{X \leq x\}$ and $\{Y \leq y\}$ are in the respective σ-fields)

$$P(X \leq x, Y \leq y) = P(X \leq x)P(Y \leq y),$$

and can be seen that it is an equivalent property.

One can formulate the independence property in terms of expectations by writing above in terms of indicators

$$\mathrm{E}\Big(I(X \leq x)I(Y \leq y)\Big) = \mathrm{E}\Big(I(X \leq x)\Big)\mathrm{E}\Big(I(Y \leq y)\Big).$$

Since it is possible to approximate indicators by continuous bounded functions, X and Y are independent if and only if for any bounded continuous functions f and g,

$$\mathrm{E}(f(X)g(Y)) = \mathrm{E}(f(X))\mathrm{E}(g(Y)).$$

$X_1, X_2, \ldots, X_n$ are called independent if for any choice of random variables $X_{i_1}, X_{i_2}, \ldots X_{i_k}$ their joint distribution is given by the product of their marginal distributions (alternatively, if the σ-fields they generate are independent).

Covariance

The covariance of two integrable random variables X and Y is defined, provided XY is integrable, by

$$Cov(X,Y) = \mathrm{E}\,(X - \mathrm{E}X)\,(Y - \mathrm{E}Y) = \mathrm{E}\,(XY) - \mathrm{E}X\mathrm{E}Y. \qquad (2.10)$$

The variance of X is the covariance of X with itself, $Var(X) = Cov(X,X)$. The Cauchy-Schwarz inequality

$$(\mathrm{E}|XY|)^2 \leq \mathrm{E}(X^2)\mathrm{E}(Y^2), \qquad (2.11)$$

assures that covariance exists for square integrable random variables. Covariance is symmetric,

$$Cov(X,Y) = Cov(Y,X),$$

and is linear in both variables (bilinear)

$$Cov(aX + bY, Z) = aCov(X,Z) + bCov(Y,Z).$$

Using this property with $X + Y$ we obtain the formula for the variance of the sum. The following property of the covariance holds

$$Var(X + Y) = Var(X) + Var(Y) + 2Cov(X,Y). \qquad (2.12)$$

Random variables X and Y are called *uncorrelated* if $Cov(X,Y) = 0$. It is easy to see from the definitions that for independent random variables

$$\mathrm{E}(XY) = \mathrm{E}X\mathrm{E}Y,$$

which implies that they are uncorrelated. The opposite implication is not true in general. The important exception is the Gaussian case.

Theorem 2.19 *If the random variables have a joint Gaussian distribution, then they are independent if and only if they are uncorrelated.*

Definition 2.20 *The covariance matrix of a random vector* $\boldsymbol{X} = (X_1, X_2, \ldots, X_n)$ *is the* $n \times n$ *matrix with the elements* $Cov(X_i, X_j)$.

2.6 Normal (Gaussian) Distributions

The Normal (Gaussian) probability density is given by

$$f(x; \mu, \sigma^2) = \frac{1}{\sqrt{2\pi}\sigma} e^{-\frac{(x-\mu)^2}{2\sigma^2}}.$$

It is completely specified by its mean μ and its standard deviation σ. The Normal family $N(\mu, \sigma^2)$ is obtained from the Standard Normal Distribution, $N(0, 1)$ by a linear transformation.

If X is $N(\mu, \sigma^2)$ then $Z = \dfrac{X - \mu}{\sigma}$ is $N(0, 1)$ and $X = \mu + \sigma Z$.

An important property of the Normal family is that a linear combination of independent Normal variables results in a Normal variable, that is, if $X_1 \sim N(\mu_1, \sigma_1^2)$ and $X_2 \sim N(\mu_2, \sigma_2^2)$ are independent then $\alpha X_1 + \beta X_2 \sim N(\alpha\mu_1 + \beta\mu_2, \alpha^2\sigma_1^2 + \beta^2\sigma_2^2)$. The moment generating function of X with $N(\mu, \sigma^2)$ distribution is given by

$$m(t) = \mathrm{E}e^{tX} = \int_{-\infty}^{\infty} e^{tx} f(x; \mu, \sigma^2) dx = e^{\mu t} e^{(\sigma t)^2/2} = e^{\mu t + (\sigma t)^2/2}.$$

A random vector $\boldsymbol{X} = (X_1, X_2, \ldots, X_n)$ has an n-variate Normal (Gaussian) distribution with mean vector $\boldsymbol{\mu}$ and covariance matrix $\boldsymbol{\Sigma}$ if there exist an $n \times n$ matrix A such that its determinant $|A| \neq 0$, and $\boldsymbol{X} = \boldsymbol{\mu} + A\boldsymbol{Z}$, where $\boldsymbol{Z} = (Z_1, Z_2, \ldots, Z_n)$ is the vector with independent standard Normal components, and $\boldsymbol{\Sigma} = AA^T$. Vectors are taken as column vectors here and in the sequel.

The probability density of $\boldsymbol{Z}$ is obtained by using independence of its components, for independent random variables the densities multiply. Then performing a change of variables in the multiple integral we find the probability density of $\boldsymbol{X}$

$$f_{\boldsymbol{X}}(\boldsymbol{x}) = \frac{1}{(2\pi)^{n/2}|\boldsymbol{\Sigma}|^{1/2}} e^{-\frac{1}{2}(\boldsymbol{x}-\boldsymbol{\mu})\boldsymbol{\Sigma}^{-1}(\boldsymbol{x}-\boldsymbol{\mu})^T}.$$

Example 2.15: Let a bivariate Normal have $\boldsymbol{\mu} = \boldsymbol{0}$ and $\boldsymbol{\Sigma} = \begin{bmatrix} 1 & \rho \\ \rho & 1 \end{bmatrix}$. Let $\boldsymbol{X} = (X, Y)$ and $\boldsymbol{x} = (x, y)$. Then $\boldsymbol{X}$ can be obtained from $\boldsymbol{Z}$ by the transformation

$X = AZ$ with $A = \begin{bmatrix} 1 & 0 \\ \rho & \sqrt{1-\rho^2} \end{bmatrix}$.

Since $\begin{cases} x &= & z_1 \\ y &= & \rho z_1 + \sqrt{1-\rho^2}z_2 \end{cases}$, the inverse transformation

$\begin{cases} z_1 &= & x \\ z_2 &= & (y - \rho x)/\sqrt{1-\rho^2} \end{cases}$ has the Jacobian

$$J = det \begin{bmatrix} \frac{\partial z_1}{\partial x} & \frac{\partial z_1}{\partial y} \\ \frac{\partial z_2}{\partial x} & \frac{\partial z_2}{\partial y} \end{bmatrix} = det \begin{bmatrix} 1 & 0 \\ \frac{1}{\sqrt{1-\rho^2}} \end{bmatrix} = \frac{1}{\sqrt{1-\rho^2}}.$$

The density of Z is given by the product of standard Normal densities, by independence, $f_Z(z_1, z_2) = \frac{1}{2\pi}e^{-\frac{1}{2}(z_1^2+z_2^2)}$. Using the formula (2.4) we obtain the joint density of the bivariate Normal

$$f_X(x,y) = \frac{1}{2\pi\sqrt{1-\rho^2}}e^{-\frac{1}{2(1-\rho^2)}[x^2 - 2\rho xy + y^2]}.$$

It follows from the definition that if X has a multivariate Normal distribution and $\mathbf{a}$ is a non-random vector then $\mathbf{a}X = \mathbf{a}(\mu + AZ) = \mathbf{a}\mu + \mathbf{a}AZ$. Since a linear combination of independent Normal random variables is a Normal random variable, $\mathbf{a}AZ$ is a Normal random variable. Hence $\mathbf{a}X$ has Normal distribution with mean $\mathbf{a}\mu$ and variance $(\mathbf{a}A)(\mathbf{a}A)^T = \mathbf{a}\Sigma\mathbf{a}^T$. Thus we have

Theorem 2.21 *A linear combination of jointly Gaussian random variables is a Gaussian random variable.*

Similarly it can be shown that if $X \sim N(\mu, \Sigma)$ and B is a non-random matrix, then $BX \sim N(B\mu, B\Sigma B^T)$.

The moment generating function of a vector X is defined as

$$E(e^{tX}) = E(e^{\sum_{i=1}^{n} t_i X_i}),$$

where $t = (t_1, t_2, \ldots, t_n)$, and tX is the scalar product of vectors t and X.

It is not hard to show that the moment generating function of a Gaussian vector $X \sim N(\mu, \Sigma)$ is given by

$$M_X(t) = e^{\mu t^T - \frac{1}{2}t\Sigma t^T}.$$

Definition 2.22 *A collection of random variables is called a Gaussian process, if the joint distribution of any finite number of its members is Gaussian.*

Theorem 2.23 *Let $X(t)$ be a process with independent Gaussian increments, that is, for any $s < t$, $X(t) - X(s)$ has a Normal distribution, and is independent of the values $X(u), u \le s$ (the σ-field $\mathcal{F}_s$ generated by the process up to time s). Then $X(t)$ is a Gaussian process.*

See Section 3.1, Example 3.3 for the proof.

2.7 Conditional Expectation

Conditional Expectation and Conditional Distribution

The conditional distribution function of X given $Y = y$ is defined by

$$P(X \leq x | Y = y) = \frac{P(X \leq x, Y = y)}{P(Y = y)},$$

provided $P(Y = y) > 0$. However, such an approach fails if the event we condition on has zero probability, $P(Y = y) = 0$. This difficulty can be overcome if X, Y have a joint density $f(x, y)$. In this case it follows that both X and Y possess densities $f_X(x)$, and $f_Y(y)$; $f_X(x) = \int_{-\infty}^{\infty} f(x, y) dy$, and $f_Y(y) = \int_{-\infty}^{\infty} f(x, y) dx$. The conditional distribution of X given $Y = y$ is defined by the conditional density

$$f(x|y) = \frac{f(x, y)}{f_Y(y)},$$

at any point where $f_Y(y) > 0$. It is easy to see that so defined $f(x|y)$ is indeed a probability density for any y, as it is non-negative and integrates to unity.

The expectation of this distribution, when it exists, is called the conditional expectation of X given $Y = y$,

$$E(X|Y = y) = \int_{-\infty}^{\infty} x f(x|y) dx. \tag{2.13}$$

The conditional expectation $E(X|Y = y)$ is a function of y. Let g denote this function, $g(y) = E(X|Y = y)$, then by replacing y by Y we obtain a random variable $g(Y)$, which is the conditional expectation of X given Y, $E(X|Y) = g(Y)$.

Example 2.16: Let X and Y have a standard bivariate Normal distribution with parameter ρ. Then

$f(x, y) = \frac{1}{2\pi\sqrt{1-\rho^2}} \exp\left\{-\frac{1}{2(1-\rho^2)}[x^2 - 2\rho xy + y^2]\right\}$, and $f_Y(y) = \frac{1}{\sqrt{2\pi}} e^{-y^2/2}$, so that

$f(x|y) = \frac{f(x,y)}{f_Y(y)} = \frac{1}{\sqrt{2\pi(1-\rho^2)}} \exp\left\{-\frac{(x-\rho y)^2}{2(1-\rho^2)}\right\}$, which is the $N(\rho y, 1 - \rho^2)$ distribution. Its mean is ρy, therefore $E(X|Y = y) = \rho y$, and $E(X|Y) = \rho Y$.

Similarly, it can be seen that in the multivariate Normal case the conditional expectation is also a linear function of Y.

The conditional distribution and the conditional expectation are defined only at the points where $f_Y(y) > 0$. Both can be defined arbitrarily on the set $\{y : f_Y(y) = 0\}$. Since there are many functions which agree on the set $\{y : f_Y(y) > 0\}$, any one of them is called a version of the conditional distribution (the conditional expectation) of X given $Y = y$. The different versions of $f(x|y)$ and $E(X|Y = y)$ differ only on the set $\{y : f_Y(y) = 0\}$, which has zero probability under the distribution of Y; $f(x|y)$ and $E(X|Y = y)$ are defined uniquely Y-almost surely.

General Conditional Expectation

The conditional expectation in a more general form is defined as follows. Let X be an integrable random variable. $E(X|Y) = G(Y)$ a function of Y such that for any bounded function h,

$$E(Xh(Y)) = E(Yh(Y)), \tag{2.14}$$

or $E\big((X - G(Y))h(Y)\big) = 0$. Existence of such a function is assured by the Radon-Nikodym theorem from functional analysis. But uniqueness is easy to prove. If there are two such functions, G_1, G_2, then $E((G_1(Y)-G_2(Y))h(Y)) = 0$. Take $h(y) = \text{sign}(G_1(y) - G_2(y))$. Then we have $E|G_1(Y) - G_2(Y)| = 0$. Thus $P(G_1(Y) = G_2(Y)) = 1$, and they coincide with (Y) probability one.

A more general conditional expectation of X given a σ-field $\mathcal{G}$, $E(X|\mathcal{G})$ is a $\mathcal{G}$-measurable random variable such that for any bounded $\mathcal{G}$-measurable ξ

$$E(\xi E(X|\mathcal{G})) = E(\xi X). \tag{2.15}$$

In the literature, $\xi = I_B$ is taken as indicator function of a set $B \in \mathcal{G}$, which is an equivalent condition: for any set $B \in \mathcal{G}$

$$\int_B X\,d\mathrm{P} = \int_B E(X|\mathcal{G})d\mathrm{P}, \ \text{ or } \ E\big(XI(B)\big) = E\Big(E(X|\mathcal{G})I(B)\Big). \tag{2.16}$$

The Radon-Nikodym theorem (see Theorem 10.6) implies that such a random variable exists and is almost surely unique, in the sense that any two versions differ only on a set of probability zero.

The conditional expectation $E(X|Y)$ is given by $E(X|\mathcal{G})$ with $\mathcal{G} = \sigma(Y)$, the σ-field generated by Y. Often the Equations (2.15) or (2.16) are not used, because easier calculations are possible to various specific properties, but they are used to establish the fundamental properties given below. In particular, the conditional expectation defined in (2.13) by using densities satisfies (2.15) or (2.16).

Properties of Conditional Expectation

Conditional expectations are random variables. Their properties are stated as equalities of two random variables. Random variables X and Y, defined on the same space, are equal if $P(X = Y) = 1$. This is also written $X = Y$

almost surely (a.s.). If not stated otherwise, whenever the equality of random variables is used it is understood in the almost sure sense, and often writing "almost surely" is omitted.

1. If $\mathcal{G}$ is the trivial field $\{\emptyset, \Omega\}$, then

$$\mathrm{E}(X|\mathcal{G}) = \mathrm{E}X, \qquad (2.17)$$

2. If X is $\mathcal{G}$-measurable, then

$$\mathrm{E}(XY|\mathcal{G}) = X\mathrm{E}(Y|\mathcal{G}). \qquad (2.18)$$

This means that if $\mathcal{G}$ contains all the information about X, then given $\mathcal{G}$, X is known, and therefore it is treated as a constant.

3. If $\mathcal{G}_1 \subset \mathcal{G}_2$ then

$$\mathrm{E}(\mathrm{E}(X|\mathcal{G}_2)|\mathcal{G}_1) = \mathrm{E}(X|\mathcal{G}_1). \qquad (2.19)$$

This is known as the smoothing property of conditional expectation. In particular by taking $\mathcal{G}_1$ to be the trivial field, we obtain the law of double expectation

$$\mathrm{E}(\mathrm{E}(X|\mathcal{G})) = \mathrm{E}(X). \qquad (2.20)$$

4. If $\sigma(X)$ and $\mathcal{G}$ are independent, then

$$\mathrm{E}(X|\mathcal{G}) = \mathrm{E}X, \qquad (2.21)$$

that is, if the information we know provides no clues about X, then the conditional expectation is the same as the expectation. The next result is an important generalization.

5. If $\sigma(X)$ and $\mathcal{G}$ are independent, and $\mathcal{F}$ and $\mathcal{G}$ are independent, and $\sigma(\mathcal{F}, \mathcal{G})$ denotes the smallest σ-field containing both of them, then

$$\mathrm{E}(X|\sigma(\mathcal{F}, \mathcal{G})) = \mathrm{E}(X|\mathcal{F}). \qquad (2.22)$$

6. Jensen's inequality. If $g(x)$ is a convex function on I, that is, for all $x, y, \in I$ and $\lambda \in (0, 1)$

$$g(\lambda x + (1 - \lambda)y) \leq \lambda g(x) + (1 - \lambda)g(y),$$

and X is a random variable with range I, then

$$g(\mathrm{E}(X|\mathcal{G})) \leq \mathrm{E}(g(X)|\mathcal{G}). \qquad (2.23)$$

In particular, with $g(x) = |x|$

$$|\mathrm{E}(X|\mathcal{G})| \leq \mathrm{E}(|X||\mathcal{G}). \qquad (2.24)$$

7. Monotone convergence. If $0 \leq X_n$, and $X_n \uparrow X$ with $E|X| < \infty$, then

$$E(X_n|\mathcal{G}) \uparrow E(X|\mathcal{G}). \tag{2.25}$$

8. Fatous' lemma. If $0 \leq X_n$, then

$$E\big(\liminf_n X_n|\mathcal{G}\big) \leq \liminf_n E(X_n|\mathcal{G}). \tag{2.26}$$

9. Dominated convergence. If $\lim_{n\to\infty} X_n = X$ almost surely and $|X_n| \leq Y$ with $EY < \infty$, then

$$\lim_{n\to\infty} E(X_n|\mathcal{G}) = E(X|\mathcal{G}). \tag{2.27}$$

For results on conditional expectations see e.g. Breiman (1968), Chapter 4. The conditional probability $P(A|\mathcal{G})$ is defined as the conditional expectation of the indicator function,

$$P(A|\mathcal{G}) = E(I_A|\mathcal{G}),$$

and it is a $\mathcal{G}$-measurable random variable, defined P-almost surely.

The following results are often used.

Theorem 2.24 *Let X and Y be two independent random variables and $\phi(x,y)$ be such that $E|\phi(X,Y)| < +\infty$. Then*

$$E(\phi(X,Y)|Y) = G(Y),$$

where $G(y) = E(\phi(X,y))$.

Theorem 2.25 *Let (X,Y) be a Gaussian vector. Then the conditional distribution of X given Y is also Gaussian. Moreover, provided the matrix $Cov(Y,Y)$ is non-singular (has the inverse),*

$$E(X|Y) = E(X) + Cov(X,Y)Cov^{-1}(Y,Y)(Y - E(Y)).$$

In the case when $Cov(Y,Y)$ is singular, the same formula holds with the inverse replaced by the generalized inverse, the Moore-Penrose pseudoinverse matrix.

If one wants to predict/estimate X by using observations on Y, then a predictor is some function of Y. For a square-integrable X, $E(X^2) < \infty$, the best predictor $\hat{X}$, by definition, minimizes the mean-square error. It is easy to show

Theorem 2.26 (Best Estimator/Predictor) *Let $\hat{X}$ be such that for any Y-measurable random variable Z,*

$$E(X - \hat{X})^2 \leq E(X - Z)^2.$$

Then $\hat{X} = E(X|Y)$.

2.8 Stochastic Processes in Continuous Time

The construction of a mathematical model of uncertainty and of flow of information in continuous time follows the same ideas as in the discrete time, but it is much more complicated. Consider constructing a probability model for a process $S(t)$ when time changes continuously between 0 and T. Take for the sample space the set of all possibilities of movements of the process. If we make a simplifying assumption that the process changes continuously, we obtain the set of all continuous functions on $[0, T]$, denoted by $C[0, T]$. This is a very rich space. In a more general model it is assumed that the observed process is a right-continuous function with left limits (regular right-continuous (RRC, càdlàg)) function.

Let the sample space $\Omega = D[0, T]$ be the set of all RRC functions on $[0, T]$. An element of this set, ω is a RRC function from $[0, T]$ into $\mathbb{R}$. First we must decide what kind of sets of these functions are measurable. The simplest sets for which we would like to calculate the probabilities are sets of the form $\{a \leq S(t_1) \leq b\}$ for some t_1. If $S(t)$ represents the price of a stock at time t, then the probability of such a set gives the probability that the stock price at time t_1 is between a and b. We are also interested in how the price of stock at time t_1 affects the price at another time t_2. Thus we need to talk about the joint distribution of stock prices $S(t_1)$ and $S(t_2)$. This means that we need to define probability on the sets of the form $\{S(t_1) \in B_1, S(t_2) \in B_2\}$ where B_1 and B_2 are intervals on the line. More generally we would like to have all *finite-dimensional distributions* of the process $S(t)$, that is, probabilities of the sets: $\{S(t_1) \in B_1, \ldots, S(t_n) \in B_n\}$, for any choice of $0 \leq t_1 \leq t_2, \ldots \leq t_n \leq T$. The sets of the form $\{\omega(\cdot) \in D[0, T] : \omega(t_1) \in B_1, \ldots, \omega(t_n) \in B_n\}$, where B_i's are intervals on the line, are called *cylinder sets* or *finite-dimensional rectangles*. The stochastic process $S(t)$ on this sample space is just $s(t)$, the value of the function s at t. Probability is defined first on the cylinder sets, and then extended to the σ-field $\mathcal{F}$ generated by the cylinders, that is, the smallest σ-field containing all cylinder sets. One needs to be careful with consistency of probability defined on cylinder sets, so that when one cylinder contains another no contradiction of probability assignment is obtained. The result that shows that a consistent family of distributions defines a probability function, continuous at $\emptyset$ on the field of cylinder sets is known as Kolmogorov's extension theorem. Once a probability is defined the field of cylinder sets, it can be extended in a unique way (by Caratheodory's theorem) to $\mathcal{F}$ (see for example, Breiman (1968), Durrett (1991) or Dudley (1989) for details).

It follows immediately from this construction that: a) for any choice of $0 \leq t_1 \leq t_2, \ldots \leq t_n \leq T$, $S(t_1), S(t_2), \ldots, S(t_n)$ is a random vector, and b) that the process is determined by its finite-dimensional distributions.

Continuity and Regularity of Paths

As discussed in the previous section, a stochastic process is determined by its finite-dimensional distributions. In studying stochastic processes it is often natural to think of them as random functions in t. Let $S(t)$ be defined for $0 \leq t \leq T$, then for a fixed ω it is a function in t, called the sample path or a realization of S. Finite-dimensional distributions do not determine the continuity property of sample paths. The following example illustrates this.

Example 2.17: Let $X(t) = 0$ for all t, $0 \leq t \leq 1$, and τ be a uniformly distributed random variable on $[0, 1]$. Let $Y(t) = 0$ for $t \neq \tau$ and $Y(t) = 1$ if $t = \tau$. Then for any fixed t, $P(Y(t) \neq 0) = P(\tau = t) = 0$, hence $P(Y(t) = 0) = 1$. So that all one-dimensional distributions of $X(t)$ and $Y(t)$ are the same. Similarly all finite-dimensional distributions of X and Y are the same. However, the sample paths of the process X, that is, the functions $X(t)_{0 \leq t \leq 1}$ are continuous in t, whereas every sample path $Y(t)_{0 \leq t \leq 1}$ has a jump at the point τ. Notice that $P(X(t) = Y(t)) = 1$ for all t, $0 \leq t \leq 1$.

Definition 2.27 *Two stochastic processes are called versions (modifications) of one another if*

$$P(X(t) = Y(t)) = 1 \text{ for all } t, 0 \leq t \leq T.$$

Thus the two processes in the Example 2.17 are versions of one another, one has continuous sample paths and the other does not. If we agree to pick any version of the process we want, then we can pick the continuous version when it exists. In general we choose the smoothest possible version of the process.

For two processes X and Y, denote by $N_t = \{X(t) \neq Y(t)\}$, $0 \leq t \leq T$. In the above Example 2.17, $P(N_t) = P(\tau = t) = 0$ for any t, $0 \leq t \leq 1$. However, $P(\bigcup_{0 \leq t \leq 1} N_t) = P(\tau = t$ for some t in $[0,1]) = 1$. Although, each of N_t is a P-null set, the union $N = \bigcup_{0 \leq t \leq 1} N_t$ contains uncountably many null sets, and in this case it is a set of probability one.

If it happens that $P(N) = 0$, then N is called an *evanescent* set, and the processes X and Y are called *indistinguishable*. Note that in this case $P(\{\omega : \exists t : X(t) \neq Y(t)\}) = P(\bigcup_{0 \leq t \leq 1} \{X(t) \neq Y(t)\}) = 0$, and $P(\bigcap_{0 \leq t \leq 1} \{X(t) = Y(t)\}) = P(X(t) = Y(t)$ for all $t \in [0,T]) = 1$. It is clear that if the time is discrete then any two versions of the process are indistinguishable. It is also not hard to see that if $X(t)$ and $Y(t)$ are versions of one another and they both are right-continuous, then they are indistinguishable.

Conditions for the existence of the continuous and the regular (paths with only jump discontinuities) versions of a stochastic process are given below.

Theorem 2.28 $S(t)$, $0 \leq t \leq T$ *is $\mathbb{R}$-valued stochastic process.*

1. *If there exist $\alpha > 0$ and $\epsilon > 0$, so that for any $0 \leq u \leq t \leq T$,*

$$E|S(t) - S(u)|^\alpha \leq C(t - u)^{1+\epsilon}, \tag{2.28}$$

for some constant C, then there exists a version of S with continuous sample paths, which are Hölder continuous of order $h < \epsilon/\alpha$.

2. *If there exist $C > 0$, $\alpha_1 > 0$, $\alpha_2 > 0$ and $\epsilon > 0$, so that for any $0 \le u \le v \le t \le T$,*

$$\mathrm{E}\big(|S(v) - S(u)|^{\alpha_1}|S(t) - S(v)|^{\alpha_2}\big) \le C(t - u)^{1+\epsilon}, \qquad (2.29)$$

then there exists a version of S with paths that may have discontinuities of the first kind only (which means that at any interior point both right and left limits exist, and one-sided limits exist at the boundaries).

Note that the above result allows to decide on the existence of the continuous (regular) version by means of the joint bivariate (trivariate) distributions of the process. The same result applies when the process takes values in $\mathbb{R}^d$, except that the Eucledean distance replaces the absolute value in the above conditions.

Functions without discontinuities of the second kind are considered to be the same if at all points of the domain they have the same right and left limits. In this case it is possible to identify any such function with its right-continuous version.

The following result gives a condition for the existence of a regular right-continuous version of a stochastic process.

Theorem 2.29 *If the stochastic process $S(t)$ is right-continuous in probability (that is, for any t the limit in probability $\lim_{u \downarrow t} S(u) = S(t)$) and it does not have discontinuities of the second kind, then it has a right-continuous version.*

Other conditions for the regularity of path can be given if we know some particular properties of the process. For example, later we give such conditions for processes that are martingales and supermartingales.

σ-field Generated by a Stochastic Process

$\mathcal{F}_t = \sigma(S_u, u \le t)$ is the smallest σ-field that contains sets of the form $\{a \le S_u \le b\}$ for $0 \le u \le t$, $a, b \in \mathbb{R}$. It is the information available to an observer of the process S up to time t.

Filtered Probability Space and Adapted Processes

A filtration $\mathbb{F}$ is a family $\{\mathcal{F}_t\}$ of increasing σ-fields on $(\Omega, \mathcal{F})$, $\mathcal{F}_t \subset \mathcal{F}$. $\mathbb{F}$ specifies how the information is revealed in time. The property that a filtration is increasing corresponds to the fact the information is not forgotten.

If we have a set Ω, a σ-field of subsets of Ω, $\mathcal{F}$, a probability P defined on elements of $\mathcal{F}$, and a filtration $\mathbb{F}$ such that

$$\mathcal{F}_0 \subset \mathcal{F}_t \subset \ldots \subset \mathcal{F}_T = \mathcal{F},$$

then $(\Omega, \mathcal{F}, \mathbb{F}, P)$ is called a filtered probability space.

A stochastic process on this space $\{S(t), 0 \leq t \leq T\}$ is called *adapted* if for all t, $S(t)$ is $\mathcal{F}_t$-measurable, that is, if for any t, $\mathcal{F}_t$ contains all the information about $S(t)$ (and may contain extra information).

The Usual Conditions

Filtration is called right-continuous if $\mathcal{F}_{t+} = \mathcal{F}_t$, where

$$\mathcal{F}_{t+} = \bigcap_{s>t} \mathcal{F}_s.$$

The standard assumption (referred to as the usual condition) is that filtration is right-continuous, for all t, $\mathcal{F}_t = \mathcal{F}_{t+}$. It has the following interpretation: any information known immediately after t is also known at t.

Remark 2.2: Note that if $S(t)$ is a process adapted to $\mathbb{F}$, then we can always take a right-continuous filtration to which $S(t)$ is adapted by taking $\mathcal{G}_t = \mathcal{F}_{t+} = \bigcap_{s>t} \mathcal{F}_s$. Then S_t is $\mathcal{G}_t$ adapted.

The assumption of right-continuous filtration has a number of important consequences. For example, it allows to assume that martingales, submartingales and supermartingales have a regular right-continuous version.

It is also assumed that any set which is a subset of a set of zero probability is $\mathcal{F}_0$-measurable. Of course, such a set must have zero probability. *A priori* such sets need not be measurable, and we enlarge the σ-fields to include such sets. This procedure is called the completion by the null sets.

Martingales, Supermartingales, Submartingales

Definition 2.30 *A stochastic process $\{X(t), t \geq 0\}$ adapted to a filtration $\mathbb{F}$ is a supermartingale (submartingale) if for any t it is integrable, $E|X(t)| < \infty$, and for any $s < t$*

$$E(X(t)|\mathcal{F}_s) \leq X(s), \ \big(E(X(t)|\mathcal{F}_s) \geq X(s)\big).$$

If $E(X(t)|\mathcal{F}_s) = X(s)$, then the process $X(t)$ is called a martingale.

An example of a martingale is given by the following

Theorem 2.31 (Doob-Levy martingale) *Let Y be an integrable random variable, that is, $\mathrm{E}|Y| < \infty$, then*

$$M(t) = \mathrm{E}(Y|\mathcal{F}_t). \tag{2.30}$$

is a martingale.

PROOF: By the law of double expectation

$$\mathrm{E}(M(t)|\mathcal{F}_s) = \mathrm{E}\big(\mathrm{E}(Y|\mathcal{F}_t)|\mathcal{F}_s\big) = \mathrm{E}(Y|\mathcal{F}_s) = M(s).$$

$\square$

Using the law of double expectation, it is easy to see that the mean of a martingale is a constant in t, the mean of a supermartingale is non-increasing in t, and the mean of a submartingale is non-decreasing in t.

If $X(t)$ is a supermartingale, then $-X(t)$ is a submartingale, directly from the definition.

We have the following result for the existence of the right-continuous version for super or submartingales, without the assumption of continuity in probability imposed on the process (see for example, Liptser and Shiryaev (1974), Vol I, p.55).

Theorem 2.32 *Let the filtration $\mathbb{F}$ be right-continuous and each of the σ-fields $\mathcal{F}_t$ be completed by the P-null sets from $\mathcal{F}$. In order that the supermartingale $X(t)$ has a right-continuous version, it is necessary and sufficient that its mean function $\mathrm{E}X(t)$ is right-continuous. In particular, any martingale with right-continuous filtration admits a regular right-continuous version.*

In view of these results, it will often be assumed that the version of the process under consideration is regular and right-continuous (càdlàg).

Stopping Times

Definition 2.33 *A non-negative random variable τ, which is allowed to take the value ∞, is called a stopping time (with respect to filtration $\mathbb{F}$) if for each t, the event*

$$\{\tau \leq t\} \in \mathcal{F}_t.$$

It is immediate that for all t, the complimentary event $\{\tau > t\} \in \mathcal{F}_t$. If $\tau < t$, then for some n, $\tau \leq t - 1/n$. Thus

$$\{\tau < t\} = \bigcup_{n=1}^{\infty} \{\tau \leq t - 1/n\}.$$

The event $\{\tau \leq t-1/n\} \in \mathcal{F}_{t-1/n}$. Since $\mathcal{F}_t$ are increasing, $\{\tau \leq t-1/n\} \in \mathcal{F}_t$, therefore $\{\tau < t\} \in \mathcal{F}_t$.

Introduce

$$\mathcal{F}_{t-} := \bigvee_{s<t} \mathcal{F}_s := \sigma\Big(\bigcup_{s<t} \mathcal{F}_s \Big).$$

The above argument shows that $\{\tau < t\} \in \mathcal{F}_{t-}$.

Theorem 2.34 *If filtration is right-continuous, then τ is a stopping time if and only if for each t, the event $\{\tau < t\} \in \mathcal{F}_t$.*

PROOF: One direction has just been established, the other one is seen as follows.

$$\{\tau \le t\} = \bigcap_{n=1}^{\infty} \{\tau < t + 1/n\}.$$

Since $\{\tau < t + 1/n\} \in \mathcal{F}_{t+1/n}$, by right-continuity of $\mathbb{F}$, $\{\tau \le t\} \in \mathcal{F}_t$.

$\square$

The assumption of right-continuity of $\mathbb{F}$ is important when studying exit times and hitting times of a set by a process. If $S(t)$ is a random process on $\mathbb{R}$ adapted to $\mathbb{F}$, then the *hitting time* of set A is defined as

$$T_A = \inf\{t \ge 0 : S(t) \in A\}. \tag{2.31}$$

The first exit time from a set D is defined as

$$\tau_D = \inf\{t \ge 0 : S(t) \notin D\}. \tag{2.32}$$

Note that $\tau_D = T_{\mathbb{R} \setminus D}$

Theorem 2.35 *Let $S(t)$ be continuous and adapted to $\mathbb{F}$. If $D = (a, b)$ is an open interval, or any other open set on $\mathbb{R}$ (a countable union of open intervals), then τ_D is a stopping time. If $A = [a, b]$ is a closed interval, or any other closed set on $\mathbb{R}$ (its complement is an open set), then T_A is a stopping time. If in addition the filtration $\mathbb{F}$ is right-continuous then also for closed sets D and open sets A, τ_D and T_A are stopping times.*

PROOF: $\{\tau_D > t\} = \{S(u) \in D,$ for all $u \le t\} = \cap_{0 \le u \le t}\{S(u) \in D\}$. This event is an uncountable intersection over all $u \le t$ of events in $\mathcal{F}_u$. The point of the proof is to represent this event as a countable intersection. Due to continuity of $S(u)$ and D being open, for any irrational u with $S(u) \in D$ there is a rational q with $S(q) \in D$. Therefore

$$\bigcap_{0 \le u \le t} \{S(u) \in D\} = \bigcap_{0 \le q-\text{rational} \le t} \{S(q) \in D\},$$

which is now a countable intersection of the events from $\mathcal{F}_t$, and hence is itself in $\mathcal{F}_t$. This shows that τ_D is a stopping time. Since for any closed set A, $\mathbb{R} \setminus A$ is open, and $T_A = \tau_{\mathbb{R} \setminus A}$, T_A is also a stopping time.

Assume now that filtration is right-continuous. If $D = [a, b]$ is a closed interval, then $D = \cap_{n=1}^{\infty}(a - 1/n, b + 1/n)$. If $D_n = (a - 1/n, b + 1/n)$, then τ_{D_n} is a stopping time, and the event $\{\tau_{D_n} > t\} \in \mathcal{F}_t$. It is easy to see that $\cap_{n=1}^{\infty}\{\tau_{D_n} > t\} = \{\tau_D \geq t\}$, hence $\{\tau_D \geq t\} \in \mathcal{F}_t$, and also $\{\tau_D < t\} \in \mathcal{F}_t$ as its complimentary, for any t. The rest of the proof follows by Theorem 2.34.

$\square$

For general processes the following result holds.

Theorem 2.36 *Let $S(t)$ be regular right-continuous and adapted to $\mathbb{F}$, and $\mathbb{F}$ be right-continuous. If A is an open set on $\mathbb{R}$, then T_A is a stopping time. If A is a closed set, then $\inf\{t > 0 : S(t) \in A, \text{ or } S(t-) \in A\}$ is a stopping time.*

It is possible, although much harder, to show that the hitting time of a Borel set is a stopping time.

Next results give basic properties of stopping times.

Theorem 2.37 *Let S and T be two stopping times, then $\min(S, T)$, $\max(S, T)$, $S + T$ are all stopping times.*

σ-field $\mathcal{F}_T$

If T is a stopping time, events observed before or at time T are described by σ-field $\mathcal{F}_T$, defined as the collection of sets
$\mathcal{F}_T = \{A \in \mathcal{F} : \text{for any } t, \ A \cap \{T \leq t\} \in \mathcal{F}_t\}$.

Theorem 2.38 *Let S and T be two stopping times. The following properties hold: If $A \in \mathcal{F}_S$, then $A \cap \{S = T\} \in \mathcal{F}_T$, consequently $\{S = T\} \in \mathcal{F}_S \cap \mathcal{F}_T$. If $A \in \mathcal{F}_S$, then $A \cap \{S \leq T\} \in \mathcal{F}_T$, consequently $\{S \leq T\} \in \mathcal{F}_S \cap \mathcal{F}_T$.*

Fubini's Theorem

Fubini's theorem allows us to interchange integrals (sums) and expectations. We give a particular case of Fubini's theorem, it is formulated in the way we use it in applications.

Theorem 2.39 *Let $X(t)$ be a stochastic process $0 \leq t \leq T$ (for all t $X(t)$ is a random variable), with regular sample paths (for all ω at any point t, $X(t)$ has left and right limits). Then*

$$\int_0^T \mathrm{E}|X(t)|dt = \mathrm{E}\left(\int_0^T |X(t)|dt\right).$$

Furthermore if this quantity is finite, then

$$\mathrm{E}\left(\int_0^T X(t)dt\right) = \int_0^T \mathrm{E}(X(t))dt.$$

Chapter 3

Basic Stochastic Processes

This chapter is mainly about Brownian motion. It is the main process in the calculus of continuous processes. The Poisson process is the main process in the calculus of processes with jumps. Both processes give rise to functions of positive quadratic variation. For Stochastic Calculus only Section 3.1-3.5 are needed, but in applications other sections are also used.

Introduction

Observations of prices of stocks, positions of a diffusing particle and many other processes observed in time are often modelled by a stochastic process. A stochastic process is an umbrella term for any collection of random variables $\{X(t)\}$ depending on time t. Time can be discrete, for example, $t = 0, 1, 2, \ldots$, or continuous, $t \geq 0$. Calculus is suited more to continuous time processes. At any time t, the observation is described by a random variable which we denote by X_t or $X(t)$. A stochastic process $\{X(t)\}$ is frequently denoted by X or with a slight abuse of notation also by $X(t)$.

In practice, we typically observe only a single realization of this process, a single path, out of a multitude of possible paths. Any single path is a function of time t, $x_t = x(t)$, $0 \leq t \leq T$; and the process can also be seen as a random function. To describe the distribution and to be able to do probability calculations about the uncertain future, one needs to know the so-called finite-dimensional distributions. Namely, we need to specify how to calculate probabilities of the form $P(X(t) \leq x)$ for any time t, i.e. the probability distribution of the random variable $X(t)$; and probabilities of the form $P(X(t_1) \leq x_1, X(t_2) \leq x_2)$ for any times t_1, t_2, i.e. the joint bivariate distributions of $X(t_1)$ and $X(t_2)$; and probabilities of the form

$$P(X(t_1) \leq x_1, X(t_2) \leq x_2, \ldots X(t_n) \leq x_n), \tag{3.1}$$

for any choice of time points $0 \leq t_1 < t_2 \ldots < t_n \leq T$, and any $n \geq 1$ with $x_1, \ldots x_n \in \mathbb{R}$. Often one does not write the formula for (3.1), but merely points out how to compute it.

3.1 Brownian Motion

Botanist R. Brown described the motion of a pollen particle suspended in fluid in 1828. It was observed that a particle moved in an irregular, random fashion. A. Einstein, in 1905, argued that the movement is due to bombardment of the particle by the molecules of the fluid, he obtained the equations for Brownian motion. In 1900, L. Bachelier used the Brownian motion as a model for movement of stock prices in his mathematical theory of speculation. The mathematical foundation for Brownian motion as a stochastic process was done by N. Wiener in 1931, and this process is also called the Wiener process.

The Brownian Motion process $B(t)$ serves as a basic model for the cumulative effect of pure noise. If $B(t)$ denotes the position of a particle at time t, then the displacement $B(t) - B(0)$ is the effect of the purely random bombardment by the molecules of the fluid, or the effect of noise over time t.

Defining Properties of Brownian Motion

Brownian motion $\{B(t)\}$ is a stochastic process with the following properties.

1. (Independence of increments) $B(t) - B(s)$, for $t > s$, is independent of the past, that is, of B_u, $0 \leq u \leq s$, or of $\mathcal{F}_s$, the σ-field generated by $B(u), u \leq s$.

2. (Normal increments) $B(t) - B(s)$ has Normal distribution with mean 0 and variance $t - s$. This implies (taking $s = 0$) that $B(t) - B(0)$ has $N(0, t)$ distribution.

3. (Continuity of paths) $B(t)$, $t \geq 0$ are continuous functions of t.

The initial position of Brownian motion is not specified in the definition. When $B(0) = x$, then the process is Brownian motion started at x. Properties 1 and 2 above determine all the finite-dimensional distributions (see (3.4) below) and it is possible to show (see Theorem 3.3) that all of them are Gaussian. P_x denotes the probability of events when the process starts at x. The time interval on which Brownian motion is defined is $[0, T]$ for some $T > 0$, which is allowed to be infinite.

We don't prove here that a Brownian motion exists, it can be found in many books on stochastic processes, and one construction is outlined in Section 5.7.

However we can deduce continuity of paths by using normality of increments and appealing to Theorem 2.28. Since

$$E(B(t) - B(s))^4 = 3(t - s)^2,$$

a continuous version of Brownian motion exists.

Remark 3.1: A definition of Brownian motion in a more general model (that contains extra information) is given by a pair $\{B(t), \mathcal{F}_t\}$, $t \geq 0$, where $\mathcal{F}_t$ is an increasing sequence of σ-fields (a filtration), $B(t)$ is an adapted process, i.e. $B(t)$ is $\mathcal{F}_t$ measurable, such that Properties 1-3 above hold.

An important representation used for calculations in processes with independent increments is that for any $s \geq 0$

$$B(t + s) = B(s) + (B(t + s) - B(s)), \tag{3.2}$$

where two variables are independent. An extension of this representation is the process version.

Let $W(t) = B(t + s) - B(s)$. Then for a fixed s, as a process in t, $W(t)$ is a Brownian motion started at 0. This is seen by verifying the defining properties.

Other examples of Brownian motion processes constructed from other processes are given below, as well as in exercises.

Example 3.1: Although $B(t) - B(s)$ is independent of the past, $2B(t) - B(s)$ or $B(t) - 2B(s)$ is not, as, for example, $B(t) - 2B(s) = (B(t) - B(s)) - B(s)$, is a sum of two variables, with only one independent of the past and $B(s)$.

The following example illustrates calculations of some probabilities for Brownian motion.

Example 3.2: Let $B(0) = 0$.
We calculate $P(B(t) \leq 0 \text{ for } t = 2)$ and $P(B(t) \leq 0 \text{ for } t = 0, 1, 2)$.
Since $B(2)$ has Normal distribution with mean zero and variance 2, $P(B(t) \leq 0 \text{ for } t = 2) = \frac{1}{2}$.
Since $B(0) = 0$, $P(B(t) \leq 0 \text{ for } t = 0, 1, 2) = P(B(1) \leq 0, B(2) \leq 0)$. Note that $B(2)$ and $B(1)$ are not independent, therefore this probability cannot be calculated as a product $P(B(1) \leq 0)P(B(2) \leq 0) = 1/4$. Using the decomposition $B(2) = B(1) + (B(2) - B(1)) = B(1) + W(1)$, where the two random variables are independent, we have

$$\begin{aligned} P(B(1) \leq 0, B(2) \leq 0) &= P(B(1) \leq 0, B(1) + W(1) \leq 0) \\ &= P(B(1) \leq 0, W(1) \leq -B(1)). \end{aligned}$$

By conditioning and by using Theorem 2.24 and (2.20)

$$P(B(1) \leq 0, W(1) \leq -B(1)) = \int_{-\infty}^{0} P(W(1) \leq -x)f(x)dx = \int_{-\infty}^{0} \Phi(-x)d\Phi(x),$$

where $\Phi(x)$ and $f(x)$ denote the distribution and the density functions of the standard Normal distribution. By changing variables the last integral, we obtain

$$\int_0^\infty \Phi(x)f(-x)dx = \int_0^\infty \Phi(x)d\Phi(x) = \int_{1/2}^1 ydy = \frac{3}{8}.$$

Transition Probability Functions

If the process is started at x, $B(0) = x$, then $B(t)$ has the $N(x,t)$ distribution. More generally, the conditional distribution of $B(t + s)$ given that $B(s) = x$ is $N(x,t)$. The transition function $P(y,t,x,s)$ is the cumulative distribution function of this distribution,
$$P(y,t,x,s) = \mathrm{P}(B(t+s) \leq y|B(s) = x) = \mathrm{P}_x(B(t) \leq y).$$
The density function of this distribution is the transition probability density function of Brownian motion,

$$p_t(x,y) = \frac{1}{\sqrt{2\pi t}}e^{-\frac{(y-x)^2}{2t}}. \tag{3.3}$$

The finite-dimensional distributions can be computed with the help of the transition probability density function, by using independence of increments in a way similar to that exhibited in the above example.

$$\mathrm{P}_x(B(t_1) \leq x_1, B(t_2) \leq x_2, \ldots, B(t_n) \leq x_n) = \tag{3.4}$$
$$\int_{-\infty}^{x_1} p_{t_1}(x,y_1)dy_1 \int_{-\infty}^{x_2} p_{t_2-t_1}(y_1,y_2)dy_2 \ldots \int_{-\infty}^{x_n} p_{t_n-t_{n-1}}(y_{n-1},y_n)dy_n.$$

Space Homogeneity

It is easy to see that the one-dimensional distributions of Brownian motion satisfy $\mathrm{P}_0(B(t) \in A) = \mathrm{P}_x(B(t) \in x + A)$, where A is an interval on the line.

If $B^x(t)$ denotes Brownian motion started at x, then it follows from (3.4) that all finite-dimensional distributions of $B^x(t)$ and $x + B^0(t)$ are the same. Thus $B^x(t) - x$ is Brownian motion started at 0, and $B^0(t) + x$ is Brownian motion started at x, in other words

$$B^x(t) = x + B^0(t). \tag{3.5}$$

The property (3.5) is called the *space homogeneous* property of Brownian motion.

Definition 3.1 *A stochastic process is called space-homogeneous if its finite-dimensional distributions do not change with a shift in space, namely if*

$$\mathrm{P}(X(t_1) \leq x_1, X(t_2) \leq x_2, \ldots X(t_n) \leq x_n|X(0) = 0)$$
$$= \mathrm{P}(X(t_1) \leq x_1 + x, X(t_2) \leq x_2 + x, \ldots X(t_n) \leq x_n + x|X(0) = x).$$

Four realizations of Brownian motion $B = B(t)$ started at 0 are exhibited in Figure 3.1. Although it is a process governed by the pure chance with zero mean, it has regions where motion looks like it has "trends".

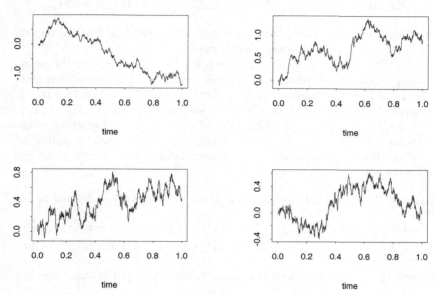

Figure 3.1: Four realizations or paths of Brownian motion B(t).

Brownian Motion as a Gaussian Process

Recall that a process is called Gaussian if all its finite-dimensional distributions are multivariate Normal.

Example 3.3: Let random variables X and Y be independent Normal with distributions $N(\mu_1, \sigma_1^2)$ and $N(\mu_2, \sigma_2^2)$. Then the distribution of $(X, X + Y)$ is bivariate Normal with mean vector $(\mu_1, \mu_1 + \mu_2)$ and covariance matrix
$$\begin{bmatrix} \sigma_1^2 & \sigma_1^2 \\ \sigma_1^2 & \sigma_1^2 + \sigma_2^2 \end{bmatrix}.$$
To see this let $\mathbf{Z} = (Z_1, Z_2)$ have standard Normal components, then it is easy to see that
$$(X, X + Y) = \boldsymbol{\mu} + A\mathbf{Z},$$
where $\boldsymbol{\mu} = (\mu_1, \mu_1 + \mu_2)$, and matrix $A = \begin{bmatrix} \sigma_1 & 0 \\ \sigma_1 & \sigma_2 \end{bmatrix}$. The result follows by the definition of the general Normal distribution as a linear transformation of standard Normals (see Section 2.6).

Similarly to the above example, the following representation

$$(B(t_1), B(t_2), \ldots, B(t_n))$$
$$= (B(t_1), B(t_1) + (B(t_2) - B(t_1)), \ldots, B(t_{n-1})) + (B(t_n)) - B(t_{n-1}))$$

shows that this vector is a linear transformation of the standard Normal vector, hence it has a multivariate Normal distribution.

Let $Y_1 = B(t_1)$, and for $k > 1$, $Y_k = B(t_k) - B(t_{k-1})$. Then by the property of independence of increments of Brownian motion, Y_k's are independent. They also have Normal distribution, $Y_1 \sim N(0, t_1)$, and $Y_k \sim N(0, t_k - t_{k-1})$. Thus $(B(t_1), B(t_2), \ldots, B(t_n))$ is a linear transformation of $(Y_1, Y_2, \ldots, Y_n)$. But $Y_1 = \sqrt{t_1}Z_1$, and $Y_k = \sqrt{t_k - t_{k-1}}Z_k$, where Z_k's are independent standard Normal. Thus $(B(t_1), B(t_2), \ldots, B(t_n))$ is a linear transformation of $(Z_1, \ldots, Z_n)$. Finding the matrix A of this transformation is left as an exercise (Exercise 3.7).

Definition 3.2 *The covariance function of the process $X(t)$ is defined by*

$$\gamma(s,t) = Cov(X(t), X(s)) \quad = \quad \mathrm{E}(X(t) - \mathrm{E}X(t))(X(s) - \mathrm{E}X(s))$$
$$= \quad \mathrm{E}(X(t)X(s)) - \mathrm{E}X(t)\mathrm{E}X(s). \qquad (3.6)$$

The next result characterizes Brownian motion as a particular Gaussian process.

Theorem 3.3 *A Brownian motion is a Gaussian process with zero mean function, and covariance function* $\min(t, s)$. *Conversely, a Gaussian process with zero mean function, and covariance function* $\min(t, s)$ *is a Brownian motion.*

PROOF: Since the mean of the Brownian motion is zero,

$$\gamma(s,t) = Cov(B(t), B(s)) = \mathrm{E}(B(t)B(s)).$$

If $t < s$ then $B(s) = B(t) + B(s) - B(t)$, and

$$\mathrm{E}(B(t), B(s)) = \mathrm{E}B^2(t) + \mathrm{E}(B(t)(B(s) - B(t))) = \mathrm{E}B^2(t) = t,$$

where we used independence of increments property. Similarly if $t > s$, $\mathrm{E}(B(t)B(s)) = s$. Therefore

$$\mathrm{E}(B(t)B(s)) = \min(t, s).$$

To show the converse, let t be arbitrary and $s \geq 0$. $X(t)$ is a Gaussian process, thus the joint distribution of $X(t), X(t+s)$ is a bivariate Normal, and by conditions has zero mean. Therefore the vector $(X(t), X(t+s) - X(t)$ is also bivariate Normal. The variables $X(t)$ and $X(t + s) - X(t)$ are uncorrelated, using that $Cov(X(t), X(t + s)) = \min(t, s)$,

$$Cov(X(t), X(t+s) - X(t)) = Cov(X(t), X(t+s)) - Cov(X(t), X(t)) = t - t = 0.$$

A property of the multivariate Normal distribution implies that these variables are independent. Thus the increment $X(t + s) - X(t)$ is independent of $X(t)$ and has $N(0, s)$ distribution. Therefore it is a Brownian motion.

$\square$

Example 3.4: We find the distribution of $B(1) + B(2) + B(3) + B(4)$.
Consider the random vector $\boldsymbol{X} = (B(1), B(2), B(3), B(4))$. Since Brownian motion is a Gaussian process, all its finite-dimensional distributions are Normal, in particular $\boldsymbol{X}$ has a multivariate Normal distribution with mean vector zero and covariance matrix given by $\sigma_{ij} = Cov(X_i, X_j)$. For example, $Cov(X_1, X_3) = Cov((B(1), B(3)) = 1$.

$$\Sigma = \begin{bmatrix} 1 & 1 & 1 & 1 \\ 1 & 2 & 2 & 2 \\ 1 & 2 & 3 & 3 \\ 1 & 2 & 3 & 4 \end{bmatrix}$$

Now, let $\mathbf{a} = (1, 1, 1, 1)$. Then

$$\boldsymbol{aX} = X_1 + X_2 + X_3 + X_4 = B(1) + B(2) + B(3) + B(4).$$

$\boldsymbol{aX}$ has a Normal distribution with mean zero and variance $a\Sigma a^T$, and in this case the variance is given by the sum of the elements of the covariance matrix. Thus $B(1) + B(2) + B(3) + B(4)$ has a Normal distribution with mean zero and variance 30. Alternatively, we can calculate the variance of the sum by the formula

$$
\begin{aligned}
&Var(X_1 + X_2 + X_3 + X_4) \\
&= \quad Cov(X_1 + X_2 + X_3 + X_4, X_1 + X_2 + X_3 + X_4) = \sum_{i,j} Cov(X_i, X_j) = 30.
\end{aligned}
$$

Example 3.5: To illustrate the use of scaling, we find the distribution of $B(\frac{1}{4}) + B(\frac{1}{2}) + B(\frac{3}{4}) + B(1)$. Consider the random vector $\boldsymbol{Y} = \left(B(\frac{1}{4}), B(\frac{1}{2}), B(\frac{3}{4}), B(1)\right)$. It is easy to see that $\boldsymbol{Y}$ and $1/2\boldsymbol{X}$, where $\boldsymbol{X} = (B(1), B(2), B(3), B(4))$ have the same law. Therefore its covariance matrix is given by $\frac{1}{4}\Sigma$, with Σ as above. Consequently, $\boldsymbol{aY}$ has a Normal distribution with mean zero and variance $30/4$

Example 3.6: We find the probability $P(\int_0^1 B(t)dt > \frac{2}{\sqrt{3}})$.
Comment first that since Brownian motion has continuous paths, the Riemann integral $\int_0^1 B(t)dt$ is well defined for any random path as we integrate path by path. To find the required probability we need to know the distribution of $\int_0^1 B(t)dt$. This can be obtain as a limit of the distributions of the approximating sums,

$$\sum B(t_i)\Delta,$$

where points t_i partition $[0, 1]$ and $\Delta = t_{i+1} - t_i$. If, for example, $t_i = i/n$, then for $n = 4$ the approximating sum is $\frac{1}{4}\left(B(\frac{1}{4}) + B(\frac{1}{2}) + B(\frac{3}{4}) + B(1)\right)$, the distribution of which was found in the previous example to be $N(0, \frac{15}{32})$. Similarly, the distribution of all of the approximating sums is Normal with zero mean. It can be shown that the limit of Gaussian distributions is a Gaussian distribution. Thus $\int_0^1 B(t)dt$ has a Normal distribution with zero mean. Therefore it only remains to compute its variance.

$$
\begin{aligned}
Var\left(\int_0^1 B(t)dt\right) &= Cov\left(\int_0^1 B(t)dt, \int_0^1 B(s)ds\right) \\
&= E\left(\int_0^1 B(t)dt \int_0^1 B(s)ds\right) = \int_0^1 \int_0^1 E\left(B(t)B(s)\right)dtds \\
&= \int_0^1 \int_0^1 Cov(B(t), B(s))dtds = \int_0^1 \int_0^1 \min(t, s)dtds = 1/3
\end{aligned}
$$

Exchanging the integrals and expectation is justified by Fubini's theorem since

$$
\int_0^1 \int_0^1 E\left|B(t)B(s)\right|dtds \leq \int_0^1 \int_0^1 \sqrt{ts}dtds < 1.
$$

Thus $\int_0^1 B(t)dt$ has $N(0, 1/3)$ distribution, and the desired probability is approximately 0.025. Later we shall prove that the distribution of the integral $\int_0^a B(t)dt$ is Normal $N(0, a^3/3)$ by considering a transformation to Itô integral, see Example 6.4.

Brownian Motion as a Random Series

The process

$$
\xi_0 \frac{t}{\sqrt{\pi}} + \frac{2}{\sqrt{\pi}} \sum_{j=1}^{\infty} \frac{\sin(jt)}{j}\xi_j, \tag{3.7}
$$

where ξ_j's $j = 0, 1, \ldots$, are independent standard Normal random variables, is Brownian motion on $[0, \pi]$. Convergence of the series is understood almost surely. This representation resembles the example of a continuous but nowhere differentiable function, Example 1.2. One can prove the assertion by showing that the partial sums converge uniformly, and verifying that the process in (3.7) is Gaussian, has zero mean, and covariance $\min(s, t)$ (see, for example, Breiman (1968), p.261, Itô and McKean (1965), p.22).

Remark 3.2: A similar, more general representation of a Brownian motion is given by using an orthonormal sequence of functions on $[0, T]$, $h_j(t)$. $B(t) = \sum_{j=0}^{\infty} \xi_j H_j(t)$, where $H_j(t) = \int_0^t h_j(s)ds$, is a Brownian motion on $[0, T]$.

3.2 Properties of Brownian Motion Paths

An occurrence of Brownian motion observed from time 0 to time T, is a random function of t on the interval $[0, T]$. It is called a *realization*, a *path* or *trajectory*.

Quadratic Variation of Brownian Motion

The quadratic variation of Brownian motion $[B, B](t)$ is defined as

$$[B, B](t) = [B, B]([0, t]) = \lim \sum_{i=1}^{n} |B(t_i^n) - B(t_{i-1}^n)|^2, \qquad (3.8)$$

where the limit is taken over all shrinking partitions of $[0, t]$, with $\delta_n = \max_i(t_{i+1}^n - t_i^n) \to 0$ as $n \to \infty$. It is remarkable that although the sums in the definition (3.8) are random, their limit is non-random, as the following result shows.

Theorem 3.4 *Quadratic variation of a Brownian motion over $[0, t]$ is t.*

PROOF: We give the proof for a sequence of partitions, for which $\sum_n \delta_n < \infty$. An example of such is when the interval is divided into two, then each subinterval is divided into two, etc. Let $T_n = \sum_i |B(t_i^n) - B(t_{i-1}^n)|^2$. It is easy to see that

$$\mathrm{E}(T_n) = \mathrm{E} \sum_i |B(t_i^n) - B(t_{i-1}^n)|^2 = \sum_{i=1}^{n} (t_i^n - t_{i-1}^n) = t - 0 = t.$$

By using the fourth moment of $N(0, \sigma^2)$ distribution is $3\sigma^4$, we obtain the variance of T_n

$$
\begin{aligned}
Var(T_n) &= Var(\sum_i |B(t_i^n) - B(t_{i-1}^n)|^2) = \sum_i Var(B(t_i^n) - B(t_{i-1}^n))^2 \\
&= \sum_i 3(t_i^n - t_{i-1}^n)^2 \leq 3 \max(t_i^n - t_{i-1}^n)t = 3t\delta_n.
\end{aligned}
$$

Therefore $\sum_{n=1}^{\infty} Var(T_n) < \infty$. Using monotone convergence theorem, we find $\mathrm{E} \sum_{n=1}^{\infty} (T_n - \mathrm{E}T_n)^2 < \infty$. This implies that the series inside the expectation converges almost surely. Hence its terms converge to zero, and $T_n - \mathrm{E}T_n \to 0$ a.s., consequently $T_n \to t$ a.s.

It is possible to show that $T_n \to t$ a.s. for any sequence of partitions which are successive refinements and satisfy $\delta_n \to 0$ as $n \to \infty$ (see for example, Loeve (1978), Vol. 2, p.253 for the proof, or Breiman (1968)).

$\square$

Varying t, the quadratic variation process of Brownian motion is t. Note that the classical quadratic variation of Brownian paths (defined as the supremum over all partitions of sums in (3.8), see Chapter 1 is infinite (e.g. Freedman (1971), p.48.)

Properties of Brownian paths

$B(t)$'s as functions of t have the following properties. Almost every sample path $B(t)$, $0 \leq t \leq T$

1. is a continuous function of t;

2. is not monotone in any interval, no matter how small the interval is;

3. is not differentiable at any point;

4. has infinite variation on any interval, no matter how small it is;

5. has quadratic variation on $[0, t]$ equal to t, for any t.

Properties 1 and 3 of Brownian motion paths state that although any realization $B(t)$ is a continuous function of t, it has increments $\Delta B(t)$ over an interval of length Δt much larger than Δt as $\Delta t \to 0$. Since $\mathrm{E}(B(t+\Delta t) - B(t))^2 = \Delta t$, it suggests that the increment is roughly like $\sqrt{\Delta t}$. This is made precise by the quadratic variation Property 5.

Note that by Theorem 1.10, a positive quadratic variation implies infinite variation, so that Property 4 follows from Property 5. Since a monotone function has finite variation, Property 2 follows from Property 4.

By Theorem 1.8 a continuous function with a bounded derivative is of finite variation. Therefore it follows from Property 4 that $B(t)$ can not have a bounded derivative on any interval, no matter how small the interval is. It is not yet the non-differentiability at any point, but it is close to it. For the proof of the result that with probability one Brownian motion paths are nowhere differentiable (due to Dvoretski, Erdös and Kakutani) see Breiman (1968) p.261. Here we show a simple statement

Theorem 3.5 *For any t almost all trajectories of Brownian motion are not differentiable at t.*

PROOF: Consider $\frac{B(t+\Delta)-B(t)}{\Delta} = \frac{\sqrt{\Delta}Z}{\Delta} = \frac{Z}{\sqrt{\Delta}}$, for some standard Normal random variable Z. Thus the ratio converges to ∞ in distribution, since

$\mathrm{P}(|\frac{Z}{\sqrt{\Delta}}| > K) \to 1$ for any K, as $\Delta \to 0$, precluding existence of the derivative at t.

$\square$

To realize the above argument on a computer take e.g. $\Delta = 10^{-20}$. Then $\Delta B(t) = 10^{-10}Z$, and $\Delta B(t)/\Delta = 10^{10}Z$, which is very large in absolute value with overwhelming probability.

3.3 Three Martingales of Brownian Motion

In this section three main martingales associated with Brownian motion are given. Recall the definition of a martingale.

Definition 3.6 *A stochastic process $\{X(t), t \geq 0\}$ is a martingale if for any t it is integrable, $\mathbb{E}|X(t)| < \infty$, and for any $s > 0$*

$$\mathbb{E}(X(t+s)|\mathcal{F}_t) = X(t), \quad a.s. \tag{3.9}$$

where $\mathcal{F}_t$ is the information about the process up to time t, and the equality holds almost surely.

The martingale property means that if we know the values of the process up to time t, and $X(t) = x$ then the expected future value at any future time is x.

Remark 3.3: $\mathcal{F}_t$ represents information available to an observer at time t. A set $A \in \mathcal{F}_t$ if and only if by observing the process up to time t one can decide whether or not A has occurred. Formally, $\mathcal{F}_t = \sigma(X(s), 0 \leq s \leq t)$ denotes the σ-field (σ-algebra) generated by the values of the process up to time t.

Remark 3.4: As the conditional expectation given a σ-field is defined as a random variable (see for example, Section 2.7), all the relations involving conditional expectations, such as equalities and inequalities, must be understood in the almost sure sense. This will always be assumed, and the almost sure "a.s." specification will be frequently dropped.

Examples of martingales constructed from Brownian motion are given in the next result.

Theorem 3.7 *Let $B(t)$ be Brownian Motion. Then*

1. *$B(t)$ is a martingale.*

2. *$B(t)^2 - t$ is a martingale.*

3. *For any u, $e^{uB(t) - \frac{u^2}{2}t}$ is a martingale.*

PROOF: The key idea in establishing the martingale property is that for any function g, the conditional expectation of $g(B(t+s) - B(t))$ given $\mathcal{F}_t$ equals to the unconditional one,

$$\mathbb{E}\big(g(B(t+s) - B(t))|\mathcal{F}_t\big) = \mathbb{E}\big(g(B(t+s) - B(t))\big), \tag{3.10}$$

due to independence of $B(t+s) - B(t)$ and $\mathcal{F}_t$. The latter expectation is just $Eg(X)$, where X Normal $N(0, s)$ random variable.

1. By definition, $B(t) \sim N(0, t)$, so that $B(t)$ is integrable with $E(B(t)) = 0$.

$$
\begin{aligned}
E(B(t+s)|\mathcal{F}_t) &= E\big(B(t) + (B(t+s) - B(t))|\mathcal{F}_t\big) \\
&= E(B(t)|\mathcal{F}_t) + E(B(t+s) - B(t)|\mathcal{F}_t) \\
&= B(t) + E(B(t+s) - B(t)) = B(t).
\end{aligned}
$$

2. By definition, $E(B^2(t)) = t < \infty$, therefore $B^2(t)$ is integrable. Since

$$
\begin{aligned}
B^2(t+s) &= \big(B(t) + B(t+s) - B(t)\big)^2 \\
&= B^2(t) + 2B(t)(B(t+s) - B(t)) + (B(t+s) - B(t))^2,
\end{aligned}
$$

$$
\begin{aligned}
&E(B^2(t+s)|\mathcal{F}_t) \\
&= B^2(t) + 2E\big(B(t)(B(t+s) - B(t))|\mathcal{F}_t\big) + E\big((B(t+s) - B(t))^2|\mathcal{F}_t\big) \\
&= B^2(t) + s,
\end{aligned}
$$

where we used that $B(t+s) - B(t)$ is independent of $\mathcal{F}_t$ and has mean 0, and (3.10) with $g(x) = x^2$. Subtracting $(t+s)$ from both sides gives the martingale property of $B^2(t) - t$.

3. Consider the moment generating function of $B(t)$,

$$
E(e^{uB(t)}) = e^{tu^2/2} < \infty,
$$

since $B(t)$ has the $N(0, t)$ distribution. This implies integrablity of $e^{uB(t) - tu^2/2}$, moreover

$$
E(e^{uB(t) - tu^2/2}) = 1.
$$

The martingale property is established by using (3.10) with $g(x) = e^{ux}$.

$$
\begin{aligned}
E\left(e^{uB(t+s)}|\mathcal{F}_t\right) &= E\left(e^{uB(t) + u(B(t+s) - B(t))}|\mathcal{F}_t\right) \\
&= e^{uB(t)} E\left(e^{u(B(t+s) - B(t))}|\mathcal{F}_t\right) \text{ (since } B(t) \text{ is } \mathcal{F}_t\text{-measurable)} \\
&= e^{uB(t)} E\left(e^{u(B(t+s) - B(t))}\right) \text{ (since increment is independent of } \mathcal{F}_t) \\
&= e^{\frac{u^2}{2}s} e^{uB(t)}.
\end{aligned}
$$

The martingale property of $e^{uB(t) - tu^2/2}$ is obtained by multiplying both sides by $e^{-\frac{u^2}{2}(t+s)}$.

$\square$

Remark 3.5: All three martingales have a central place in the theory. The martingale $B^2(t) - t$ provides a characterization (Levy's characterization) of Brownian motion. It will be seen later that if a process $X(t)$ is a continuous martingale such that $X^2(t) - t$ is also a martingale, then $X(t)$ is Brownian motion. The martingale $e^{uB(t)-tu^2/2}$ is known as the exponential martingale, and as it is related to the moment generating function, it is used for establishing distributional properties of the process.

3.4 Markov Property of Brownian Motion

The Markov Property states that if we know the present state of the process, then the future behaviour of the process is independent of its past. The process $X(t)$ has the Markov property if the conditional distribution of $X(t+s)$ given $X(t) = x$, does not depend on the past values (but it may depend on the present value x). The process "does not remember" how it got to the present state x. Let $\mathcal{F}_t$ denote the σ-field generated by the process up to time t.

Definition 3.8 *X is a Markov process if for any t and $s > 0$, the conditional distribution of $X(t+s)$ given $\mathcal{F}_t$ is the same as the conditional distribution of $X(t+s)$ given $X(t)$, that is,*

$$P(X(t+s) \leq y|\mathcal{F}_t) = P(X(t+s) \leq y|X(t)), \quad a.s. \tag{3.11}$$

Theorem 3.9 *Brownian motion $B(t)$ possesses Markov property.*

PROOF: It is easy to see by using the moment generating function that the conditional distribution of $B(t + s)$ given $\mathcal{F}_t$ is the same as that given $B(t)$. Indeed,

$$E(e^{uB(t+s)}|\mathcal{F}_t) = e^{uB(t)}E\left(e^{u(B(t+s)-B(t))}|\mathcal{F}_t\right)$$

$$= e^{uB(t)}E\left(e^{u(B(t+s)-B(t))}\right) \quad (\text{since } e^{u(B(t+s)-B(t))} \text{ is independent of } \mathcal{F}_t)$$

$$= e^{uB(t)}e^{u^2s/2} \quad (\text{since } B(t+s) - B(t) \text{ is } N(0,s))$$

$$= e^{uB(t)}E\left(e^{u(B(t+s)-B(t))}|B(t)\right) = E\left(e^{uB(t+s)}|B(t)\right).$$

□

The transition probability function of a Markov process X is defined as

$$P(y, t, x, s) = P(X(t) \leq y|X(s) = x)$$

the conditional distribution function of the process at time t, given that it is at point x at time $s < t$. It is possible to choose them so that for any fixed

x they are true probabilities on the line. In the case of Brownian motion it is given by the distribution function of the Normal $N(x, t - s)$ distribution

$$P(y, t, x, s) = \int_{-\infty}^{y} \frac{1}{\sqrt{2\pi(t-s)}} e^{\frac{(u-x)^2}{2(t-s)}} \, du.$$

The transition probability function of Brownian motion satisfies $P(y, t, x, s) = P(y, t - s, x, 0)$. In other words,

$$P(B(t) \leq y | B(s) = x) = P(B(t - s) \leq y | B(0) = x). \qquad (3.12)$$

For fixed x and t, $P(y, t, x, 0)$ has the density $p_t(x, y)$ is given by (3.3). The property (3.12) states that Brownian motion is time-homogeneous, that is, its distributions do not change with a shift in time. For example, the distribution of $B(t)$ given $B(s) = x$ is the same as that of $B(t - s)$ given $B(0) = x$. It follows from (3.12) and (3.4) that all finite-dimensional distributions of Brownian motion are time-homogeneous.

In what follows P_x denotes the conditional probability given $B(0) = x$. More information on transition functions is given in Section 5.5.

Stopping Times and Strong Markov Property

Definition 3.10 *A random time T is called a stopping time for $B(t)$, $t \geq 0$, if for any t it is possible to decide whether T has occurred or not by observing $B(s)$, $0 \leq s \leq t$. More rigorously, for any t the sets $\{T \leq t\} \in \mathcal{F}_t$, the σ-field generated by $B(s)$, $0 \leq s \leq t$.*

Example 3.7: Examples of stopping times and random times.

1. Any non-random time T is a stopping time. Formally, $\{T \leq t\}$ is either the $\emptyset$ or Ω, which are members of $\mathcal{F}_t$ for any t.

2. Let T be the first time $B(t)$ takes value (hits) 1. Then T is a stopping time. Clearly, if we know $B(s)$ for all $s \leq t$ then we know whether the Brownian motion took value 1 before or at t or not. Thus we know that $\{T \leq t\}$ has occurred or not just by observing the past of the process prior to t. Formally, $\{T \leq t\} = \{B(u) < 1, \text{for all } u \leq t\} \in \mathcal{F}_t$.

3. Similarly, the first passage time of level a, $T_a = \inf\{t > 0 : B(t) = a\}$ is a stopping time.

4. Let T be the time when Brownian motion reaches its maximum on the interval $[0, 1]$. Then clearly, to decide whether $\{T \leq t\}$ has occurred or not, it is not enough to know the values of the process prior to t, one needs to know all the values on the interval $[0, 1]$. So that T is not a stopping time.

5. Let T be the last zero of Brownian motion before time $t = 1$. Then T is not a stopping time, since if $T \leq t$, then there are no zeros in $(t, 1]$, which is the event that is decided by observing the process up to time 1, and this set does not belong to $\mathcal{F}_t$.

The strong Markov property is similar to the Markov property, except that in the definition a fixed time t is replaced by a stopping time T.

Theorem 3.11 *Brownian motion $B(t)$ has the Strong Markov property: for any finite stopping time T the regular conditional distribution of $B(T+t), t \geq 0$ given $\mathcal{F}_T$ is $P_{B(T)}$, that is,*

$$P(B(T+t) \leq y | \mathcal{F}_T) = P(B(T+t) \leq y | B(T)) \ a.s.$$

Corollary 3.12 *Let T be a finite stopping time. Define the new process in $t \geq 0$ by*

$$\hat{B}(t) = B(T+t) - B(T). \tag{3.13}$$

Then $\hat{B}(t)$ is a Brownian motion is started at zero and independent of $\mathcal{F}_T$.

We don't give the proof of the strong Markov property here, it can be found, for example in Rogers and Williams (1994) p.21, and can be done by using the exponential martingale and the Optional Stopping Theorem given in Chapter 7.

Note that the strong Markov property applies only when T is a stopping time. If T is just a random time, then $B(T+t) - B(T)$ need not be Brownian motion.

3.5 Hitting Times and Exit Times

Let T_x denote the first time $B(t)$ hits level x, $T_x = \inf\{t > 0 : B(t) = x\}$. Denote the time to exit an interval (a, b) by $\tau = \min(T_a, T_b)$.

Theorem 3.13 *Let $a < x < b$, and $\tau = \min(T_a, T_b)$. Then $P_x(\tau < \infty) = 1$ and $E_x \tau < \infty$.*

PROOF: $\{\tau > 1\} = \{a < B(s) < b, \text{ for all } 0 \leq s \leq 1\} \subset \{a < B(1) < b\}$. Therefore we have

$$P_x(\tau > 1) \leq P_x(B(1) \in (a, b)) = \frac{1}{\sqrt{2\pi}} \int_a^b e^{-(x-y)^2/2} dy.$$

The function $P_x(B(1) \in (a, b))$ is continuous in x on $[a, b]$, hence it reaches its maximum $\theta < 1$. By using the strong Markov property we can show that $P_x(\tau > n) \leq \theta^n$. For any non-negative random variable $X \geq 0$, $EX \leq \sum_{n=0}^{\infty} P(X > n)$ (see Exercise 3.2). Therefore,

$$E_x \tau \leq \sum_{n=0}^{\infty} \theta^n = \frac{1}{1-\theta} < \infty.$$

The bound on $P_x(\tau > n)$ is established as follows

$$
\begin{aligned}
P_x(\tau > n) &= P_x(B(s) \in (a, b),\ 0 \le s \le n) \\
&= P_x(B(s) \in (a, b),\ 0 \le s \le n - 1, B(s) \in (a, b),\ n - 1 \le s \le n) \\
&= P_x(\tau > n - 1, B(s) \in (a, b),\ n - 1 \le s \le n) \\
&= P_x(\tau > n - 1, B(n - 1) + \hat{B}(s) \in (a, b),\ 0 \le s \le 1) \quad \text{by (3.13)} \\
&= E(P_x(\tau > n - 1, \hat{B}^y(s) \in (a, b),\ 0 \le s \le 1 | B(n - 1) = y)) \\
&= E((P_x(\tau > n - 1 | B(n - 1) = y) P_x(\hat{B}^y(s) \in (a, b),\ 0 \le s \le 1) | B(n - 1) = y)) \\
&= E((P_x(\tau > n - 1 | B(n - 1) = y)) P_y(\hat{B}(s) \in (a, b),\ 0 \le s \le 1)) \\
&\le \ \max_y P_y(\hat{B}(s) \in (a, b),\ 0 \le s \le 1) P_x(\tau > n - 1) \\
&\le \ \theta P_x(\tau > n - 1) \le \theta^n, \quad \text{by iterations.}
\end{aligned}
$$

$\square$

The next result gives the recurrence property of Brownian motion.

Theorem 3.14

$$
P_a(T_b < \infty) = 1, \quad P_a(T_a < \infty) = 1
$$

PROOF: The second statement follows from the first, since

$$
P_a(T_a < \infty) \ge P_a(T_b < \infty) P_b(T_a < \infty) = 1.
$$

We show now that $P_0(T_1 < \infty) = 1$, for other points the proof is similar. Observe firstly that by the previous result and by symmetry, for any a and b

$$
P_{\frac{a+b}{2}}(T_a < T_b) = \frac{1}{2}
$$

Hence $P_0(T_{-1} < T_1) = \frac{1}{2}$, $P_{-1}(T_{-3} < T_1) = \frac{1}{2}$, $P_{-3}(T_{-7} < T_1) = \frac{1}{2}$, etc. Consider now $P_0(T_{-(2^n-1)} < T_1)$. Since the paths of Brownian motion are continuous, to reach $-(2^n - 1)$ the path must reach -1 first, then it must go from -1 to -3, etc. Hence we obtain

$$
\begin{aligned}
&P_0(T_{-(2^n-1)} < T_1) \\
&= \ P_0(T_{-1} < T_1) P_{-1}(T_{-3} < T_1) \ldots P_{-(2^{n-1}-1)}(T_{-(2^n-1)} < T_1) = \frac{1}{2^n}
\end{aligned}
$$

If A_n denotes the event that Brownian motion hits $-(2^n - 1)$ before it hits 1, then we showed that $P(A_n) = 2^{-n}$. Notice that $A_n \subset A_{n-1}$, as if Brownian motion hits $-(2^n - 1)$ before 1, it also hits the points bigger than $-(2^n - 1)$. Thus

$$
\bigcap_{i=1}^{n} A_i = A_n
$$

and

$$P(\bigcap_{i=1}^{\infty} A_i) = \lim_{n \to \infty} P(A_n) = \lim_{n \to \infty} 2^{-n} = 0.$$

This implies that

$$P(\bigcup_{n=1}^{\infty} A_n^c) = 1.$$

In other words, with probability 1, one of the events complimentary to A_n occurs, that is, there is n such that Brownian motion hits 1 before it hits $-(2^n - 1)$. This implies that $P_0(T_1 < \infty) = 1$. Another proof of this fact, that uses properties of martingales, is given in Chapter 7.

$\square$

3.6 Maximum and Minimum of Brownian Motion

In this section we establish the distribution of the maximum and the minimum of Brownian motion on $[0, t]$,

$$M(t) = \max_{0 \le s \le t} B(s) \quad \text{and} \quad m(t) = \min_{0 \le s \le t} B(s),$$

as well as the distribution of the first hitting (passage) time of x, $T_x = \inf\{t > 0 : B(t) = x\}$.

Theorem 3.15 *For any $x > 0$,*

$$P_0(M(t) \ge x) = 2P_0(B(t) \ge x) = 2(1 - \Phi(\frac{x}{\sqrt{t}})),$$

where $\Phi(x)$ stands for the standard Normal distribution function.

PROOF: Notice that the events $\{M(t) \ge x\}$ and $\{T_x \le t\}$ are the same. Indeed, if the maximum at time t is greater than x, then at some time before t Brownian motion took value x, and if Brownian motion took value x at some time before t, then the maximum will be at least x. Since

$$\{B(t) \ge x\} \subset \{T_x \le t\}$$

$$P(B(t) \ge x) = P(B(t) \ge x, T_x \le t).$$

As $B(T_x) = x$,

$$P(B(t) \ge x) = P(T_x \le t, B(T_x + (t - T_x)) - B(T_x) \ge 0).$$

By Theorem 3.14, T_x is a finite stopping time, and by the strong Markov property (3.13), the random variable $\hat{B}(s) = B(T_x + s) - B(T_x)$ is independent of $\mathcal{F}_{T_x}$ and has a Normal distribution, so we have

$$P(B(t) \geq x) = P(T_x \leq t, \hat{B}(t - T_x) \geq 0). \tag{3.14}$$

If we had s independent of T_x, then

$$
\begin{aligned}
P(T_x \leq t, \hat{B}(s) \geq 0) &= P(T_x \leq t)P(\hat{B}(s \geq 0) \\
&= P(T_x \leq t)\frac{1}{2} = P(M(t) \geq x)\frac{1}{2}, \tag{3.15}
\end{aligned}
$$

and we are done. But in (3.14) $s = t - T_x$, and is clearly dependent on T_x. It is not easy to show that

$$
\begin{aligned}
P(B(t) \geq x) &= P(T_x \leq t, \hat{B}(t - T_x) \geq 0) \\
&= P(T_x \leq t)\frac{1}{2} = P(M(t) \geq x)\frac{1}{2}.
\end{aligned}
$$

The proof can be found, for example, in Dudley (1989) p.361.

$\square$

A simple application of the result is given in the following example, from which it follows that Brownian motion changes sign in $(0, \varepsilon)$, for any ε however small.

Example 3.8:   We find the probability $P(B(t) \leq 0$ for all $t, 0 \leq t \leq 1)$. Note that the required probability involves uncountably many random variables: all $B(t)$'s are less than or equal to zero, $0 \leq t \leq 1$, we want to know the probability that the entire path from 0 to 1 will stay below 0. We could calculate the desired probability for n values of the process and then take the limit as $n \to \infty$. But it is simpler in this case to express this probability as a function of the whole path. All $B(t)$'s are less or equal zero, if and only if their maximum is less than or equal to zero.

$$\{B(t) \leq 0 \text{ for all } t, 0 \leq t \leq 1\} = \{\max_{0 \leq t \leq 1} B(t) \leq 0\},$$

and consequently these events have same probabilities. Now,

$$P(\max_{0 \leq t \leq 1} B(t) \leq 0) = 1 - P(\max_{0 \leq t \leq 1} B(t) > 0).$$

By the law of the maximum of Brownian motion,

$$P(\max_{0 \leq t \leq 1} B(t) > 0) = 2P(B(1) > 0) = 1.$$

Hence $P(B(t) \leq 0$ for all $t, 0 \leq t \leq 1) = 0$.

To find the distribution of the minimum of Brownian motion $m(t) = \min_{0 \leq s \leq t} B(s)$ we use the symmetry argument, and that

$$-\min_{0 \leq s \leq t} B(s) = \max_{0 \leq s \leq t} (-B(s)).$$

Theorem 3.16 *If $B(t)$ is a Brownian Motion with $B(0) = 0$, then $\hat{B}(t) = -B(t)$ is also a Brownian motion with $\hat{B}(0) = 0$.*

PROOF: The process $\hat{B}(t) = -B(t)$ has independent and normally distributed increments. It also has continuous paths, therefore it is Brownian motion.

□

Theorem 3.17 *For any $x < 0$*

$$P_0(\min_{0 \le s \le t} B(s) \le x) = 2P_0(B(t) \ge -x) = 2P_0(B(t) \le x)$$

The proof is straightforward and is left as an exercise.

3.7 Distribution of Hitting Times

T_x is finite by Theorem 3.14. The result below gives the distribution of T_x and establishes that T_x has infinite mean.

Theorem 3.18 *The probability density of T_x is given by*

$$f_{T_x}(t) = \frac{|x|}{\sqrt{2\pi}} t^{-\frac{3}{2}} e^{-\frac{x^2}{2t}},$$

which is the Inverse Gamma density with parameters $\frac{1}{2}$ and $\frac{x^2}{2}$. $E_0 T_x = +\infty$.

PROOF: Take $x > 0$. The events $\{M(t) \ge x\}$ and $\{T_x \le t\}$ are the same, so that

$$
\begin{aligned}
P(T_x \le t) &= P(M(t) \ge x) \\
&= 2P(B(t) \ge x) = \int_x^\infty \sqrt{\frac{2}{\pi t}} e^{-\frac{y^2}{2t}} \, dy.
\end{aligned}
$$

The formula for the density of T_x is obtained by differentiation after the change of variables $u = \frac{y}{\sqrt{t}}$ in the integral. Finally,

$$E_0 T_x = \frac{|x|}{\sqrt{2\pi}} \int_0^\infty t^{-\frac{1}{2}} e^{-\frac{x^2}{2t}} \, dt = \infty, \quad \text{since} \quad t^{-\frac{1}{2}} e^{-\frac{x^2}{2t}} \sim 1/\sqrt{t}, \ t \to \infty.$$

For $x < 0$ the proof is similar.

□

Remark 3.6: The property $P(T_x < \infty) = 1$ is called the *recurrence* property of Brownian motion. Although $P(T_x < \infty) = 1$, $E(T_x) = \infty$, even though x is visited with probability one, the expected time for it to happen is infinite.

The next result looks at hitting times T_x as a process in x.

Theorem 3.19 *The process of hitting times* $\{T_x\}$, $x \geq 0$, *has increments independent of the past, that is, for any* $0 < a < b$, $T_b - T_a$ *is independent of* $B(t), t \leq T_a$, *and the distribution of the increment* $T_b - T_a$ *is the same as that of* T_{b-a} *and it is given by the density*

$$f_{T_b - T_a}(t) = \frac{b - a}{\sqrt{2\pi}} t^{-\frac{3}{2}} e^{-\frac{(b-a)^2}{2t}}.$$

PROOF: By the strong Markov property $\hat{B}(t) = B(T_a + t) - B(T_a)$ is Brownian motion started at zero, and independent of the past $B(t), t \leq T_a$. $T_b - T_a = \inf\{t \geq 0 : \hat{B}(t) = b - a\}$. Hence $T_b - T_a$ is the same as first hitting time of $b - a$ by $\hat{B}$.

□

3.8   Reflection Principle and Joint Distributions

Let $B(t)$ be a Brownian motion started at x, and $\hat{B}(t) = -B(t)$. Then $\hat{B}(t)$ is a Brownian motion started at $-x$. The proof is straightforward by checking the defining properties. This is the simplest form of the reflection principle. Here the Brownian motion is reflected about the horizontal axis.

In greater generality, the process that is obtained by reflection of a Brownian motion about the horizontal line passing through $(T, B(T))$, for a stopping time T, is also a Brownian motion.

Note that for $t \geq T$ the reflected path is, $\hat{B}(t) - B(T) = -(B(t) - B(T))$, giving $\hat{B}(t) = 2B(T) - B(t)$.

Theorem 3.20 (Reflection Principle) *Let* T *be a stopping time. Define* $\hat{B}(t) = B(t)$ *for* $t \leq T$, *and* $\hat{B}(t) = 2B(T) - B(t)$ *for* $t \geq T$. *Then* $\hat{B}$ *is also Brownian motion.*

The proof is beyond the scope of this book, but a heuristic justification is that for $t \geq T$ the process $-(B(t) - B(T))$ is also a Brownian motion by the strong Markov property, so that $\hat{B}$, constructed from the Brownian motion before the stopping time, and another Brownian motion after the stopping time, is again Brownian motion. For a rigorous proof see Freedman (1971). Using the Reflection Principle, the joint distribution of the Brownian motion with its maximum can be obtained.

Theorem 3.21 *The joint distribution of* $(B(t), M(t))$ *has the density*

$$f_{B,M}(x, y) = \sqrt{\frac{2}{\pi}} \frac{(2y - x)}{t^{3/2}} e^{\frac{-(2y-x)^2}{2t}}, \; for \; y \geq 0, x \leq y. \qquad (3.16)$$

PROOF: Let, for $y > 0$ and $y > x$, $\hat{B}(t)$ be $B(t)$ reflected at T_y. Then

$$
\begin{aligned}
&P(B(t) \leq x, M(t) \geq y) \\
&= \quad P(T_y \leq t, B(t) \leq x) \quad (\text{ since } \{M(t) \geq y\} = \{T_y \leq t\}) \\
&= \quad P(T_y \leq t, \hat{B}(t) \geq 2y - x) \quad (\text{ on } \{T_y \leq t\}, \ \hat{B}(t) = 2y - B(t)) \\
&= \quad P(T_y \leq t, B(t) \geq 2y - x) \quad (\text{ since } T_y \text{ is the same for } B \text{ and } \hat{B}) \\
&= \quad P(B(t) \geq 2y - x) \quad (\text{ since } y - x > 0, \text{ and } \{B(t) \geq 2y - x\} \subset \{T_y \leq t\}) \\
&= \quad 1 - \Phi\left(\frac{2y - x}{\sqrt{t}}\right).
\end{aligned}
$$

The density is obtained by differentiation.

$\square$

It is possible to show (see for example, Karatzas and Shreve 1988, p.123-24) that $|B(t)|$ and $M(t) - B(t)$ have the same distribution.

Theorem 3.22 *The two processes $|B(t)|$ and $M(t) - B(t)$ are both Markov processes with transition probability density function $p_t(x, y) + p_t(x, -y)$, where $p_t(x, y) = \frac{1}{\sqrt{2\pi t}} e^{-\frac{(y-x)^2}{2t}}$ is the transition probability function of Brownian motion. Consequently they have same finite-dimensional distributions.*

The next result gives the joint distribution of $B(t)$, $M(t)$ and $m(t)$, for a proof see Freedman 1971, p.26-27.

Theorem 3.23

$$
P(a < m(t) \leq M(t) < b, \ and \ B(t) \in A) = \int_A k(y)\,dy, \tag{3.17}
$$

where $k(y) = \sum_{n=-\infty}^{\infty} p_t(2n(b - a), y) - p_t(2a, 2n(b - a) + y)$ for $t > 0$ and $a < 0 < b$.

Remark 3.7: Joint distributions given above are used in pricing of the so-called barrier options, see Chapter 11.

3.9 Zeros of Brownian Motion. Arcsine Law

A time point τ is called a zero of Brownian motion if $B(\tau) = 0$. As an application of the distribution of the maximum we obtain the following information about zeros of Brownian motion. Below $\{B^x(t)\}$ denotes Brownian motion started at x.

Theorem 3.24 *For any $x \neq 0$, the probability that $\{B^x(t)\}$ has at least one zero in the time interval $(0, t)$, is given by*

$$\frac{|x|}{\sqrt{2\pi}} \int_0^t u^{-\frac{3}{2}} e^{-\frac{x^2}{2u}} du.$$

PROOF: If $x < 0$, then due to continuity of $B^x(t)$, (draw a picture of this event)

$$P(B^x \text{ has at least one zero between 0 and } t) = P(\max_{0 \leq s \leq t} B^x(t) \geq 0).$$

Since $B^x(t) = B(t) + x$, where $B(t)$ is Brownian motion started at zero at time zero,

$$P_x(B \text{ has a zero between 0 and } t) = P(\max_{0 \leq s \leq t} B^x(t) \geq 0)$$

$$= P_0(\max_{0 \leq s \leq t} B(t) + x \geq 0) = P_0(\max_{0 \leq s \leq t} B(t) \geq -x)$$

$$= 2P_0(B(t) \geq -x) = P_0(T_x \leq t)$$

$$= \int_0^t f_{T_x}(u) du = \frac{-x}{\sqrt{2\pi}} \int_0^t u^{-\frac{3}{2}} e^{-\frac{x^2}{2u}} du.$$

For $x > 0$ the proof is similar, and is based on the distribution of the minimum of Brownian motion.

$\square$

Using this result we can establish

Theorem 3.25 *The probability that Brownian motion $B(t)$ has at least one zero in the time interval (a, b) is given by*

$$\frac{2}{\pi} \arccos \sqrt{\frac{a}{b}}.$$

PROOF: Denote by $h(x) = P(B \text{ has at least one zero in } (a, b)|B_a = x)$. By the Markov property $P(B \text{ has at least one zero in } (a, b)|B_a = x)$ is the same as $P(B^x \text{ has at least one zero in } (0, b - a))$. By conditioning

$$P(B \text{ has at least one zero in } (a, b))$$

$$= \int_{-\infty}^{\infty} P(B \text{ has at least one zero in } (a, b)|B_a = x)P(B_a \in dx)$$

$$= \int_{-\infty}^{\infty} h(x)P(B_a \in dx) = \sqrt{\frac{2}{\pi a}} \int_0^{\infty} h(x)e^{-\frac{x^2}{2a}} dx.$$

Putting in the expression for $h(x)$ from the previous example and performing the necessary calculations we obtain the result.

$\square$

The Arcsine law now follows:

Theorem 3.26 *The probability that Brownian motion $\{B(t)\}$ has no zeros in the time interval (a, b) is given by $\frac{2}{\pi} \arcsin \sqrt{\frac{a}{b}}$.*

The next result gives distributions of the last zero before t, and the first zero after t. Let

$$\gamma_t = \sup\{s \leq t : B(s) = 0\} = \text{last zero before } t. \tag{3.18}$$

$$\beta_t = \inf\{s \geq t : B(s) = 0\} = \text{first zero after } t. \tag{3.19}$$

Note that β_t is a stopping time but γ_t is not.

Theorem 3.27

$$P(\gamma_t \leq x) = \frac{2}{\pi} \arcsin \sqrt{\frac{x}{t}}. \tag{3.20}$$

$$P(\beta_t \geq y) = \frac{2}{\pi} \arcsin \sqrt{\frac{t}{y}}. \tag{3.21}$$

$$P(\gamma_t \leq x, \beta_t \geq y) = \frac{2}{\pi} \arcsin \sqrt{\frac{x}{y}}. \tag{3.22}$$

PROOF: All of these follow from the previous result. For example, $P(\gamma_t \leq x) = P(B \text{ has no zeros in } (x, t))$.
$P(\gamma_t \leq x, \beta_t \geq y) = P(B \text{ has no zeros in } (x, y))$.

□

Since Brownian motion is continuous, and it has no zeros on the interval (γ_t, β_t) it keeps the same sign on this interval, either positive or negative. When Brownian motion is entirely positive or entirely negative on an interval, it is said that it is an *excursion* of Brownian motion. Thus the previous result states that excursions have the arcsine law. To picture a Brownian path consider for every realization $B = \{B(t), 0 \leq t \leq 1\}$, the set of its zeros on the interval $[0, 1]$, that is, the random set $L_0 = L_0(B) = \{t : B(t) = 0, 0 \leq t \leq 1\}$.

Theorem 3.28 *The set of zeros of Brownian motion is a random uncountable closed set without isolated points and has Lebesgue measure zero.*

PROOF: According to the Example 3.8, the probability that Brownian motion stays below zero on the interval $[0, 1]$ is zero. Therefore it changes sign on this interval. This implies, since Brownian motion is continuous, that it has a zero inside $[0, 1]$. The same reasoning leads to the conclusion that for any positive t, the probability that Brownian motion has the same sign on the interval $[0, t]$ is zero. Therefore it has a zero inside $[0, t]$ for any t, no matter how small it is. This implies that the set of zeros is an infinite set, moreover time $t = 0$ is a limit of zeros from the right.

Observe next, that the set of zeros is closed, that is, if $B(\tau_n) = 0$ and $\lim_{n\to\infty} \tau_n = \tau$ then $B(\tau) = 0$. This is true since $B(t)$ is a continuous function of t.

By using the strong Markov property, it is possible to see that any zero of Brownian motion is a limit of other zeros. If $B(\tau) = 0$, and τ is a stopping time, then by (3.13) $\hat{B}(t) = B(t + \tau) - B(\tau) = B(t + \tau)$ is again Brownian motion started anew at time τ. Therefore time $t = 0$ for the new Brownian motion $\hat{B}$ is a limit from the right of zeros of $\hat{B}$. But $\hat{B}(t) = B(t + \tau)$, so that τ a limit from the right of zeros of B. However, not every zero of Brownian motion is a stopping time. For example, for a fixed t, γ_t, the last zero before t is not a stopping time. Nevertheless, using a more intricate argument, one can see that any zero is a limit of other zeros. A sketch is given below. If τ is the first zero after t, then τ is a stopping time. Thus the set of all sample paths such that τ is a limit point of zeros from the right has probability one. The intersection of such sets over all rational t's is again a set of probability one. Therefore for almost all sample paths the first zero that follows any rational number is a limit of zeros from the right. This implies that any point of L_0 is a limit of points from L_0 (it is a perfect set). A general result from the set theory, which is not hard to prove, states that if an infinite set coincides with the set of its limit points, then it is uncountable.

Although uncountable, L_0 has Lebesgue measure zero. This is seen by writing the Lebesgue measure of L_0 as $|L_0| = \int_0^1 I(B(t) = 0)dt$. It is a non-negative random variable. Taking the expectation, and interchanging the integrals by Fubini's theorem

$$\mathrm{E}|L_0| = \mathrm{E} \int_0^1 I(B(t) = 0)dt = \int_0^1 \mathrm{P}(B(t) = 0)dt = 0.$$

This implies $\mathrm{P}(|L_0| = 0) = 1$.

$\square$

Theorem 3.29 *Any level set $L_a = \{t : B(t) = a, 0 \leq t \leq 1\}$ has the same properties as L_0.*

PROOF: Let T_a be the first time with $B(t) = a$. Then by the strong Markov property, $\hat{B}(t) = B_{T_a+t} - B_{T_a} = B_{T_a+t} - a$ is a Brownian motion. The set of zeros of $\hat{B}$ is the level a set of B.

$\square$

3.10   Size of Increments of Brownian Motion

Increments over large time intervals satisfy the Law of Large Numbers and the Law of the Iterated Logarithm. For proofs see, for example, Karatzas and Shreve (1988).

Theorem 3.30 (Law of Large Numbers)

$$\lim_{t \to \infty} \frac{B(t)}{t} = 0 \ a.s.$$

A more precise result is provided by the Law of Iterated Logarithm.

Theorem 3.31 (Law of Iterated Logarithm)

$$\limsup_{t \to \infty} \frac{B(t)}{\sqrt{2t \ln \ln t}} = 1, \ a.s.$$

$$\liminf_{t \to \infty} \frac{B(t)}{\sqrt{2t \ln \ln t}} = -1 \ a.s.$$

To obtain the behaviour for small t near zero the process $W(t) = tB(1/t)$ is considered, which is also Brownian motion.

Example 3.9: Let $B(t)$ be Brownian motion. The process $W(t)$ defined as $W(t) = tB(1/t)$, for $t > 0$, and $W(0) = 0$, is also Brownian motion. Indeed, $W(t)$ has continuous paths. Continuity at zero follows from the Law of Large Numbers. It is, clearly, a Gaussian process, and has zero mean. Its covariance is given by

$$Cov(W(t), W(s)) = \mathrm{E}(W(t)W(s)) = ts\mathrm{E}(B(1/t)B(1/s)) = ts(1/t) = s, \text{ for } s < t.$$

Since $W(t)$ is a Gaussian process with zero mean, and the covariance of Brownian motion, it is Brownian motion.

This result allows us to transfer results on the behaviour of paths of Brownian motion for large t to that of small t. For example, we have immediately the Law of Iterated Logarithm near zero, from the same law near infinity.

Graphs of some Functions of Brownian Motion

Graphs of some Functions of Brownian motion are given in order to visualize these processes. To obtain these 1000 independent Normal random variables with mean zero and variance 0.001 were generated. Time is taken to be discrete (as any other variable on a computer) varying from 0 to 1 with steps of 0.001. The first two pictures in Figure 3.10 are realizations of White noise. Pictures of $0.1B(t) + t$ and $B(t) + 0.1t$ (second row of Figure 3.10) demonstrate that when noise is small in comparison with drift, the drift dominates, and if drift is small, then the noise dominates in the behaviour of the process. The next two are realizations of the martingale $B^2(t) - t$ which has zero mean. By the recurrence property of Brownian motion, $B(t)$ will always come back to zero. Thus $B^2(t) - t$ will always come back to $-t$ in the long run. The last two pictures are realization of the exponential martingale $e^{B(t)-t/2}$. Although this martingale has mean 1, $\lim_{t \to \infty} e^{B(t)-t/2} = 0$, which can be seen by the Law of Large Numbers. Therefore a realization of this martingale will approach zero in the long run.

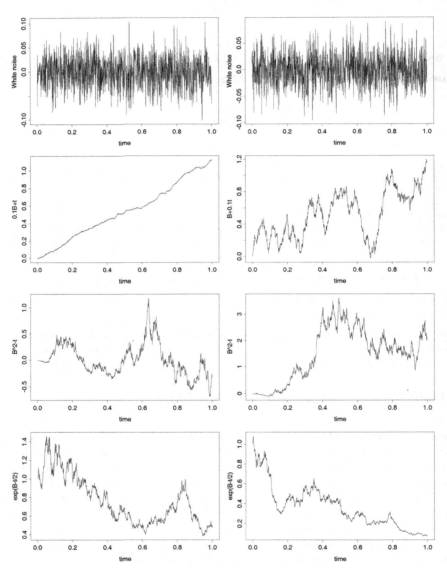

Figure 3.2: *White noise and Functions of Brownian motion.*

3.11 Brownian Motion in Higher Dimensions

Definition 3.32 *Define Brownian motion in dimension two and higher as a random vector $B(t) = (B^1(t), B^2(t), \ldots, B^n(t))$ with all coordinates $B^i(t)$ being independent one-dimensional Brownian motions.*

Alternatively Brownian motion in $\mathbb{R}^n$ can be defined as a process with independent multivariate Gaussian increments. It follows from the definitions, similarly to the one-dimensional case, that Brownian motion in $\mathbb{R}^n$ is a Markov Gaussian process homogeneous both in space and time. Its transition probability density is given by

$$p_t(\boldsymbol{x}, \boldsymbol{y}) = \frac{1}{\sqrt{2\pi t}} e^{-(|\boldsymbol{x}-\boldsymbol{y}|^2)/2t}, \tag{3.23}$$

where $\boldsymbol{x}, \boldsymbol{y}$ are n-dimensional vectors and $|\boldsymbol{x}|^2$ is the length of $\boldsymbol{x}$.

Remark 3.8: In dimensions one and two Brownian motion is recurrent, that is, it will come back to a neighbourhood, however small, of any point infinitely often. In dimensions three and higher Brownian motion is transient, it will leave a ball, however large, around any point never to return (Polya (1922)).

3.12 Random Walk

The analogue of Brownian motion process in discrete time $t = 0, 1, 2, \ldots, n, \ldots$ is the Random Walk process. Brownian motion can be constructed as the limit of Random Walks, when step sizes get smaller and smaller. Random Walks occur in many applications, including Insurance, Finance and Biology.

A model of pure chance is served by an ideal coin being tossed with equal probabilities for the Heads and Tails to come up. Introduce a random variable ξ taking values $+1$ (Heads) and -1 (Tails) with probability $\frac{1}{2}$. If the coin is tossed n times then a sequence of random variables $\xi_1, \xi_2, \ldots, \xi_n$ describes this experiment. All ξ_i have exactly the same distribution as ξ_1, moreover they are all independent. The process S_n is a Random walk, defined by $S_0 = 0$ and

$$S_n = \xi_1 + \xi_2 + \ldots + \xi_n \tag{3.24}$$

S_n gives the amount of money after n plays when betting \$1 on the outcomes of a coin when \$1 is won if Heads come up, but lost otherwise.

Since $\mathrm{E}(\xi_i) = 0$, and $Var(\xi_i) = \mathrm{E}(\xi^2) = 1$, the mean and the variance of the random walk are given by

$$\mathrm{E}(S_n) = \mathrm{E}(\xi_1 + \xi_2 + \ldots \xi_n) = \mathrm{E}(\xi_1) + \mathrm{E}(\xi_2) + \ldots \mathrm{E}(\xi_n) = 0,$$

$$Var(S_n) = Var(\xi_1) + Var(\xi_2) + \ldots + Var(\xi_n) = nVar(\xi_1) = n,$$

as the variance of a sum of independent variables equals to the sum of variances. More generally, a random walk is the process

$$S_n = S_0 + \sum_{i=1}^{n} \xi_i, \tag{3.25}$$

where ξ_i's are independent and identically distributed random variables. In particular, this model contains gambling on the outcomes of a biased coin $P(\xi_i = 1) = p$, $P(\xi_i = -1) = q = 1 - p$.

Martingales in Random Walks

Some interesting questions about Random Walks, such as ruin probabilities and the like, can be answered with the help of martingales.

Theorem 3.33 *The following are martingales.*

1. $S_n - \mu n$, *where* $\mu = E(\xi_1)$. *In particular, if the Random Walk is unbiased* $(\mu = 0)$, *then it is itself is a martingale.*

2. $(S_n - \mu n)^2 - \sigma^2 n$, *where* $\sigma^2 = E(\xi_1 - \mu)^2 = Var(\xi_1)$.

3. *For any* u, $e^{uS_n - nh(u)}$, *where* $h(u) = \ln E(e^{u\xi_1})$. *In particular, in the case* $P(\xi_1 = 1) = p$, $P(\xi_1 = -1) = q = 1 - p$, $(\frac{q}{p})^{S_n}$ *is a martingale.*

PROOF:
1. Since, by the triangle inequality,

$$E|S_n - n\mu| = E|S_0 + \sum_{i=1}^{n} \xi_i - n\mu| \leq E|S_0| + \sum_{i=1}^{n} E|\xi_i| + n|\mu| = E|S_0| + n(E|\xi_1| + |\mu|),$$

$S_n - n\mu$ is integrable provided $E|\xi_1| < \infty$, and $E|S_0| < \infty$. To establish the martingale property consider for any n

$$E(S_{n+1}|S_n) = S_n + E(\xi_{n+1}|S_n).$$

Since ξ_{n+1} is independent of the past, and S_n is determined by the first n variables, ξ_{n+1} is independent of S_n. Therefore, $E(\xi_{n+1}|S_n) = E(\xi_{n+1})$. It now follows that

$$E(S_{n+1}|S_n) = S_n + E(\xi_{n+1}|S_n) = S_n + \mu,$$

and subtracting $(n + 1)\mu$ from both sides of the equation, the martingale property is obtained,

$$E(S_{n+1} - (n + 1)\mu|S_n) = S_n - n\mu.$$

2. This is left as an exercise.

3. Put $M_n = e^{uS_n - nh(u)}$. Since $M_n \geq 0$, $\mathrm{E}|M_n| = \mathrm{E}(M_n)$, which is given by

$$
\begin{aligned}
\mathrm{E}(M_n) &= \mathrm{E}e^{uS_n - nh(u)} = e^{-nh(u)}\mathrm{E}e^{uS_n} = e^{-nh(u)}\mathrm{E}e^{u(S_0 + \sum_{i=1}^n \xi_i)} \\
&= e^{uS_0}e^{-nh(u)}\mathrm{E}\prod_{i=1}^n e^{u\xi_i} = e^{uS_0}e^{-nh(u)}\prod_{i=1}^n \mathrm{E}(e^{u\xi_i}) \text{ by independence} \\
&= e^{uS_0}e^{-nh(u)}\prod_{i=1}^n e^{h(u)} = e^{uS_0} < \infty.
\end{aligned}
$$

The martingale property is shown by using the fact that

$$
S_{n+1} = S_n + \xi_{n+1}, \tag{3.26}
$$

with ξ_{n+1} independent of S_n and of all previous ξ_i's $i \leq n$, or independent of $\mathcal{F}_n$. Using the properties of conditional expectation, we have

$$
\begin{aligned}
\mathrm{E}(e^{uS_{n+1}}|\mathcal{F}_n) &= \mathrm{E}(e^{uS_n + u\xi_{n+1}}|\mathcal{F}_n) \\
&= e^{uS_n}\mathrm{E}(e^{u\xi_{n+1}}|\mathcal{F}_n) = e^{uS_n}\mathrm{E}(e^{u\xi_{n+1}}) \\
&= e^{uS_n + h(u)}.
\end{aligned}
$$

Multiplying both sides of the above equation by $e^{-(n+1)h(u)}$, the martingale property is obtained, $\mathrm{E}(M_{n+1}|\mathcal{F}_n) = M_n$.

In the special case when $\mathrm{P}(\xi_i = 1) = p$, $\mathrm{P}(\xi_i = -1) = q = 1 - p$ choosing $u = \ln(q/p)$ in the previous martingale, we have $e^{u\xi_1} = (q/p)^{\xi_1}$ and $\mathrm{E}(e^{u\xi_1}) = 1$. Thus $h(u) = \ln \mathrm{E}(e^{u\xi_1}) = 0$, and $e^{uS_n - nh(u)} = (q/p)^{S_n}$. Alternatively in this case, the martingale property of $(q/p)^{S_n}$ is easy to verify directly.

$\square$

3.13 Stochastic Integral in Discrete Time

Let S_n be an unbiased Random Walk representing the capital of a player when betting on a fair coin. Let H_n can be the amount of money (the number of betting units) a gambler will bet at time n. This can be based on the outcomes of the game at times $1, \ldots, n - 1$ but not on the outcome at time n. This is an example of a predictable process. The concept of predictable processes plays a most important role in stochastic calculus. The process $\{H_n\}$ is called predictable if H_n can be predicted with certainty from the information available at time $n - 1$. Rigorously, let $\mathcal{F}_{n-1}$ be the σ-field generated by $S_0, S_1, \ldots, S_{n-1}$.

Definition 3.34 $\{H_n\}$ *is called predictable if for all $n \geq 1$, H_n is $\mathcal{F}_{n-1}$ measurable.*

If a betting strategy $\{H_n\}_{n=0}^t$ is used in the game of coin tossing, then the *gain* at time t is given by

$$(H \cdot S)_t = \sum_{n=1}^{t} H_n(S_n - S_{n-1}), \tag{3.27}$$

since $S_n - S_{n-1} = +1$ or -1 when the n-th toss results in a win or loss respectively. More generally, $S_n - S_{n-1} = \xi_n$ represents the amount of money lost or won on one betting unit on the n-th bet. If H_n units are placed, then the amount of money lost or won on the n-th bet is $H_n(S_n - S_{n-1})$. The gain at time t is obtained by adding up monies lost and won on all the bets.

Definition 3.35 *The stochastic integral in discrete time of a predictable process H with respect to the process S is defined by*

$$(H \cdot S)_t := H_0 S_0 + \sum_{n=1}^{t} H_n(S_n - S_{n-1}) \tag{3.28}$$

The stochastic integral gives the gain in a game of chance when betting on S and the betting strategy H is used. For a martingale the stochastic integral (3.28) is also called a martingale transform. The next result states that a betting system used on a martingale will result again in a martingale.

Theorem 3.36 *If M_n is a martingale, H_n is predictable and the random variables $(H \cdot M)_t$ are integrable, then $(H \cdot M)_t$ is a martingale.*

PROOF:

$$\mathrm{E}[(H \cdot M)_{t+1}|\mathcal{F}_t] = \mathrm{E}[(H \cdot M)_t|\mathcal{F}_t] + \mathrm{E}[H_{t+1}(M_{t+1} - M(t))|\mathcal{F}_t]$$

$$= (H \cdot M)_t + H_{t+1}\mathrm{E}[(M_{t+1} - M(t))|\mathcal{F}_t] = (H \cdot M)_t.$$

$\square$

As a corollary we obtain

Theorem 3.37 *If M_n is a martingale, H_n is predictable and bounded, then then $(H \cdot M)_t$ is a martingale.*

PROOF: The assumption of bounded H_n implies that

$$\mathrm{E}|(H \cdot M)_t| = \mathrm{E}|\sum_{n=1}^{t} H_n(M_n - M_{n-1})|$$

$$\leq \sum_{n=1}^{t} \mathrm{E}|H_n(M_n - M_{n-1})| \leq 2C \sum_{n=1}^{t} \mathrm{E}|M_n| < \infty.$$

So that $(H \cdot M)_t$ is integrable, and the condition of the previous theorem is fulfilled.

$\square$

Stopped Martingales

Let $(M_n, \mathcal{F}_n)$ be a martingale and τ be a stopping time. Recall the definition

Definition 3.38 *A random time is a stopping time if for any n,*
$\{\tau > n\} \in \mathcal{F}_n$.

Take $H_n = 1$ if $n \leq \tau$, and $H_n = 0$ if $n > \tau$, in other words, $H_n = I(\tau \geq n)$. Then H_n is predictable, because $\{\tau \geq n\} = \{\tau > n+1\} \in \mathcal{F}_{n+1}$. The stochastic integral gives the martingale stopped at τ,

$$(H \cdot M)_n = H_0 M_0 + H_1(M_1 - M_0) + \cdots + H_n(M_n - M_{n-1})$$

$$= M_{\tau \wedge n} = M_\tau I(\tau \leq n) + M_n I(\tau > n).$$

Since $H_n = I(\tau \geq n)$ is bounded by 1, Theorem 3.37 implies that the process $(H \cdot M)_n = M_{\tau \wedge n}$ is a martingale. Thus we have shown

Theorem 3.39 *A martingale stopped at a stopping time τ, $M_{\tau \wedge n}$ is a martingale. In particular,*

$$\mathbb{E} M_{\tau \wedge n} = \mathbb{E} M_0. \tag{3.29}$$

Comment here that Theorem 3.39 holds also in continuous time, see Theorem 7.14. It is a Basic Stopping result, which is harder to prove.

Example 3.10: (Doubling bets strategy).
Consider the doubling strategy when betting on the Heads in tosses of a fair coin. Bet $H_1 = 1$. If Heads comes up then stop. The profit is $G_1 = 1$. If the outcome is Tails, then bet $H_2 = 2$ on the second toss. If the second toss comes up heads, then stop. The profit is $G_2 = 4 - 3 = 1$. If the game continues for n steps, (meaning that the $n - 1$ tosses did not result in a win) then bet $H_n = 2^{n-1}$ on the n-th toss. If the n-th toss comes up Heads then stop. The profit is $G_n = 2 \times 2^{n-1} - (1 + 2 + \ldots 2^{n-1}) = 2^n - (2^n - 1) = 1$. The probability that the game will stop at a finite number of steps is one minus the probability that Heads never come up. Probability of only Tails on the first n tosses is, by independence, 2^{-n}. The probability that Heads never comes up is the limit $\lim_{n \to \infty} 2^{-n} = 0$, thus the game will stop for sure. For any non-random time T the gain process G_t, $t \leq T$ is a martingale with zero mean. The doubling strategy does not contradict the result above, because the strategy uses an unbounded stopping time, the first time one dollar is won.

Further information on discrete time martingales and on their stopping is given in Chapter 7.

3.14   Poisson Process

If Brownian motion process is a basic model for cumulative small noise present continuously, the Poisson process is a basic model for cumulative noise that occurs as a shock.

Let $\lambda > 0$. A random variable X has a Poisson distribution with parameter λ, denoted $Pn(\lambda)$, if it takes non-negative integer values $k \geq 0$ with probabilities

$$P(X = k) = e^{-\lambda}\frac{\lambda^k}{k!}, \quad k = 0, 1, 2, \ldots \ldots \tag{3.30}$$

The moment generating function of this distribution is given by

$$E(e^{uX}) = e^{\lambda(e^u - 1)}. \tag{3.31}$$

Defining Properties of Poisson process

A Poisson process $N(t)$ is a stochastic process with the following properties.

1. (Independence of increments) $N(t) - N(s)$ is independent of the past, that is, of $\mathcal{F}_s$, the σ-field generated by $N(u), u \leq s$.

2. (Poisson increments) $N(t) - N(s), t > s$, has a Poisson distribution with parameter $\lambda(t - s)$. If $N(0) = 0$, then $N(t)$ has the $Pn(\lambda t)$ distribution.

3. (Step function paths) The paths $N(t), t \geq 0$, are increasing functions of t changing only by jumps of of size 1.

Remark 3.9: A definition of a Poisson process in a more general model (that contains extra information) is given by a pair $\{N(t), \mathcal{F}_t\}, t \geq 0$, where $\mathcal{F}_t$ is an increasing sequence of σ-fields (a filtration), $N(t)$ is an adapted process, i.e. $N(t)$ is $\mathcal{F}_t$ measurable, such that Properties 1-3 above hold.

Consider a model for occurrence of independent events. Define the rate λ as the average number of events per unit of time. Let $N(t)$ be the number of events that occur up to time t, i.e. in the time interval $(0, t]$. Then $N(t) - N(s)$ gives the number of events that occur in the time interval $(s, t]$.

A Poisson process $N(t)$ can be constructed as follows. Let $\tau_1, \tau_2, \ldots$ be independent random variables with the exponential $\exp(\lambda)$ distribution, that is, $P(\tau_1 > t) = e^{-\lambda t}$. τ's represent the times between occurrence of successive events. Let $T_n = \sum_{i=1}^n \tau_i$, be the time of the n-th event. Then

$$N(t) = \sup\{n : T_n \leq t\}$$

counts the number of events up to time t. It is not hard to verify the defining properties of the Poisson process for this construction. $N(t)$ has the Poisson distribution with parameter λt. Consequently,

$$P(N(t) = k) = e^{-\lambda t} \frac{(\lambda t)^k}{k!}, \quad k = 0, 1, 2, \ldots,$$

$$EN(t) = \lambda t, \text{ and } Var(N(t)) = \lambda t.$$

Variation and Quadratic Variation of the Poisson Process

Let $0 = t_0^n < t_1^n < \ldots < t_n^n = t$ be a partition of $[0, t]$. Then it is easy to see that variation of a Poisson path is

$$V_N(t) = \lim \sum_{i=1}^{n} |N(t_i^n) - N(t_{i-1}^n)| = N(t) - N(0) = N(t), \quad (3.32)$$

where the limit is taken when $\delta_n = \max_i(t_i^n - t_{i-1}^n) \to 0$ and $n \to \infty$. Recall that variation of a pure jump function is the sum of absolute values of the jumps, see Example 1.6. Since the Poisson process has only positive jumps of size one (3.32) follows.

To calculate its quadratic variation, observe that $N(t_i^n) - N(t_{i-1}^n)$ takes only two values 0 and 1 for small $t_i^n - t_{i-1}^n$, hence it is the same as its square, $N(t_i^n) - N(t_{i-1}^n) = (N(t_i^n) - N(t_{i-1}^n))^2$. Thus the quadratic variation of N is the same as its variation

$$[N, N](t) = \lim \sum_{i=1}^{n} (N(t_i^n) - N(t_{i-1}^n))^2 = N(t) - N_0 = N(t).$$

Thus for a Poisson process both the variation and quadratic variation are positive and finite.

Poisson Process Martingales

The process $N(t)$ is increasing, hence it can not be a martingale. However, the compensated process $N(t) - \lambda t$ is a martingale. This martingale is analogous to the Brownian motion.

Theorem 3.40 *The following are martingales.*

1. $N(t) - \lambda t$.

2. $(N(t) - \lambda t)^2 - \lambda t$.

3. $e^{\ln(1-u)N(t)+u\lambda t}$, for any $0 < u < 1$.

PROOF: The martingale property follows from independence of increments, the Poisson distribution of increments and the expressions for the mean and the variance of Poisson distribution. We show the martingale property for the exponential martingale.

$$\mathrm{E}\left(e^{\ln(1-u)N(t+s)}|\mathcal{F}_t\right) = \mathrm{E}\left(e^{\ln(1-u)N(t)+\ln(1-u)(N(t+s)-N(t))}|\mathcal{F}_t\right)$$

$$= e^{\ln(1-u)N(t)}\mathrm{E}\left(e^{\ln(1-u)(N(t+s)-N(t))})|\mathcal{F}_t\right) \text{ (since } N(t) \text{ is } \mathcal{F}_t\text{-measurable)}$$

$$= e^{\ln(1-u)N(t)}\mathrm{E}\left(e^{\ln(1-u)(N(t+s)-N(t))}\right) \text{ (increment is independent of } \mathcal{F}_t\text{)}$$

$$= e^{\ln(1-u)N(t)}e^{-u\lambda s} \text{ by (3.31), since } N(t+s) - N(t) \text{ is Poisson}(\lambda s).$$

Multiplying both sides by $e^{u\lambda(t+s)}$, the martingale property follows.

$\square$

Using the exponential martingale, it can be shown that the Poisson process has the strong Markov property.

3.15   Exercises

Exercise 3.1: Derive the moment generating function of the multivariate Normal distribution $N(\boldsymbol{\mu}, \boldsymbol{\Sigma})$.

Exercise 3.2: Show that for a non-negative random variable X, $\mathrm{E}X = \int_0^\infty \mathrm{P}(X \geq x)dx$. Hint: use $\mathrm{E}X = \int_0^\infty x dF(x) = \int_0^\infty \int_0^x dt dF(x)$ and change the order of integration.

Exercise 3.3: Show that if $X \geq 0$, $\mathrm{E}X \leq \sum_{n=0}^\infty \mathrm{P}(X > n)$.

Exercise 3.4: Let $B(t)$ be a Brownian motion. Show that the following processes are Brownian motions on $[0, T]$.

1. $X(t) = -B(t)$.

2. $X(t) = B(T - t) - B(T)$, where $T < +\infty$.

3. $X(t) = cB(t/c^2)$, where $T \leq +\infty$.

4. $X(t) = tB(1/t)$, $t > 0$, and $X(0) = 0$.

Hint: Check the defining properties. Alternatively, show that the process is a Gaussian process with correlation function $\min(s, t)$. Alternatively, show that the process is a continuous martingale with quadratic variation t (this is the Levy's characterization, and will be proven later.)

Exercise 3.5: Let $B(t)$ and $W(t)$ be two independent Brownian motions. Show that $X(t) = (B(t) + W(t))/\sqrt{2}$ is also a Brownian motion. Find correlation between $B(t)$ and $X(t)$.

Exercise 3.6: Let $\boldsymbol{B}(t)$ be an n-dimensional Brownian motion, and $\boldsymbol{x}$ is a non-random vector in $\mathbb{R}^n$ with length 1, $|\boldsymbol{x}|^2 = 1$. Show that $W(t) = \boldsymbol{x} \cdot \boldsymbol{B}(t)$ is a (one-dimensional) Brownian motion.

Exercise 3.7: Let $B(t)$ be a Brownian motion and $0 \le t_1, \ldots \le t_n$. Give a matrix A, such that $(B(t_1), B(t_2), \ldots, B(t_n))^T = A(Z_1, \ldots, Z_n)^T$, where Z_i's are standard Normal variables. Hence give the covariance matrix of $(B(t_1), B(t_2), \ldots, B(t_n))$. Here ^{T} stands for transpose and the vectors are column vectors.

Exercise 3.8: Let $B(t)$ be a Brownian motion and $0 \le s < t$. Show that the conditional distribution of $B(s)$ given $B(t) = b$ is Normal and give its mean and variance.

Exercise 3.9: Show that the random variables $M(t)$ and $|B(t)|$ have the same distribution.

Exercise 3.10: Show that the moments of order r of the hitting time T_x are finite $E(T_x^r) < \infty$ if and only if $r < 1/2$.

Exercise 3.11: Derive the distribution of the maximum $M(t)$ from the joint distribution of $(B(t), M(t))$.

Exercise 3.12: By considering $-B(t)$, derive the joint distribution of $B(t)$ and $m(t) = \min_{s \le t} B(s)$.

Exercise 3.13: Show that the random variables $M(t)$, $|B(t)|$ and $M(t) - B(t)$ have the same distributions, cf. Exercise 3.9.

Exercise 3.14: The first zero of Brownian motion started at zero is 0. What is the second zero?

Exercise 3.15: Let T be the last time before 1 a Brownian motion visits 0. Explain why $X(t) = B(t + T) - B(T) = B(t + T)$ is not a Brownian motion.

Exercise 3.16: Formulate the Law of Large Numbers and the Law of Iterated Logarithm for Brownian motion near zero.

Exercise 3.17: Let $B(t)$ be Brownian motion. Show that $e^{-\alpha t} B(e^{2\alpha t})$ is a Gaussian process. Find its mean and covariance functions.

Exercise 3.18: Let $X(t)$ be a Gaussian process with zero mean and covariance $\gamma(s,t) = e^{-\alpha|t-s|}$. Show that X has a version with continuous paths.

Exercise 3.19: Show that in a Normal Random Walk, $S_n = S_0 + \sum_{i=1}^{n} \xi_i$ when ξ_i's are standard Normal random variables, $e^{uS_n - nu^2/2}$ is a martingale.

Exercise 3.20: Let $S_n = S_0 + \sum_{i=1}^{n} \xi_i$ be a Random Walk, with $P(\xi_1 = 1) = p$, $P(\xi_1 = -1) = 1 - p$. Show that for any λ, $e^{\gamma S_n - \lambda n}$ is a martingale for the appropriate value of γ.

Exercise 3.21: The process X_t, is defined for discrete times $t = 1, 2, \ldots$. It can take only three values 1, 2 and 3. Its behaviour is defined by the rule: from state 1 it goes to 2, from 2 it goes to 3 and from 3 it goes back to 1. X_1 takes values 1, 2, 3 with equal probabilities. Show that this process is Markov. Show also that

$$P(X_3 = 3 | X_2 = 1 \text{ or } 2, X_1 = 3) \neq P(X_3 = 3 | X_2 = 1 \text{ or } 2).$$

This demonstrates that to apply Markov property we must know the present state of the process exactly, it is not enough to know that it can take one of the two (or more) possible values.

Exercise 3.22: A discrete-time process $X(t)$, $t = 0, 1, 2, \ldots$, is said to be autoregressive of order p (AR(p)) if there exists $a_1, \ldots, a_p \in \mathbb{R}$, and a white noise $Z(t)$ ($E(Z(t)) = 0$, $E(Z^2(t)) = \sigma^2$ and, for $s > 0$, $E(Z(t)Z(t+s)) = 0$) such that

$$X(t) = \sum_{s=1}^{p} a_s X(t-s) + Z(t).$$

1. Show that $X(t)$ is Markovian if and only if $p = 1$.

2. Show that if $X(t)$ is AR(2), then $\boldsymbol{Y}(t) = (X(t), X(t+1))$ is Markovian.

3. Suppose that $Z(t)$ is a Gaussian process. Write the transition probability function of an AR(1) process $X(t)$.

Exercise 3.23: The distribution of a random variable τ has the lack of memory property if $P(\tau > a + b | \tau > a) = P(\tau > b)$. Verify the lack of memory property for the exponential $\exp(\lambda)$ distribution. Show that if τ has the lack of memory property and a density, then it has an exponential distribution.

Chapter 4

Brownian Motion Calculus

In this chapter stochastic integrals with respect to Brownian motion are introduced and their properties are given. They are also called Itô integrals, and the corresponding calculus Itô calculus.

4.1 Definition of Itô Integral

Our goal is to define the stochastic integral $\int_0^T X(t)dB(t)$, also denoted $\int X\,dB$ or $X \cdot B$. This integral should have the property that if $X(t) = 1$ then $\int_0^T dB(t) = B(T) - B(0)$. Similarly, if $X(t)$ is a constant c, then the integral should be $c(B(T) - B(0))$. In this way we can integrate constant processes with respect to B. The integral over $(0, T]$ should be the sum of integrals over two subintervals $(0, a]$ and $(a, T]$. Thus if $X(t)$ takes two values c_1 on $(0, a]$, and c_2 on $(a, T]$, then the integral of X with respect to B is easily defined. In this way the integral is defined for *simple* processes, that is, processes which are constant on finitely many intervals. By the limiting procedure the integral is then defined for more general processes.

Itô Integral of Simple Processes

Consider first integrals of a *non-random* simple process $X(t)$, which is a function of t and does not depend on $B(t)$. By definition a simple non-random process $X(t)$ is a process for which there exist times $0 = t_0 < t_1 < \ldots < t_n = T$ and constants $c_0, c_1, \ldots, c_{n-1}$, such that

$$X(t) = c_0 I_0(t) + \sum_{i=0}^{n-1} c_i I_{(t_i, t_{i+1}]}(t). \tag{4.1}$$

The Itô integral $\int_0^T X(t)dB(t)$ is defined as a sum

$$\int_0^T X(t)dB(t) = \sum_{i=0}^{n-1} c_i\big(B(t_{i+1}) - B(t_i)\big). \tag{4.2}$$

It is easy to see by using the independence property of Brownian increments that the integral, which is the sum in (4.2) is a Gaussian random variable with mean zero and variance

$$Var\left(\int_0^T X(t)dB(t)\right) = Var\left(\sum_{i=0}^{n-1} c_i\left(B(t_{i+1}) - B(t_i)\right)\right)$$

$$= \sum_{i=0}^{n-1} Var\left(c_i(B(t_{i+1}) - B(t_i))\right) = \sum_{i=0}^{n-1} c_i^2(t_{i+1} - t_i).$$

Example 4.1: Let $X(t) = -1$ for $0 \le t \le 1$, $X(t) = 1$ for $1 < t \le 2$, and $X(t) = 2$ for $2 < t \le 3$. Then (note that $t_i = 0, 1, 2, 3$, $c_i = X(t_{i+1})$, $c_0 = -1$, $c_1 = 1$, $c_2 = 2$)

$$\int_0^3 X(t)dB(t) = c_0\big(B(1) - B(0)\big) + c_2\big(B(2) - B(1)\big) + c_3\big(B(3) - B(2)\big)$$

$$= -B(1) + \big(B(2) - B(1)\big) + 2\big(B(3) - B(2)\big) = 2B(3) - B(2) - 2B(1).$$

Its distribution is $N(0,6)$, either directly as a sum of independent $N(0,1) + N(0,1) + N(0,4)$ or by using the result above.

By taking limits of simple non-random processes, more general but still only non-random processes can be integrated with respect to Brownian motion.

To integrate random processes, it is important to allow for constants c_i in (4.1) to be random. If c_i's are replaced by random variables ξ_i's, then in order to have convenient properties of the integral the random variable ξ_i's are allowed to depend on the values of $B(t)$ for $t \le t_i$, but not on future values of $B(t)$ for $t > t_i$. If $\mathcal{F}_t$ is the σ-field generated by Brownian motion up to time t, then ξ_i is $\mathcal{F}_{t_i}$-measurable. The approach of defining the integral by approximation can be carried out for the class of *adapted* processes $X(t)$, $0 \le t \le T$.

Definition 4.1 *A process X is called adapted to the filtration $\mathbb{F} = (\mathcal{F}_t)$, if for all t, $X(t)$ is $\mathcal{F}_t$-measurable.*

Remark 4.1: In order that the integral has desirable properties, in particular that the expectation and the integral can be interchanged (by Fubini's theorem), the requirement that X is adapted is too weak, and a stronger condition, that of a *progressive* (progressively measurable) process is needed. X

is progressive if it is a measurable function in the pair of variables (t, ω), i.e. $\mathcal{B}([0, t]) \times \mathcal{F}_t$ measurable as a map from $[0, t] \times \Omega$ into $\mathbb{R}$. It can be seen that every adapted right-continuous with left limits or left-continuous with right limits (regular, cadlag) process is progressive. Since it is easier to understand what is meant by a regular adapted process, we use 'regular adapted' terminology without further reference to progressive or measurable in (t, ω) processes.

Definition 4.2 *A process* $X = \{X(t), \ 0 \leq t \leq T\}$ *is called a simple adapted process if there exist times* $0 = t_0 < t_1 < \ldots < t_n = T$ *and random variables* $\xi_0, \xi_1, \ldots, \xi_{n-1}$, *such that* ξ_0 *is a constant,* ξ_i *is* $\mathcal{F}_{t_i}$-*measurable (depends on the values of* $B(t)$ *for* $t \leq t_i$, *but not on values of* $B(t)$ *for* $t > t_i$), *and* $\mathrm{E}(\xi_i^2) < \infty$, $i = 0, \ldots, n-1$; *such that*

$$X(t) = \xi_0 I_0(t) + \sum_{i=0}^{n-1} \xi_i I_{(t_i, t_{i+1}]}(t). \tag{4.3}$$

For simple adapted processes Itô integral $\int_0^T X dB$ is defined as a sum

$$\int_0^T X(t) dB(t) = \sum_{i=0}^{n-1} \xi_i \big(B(t_{i+1}) - B(t_i) \big). \tag{4.4}$$

Note that when ξ_i's are random, the integral need not have a Normal distribution, as in the case of non-random c_i's.

Remark 4.2: Simple adapted processes are defined as left-continuous step functions. One can take right-continuous functions. However, when the stochastic integral is defined with respect to general martingales, other than the Brownian motion, only left-continuous functions are taken.

Properties of the Itô Integral of Simple Adapted Processes

Here we establish main properties of the Itô integral of simple processes. These properties carry over to the Itô integral of general processes.

1. Linearity. If $X(t)$ and $Y(t)$ are simple processes and α and β are some constants then

$$\int_0^T \big(\alpha X(t) + \beta Y(t) \big) \, dB(t) = \alpha \int_0^T X(t) dB(t) + \beta \int_0^T Y(t) dB(t).$$

2. For the indicator function of an interval $I_{(a,b]}(t)$ ($I_{(a,b]}(t) = 1$ when $t \in (a, b]$, and zero otherwise)

$$\int_0^T I_{(a,b]}(t) dB(t) = B(b) - B(a), \quad \int_0^T I_{(a,b]}(t) X(t) dB(t) = \int_a^b X(t) dB(t).$$

3. Zero mean property. $\mathrm{E} \int_0^T X(t)dB(t) = 0$

4. Isometry property.

$$\mathrm{E}\left(\int_0^T X(t)dB(t) \right)^2 = \int_0^T \mathrm{E}(X^2(t))dt \qquad (4.5)$$

PROOF: Properties 1 and 2 are verified directly from the definition. Proof of linearity of the integral follows from the fact that a linear combination of simple processes is again a simple process, and so is $I_{(a,b]}(t)X(t)$.

Since ξ_i's are square integrable, then by the Cauchy-Schwarz inequality

$$\mathrm{E}\big|\xi_i(B(t_{i+1}) - B(t_i))\big| \leq \sqrt{\mathrm{E}(\xi_i^2)\mathrm{E}(B(t_{i+1}) - B(t_i))^2} < \infty,$$

which implies that

$$\mathrm{E}\Big| \sum_{i=0}^{n-1} \xi_i\big(B(t_{i+1}) - B(t_i)\big)\Big| \leq \sum_{i=0}^{n-1} \mathrm{E}\big|\xi_i\big(B(t_{i+1}) - B(t_i)\big)\big| < \infty, \qquad (4.6)$$

and the stochastic integral has expectation. By the martingale property of Brownian motion, using that ξ_i's are $\mathcal{F}_{t_i}$-measurable

$$\mathrm{E}\big(\xi_i(B(t_{i+1}) - B(t_i))|\mathcal{F}_{t_i}\big) = \xi_i\mathrm{E}\big((B(t_{i+1}) - B(t_i))|\mathcal{F}_{t_i}\big) = 0, \qquad (4.7)$$

and it follows that $\mathrm{E}\big(\xi_i(B(t_{i+1}) - B(t_i))\big) = 0$, which implies Property 3.

To prove Property 4, write the square as the double sum

$$\mathrm{E}\left(\sum_{i=0}^{n-1} \xi_i(B(t_{i+1}) - B(t_i)) \right)^2 = \sum_{i=0}^{n-1} \mathrm{E}\big(\xi_i^2\,(B(t_{i+1}) - B(t_i))^2\big)$$
$$+2 \sum_{i<j} \mathrm{E}\big(\xi_i\xi_j\,(B(t_{i+1}) - B(t_i))\,(B(t_{j+1}) - B(t_j))\big). \qquad (4.8)$$

Using the martingale property of Brownian motion,

$$\sum_{i=0}^{n-1} \mathrm{E}\big(\xi_i^2\,(B(t_{i+1}) - B(t_i))^2\big) = \sum_{i=0}^{n-1} \mathrm{E}\mathrm{E}\left(\xi_i^2\,(B(t_{i+1}) - B(t_i))^2\,|\mathcal{F}_{t_i} \right)$$
$$= \sum_{i=0}^{n-1} \mathrm{E}\left(\xi_i^2\mathrm{E}\left((B(t_{i+1}) - B(t_i))^2\,|\mathcal{F}_{t_i} \right) \right) = \sum_{i=0}^{n-1} \mathrm{E}(\xi_i^2)(t_{i+1} - t_i).$$

The last sum is exactly $\int_0^T \mathrm{E}(X^2(t))dt$, since $X(t) = \xi_i$ on $(t_i, t_{i+1}]$. By conditioning, in a similar way, we obtain for $i < j$,

$$\mathrm{E}\big(\xi_i\xi_j\,(B(t_{i+1}) - B(t_i))\,(B(t_{j+1}) - B(t_j))\big) = 0,$$

so that the sum $\sum_{i<j}$ in (4.8) vanishes, and Property 4 is proved.

□

Itô Integral of Adapted Processes

Let $X^n(t)$ be a sequence of simple processes convergent in probability to the process $X(t)$. Then, under some conditions, the sequence of their integrals $\int_0^T X^n(t)dB(t)$ also converges in probability to a limit J. The random variable J is taken to be the integral $\int_0^T X(t)dB(t)$.

Example 4.2: We find $\int_0^T B(t)dB(t)$.
Let $0 = t_0^n < t_1^n < t_2^n < \ldots < t_n^n = T$ be a partition of $[0, T]$, and let

$$X^n(t) = \sum_{i=0}^{n-1} B(t_i^n) I_{(t_i^n, t_{i+1}^n]}(t).$$

Then for any n, $X^n(t)$ is a simple adapted process. (Here $\xi_i^n = B(t_i^n)$.) By the continuity of $B(t)$, $\lim_{n \to \infty} X^n(t) = B(t)$ almost surely as $\max_i(t_{i+1}^n - t_i^n) \to 0$. The Itô integral of the simple function $X^n(t)$ is given by

$$\int_0^T X^n(t)dB(t) = \sum_{i=0}^{n-1} B(t_i^n)(B(t_{i+1}^n) - B(t_i^n)).$$

We show that this sequence of integrals converges in probability to $J = \frac{1}{2}B^2(T) - \frac{1}{2}T$. Adding and subtracting $B^2(t_{i+1}^n)$, we obtain

$$B(t_i^n)\big(B(t_{i+1}^n) - B(t_i^n)\big) = \frac{1}{2}\Big(B^2(t_{i+1}^n) - B^2(t_i^n) - \big(B(t_{i+1}^n) - B(t_i^n)\big)^2\Big),$$

and

$$\begin{aligned}
\int_0^T X^n(t)dB(t) &= \frac{1}{2}\sum_{i=0}^{n-1}\big(B^2(t_{i+1}^n) - B^2(t_i^n)\big) - \frac{1}{2}\sum_{i=0}^{n-1}\big(B(t_{i+1}^n) - B(t_i^n)\big)^2 \\
&= \frac{1}{2}B^2(T) - \frac{1}{2}B^2(0) - \frac{1}{2}\sum_{i=0}^{n-1}\big(B(t_{i+1}^n) - B(t_i^n)\big)^2,
\end{aligned}$$

since the first sum is a telescopic one. By the definition of the quadratic variation of Brownian motion the second sum converges in probability to T. Therefore $\int_0^T X^n(t)dB(t)$ converges in probability to the limit J

$$\int_0^T B(t)dB(t) = J = \lim_{n \to \infty} \int_0^T X^n(t)dB(t) = \frac{1}{2}B^2(T) - \frac{1}{2}T. \tag{4.9}$$

Remark 4.3:

- If $X(t)$ is a differentiable function (more generally, a function of finite variation), then the stochastic integral $\int_0^T X(t)dB(t)$ can be defined by formally using the integration by parts:

$$\int_0^T X(t)dB(t) = X(T)B(T) - X(0)B(0) - \int_0^T B(t)dX(t),$$

(Paley, Wiener and Zygmund *Math. Z.* 37, 1933, 647-668.)

But this approach fails when $X(t)$ depends on $B(t)$.

- Brownian motion has no derivative, but it has a generalized derivative as a Schwartz distribution. It is defined by the following relation. For a smooth function g with a compact support (zero outside a finite interval)

$$\int g(t)B'(t)dt := -\int B(t)g'(t)dt.$$

But this approach fails when $g(t)$ depends on $B(t)$.

- For simple processes the Itô integral is defined for each ω, path by path, but in general, this is not possible. For example, $\int_0^1 B(\omega, t)dB(\omega, t)$ is not defined, whereas $\left(\int_0^1 B(t)dB(t)\right)(\omega) = J(\omega)$ is defined as a limit in probability of integrals (sums) of simple processes.

Theorem 4.3 *Let $X(t)$ be a regular adapted process such that with probability one $\int_0^T X^2(t)dt < \infty$. Then Itô integral $\int_0^T X(t)dB(t)$ is defined and has the following properties.*

1. *Linearity. If Itô integrals of $X(t)$ and $Y(t)$ are defined and α and β are some constants then*

$$\int_0^T \left(\alpha X(t) + \beta Y(t)\right) dB(t) = \alpha \int_0^T X(t)dB(t) + \beta \int_0^T Y(t)dB(t).$$

2.

$$\int_0^T X(t)I_{(a,b]}(t)dB(t) = \int_a^b X(t)dB(t).$$

The following two properties hold when the process satisfies an additional assumption

$$\int_0^T E(X^2(t))dt < \infty. \tag{4.10}$$

3. *Zero mean property. If condition (4.10) holds then*

$$E\left(\int_0^T X(t)dB(t)\right) = 0. \tag{4.11}$$

4. *Isometry property. If condition (4.10) holds then*

$$E\left(\int_0^T X(t)dB(t)\right)^2 = \int_0^T E(X^2(t))dt. \tag{4.12}$$

PROOF: Only an outline of the proof is given. Firstly it is shown that the Itô integral is well defined for adapted processes that satisfy an additional assumption (4.10). Such processes can be approximated by simple processes

$$X^n(t) = X(0) + \sum_{i=0}^{n-1} X(t_i^n) I_{(t_i^n, t_{i+1}^n]}(t), \tag{4.13}$$

where $\{t_i^n\}$ is a partition of $[0, T]$ with $\delta_n = \max_i(t_{i+1}^n - t_i^n) \to 0$ as $n \to \infty$. In this sum only one term is different from zero, corresponding to the interval in the partition containing t. The process $X^n(t)$ equals $X(t)$ for all the points in the partition, but may differ in each small interval (t_i^n, t_{i+1}^n). Now,

$$\int_0^T (X^n(t))^2 dt = \sum_{k=0}^{n-1} X^2(t_i^n)(t_{i+1}^n - t_i^n) \longrightarrow \int_0^T X^2(t) dt, \tag{4.14}$$

$$\int_0^T \mathrm{E}(X^n(t))^2 dt = \sum_{k=0}^{n-1} \mathrm{E}X^2(t_i^n)(t_{i+1}^n - t_i^n) \longrightarrow \int_0^T \mathrm{E}X^2(t) dt, \tag{4.15}$$

as $n \to \infty$, because the sums are Riemann sums for the corresponding integrals. Moreover, it is possible to show that,

$$\lim_{n \to \infty} \int_0^T \mathrm{E}(X^n(t) - X(t))^2 dt = 0. \tag{4.16}$$

Denote the Itô integral of the simple process by

$$J_n = \int_0^T X^n(t) dB(t) = \sum_{k=0}^{n-1} X(t_i^n)(B(t_{i+1}^n) - B(t_i^n)). \tag{4.17}$$

The condition (4.10) for J_n holds, so that $\mathrm{E}(J_n) = 0$ and by the isometry property $\mathrm{E}(J_n^2)$ is given by the sum in (4.15). Using the isometry property (4.16) implies that

$$\mathrm{E}(J_n - J_m)^2 = \mathrm{E}\left(\int_0^T X^m(t) dB(t) - \int_0^T X^n(t) dB(t)\right)^2$$

$$= \mathrm{E}\left(\int_0^T (X^m(t) - X^n(t)) dB(t)\right)^2 = \mathrm{E}\int_0^T (X^m(t) - X^n(t))^2 dt$$

$$\leq 2\mathrm{E}\int_0^T (X^m(t) - X(t))^2 dt + 2\mathrm{E}\int_0^T (X^n(t) - X(t))^2 dt \longrightarrow 0, \tag{4.18}$$

as $n, m \to \infty$. The space L^2 of random variables with zero mean, finite second moments and convergence in the mean-square is complete, and (4.18) shows that J_n form a Cauchy sequence in this space. This implies that there is an element J, such that $J_n \to J$ in L^2. This limit J is taken to be the Itô integral $\int_0^T X(t)dB(t)$. If we were to use another approximating sequence then, it is not hard to check, this limit does not change.

Now consider adapted processes with finite integral $\int_0^T X^2(t)dt$ but not necessarily of finite expectation. It can be shown, using the previous result, that such processes can be approximated by simple processes by taking limit in probability rather than in mean square. The sequence of corresponding Itô integrals is a Cauchy sequence in probability. It converges in probability to a limit $\int_0^T X(t)dB(t)$.

For details of proof see for example, Gihman and Skorohod (1972), Liptser and Shiryaev (1977), Karatzas and Shreve (1988).

□

Note that Itô integrals need not have mean and variance, but when they do, the mean is zero and the variance is given by (4.12).

Corollary 4.4 *If X is a continuous adapted process then the Itô integral $\int_0^T X(t)dB(t)$ exists. In particular, $\int_0^T f(B(t))dB(t)$, where f is a continuous function on $\mathbb{R}$ is well defined.*

PROOF: Since any path of $X(t)$ is a continuous function, $\int_0^T X^2(t)dt < \infty$, and the result follows by Theorem 4.3. If f is continuous on $\mathbb{R}$, then $f(B(t))$ is continuous on $[0, T]$.

□

Remark 4.4:   It follows from the proof that the sums (4.17) approximate the Itô integral $\int_0^T X(t)dB(t)$

$$\sum_{k=0}^{n-1} X(t_i^n)(B(t_{i+1}^n) - B(t_i^n)).$$

In approximation of the Stieltjes integral by sums, the function f on the interval $[t_i, t_{i+1}]$ is replaced by its value at some middle point $\theta_i \in [t_i, t_{i+1}]$, whereas in the above approximations for the Itô integral, the left most point must be taken for $\theta_i = t_i$, otherwise the process may not be adapted.

It is possible to define an integral (different to Itô integral) when θ_i is chosen to be an interior point of the interval, $\theta_i = \lambda t_i + (1 - \lambda)t_{i+1}$, for some $\lambda \in (0, 1)$. The resulting integral may depend on the choice of λ. When $\lambda = 1/2$, the Stratanovich stochastic integral results. Calculus with such integrals is closely related to the Itô calculus.

Remark 4.5:   Note that the Itô integral does not have the monotonicity property: $X(t) \leq Y(t)$ does not imply $\int_0^T X(t)dB(t) \leq \int_0^T Y(t)dB(t)$. A simple counter-example is $\int_0^1 1 \times dB(t) = B(1)$. With probability half this is smaller than 0, the Itô integral of 0.

We give examples of Itô integrals of the form $\int_0^1 f(B(t))dB(t)$ with and without the first two moments.

Example 4.3: Take $f(t) = e^t$. $\int_0^1 e^{B(t)}dB(t)$ is well defined as e^x is continuous on $\mathbf{R}$. Since $\mathrm{E}(\int_0^1 e^{2B(t)}dt) = \int_0^1 \mathrm{E}(e^{2B(t)})dt = \int_0^1 e^{2t}dt = \frac{1}{2}(e^2 - 1) < \infty$, $\mathrm{E}(\int_0^1 e^{B(t)}dB(t)) = 0$, and $\mathrm{E}(\int_0^1 e^{B(t)}dB(t))^2 = \frac{1}{2}(e^2 - 1)$.

Example 4.4: Take $f(t) = t$, that is, consider $\int_0^1 B(t)dB(t)$. Then the condition (4.10) is satisfied, since $\int_0^1 \mathrm{E}(B^2(t))dt = \int_0^1 tdt = 1/2 < \infty$. Thus $\int_0^1 B(t)dB(t)$ has mean zero and variance $1/2$.

Example 4.5: Take $f(t) = e^{t^2}$, that is, consider $\int_0^1 e^{B^2(t)}dB(t)$. Although this integral is well defined, the condition (4.10) fails, as $\int_0^1 \mathrm{E}(e^{2B^2(t)})dt = \infty$, due to the fact that $\mathrm{E}(e^{2B^2(t)}) = \int e^{2x^2} \frac{1}{\sqrt{2\pi t}} e^{-\frac{x^2}{2t}} = \infty$ for $t \geq 1/4$. Therefore we can not claim that this Itô integral has finite moments. By using martingale inequalities given in the sequel, it can be shown that the expectation of the Itô integral does not exist.

Example 4.6: Let $J = \int_0^1 tdB(t)$. We calculate $\mathrm{E}(J)$ and $Var(J)$.

Since $\int_0^1 t^2dt < \infty$, the Itô integral is defined. Since the integrand t is non-random, condition (4.10) holds and the integral has the first two moments, $\mathrm{E}(J) = 0$, and $\mathrm{E}(J^2) = \int_0^1 t^2dt = 1/3$.

Example 4.7: For what values of α is the integral $\int_0^1 (1 - t)^{-\alpha}dB(t)$ defined?

For the Itô integral to be defined it must have $\int_0^1 (1 - t)^{-2\alpha}dt < \infty$. This gives $\alpha < 1/2$.

A consequence of the isometry property is the expectation of the product of two Itô integrals.

Theorem 4.5 *Let $X(t)$ and $Y(t)$ be regular adapted processes, such that* $\mathrm{E}\int_0^T X(t)^2dt < \infty$ *and* $\mathrm{E}\int_0^T Y(t)^2dt < \infty$. *Then*

$$\mathrm{E}\left(\int_0^T X(t)dB(t) \int_0^T Y(t)dB(t)\right) = \int_0^T \mathrm{E}(X(t)Y(t))dt. \qquad (4.19)$$

PROOF: Denote the Itô integrals $I_1 = \int_0^T X(t)dB(t)$, $I_2 = \int_0^T Y(t)dB(t)$. Write their product by using the identity $I_1 I_2 = (I_1 + I_2)^2/2 - I_1^2/2 - I_2^2/2$. Then use the isometry property. $\qquad \square$

4.2   Itô Integral Process

Let X be a regular adapted process, such that $\int_0^T X^2(s)ds < \infty$ with probability one, so that $\int_0^t X(s)dB(s)$ is defined for any $t \leq T$. Since it is a random variable for any fixed t, $\int_0^t X(s)dB(s)$ as a function of the upper limit t defines a stochastic process

$$Y(t) = \int_0^t X(s)dB(s). \qquad (4.20)$$

It is possible to show that there is a version of the Itô integral $Y(t)$ with continuous sample paths. It is always assumed that the continuous version of the Itô integral is taken. It will be seen later in this section that the Itô integral has a positive quadratic variation and infinite variation.

Martingale Property of the Itô Integral

It is intuitively clear from the construction of Itô integrals that they are adapted. To see this more formally, Itô integrals of simple processes are clearly adapted, and also continuous. Since $Y(t)$ is a limit of integrals of simple processes, it is itself adapted.

Suppose that in addition to the condition $\int_0^T X^2(s)ds < \infty$, condition (4.10) holds, $\int_0^T \mathrm{E}X^2(s)ds < \infty$. (The latter implies the former by Fubini's theorem.) Then $Y(t) = \int_0^t X(s)dB(s)$, $0 \leq t \leq T$, is defined and possesses first two moments. It can be shown, first for simple processes and then in general, that for $s < t$,

$$\mathrm{E}\left(\int_s^t X(u)dB(u)|\mathcal{F}_s\right) = 0.$$

Thus

$$
\begin{aligned}
\mathrm{E}(Y(t)|\mathcal{F}_s) &= \mathrm{E}\left(\int_0^t X(u)dB(u)|\mathcal{F}_s\right) \\
&= \int_0^s X(u)dB(u) + \mathrm{E}\left(\int_s^t X(u)dB(u)|\mathcal{F}_s\right) \\
&= \int_0^s X(u)dB(u) = Y(s).
\end{aligned}
$$

Therefore $Y(t)$ is a martingale. The second moments of $Y(t)$ are given by the isometry property,

$$\mathrm{E}\left(\int_0^t X(s)dB(s)\right)^2 = \int_0^t \mathrm{E}X^2(s)ds. \qquad (4.21)$$

This shows that $\sup_{t \leq T} \mathrm{E}(Y^2(t)) = \int_0^T \mathrm{E}X^2(s)ds < \infty$.

Definition 4.6 *A martingale is called square integrable on* $[0, T]$ *if its second moments are bounded.*

Thus we have

Theorem 4.7 *Let* $X(t)$ *be an adapted process such that* $\int_0^T EX^2(s)ds < \infty$. *Then* $Y(t) = \int_0^t X(s)dB(s)$, $0 \le t \le T$, *is a continuous zero mean square integrable martingale.*

Remark 4.6: If $\int_0^T EX^2(s)ds = \infty$, then the Itô integral $\int_0^t X(s)dB(s)$ may fail to be a martingale, but it is always a *local martingale* (see Chapter 7) for definition and properties.

Theorem 4.7 above provides a way of constructing martingales.

Corollary 4.8 *For any bounded function* f *on* $\mathbb{R}$, $\int_0^t f(B(s))dB(s)$ *is a square integrable martingale.*

PROOF: $X(t) = f(B(t))$ is adapted, and since $|f(x)| < K$, for some constant $K > 0$, $\int_0^T Ef^2(B(s))ds \le KT$. The result follows by Theorem 4.7.

$\square$

Quadratic Variation and Covariation of Itô Integrals

The Itô integral $Y(t) = \int_0^t X(s)dB(s)$, $0 \le t \le T$, is a random function of t. It is continuous and adapted. The quadratic variation of Y is defined by (see (1.13))

$$[Y, Y](t) = \lim \sum_{i=0}^{n-1} (Y(t_{i+1}^n) - Y(t_i^n))^2, \tag{4.22}$$

where for each n, $\{t_i^n\}_{i=0}^n$, is a partition of $[0, t]$, and the limit is in probability, taken over all partitions with $\delta_n = \max_i(t_{i+1}^n - t_i^n) \to 0$ as $n \to \infty$.

Theorem 4.9 *The quadratic Variation of the Itô integral* $\int_0^t X(s)dB(s)$ *is given by*

$$\left[\int_0^t X(s)dB(s), \int_0^t X(s)dB(s)\right](t) = \int_0^t X^2(s)ds. \tag{4.23}$$

It is easy to verify the result for simple processes, see the Example below. The general case can be proved by approximations by simple processes.

Example 4.8: For simplicity, suppose that X takes only two different values on $[0, 1]$: ξ_0 on $[0, 1/2]$ and ξ_1 on $[1/2, 1]$

$$X_t = \xi_0 I_{[0,1/2]}(t) + \xi_1 I_{(1/2,1]}(t).$$

It is easy to see that

$$Y(t) = \int_0^t X(s)ds = \begin{cases} \xi_0 B(t) & \text{if } t \le 1/2 \\ \xi_0 B(1/2) + \xi_1\big(B(t) - B(1/2)\big) & \text{if } t > 1/2. \end{cases}$$

Thus for any partition of $[0, t]$,

$$Y(t_{i+1}^n) - Y(t_i^n) = \begin{cases} \xi_0\big(B(t_{i+1}^n) - B(t_i^n)\big) & \text{if } t_i^n < t_{i+1}^n \le 1/2 \\ \xi_1\big(B(t_{i+1}^n) - B(t_i^n)\big) & \text{if } 1/2 \le t_i^n < t_{i+1}^n. \end{cases}$$

Including $1/2$ in a partition, one can verify that: for $t \le 1/2$

$$[Y, Y](t) = \lim \sum_{i=0}^{n-1} (Y(t_{i+1}^n) - Y(t_i^n))^2$$

$$= \xi_0^2 \lim \sum_{i=0}^{n-1} (B(t_{i+1}^n) - B(t_i^n))^2 = \xi_0^2[B, B](t) = \xi_0^2 t = \int_0^t X^2(s)ds;$$

and for $t > 1/2$

$$[Y, Y](t) = \lim \sum_{i=0}^{n-1} (Y(t_{i+1}) - Y(t_i))^2$$

$$= \xi_0^2 \lim \sum_{t_i < 1/2} (B(t_{i+1}) - B(t_i))^2 + \xi_1^2 \lim \sum_{t_i > 1/2} (B(t_{i+1}) - B(t_i))^2$$

$$= \xi_0^2[B, B](1/2) + \xi_1^2[B, B]((1/2, t]) = \int_0^t X^2(s)ds.$$

The limits above are limits in probability when $\delta_n = \max_i\{(t_{i+1}^n - t_i^n)\} \to 0$. In the same way (4.23) is verified for any simple function.

Example 4.9: Using the formula (4.23), quadratic variation of the Itô integral

$$\left[\int_0^{\cdot} B(s)dB(s)\right](t) = \int_0^t B^2(s)ds.$$

Corollary 4.10 *If $\int_0^t X^2(s)ds > 0$, for all $t \le T$, then the Itô integral $Y(t) = \int_0^t X(s)dB(s)$ has infinite variation on $[0, t]$ for all $t \le T$.*

PROOF: If $Y(t)$ were of finite variation, its quadratic variation would be zero, leading to a contradiction.

$\square$

Like Brownian motion, the Itô integral $Y(t)$ is a continuous but nowhere differentiable function of t.

Let now $Y_1(t)$ and $Y_2(t)$ be Itô integrals of $X_1(t)$ and $X_2(t)$ with respect to the *same* Brownian motion $B(t)$. Then, clearly, the process $Y_1(t) + Y_2(t)$ is also an Itô integral of $X_1(t) + X_2(t)$ with respect to $B(t)$.

Quadratic covariation of Y_1 and Y_2 on $[0, t]$ is defined by

$$[Y_1, Y_2](t) = \frac{1}{2} ([Y_1 + Y_2, Y_1 + Y_2](t) - [Y_1, Y_1](t) - [Y_2, Y_2](t)). \qquad (4.24)$$

By (4.23) it follows that

$$[Y_1, Y_2](t) = \int_0^t X_1(s)X_2(s)ds. \qquad (4.25)$$

It is clear that $[Y_1, Y_2](t) = [Y_2, Y_1](t)$, and it can be seen that quadratic covariation is given by the limit in probability of products of increments of the processes Y_1 and Y_2 when partitions $\{t_i^n\}$ of $[0, t]$ shrink,

$$[Y_1, Y_2](t) = \lim \sum_{i=0}^{n-1} (Y_1(t_{i+1}^n) - Y_1(t_i^n)) (Y_2(t_{i+1}^n) - Y_2(t_i^n)).$$

4.3 Itô Integral and Gaussian Processes

We have seen in Section 4.1 that the Itô integral of simple non-random processes is a Normal random variable. It is easy to see by using moment generating functions (see Exercise 4.3) that a limit in probability of such a sequence is also Gaussian. This implies the following result.

Theorem 4.11 *If $X(t)$ is non-random such that $\int_0^T X^2(s)ds < \infty$, then its Itô integral $Y(t) = \int_0^t X(s)dB(s)$ is a Gaussian process with zero mean and covariance function given by*

$$Cov(Y(t), Y(t+u)) = \int_0^t X^2(s)ds, \ u \geq 0. \qquad (4.26)$$

Moreover, $Y(t)$ is a square integrable martingale.

PROOF: Since the integrand is non-random, $\int_0^t EX^2(s)ds = \int_0^t X^2(s)ds < \infty$. By the zero mean property of Itô integral, Y has zero mean. To compute the covariance function, write $\int_0^{t+u}$ as $\int_0^t + \int_t^{t+u}$ and use the martingale property of $Y(t)$ to obtain

$$E \left(\int_0^t X(s)dB(s) E \left(\int_t^{t+u} X(s)dB(s) \Big| \mathcal{F}_t \right) \right) = 0.$$

Hence

$$Cov(Y(t), Y(t+u)) = \mathrm{E}\left(\int_0^t X(s)dB(s) \int_0^{t+u} X(s)dB(s)\right)$$

$$= \mathrm{E}\left(\int_0^t X(s)dB(s)\right)^2 = \int_0^t \mathrm{E}X^2(s)ds = \int_0^t X^2(s)ds.$$

□

A proof of normality of integrals of non-random processes will be done later by using Itô's formula.

Example 4.10: According to Theorem 4.11 $J = \int_0^t sdB(s)$ has a Normal $N(0, t^3/3)$ distribution.

Example 4.11:

Let $X(t) = 2I_{[0,1]}(t) + 3I_{(1,3]}(t) - 5I_{(3,4]}(t)$. Give the Itô integral $\int_0^4 X(t)dB(t)$ as a sum of random variables, give its distribution, mean and variance. Show that the process $M(t) = \int_0^t X(s)dB(s)$, $0 \leq t \leq 4$, is a Gaussian process and a martingale.

$$\int_0^4 X(t)dB(t) = \int_0^1 X(t)dB(t) + \int_1^3 X(t)dB(t) + \int_3^4 X(t)dB(t)$$

$$= \int_0^1 2dB(t) + \int_1^3 3dB(t) + \int_3^4 (-5)dB(t)$$

$$= 2(B(1) - B(0)) + 3(B(3) - B(1)) - 5(B(4) - B(3)).$$

The Itô integral is a sum of 3 independent Normal random variables (by independence of increments of Brownian motion), $2N(0,1) + 3N(0,2) - 5N(0,1)$. Its distribution is $N(0, 47)$.

The martingale property and the Gaussian property of $M(t) = \int_0^t X(s)dB(s)$, $0 \leq t \leq 4$, follow from the independence of the increments of $M(t)$, zero mean increments and the Normality of the increments. $M(t) - M(s) = \int_s^t X(u)dB(u)$. Take for example $0 < s < t < 1$, then $M(t) - M(s) = \int_s^t X(u)dB(u) = 2(B(t) - B(s))$, which is independent of the Brownian motion up to time s, and has $N(0, 4(t-s))$ distribution.

If $0 < s < 1 < t < 3$, then $M(t) - M(s) = \int_s^t X(u)dB(u) = \int_s^1 X(u)dB(u) + \int_1^t X(u)dB(u) = 2(B(1) - B(s)) + 3(B(t) - B(1))$, which is independent of the Brownian motion up to time s, $B(u), u \leq s$ (and also $M(u), u \leq s$), and has $N(0, 4(1-s) + 9(t-1))$ distribution. Other cases are similar. By Theorem 2.23 the process $M(t)$ is Gaussian.

Independence of increments plus zero mean of increments imply the martingale property of $M(t)$. For example, If $0 < s < 1 < t < 3$, $\mathrm{E}(M(t)|M(u), u \leq s) = \mathrm{E}(M(s) + M(t) - M(s)|M(u), u \leq s) = M(s) + \mathrm{E}(M(t) - M(s)|M(u), u \leq s) = M(s)$.

If $Y(t) = \int_0^t X(t, s)dB(s)$ where $X(t, s)$ depends on the upper integration limit t, then $Y(t)$ need not be a martingale, but remains a gaussian process for non-random $X(t, s)$.

Theorem 4.12 *For any* $t \leq T$, *let* $X(t, s)$ *be a regular non-random function with* $\int_0^t X^2(t, s)ds < \infty$. *Then the process* $Y(t) = \int_0^t X(t, s)dB(s)$ *is a Gaussian process with mean zero and covariance function* $t, u \geq 0$

$$Cov(Y(t), Y(t + u)) = \int_0^t X(t, s)X(t + u, s)ds. \qquad (4.27)$$

PROOF: For a fixed t, the distribution of $Y(t)$, as that of an Itô integral of a non-random function, is Normal with mean 0 and variance $\int_0^t X^2(t, s)ds$. We don't prove the process is Gaussian (it can be seen by approximating $X(t, s)$ by functions of the form $f(t)g(s)$), but calculate the covariance. For $u > 0$

$$Y(t + u) = \int_0^t X(t + u, s)dB(s) + \int_t^{t+u} X(t + u, s)dB(s).$$

Since $X(t + u, s)$ is non-random, the Itô integral $\int_t^{t+u} X(t + u, s)dB(s)$ is independent of $\mathcal{F}_t$. Therefore

$$\mathrm{E}\left(\int_0^t X(t, s)dB(s) \int_t^{t+u} X(t + u, s)dB(s)\right) = 0,$$

and

$$
\begin{aligned}
Cov(Y(t), Y(t + u)) &= \mathrm{E}\left(Y(t)Y(t + u)\right) \\
&= \mathrm{E}\left(\int_0^t X(t, s)dB(s) \int_0^t X(t + u, s)dB(s)\right) \\
&= \int_0^t X(t, s)X(t + u, s)ds, \qquad (4.28)
\end{aligned}
$$

where the last equality is obtained by the expectation of a product of Itô integrals, Equation (4.19).

$\square$

4.4   Itô's Formula for Brownian Motion

Itô's formula, also known as the *change of variable* and the *chain rule*, is one of the main tools of stochastic calculus. It gives rise to many others, such as Dynkin, Feynman-Kac, and integration by parts formulae.

Theorem 4.13 *If* $B(t)$ *is a Brownian motion on* $[0, T]$ *and* $f(x)$ *is a twice continuously differentiable function on* $\mathbb{R}$, *then for any* $t \leq T$

$$f(B(t)) = f(0) + \int_0^t f'(B(s))dB(s) + \frac{1}{2}\int_0^t f''(B(s))ds. \qquad (4.29)$$

PROOF: Note first that both integrals in (4.29) are well defined, the Itô integral by Corollary 4.4. Let $\{t_i^n\}$ be a partition of $[0, t]$. Clearly,

$$f(B(t)) = f(0) + \sum_{i=0}^{n-1} \left(f(B(t_{i+1}^n)) - f(B(t_i^n)) \right).$$

Apply Taylor's formula to $f(B(t_{i+1}^n)) - f(B(t_i^n))$ to obtain

$$f(B(t_{i+1}^n)) - f(B(t_i^n)) = f'(B(t_i^n))(B(t_{i+1}^n) - B(t_i^n)) + \frac{1}{2}f''(\theta_i^n)(B(t_{i+1}^n) - B(t_i^n))^2,$$

where $\theta_i^n \in (B(t_i^n), B(t_{i+1}^n))$. Thus,

$$
\begin{aligned}
f(B(t)) \quad = \quad & f(0) + \sum_{i=0}^{n-1} f'(B(t_i^n))(B(t_{i+1}^n) - B(t_i^n)) \\
& + \frac{1}{2}\sum_{i=0}^{n-1} f''(\theta_i^n)(B(t_{i+1}^n) - B(t_i^n))^2.
\end{aligned}
\tag{4.30}
$$

Taking limits as $\delta_n \to 0$, the first sum in (4.30) converges to the Itô integral $\int_0^t f'(B(s))dB(s)$. By the theorem below the second sum in (4.30) converges to $\int_0^t f''(B(s))ds$ and the result follows.

$\square$

Theorem 4.14 *If g is a bounded continuous function and $\{t_i^n\}$ represents partitions of $[0, t]$, then for any $\theta_i^n \in (B(t_i^n), B(t_{i+1}^n))$, the limit in probability*

$$\lim_{\delta_n \to 0} \sum_{i=0}^{n-1} g(\theta_i^n)\left(B(t_{i+1}^n) - B(t_i^n)\right)^2 = \int_0^t g(B(s))ds. \tag{4.31}$$

PROOF: Take first $\theta_i^n = B(t_i^n)$ to be the left end of the interval $(B(t_i^n), B(t_{i+1}^n))$. We show that the sums converge in probability

$$\sum_{i=0}^{n-1} g(B(t_i^n))\left(B(t_{i+1}^n) - B(t_i^n)\right)^2 \to \int_0^t g(B(s))ds. \tag{4.32}$$

By continuity of $g(B(t))$ and definition of the integral, it follows that

$$\sum_{i=0}^{n-1} g(B(t_i^n))(t_{i+1}^n - t_i^n) \to \int_0^t g(B(s))ds. \tag{4.33}$$

Next we show that the difference between the sums converges to zero in L^2,

$$\sum_{i=0}^{n-1} g(B(t_i^n))\left(B(t_{i+1}^n) - B(t_i^n)\right)^2 - \sum_{i=0}^{n-1} g(B(t_i))(t_{i+1} - t_i) \to 0. \tag{4.34}$$

With $\Delta B_i = B(t_{i+1}^n) - B(t_i^n)$ and $\Delta t_i = t_{i+1}^n - t_i^n$, by using conditioning it is seen that the cross-product term in the following expression vanishes and

$$
\mathrm{E}\left(\sum_{i=0}^{n-1} g(B(t_i^n))\left((\Delta B_i)^2 - \Delta t_i\right)\right)^2 = \mathrm{E}\sum_{i=0}^{n-1} g^2(B(t_i^n))\mathrm{E}\left(\left((\Delta B_i)^2 - \Delta t_i\right)^2 | \mathcal{F}_{t_i}\right)
$$

$$
= 2\mathrm{E}\sum_{i=0}^{n-1} g^2(B(t_i^n))(\Delta t_i)^2 \leq \delta 2\mathrm{E}\sum_{i=0}^{n-1} g^2(B(t_i^n))\Delta t_i \to 0 \text{ as } \delta \to 0.
$$

It follows that

$$
\sum_{i=0}^{n-1} g(B(t_i^n))\left((\Delta B_i)^2 - \Delta t_i\right) \to 0,
$$

in the square mean (L^2), implying (4.34) and that both sums in (4.33) and (4.32) have the same limit, and (4.32) is established. Now for any choice of θ_i^n, we have as $\delta_n \to 0$,

$$
\sum_{i=0}^{n-1}\left(g(\theta_i^n) - g(B(t_i^n))\right)\left(B(t_{i+1}^n) - B(t_i^n)\right)^2
$$

$$
\leq \max_i \left(g(\theta_i^n) - g(B(t_i^n))\right)\sum_{i=0}^{n-1}\left(B(t_{i+1}^n) - B(t_i^n)\right)^2 \to 0. \quad (4.35)
$$

The first term converges to zero almost surely by continuity of g and B, and the second converges in probability to the quadratic variation of Brownian motion, t, implying convergence to zero in probability in (4.35). This implies that both sums $\sum_{i=0}^{n-1} g(\theta_i^n)(\Delta B_i)^2$ and $\sum_{i=0}^{n-1} g(B(t_i))(\Delta B_i)^2$ have the same limit in probability, and the result follows by (4.32).

$\square$

Example 4.12: Taking $f(x) = x^m$, $m \geq 2$, we have

$$
B^m(t) = m\int_0^t B^{m-1}(s)dB(s) + \frac{m(m-1)}{2}\int_0^t B^{m-2}(s)ds.
$$

With $m = 2$,

$$
B^2(t) = 2\int_0^t B(s)dB(s) + t.
$$

Rearranging, we recover the result on the stochastic integral

$$
\int_0^t B(s)dB(s) = \frac{1}{2}B^2(t) - \frac{1}{2}t.
$$

Example 4.13: Taking $f(x) = e^x$, we have

$$
e^{B(t)} = 1 + \int_0^t e^{B(s)}dB(s) + \frac{1}{2}\int_0^t e^{B(s)}ds.
$$

4.5   Itô Processes and Stochastic Differentials

Definition of Itô Processes

An Itô process has the form

$$Y(t) = Y(0) + \int_0^t \mu(s)ds + \int_0^t \sigma(s)dB(s), \ \ 0 \leq t \leq T, \quad (4.36)$$

where $Y(0)$ is $\mathcal{F}_0$-measurable, processes $\mu(t)$ and $\sigma(t)$ are $\mathcal{F}_t$-adapted, such that $\int_0^T |\mu(t)|dt < \infty$ and $\int_0^T \sigma^2(t)dt < \infty$.

It is said that the process $Y(t)$ has the *stochastic differential* on $[0, T]$

$$dY(t) = \mu(t)dt + \sigma(t)dB(t), \qquad 0 \leq t \leq T. \quad (4.37)$$

We emphasize that a representation (4.37) only has meaning by the way of (4.36), and no other.

Note that the processes μ and σ in (4.36) may (and often do) depend on $Y(t)$ or $B(t)$ as well, or even on the whole past path of $B(s), s \leq t$; for example they may depend on the maximum of Brownian motion $\max_{s \leq t} B(s)$.

Example 4.14: Example 4.12 shows that

$$B^2(t) = t + 2 \int_0^t B(s)dB(s). \quad (4.38)$$

In other words, with $Y(t) = B^2(t)$ we can write $Y(t) = \int_0^t ds + \int_0^t 2B(s)dB(s)$. Thus $\mu(s) = 1$ and $\sigma(s) = 2B(s)$. The stochastic differential of $B^2(t)$

$$d(B^2(t)) = 2B(t)dB(t) + dt.$$

The only meaning this has is the integral relation (4.38).

Example 4.15: Example 4.13 shows that $Y(t) = e^{B(t)}$ has stochastic differential

$$de^{B(t)} = e^{B(t)}dB(t) + \frac{1}{2}e^{B(t)}dt,$$

or

$$dY(t) = Y(t)dB(t) + \frac{1}{2}Y(t)dt.$$

Itô's formula (4.29) in differential notation becomes: for a C^2 function f

$$d(f(B(t)) = f'(B(t))dB(t) + \frac{1}{2}f''(B(t))dt. \quad (4.39)$$

Example 4.16: We find $d(\sin(B(t)))$.
$f(x) = \sin(x)$, $f'(x) = \cos(x)$, $f''(x) = -\sin(x)$. Thus

$$d(\sin(B(t))) = \cos(B(t))dB(t) - \frac{1}{2}\sin(B(t))dt.$$

Similarly,

$$d(\cos(B(t))) = -\sin(B(t))dB(t) - \frac{1}{2}\cos(B(t))dt.$$

Example 4.17: We find $d(e^{iB(t)})$ with $i^2 = -1$.
The application of Itô's formula to a complex-valued function means its application
to the real and complex parts of the function. A formal application by treating i as
another constant gives the same result. Using the above example, we can calculate
$d(e^{iB(t)}) = d\cos(B(t)) + id\sin(B(t))$, or directly by using Itô's formula with
$f(x) = e^{ix}$, we have $f'(x) = ie^{ix}$, $f''(x) = -e^{ix}$ and

$$d\left(e^{iB(t)}\right) = ie^{iB(t)}dB(t) - \frac{1}{2}e^{iB(t)}dt.$$

Thus $X(t) = e^{iB(t)}$ has stochastic differential

$$dX(t) = iX(t)dB(t) - \frac{1}{2}X(t)dt.$$

Quadratic Variation of Itô Processes

Let $Y(t)$ be an Itô process

$$Y(t) = Y(0) + \int_0^t \mu(s)ds + \int_0^t \sigma(s)dB(s), \qquad (4.40)$$

where it is assumed that μ and σ are such that the integrals in question are
defined. Then by the properties of the integrals, $Y(t)$, $0 \le t \le T$, is a (random)
continuous function, the integral $\int_0^t \mu(s)ds$ is a continuous function of t and is
of finite variation (it is differentiable almost everywhere), and the Itô integral
$\int_0^t \sigma(s)dB(s)$ is continuous. Quadratic variation of Y on $[0,t]$ is defined by
(see (1.13))

$$[Y](t) = [Y,Y]([0,t]) = \lim_{\delta_n \to 0} \sum_{i=0}^{n-1} \left(Y(t_{i+1}^n) - Y(t_i^n)\right)^2, \qquad (4.41)$$

where for each n, $\{t_i^n\}$, is a partition of $[0,t]$, and the limit is in probability
taken over partitions with $\delta_n = \max_i(t_{i+1}^n - t_i^n) \to 0$ as $n \to \infty$, and is given
by

$$[Y](t) = \left[\int_0^{\cdot} \mu(s)ds + \int_0^{\cdot} \sigma(s)dB(s)\right](t)$$

$$= \left[\int_0^{\cdot} \mu(s)ds\right](t) + 2\left[\int_0^{\cdot} \mu(s)ds, \int_0^{\cdot} \sigma(s)dB(s)\right](t) + \left[\int_0^{\cdot} \sigma(s)dB(s)\right](t).$$

The result on the covariation, Theorem 1.11, states that the quadratic covariation of a continuous function with a function of finite variation is zero. This implies that the quadratic covariation of the integral $\int_0^t \mu(s)ds$ with terms above is zero, and we obtain by using the result on the quadratic variation of Itô integrals (Theorem 4.9)

$$[Y](t) = \left[\int_0^\cdot \sigma(s)dB(s)\right](t) = \int_0^t \sigma^2(s)ds. \qquad (4.42)$$

If $Y(t)$ and $X(t)$ have stochastic differentials with respect to the same Brownian motion $B(t)$, then clearly process $Y(t) + X(t)$ also has a stochastic differential with respect to the same Brownian motion. It follows that covariation of X and Y on $[0, t]$ exists and is given by

$$[X, Y](t) = \frac{1}{2}\Big([X + Y, X + Y](t) - [X, X](t) - [Y, Y](t)\Big). \qquad (4.43)$$

Theorem 1.11 has an important corollary

Theorem 4.15 *If X and Y are Itô processes and X is of finite variation, then covariation $[X, Y](t) = 0$.*

Example 4.18: Let $X(t) = \exp(t)$, $Y(t) = B(t)$, then $[X, Y](t) = [\exp, B](t) = 0$.

Introduce a convention that allows a formal manipulation with stochastic differentials.

$$dY(t)dX(t) = d[X, Y](t), \qquad (4.44)$$

and in particular

$$(dY(t))^2 = d[Y, Y](t). \qquad (4.45)$$

Since $X(t) = t$ is a continuous function of finite variation and $Y(t) = B(t)$ is continuous with quadratic variation t, the following rules follow

$$dB(t)dt = 0, \quad (dt)^2 = 0, \qquad (4.46)$$

but

$$(dB(t))^2 = d[B, B](t) = dt. \qquad (4.47)$$

Remark 4.7: In some texts, for example, Protter (1992), quadratic variation is defined by adding the value $Y^2(0)$ to (4.41). The definition given here gives a more familiar looking formula for integration by parts, and it is used in many texts, for example, Rogers and Williams (1987) p.59, Metivier (1982) p.175.

Integrals with respect to Itô processes

It is necessary to extend integration with respect to processes obtained from Brownian motion. Let the Itô integral process $Y(t) = \int_0^t X(s)dB(s)$ be defined for all $t \leq T$, where $X(t)$ is an adapted process, such that $\int_0^T X^2(s)ds < \infty$ with probability one. Let an adapted process $H(t)$ satisfy $\int_0^T H^2(s)X^2(s)ds < \infty$ with probability one. Then the Itô integral process $Z(t) = \int_0^t H(s)X(s)dB(s)$ is also defined for all $t \leq T$. In this case one can formally write by identifying $dY(t)$ and $X(t)dB(t)$,

$$Z(t) = \int_0^t H(s)dY(s) := \int_0^t H(s)X(s)dB(s). \tag{4.48}$$

In Chapter 8 integrals with respect to $Y(t)$ will be introduced in a direct way, but the result agrees with the one above.

More generally, if Y is an Itô process satisfying

$$dY(t) = \mu(t)dt + \sigma(t)dB(t), \tag{4.49}$$

and H is adapted and satisfies $\int_0^t H^2(s)\sigma^2(s)ds < \infty$, $\int_0^t |H(s)\mu(s)|ds < \infty$, then $Z(t) = \int_0^t H(s)dY(s)$ is defined as

$$Z(t) = \int_0^t H(s)dY(s) := \int_0^t H(s)\mu(s)ds + \int_0^t H(s)\sigma(s)dB(s). \tag{4.50}$$

Example 4.19: If $a(t)$ denotes the number of shares held at time t, then the gain from trading in shares during the time interval $[0, T]$ is given by $\int_0^T a(t)dS(t)$.

4.6 Itô's Formula for Itô processes

Theorem 4.16 (Itô's formula for $f(X(t))$) *Let $X(t)$ have a stochastic differential for $0 \leq t \leq T$*

$$dX(t) = \mu(t)dt + \sigma(t)dB(t). \tag{4.51}$$

If $f(x)$ is twice continuously differentiable (C^2 function), then the stochastic differential of the process $Y(t) = f(X(t))$ exists and is given by

$$
\begin{aligned}
df(X(t)) &= f'(X(t))dX(t) + \frac{1}{2}f''(X(t))d[X,X](t) \\
&= f'(X(t))dX(t) + \frac{1}{2}f''(X(t))\sigma^2(t)dt \\
&= \left(f'(X(t))\mu(t) + \frac{1}{2}f''(X(t))\sigma^2(t) \right) dt + f'(X(t))\sigma(t)dB(t).
\end{aligned}
\tag{4.52}
$$

The meaning of the above is

$$f(X(t)) = f(X(0)) + \int_0^t f'(X(s))dX(s) + \frac{1}{2}\int_0^t f''(X(s))\sigma^2(s)ds, \quad (4.53)$$

where the first integral is an Itô integral with respect to the stochastic differential. Existence of the integrals in the formula (4.53) is assured by the arguments following Theorem 4.13. The proof also follows the same ideas as Theorem 4.13, and is omitted. Proofs of Itô's formula can be found in Liptser and Shiryaev (2001), p. 124, Revuz and Yor (2001) p. 146, Protter (1992), p. 71, Rogers and Williams (1990), p. 60.

Example 4.20: Let $X(t)$ have stochastic differential

$$dX(t) = X(t)dB(t) + \frac{1}{2}X(t)dt. \quad (4.54)$$

We find a process X satisfying (4.54). Let's look for a *positive* process X. Using Itô's formula for $\ln X(t)$ $((\ln x)' = 1/x$ and $(\ln x)'' = -1/x^2)$,

$$\begin{aligned} d\ln X(t) &= \frac{1}{X(t)}dX(t) - \frac{1}{2X_t^2}X_t^2 dt \text{ by using } \sigma(t) = X(t) \\ &= dB(t) + \frac{1}{2}dt - \frac{1}{2}dt = dB(t). \end{aligned}$$

So that $\ln X(t) = \ln X(0) + B(t)$, and we find

$$X(t) = X(0)e^{B(t)}. \quad (4.55)$$

Using Itô's formula we verify that this $X(t)$ indeed satisfies (4.54). We don't claim at this stage that (4.55) is the only solution.

Integration by Parts

We give a representation of the quadratic covariation $[X, Y](t)$ of two Itô processes $X(t)$ and $Y(t)$ in terms of Itô integrals. This representation gives rise to the integration by parts formula.

Quadratic covariation is a limit over decreasing partitions of $[0, t]$,

$$[X, Y](t) = \lim_{\delta_n \to 0} \sum_{i=0}^{n-1} \left(X(t_{i+1}^n) - X(t_i^n)\right)\left(Y(t_{i+1}^n) - Y(t_i^n)\right). \quad (4.56)$$

The sum on the right above can be written as

$$= \sum_{i=0}^{n-1} \left(X(t_{i+1}^n)Y(t_{i+1}^n) - X(t_i^n)Y(t_i^n)\right)$$

$$-\sum_{i=0}^{n-1} X(t_i^n)\big(Y(t_{i+1}^n) - Y(t_i^n)\big) - \sum_{i=0}^{n-1} Y(t_i^n)\big(X(t_{i+1}^n) - X(t_i^n)\big)$$

$$= \quad X(t)Y(t) - X(0)Y(0)$$

$$-\sum_{i=0}^{n-1} X(t_i^n)\big(Y(t_{i+1}^n) - Y(t_i^n)\big) - \sum_{i=0}^{n-1} Y(t_i^n)\big(X(t_{i+1}^n) - X(t_i^n)\big).$$

The last two sums converge in probability to Itô integrals $\int_0^t X(s)dY(s)$ and $\int_0^t Y(s)dX(s)$, cf. Remark (4.4). Thus the following expression is obtained

$$[X, Y](t) = X(t)Y(t) - X(0)Y(0) - \int_0^t X(s)dY(s) - \int_0^t Y(s)dX(s). \quad (4.57)$$

The formula for integration by parts (stochastic product rule) is given by

$$X(t)Y(t) - X(0)Y(0) = \int_0^t X(s)dY(s) + \int_0^t Y(s)dX(s) + [X, Y](t). \quad (4.58)$$

In differential notations this reads

$$d\big(X(t)Y(t)\big) = X(t)dY(t) + Y(t)dX(t) + d[X, Y](t). \quad (4.59)$$

If

$$dX(t) = \mu_X(t)dt + \sigma_X(t)dB(t), \quad (4.60)$$

$$dY(t) = \mu_Y(t)dt + \sigma_Y(t)dB(t), \quad (4.61)$$

then, as seen earlier, their quadratic covariation can be obtained formally by multiplication of dX and dY, namely

$$d[X, Y](t) \quad = \quad dX(t)dY(t)$$
$$= \quad \sigma_X(t)\sigma_Y(t)(dB(t))^2 = \sigma_X(t)\sigma_Y(t)dt,$$

leading to the formula

$$d\big(X(t)Y(t)\big) = X(t)dY(t) + Y(t)dX(t) + \sigma_X(t)\sigma_Y(t)dt.$$

Note that if one of the processes is continuous and is of finite variation, then the covariation term is zero. Thus for such processes the stochastic product rule is the same as usual.

The integration by parts formula (4.59) can be established rigorously by making the argument above more precise, or by using Itô's formula for the function of two variables xy, or by approximations by simple processes.

Formula (4.57) provides an alternative representation for quadratic variation

$$[X, X](t) = X^2(t) - X^2(0) - 2 \int_0^t X(s)dX(s). \qquad (4.62)$$

For Brownian motion this formula was established in Example 4.2.

It follows from the definition of quadratic variation, that it is a nondecreasing process in t, and consequently it is of finite variation. It is also obvious from (4.62) that it is continuous. By the polarization identity, covariation is also continuous and is of finite variation.

Example 4.21: $X(t)$ has stochastic differential

$$dX(t) = B(t)dt + tdB(t), \ X(0) = 0.$$

We find $X(t)$, give its distribution, its mean and covariance. $X(t) = tB(t)$ satisfies the above equation, since the product rule for stochastic differentials is the same as usual, when one of the processes is continuous and of finite variation. Thus $X(t) = tB(t)$ is Gaussian, with mean zero, and covariance function

$$\begin{aligned} \gamma(t, s) &= Cov(X(t), X(s)) = \mathrm{E}\left(X(t)X(s)\right) \\ &= \mathrm{E}\left(B(t)B(s)\right) = Cov\left(B(t)B(s)\right) = \min(t, s). \end{aligned}$$

Example 4.22: Let $Y(t)$ have stochastic differential

$$dY(t) = \frac{1}{2}Y(t)dt + Y(t)dB(t), \ Y(0) = 1.$$

Let $X(t) = tB(t)$. We find $d\left(X(t)Y(t)\right)$.
$Y(t)$ is a Geometric Brownian motion $e^{B(t)}$ (see Example 4.17). For $d\left(X(t)Y(t)\right)$ use the product rule. We need the expression for $d[X, Y](t)$.

$$d[X, Y](t) = dX(t)dY(t) = (B(t)dt + tdB(t))\left(\frac{1}{2}Y(t)dt + Y(t)dB(t)\right)$$

$$= \frac{1}{2}B(t)Y(t)(dt)^2 + \left(B(t)Y(t) + \frac{1}{2}tY(t)\right)dB(t)dt + tY(t)(dB(t))^2 = tY(t)dt,$$

as $(dB(t))^2 = dt$ and all the other terms are zero. Thus

$$d\left(X(t)Y(t)\right) = X(t)dY(t) + Y(t)dX(t) + d[X, Y](t)$$

$$= X(t)dY(t) + Y(t)dX(t) + tY(t)dt,$$

and substituting the expressions for X and Y the answer is obtained.

Example 4.23: Let f be a C^2 function and $B(t)$ Brownian motion. We find quadratic covariation $[f(B), B](t)$.
We find the answer by doing formal calculations. Using Itô's formula

$$df(B(t)) = f'(B(t))dB(t) + \frac{1}{2}f''(B(t))dt,$$

and the convention
$$d[f(B), B](t) = df(B(t))dB(t),$$
we have
$$d[f(B), B](t) = df(B(t))dB(t) = f'(B(t))(dB(t))^2 + \frac{1}{2}f''(B(t))dB(t)dt = f'(B(t))dt.$$

Here we used $(dB)^2 = dt$, and $dBdt = 0$. Thus

$$[f(B), B](t) = \int_0^t f'(B(s))ds.$$

In a more intuitive way, from the definition of the covariation, taking limits over shrinking partitions

$$
\begin{aligned}
[f(B), B](t) &= \lim \sum_{i=0}^{n-1} \left(f(B(t_{i+1}^n)) - f(B(t_i^n)) \right) \left(B(t_{i+1}^n) - B(t_i^n) \right) \\
&= \lim \sum_{i=0}^{n-1} \frac{f(B(t_{i+1}^n)) - f(B(t_i^n))}{B(t_{i+1}^n) - B(t_i^n)} \left(B(t_{i+1}^n) - B(t_i^n) \right)^2 \\
&\approx \lim \sum_{i=0}^{n-1} f'(B(t_i^n)) \left(B(t_{i+1}^n) - B(t_i^n) \right)^2 = \int_0^t f'(B(s))ds,
\end{aligned}
$$

where we have used Theorem 4.14 in the last equality.

Example 4.24: Let $f(t)$ be an increasing differentiable function, and let $X(t) = B(f(t))$. We show that

$$[X, X](t) = [B(f), B(f)](t) = [B, B]f(t) = f(t). \tag{4.63}$$

By taking limits over shrinking partitions

$$
\begin{aligned}
[X, X](t) &= \lim \sum_{i=0}^{n-1} \left(B(f(t_{i+1}^n)) - B(f(t_i^n)) \right)^2 \\
&= \lim \sum_{i=0}^{n-1} \left(f(t_{i+1}^n) - f(t_i^n) \right) \left(\frac{B(f(t_{i+1}^n)) - B(f(t_i^n))}{\sqrt{f(t_{i+1}^n) - f(t_i^n)}} \right)^2 \\
&= \lim \sum_{i=0}^{n-1} \left(f(t_{i+1}^n) - f(t_i^n) \right) Z_i^2 = \lim T_n,
\end{aligned}
$$

where $Z_i = \frac{B(f(t_{i+1}^n)) - B(f(t_i^n))}{\sqrt{f(t_{i+1}^n) - f(t_i^n)}}$ are Standard Normal, and independent, by the properties of Brownian motion, and $T_n = \sum_{i=0}^{n-1} \left(f(t_{i+1}^n) - f(t_i^n) \right) Z_i^2$. Then for any n,

$$E(T_n) = \sum_{i=0}^{n-1} \left(f(t_{i+1}^n) - f(t_i^n) \right) = f(t).$$

$$Var(T_n) = Var\left(\sum_{i=0}^{n-1}(f(t_{i+1}^n) - f(t_i^n))\right) = 3\sum_{i=0}^{n-1}(f(t_{i+1}^n) - f(t_i^n))^2,$$

by independence, and $Var(Z^2) = 3$. The last sum converges to zero, since f is of finite variation and continuous, implying that

$$E(T_n - f(t))^2 \to 0.$$

This means that the limit in L^2 of T_n is $f(t)$, which implies that the limit in probability of T_n is $f(t)$, and

$$[B(f), B(f)] = f(t).$$

Itô's Formula for Functions of Two Variables

If two processes X and Y both possess a stochastic differential with respect to $B(t)$ and $f(x, y)$ has continuous partial derivatives up to order two, then $f(X(t), Y(t))$ also possesses a stochastic differential. To find its form consider formally the Taylor expansion of order two,

$$df(x, y) = \frac{\partial f(x, y)}{\partial x}dx + \frac{\partial f(x, y)}{\partial y}dy$$
$$+ \frac{1}{2}\left(\frac{\partial^2 f(x, y)}{(\partial x)^2}(dx)^2 + \frac{\partial^2 f(x, y)}{(\partial y)^2}(dy)^2 + 2\frac{\partial^2 f(x, y)}{\partial x \partial y}dxdy\right).$$

Now, $(dX(t))^2 = dX(t)dX(t) = d[X, X](t) = \sigma_X^2(X(t))dt$, $(dY(t))^2 = d[Y, Y]_t = \sigma_Y^2(Y(t))dt$, and $dX(t)dY(t) = d[X, Y]_t = \sigma_X(X(t))\sigma_Y(Y(t))dt$, where $\sigma_X(t)$, and $\sigma_Y(t)$ are the diffusion coefficients of X and Y respectively. So we have

Theorem 4.17 *Let $f(x, y)$ have continuous partial derivatives up to order two (a C^2 function) and X, Y be Itô processes, then*

$$df(X(t), Y(t)) = \frac{\partial f}{\partial x}(X(t), Y(t))dX(t) + \frac{\partial f}{\partial y}(X(t), Y(t))dY(t)$$
$$+ \frac{1}{2}\frac{\partial^2 f}{\partial x^2}(X(t), Y(t))\sigma_X^2(X(t))dt + \frac{1}{2}\frac{\partial^2 f}{\partial y^2}(X(t), Y(t))\sigma_Y^2(Y(t))dt$$
$$+ \frac{\partial^2 f}{\partial x \partial y}(X(t), Y(t))\sigma_X(X(t))\sigma_Y(Y(t))dt. \tag{4.64}$$

The proof is similar to that of Theorem 4.13, and is omitted. It is stressed that differential formulae have meaning only through their integral representation.

Example 4.25: If $f(x, y) = xy$, then we obtain a differential of a product (or the product rule)which gives the integration by parts formula.

$$d(X(t)Y(t)) = X(t)dY(t) + Y(t)dX(t) + \sigma_X(t)\sigma_Y(t)(t)dt.$$

An important case of Itô's formula is for functions of the form $f(X(t), t)$.

Theorem 4.18 *Let $f(x, t)$ be twice continuously differentiable in x, and continuously differentiable in t (a $C^{2,1}$ function) and X be an Itô process, then*

$$df(X(t), t) = \frac{\partial f}{\partial x}(X(t), t)dX(t) + \frac{\partial f}{\partial t}(X(t), t)dt + \frac{1}{2}\sigma_X^2(X(t), t)\frac{\partial^2 f}{\partial x^2}(X(t), t)dt.$$

(4.65)

This formula can be obtained from Theorem 4.17 by taking $Y(t) = t$ and observing that $d[Y, Y] = 0$ and $d[X, Y] = 0$.

Example 4.26: We find stochastic differential of $X(t) = e^{B(t)-t/2}$.
Use Itô's formula with $f(x, t) = e^{x-t/2}$. $X(t) = f(B(t), t)$ satisfies

$$
\begin{aligned}
dX(t) = df(B(t), t) &= \frac{\partial f}{\partial x}dB(t) + \frac{\partial f}{\partial t}dt + \frac{1}{2}\frac{\partial^2 f}{\partial^2 x}dt \\
&= f(B(t), t)dB(t) - \frac{1}{2}f(B(t), t)dt + \frac{1}{2}f(B(t), t)dt \\
&= f(B(t), t)dB(t) = X(t)dB(t).
\end{aligned}
$$

So that

$$dX(t) = X(t)dB(t).$$

4.7 Itô Processes in Higher Dimensions

Let $\boldsymbol{B}(t) = (B_1(t), B_2(t), \ldots, B_d(t))$ be Brownian motion in $\mathbb{R}^d$, that is, all coordinates $B_i(t)$ are independent one-dimensional Brownian motions. Let $\mathcal{F}_t$ be the σ-field generated by $\boldsymbol{B}(s)$, $s \leq t$. Let $\boldsymbol{H}(t)$ be a regular adapted process d-dimensional vector process, i.e. each of its coordinates is such. If for each j, $\int_0^T H_j^2(t)dt < \infty$, then the Itô integrals $\int_0^T H_j(t)dB_j(t)$ are defined. A single equivalent condition in terms of the length of the vector $|\boldsymbol{H}|^2 = \sum_{i=1}^d H_i^2$ is

$$\int_0^T |\boldsymbol{H}(t)|^2 dt < \infty.$$

It is customary to use a scalar product notation (even suppressing $\cdot$)

$$\boldsymbol{H}(t)\cdot d\boldsymbol{B}(t) = \sum_{j=1}^d H_j(t)dB_j(t), \quad \text{and} \quad \int_0^T \boldsymbol{H}(t)\cdot d\boldsymbol{B}(t) = \sum_{j=1}^d \int_0^T H_j(t)dB_j(t).$$

(4.66)

If $b(t)$ is an integrable function then the process

$$dX(t) = b(t)dt + \sum_{j=1}^d H_j(t)dB_j(t)$$

is well defined. It is a scalar Itô process driven by a d-dimensional Brownian motion. More generally, we can have any number n of process driven by a d-dimensional Brownian motion, (the vector $\boldsymbol{H}_i = (\sigma_{i1}, \ldots \sigma_{id})$)

$$dX_i(t) = b_i(t)dt + \sum_{j=1}^{d} \sigma_{ij}(t)dB_j(t), \quad , i = 1, \ldots, n \qquad (4.67)$$

where σ is $n \times d$ matrix valued function, $\boldsymbol{B}$ is d-dimensional Brownian motion, $\boldsymbol{X}, \boldsymbol{b}$ are n-dim vector-valued functions, the integrals with respect to Brownian motion are Itô integrals. Then $\boldsymbol{X}$ is called an Itô process. In vector form (4.67) becomes

$$d\boldsymbol{X}(t) = \boldsymbol{b}(t)dt + \sigma(t)d\boldsymbol{B}_t. \qquad (4.68)$$

The dependence of $\boldsymbol{b}(t)$ and $\sigma(t)$ on time t can be via the whole path of the process up time t, path of $\boldsymbol{B}_s, s \leq t$. The only restriction is that this dependence results in:

for any $i = 1, 2, \ldots n$, $b_i(t)$ is adapted and $\int_0^T |b_i(t)|dt < \infty$ a.s.

for any $i = 1, 2, \ldots n$, $\sigma_{ij}(t)$ is adapted and $\int_0^T \sigma_{ij}^2(t)dt < \infty$ a.s., which assure existence of the required integrals.

An important case is when this dependence is of the form $\boldsymbol{b}(t) = \boldsymbol{b}(\boldsymbol{X}(t), t)$, $\sigma(t) = \sigma(\boldsymbol{X}(t), t)$. In this case the stochastic differential is written as

$$d\boldsymbol{X}(t) = \boldsymbol{b}(\boldsymbol{X}(t), t)dt + \sigma(\boldsymbol{X}(t), t)d\boldsymbol{B}(t), \qquad (4.69)$$

and $\boldsymbol{X}(t)$ is then a diffusion process, see Chapters 5 and 6.

For Itô's formula we need the quadratic variation of a multi-dimensional Itô processes. It is not hard to see that quadratic covariation of two independent Brownian motions is zero.

Theorem 4.19 *Let $B_1(t)$ and $B_2(t)$ be independent Brownian motions. Then their covariation process exists and is identically zero.*

PROOF: Let $\{t_i^n\}$ be a partition of $[0, t]$ and consider

$$T_n = \sum_{i=0}^{n-1} \big(B_1(t_{i+1}^n) - B_1(t_i^n)\big)\big(B_2(t_{i+1^n}) - B_2(t_i^n)\big).$$

Using independence of B_1 and B_2, $\mathrm{E}(T_n) = 0$. Since increments of Brownian motion are independent, the variance of the sum is sum of variances, and we have

$$
\begin{aligned}
Var(T_n) &= \sum_{i=0}^{n-1} \mathrm{E}\big(B_1(t_{i+1}^n) - B_1(t_i^n)\big)^2 \mathrm{E}\big(B_2(t_{i+1}^n) - B_2(t_i^n)\big)^2 \\
&= \sum_{i=0}^{n-1} (t_{i+1}^n - t_i^n)^2 \leq \max_i (t_{i+1}^n - t_i^n)t.
\end{aligned}
$$

Thus $Var(T_n) = \mathrm{E}(T_n^2) \to 0$ as $\delta_n = \max_i(t_{i+1}^n - t_i^n) \to 0$. This implies that $T_n \to 0$ in probability, and the result is proved.

□

Thus for $k \neq l$, $k, l = 1, 2, \ldots d$,

$$[B_k, B_l](t) = 0. \tag{4.70}$$

Using (4.70), and the bi-linearity of covariation, it is easy to see from (4.67)

$$d[X_i, X_j](t) = dX_i(t)dX_j(t) = a_{ij}dt, \text{ for } i, j = 1, \ldots n. \tag{4.71}$$

where a, called the diffusion matrix, is given by

$$a = \sigma\sigma^{Tr}, \tag{4.72}$$

with σ^{Tr} denoting the transposed matrix of σ.

Itô's Formula for Functions of Several Variables

If $X(t) = (X_1(t), X_2(t), \ldots, X_n(t))$ is a vector Itô process and $f(x_1, x_2, \ldots, x_n)$ is a C^2 function of n variables, then $f(X_1(t), X_2(t), \ldots, X_n(t))$ is also an Itô process, moreover its stochastic differential is given by

$$
\begin{aligned}
&df(X_1(t), X_2(t), \ldots, X_n(t)) \\
&= \sum_{i=1}^n \frac{\partial}{\partial x_i} f(X_1(t), X_2(t), \ldots, X_n(t))dX_i(t) \\
&\quad + \frac{1}{2}\sum_{i=1}^n \sum_{j=1}^n \frac{\partial^2}{\partial x_i \partial x_j} f(X_1(t), X_2(t), \ldots, X_n(t))d[X_i, X_j](t). \tag{4.73}
\end{aligned}
$$

When there is only one Brownian motion, $d = 1$, this formula is a generalization of Itô's formula for a function of two variables (Theorem 4.17).

For examples and applications see multi-dimensional diffusions in Chapters 5 and 6. We comment here on the integration by parts formula.

Remark 4.8: (Integration by Parts)
Let $X(t)$ and $Y(t)$ be two Itô processes that are adapted to *independent* Brownian motions B_1 and B_2. Take $f(x, y) = xy$ and note that only one of the second derivatives is different from zero, $\frac{\partial^2 xy}{\partial x \partial y}$, but then the term it multiplies is zero, $d[B_1, B_2](t) = 0$ by Theorem 4.19. So the covariation of $X(t)$ and $Y(t)$ is zero, and one obtains from (4.73)

$$d(X(t)Y(t)) = X(t)dY(t) + Y(t)dX(t), \tag{4.74}$$

which is the usual integration by parts formula.

Remark 4.9: In some applications correlated Brownian motions are used. These are obtained by a linear transformation of independent Brownian motions. If B_1 and B_2 W are independent, then the pair of processes B_1 and $W = \rho B_1 + \sqrt{1 - \rho^2} B_2$ are correlated Brownian motions. It is easy to see that W is indeed a Brownian motion, and that $d[B_1, W](t) = \rho dt$.

More results about Itô processes in higher dimensions are given in Chapter 6.

Remark 4.10: Itô's formula can be generalized to functions less smooth than C^2, in particular for $f(x) = |x|$. Itô's formula for $f(x) = |x|$ becomes Tanaka's formula, and leads to the concept of local time. This development requires additional concepts, which are given later, see Section 8.7 in the general theory for semimartingales.

Notes. Material in this chapter can be found in Gihman and Skorohod (1972), Liptser and Shiryaev (1977), (1989), Karatzas and Shreve (1988), Gard (1988), Rogers and Williams (1990), (1994).

4.8   Exercises

Exercise 4.1: Give values of α for which the following process is defined $Y(t) = \int_0^t (t-s)^{-\alpha} dB(s)$. (This process is used in the definition of the so-called Fractional Brownian motion.)

Exercise 4.2: Show that if X is a simple bounded adapted process, then $\int_0^t X(s) dB(s)$ is continuous.

Exercise 4.3: Let X_n be a Gaussian sequence convergent in distribution to X. Show that the distribution of X is either Normal or degenerate. Deduce that if $\mathbb{E}X_n \to \mu$ and $Var(X_n) \to \sigma^2 > 0$ then the limit is $N(\mu, \sigma^2)$. Since convergence in probability implies convergence in distribution, deduce convergence of Itô integrals of simple non-random processes to a Gaussian limit.

Exercise 4.4: Show that if $X(t)$ is non-random (does not depend on $B(t)$) and is a function of t and s with $\int_0^t X^2(t, s) ds < \infty$ then $\int_0^t X(t, s) dB(s)$ is a Gaussian random variable $Y(t)$. The collection $Y(t)$, $0 \leq t \leq T$, is a Gaussian process with zero mean and covariance function for $u \geq 0$ given by $Cov(Y(t), Y(t+u)) = \int_0^t X(t, s) X(t+u, s) ds$.

Exercise 4.5: Show that a Gaussian martingale on a finite time interval $[0, T]$ is a square integrable martingale with independent increments. Deduce that if X is non-random and $\int_0^t X^2(s) ds < \infty$ then $Y(t) = \int_0^t X(s) dB(s)$ is a Gaussian square integrable martingale with independent increments.

Exercise 4.6: Obtain the alternative relation for the quadratic variation of Itô processes, Equation (4.62), by applying Itô's formula to $X^2(t)$.

Exercise 4.7: $X(t)$ has a stochastic differential with $\mu(x) = bx + c$ and $\sigma^2(x) = 4x$. Assuming $X(t) \geq 0$, find the stochastic differential for the process $Y(t) = \sqrt{X(t)}$.

Exercise 4.8: A process $X(t)$ on $(0,1)$ has a stochastic differential with coefficient $\sigma(x) = x(1 - x)$. Assuming $0 < X(t) < 1$, show that the process defined by $Y(t) = \ln(X(t)/(1 - X(t)))$ has a constant diffusion coefficient.

Exercise 4.9: $X(t)$ has a stochastic differential with $\mu(x) = cx$ and $\sigma^2(x) = x^a$, $c > 0$. Let $Y(t) = X(t)^b$. What choice of b will give a constant diffusion coefficient for Y?

Exercise 4.10: Let $X(t) = tB(t)$ and $Y(t) = e^{B(t)}$. Find $d\left(\frac{X(t)}{Y(t)}\right)$.

Exercise 4.11: Obtain the differential of a ratio formula $d\left(\frac{X(t)}{Y(t)}\right)$ by taking $f(x, y) = x/y$. Assume that the process Y stays away from 0.

Exercise 4.12: Find $d\big(M(t)\big)^2$, where $M(t) = e^{B(t)-t/2}$

Exercise 4.13: Let $M(t) = B^3(t) - 3tB(t)$. Show that M is a martingale, first directly and then by using Itô integrals.

Exercise 4.14: Show that $M(t) = e^{t/2}\sin(B(t))$ is a martingale by using Itô's formula.

Exercise 4.15: For a function of n variables and n-dimensional Brownian motion, write Itô's formula for $f(B_1(t), \ldots, B_n(t))$ by using gradient notation $\nabla f = (\frac{\partial}{\partial x_1}, \ldots, \frac{\partial}{\partial x_n})$.

Exercise 4.16: $\Phi(x)$ is the standard Normal distribution function. Show that for a fixed $T > 0$ the process $\Phi(\frac{B(t)}{\sqrt{T-t}})$, $0 \leq t \leq T$ is a martingale.

Exercise 4.17: Let $X(t) = (1 - t) \int_0^t \frac{dB(s)}{1-s}$, where $0 \leq t < 1$. Find $dX(t)$.

Exercise 4.18: Let $X(t) = tB(t)$. Find its quadratic variation $[X, X](t)$.

Exercise 4.19: Let $X(t) = \int_0^t (t - s)dB(s)$. Find $dX(t)$ and its quadratic variation $[X, X](t)$. Compare to the quadratic variation of Itô integrals.

Chapter 5

Stochastic Differential Equations

Differential equations are used to describe the evolution of a system. Stochastic Differential Equations (SDEs) arise when a random noise is introduced into ordinary differential equations (ODEs). In this chapter we define two concepts of solutions of SDEs, the strong and the weak solution.

5.1 Definition of Stochastic Differential Equations

Ordinary Differential Equations

If $x(t)$ is a differentiable function defined for $t \geq 0$, $\mu(x, t)$ is a function of x, and t, and the following relation is satisfied for all t, $0 \leq t \leq T$

$$\frac{dx(t)}{dt} = x'(t) = \mu(x(t), t), \text{ and } x(0) = x_0, \tag{5.1}$$

then $x(t)$ is a solution of the ODE with the initial condition x_0. Usually the requirement that $x'(t)$ is continuous is added. See also Theorem 1.4.

The above equation can be written in other forms.

$$dx(t) = \mu(x(t), t)dt$$

and (by continuity of $x'(t)$)

$$x(t) = x(0) + \int_0^t \mu(x(s), s)ds.$$

Before we give a rigorous definition of SDEs, we show how they arise as a randomly perturbed ODEs and give a physical interpretation.

White Noise and SDEs

The White Noise process $\xi(t)$ is formally defined as the derivative of the Brownian motion,

$$\xi(t) = \frac{dB(t)}{dt} = B'(t). \tag{5.2}$$

It does not exist as a function of t in the usual sense, since a Brownian motion is nowhere differentiable.

If $\sigma(x, t)$ is the intensity of the noise at point x at time t, then it is agreed that $\int_0^T \sigma(X(t), t)\xi(t)dt = \int_0^T \sigma(X(t), t)B'(t)dt = \int_0^T \sigma(X(t), t)dB(t)$, where the integral is Itô integral.

Stochastic Differential Equations arise, for example, when the coefficients of ordinary equations are perturbed by White Noise.

Example 5.1: Black-Scholes-Merton model for growth with uncertain rate of return. $x(t)$ is the value of \$1 after time t, invested in a savings account. By the definition of compound interest, it satisfies the ODE $dx(t)/x(t) = rdt$, or $dx(t)/dt = rx(t)$, (r is called the interest rate). If the rate is uncertain, it is taken to be perturbed by noise, $r + \xi(t)$, and following SDE is obtained

$$\frac{dX(t)}{dt} = (r + \sigma\xi(t))X(t),$$

meaning

$$dX(t) = rX(t)dt + \sigma X(t)dB(t).$$

Case $\sigma = 0$ corresponds to no noise, and recovers the deterministic equation. The solution of the deterministic equation is easily obtained by separating variables as $x(t) = e^{rt}$. The solution to the above SDE is given by a geometric Brownian motion, as can be verified by Itô's formula (see Example 5.5)

$$X(t) = e^{(r-\sigma^2/2)t+\sigma B(t)}. \tag{5.3}$$

Example 5.2: Population growth. If $x(t)$ denotes the population density, then the population growth can be described by the ODE $dx(t)/dt = ax(t)(1 - x(t))$. The growth is exponential with birth rate a, when this density is small, and slows down when the density increases. Random perturbation of the birth rate results in the equation $dX(t)/dt = (a + \sigma\xi(t))X(t)(1 - X(t))$, or the SDE

$$dX(t) = aX(t)(1 - X(t))dt + \sigma X(t)(1 - X(t))dB(t).$$

A Physical Model of Diffusion and SDEs

The physical phenomena which give rise to the mathematical model of diffusion (and of Brownian motion) is the microscopic motion of a particle suspended in a fluid. Molecules of the fluid move with various velocities and collide with the particle from every possible direction producing a constant bombardment. As a result of this bombardment the particle exhibits an ever present erratic movement. This movement intensifies with increase in the temperature of the fluid. Denote by $X(t)$ the displacement of the particle in one direction from its initial position at time t. If $\sigma(x,t)$ measures the effect of temperature at point x at time t, then the displacement due to bombardment during time $[t, t + \Delta]$ is modelled as $\sigma(x,t)\big(B(t + \Delta) - B(t)\big)$. If the velocity of the fluid at point x at time t is $\mu(x,t)$, then the displacement of the particle due to the movement of the fluid during is $\mu(x,t)\Delta$. Thus the total displacement from its position x at time t is given by

$$X(t + \Delta) - x \approx \mu(x,t)\Delta + \sigma(x,t)\Big(B(t + \Delta) - B(t)\Big). \qquad (5.4)$$

Thus we obtain from this equation, the mean displacement from x during short time Δ is given by

$$\mathrm{E}\Big((X(t + \Delta) - X(t))|X(t) = x\Big) \approx \mu(x,t) \cdot \Delta, \qquad (5.5)$$

and the second moment of the displacement from x during short time Δ is given by

$$\mathrm{E}\Big((X(t + \Delta) - X(t))^2|X(t) = x\Big) \approx \sigma^2(x,t)\Delta. \qquad (5.6)$$

The above relations show that for small intervals of time both the mean and the second moment (and variance) of the displacement of a diffusing particle at time t at point x are proportional to the length of the interval, with coefficients $\mu(x,t)$ and $\sigma^2(x,t)$ respectively.

It can be shown that, taken as asymptotic relations as $\Delta \to 0$, that is, replacing $\approx$ sign by the equality and adding terms $o(\Delta)$ to the right-hand sides, these two requirements characterize diffusion processes.

Assuming that $\mu(x,t)$ and $\sigma(x,t)$ are smooth functions, heuristic Equation (5.4) also points out that for small intervals of time Δ, diffusions are approximately Gaussian processes. Given $X(t) = x$, $X(t+\Delta) - X(t)$ is approximately normally distributed, $N\big(\mu(x,t)\Delta, \sigma^2(x,t)\Delta\big)$. Of course, for large intervals of time diffusions are not Gaussian, unless the coefficients are non-random.

A stochastic differential equation is obtained heuristically from the relation (5.4) by replacing Δ by dt, and $\Delta B = B(t + \Delta) - B(t)$ by $dB(t)$, and $X(t + \Delta) - X(t)$ by $dX(t)$.

Stochastic Differential Equations

Let $B(t)$, $t \geq 0$, be Brownian motion process. An equation of the form

$$dX(t) = \mu(X(t), t)dt + \sigma(X(t), t)dB(t), \qquad (5.7)$$

where functions $\mu(x, t)$ and $\sigma(x, t)$ are given and $X(t)$ is the unknown process, is called a stochastic differential equation (SDE) driven by Brownian motion. The functions $\mu(x, t)$ and $\sigma(x, t)$ are called the coefficients.

Definition 5.1 *A process $X(t)$ is called a strong solution of the SDE (5.7) if for all $t > 0$ the integrals $\int_0^t \mu(X(s), s)ds$ and $\int_0^t \sigma(X(s), s)dB(s)$ exist, with the second being an Itô integral, and*

$$X(t) = X(0) + \int_0^t \mu(X(s), s)ds + \int_0^t \sigma(X(s), s)dB(s). \qquad (5.8)$$

Remark 5.1:
1. A strong solution is some function (functional) $F(t, (B(s), s \leq t))$ of the given Brownian motion $B(t)$.
2. When $\sigma = 0$, the SDE becomes an ordinary differential equation (ODE).
3. Another interpretation of (5.7), called the weak solution, is a solution in distribution which will be given later.

Equations of the form (5.7) are called diffusion-type SDEs. More general SDEs have the form

$$dX(t) = \mu(t)dt + \sigma(t)dB(t), \qquad (5.9)$$

where $\mu(t)$ and $\sigma(t)$ can depend on t and the whole past of the processes $X(t)$ and $B(t)$ $(X(s), B(s), s \leq t)$, that is, $\mu(t) = \mu\big((X(s), s \leq t), t\big)$, $\sigma(t) = \sigma\big((X(s), s \leq t), t\big)$. The only restriction on $\mu(t)$ and $\sigma(t)$ is that they must be adapted processes, with respective integrals defined. Although many results (such as existence and uniqueness results) can be formulated for general SDEs, we concentrate here on diffusion-type SDEs.

Example 5.3: We have seen that $X(t) = \exp(B(t) - t/2)$ is a solution of the stochastic exponential SDE $dX(t) = X(t)dB(t)$, $X(0) = 1$.

Example 5.4: Consider the process $X(t)$ satisfying $dX(t) = a(t)dB(t)$, where $a(t)$ is a non-random. Clearly, $X(t) = X(0) + \int_0^t a(s)dB(s)$. We can represent this as a function of the Brownian motion by integrating by parts, $X(t) = X(0) + a(t)B(t) - \int_0^t B(s)a'(s)ds$, assuming $a(t)$ is differentiable. In this case the function $F(t, (x(s), s \leq t)) = X(0) + a(t)x(t) - \int_0^t x(s)a'(s)ds$.

The next two examples demonstrate how to find a strong solution by using Itô's formula and integration by parts.

Example 5.5: (Example 5.1 continued)
Consider the SDE

$$dX(t) = \mu X(t)dt + \sigma X(t)dB(t), \quad X(0) = 1. \tag{5.10}$$

Take $f(x) = \ln x$, then $f'(x) = 1/x$ and $f''(x) = -1/x^2$.

$$
\begin{aligned}
d(\ln X(t)) &= \frac{1}{X(t)}dX(t) + \frac{1}{2}\left(-\frac{1}{X^2(t)}\right)\sigma^2 X^2(t)dt \\
&= \frac{1}{X(t)}\left(\mu X(t)dt + \sigma X(t)dB(t)\right) - \frac{1}{2}\sigma^2 dt \\
&= (\mu - \frac{1}{2}\sigma^2)dt + \sigma dB(t)
\end{aligned}
$$

So that $Y(t) = \ln X(t)$ satisfies

$$dY(t) = (\mu - \frac{1}{2}\sigma^2)dt + \sigma dB(t).$$

Its integral representation gives

$$Y(t) = Y(0) + (\mu - \frac{1}{2}\sigma^2)t + \sigma B(t),$$

and

$$X(t) = X(0)e^{(\mu - \frac{1}{2}\sigma^2)t + \sigma B(t)}. \tag{5.11}$$

Example 5.6: (Langevin equation and Ornstein-Uhlenbeck process)
Consider the SDE

$$dX(t) = -\alpha X(t)dt + \sigma dB(t), \tag{5.12}$$

where α and σ are some non-negative constants.

Note that in the case $\sigma = 0$, the solution to the ODE is $x_0 e^{-\alpha t}$, or in other words $x(t)e^{\alpha t}$ is a constant. To solve the SDE consider the process $Y(t) = X(t)e^{\alpha t}$. Use the differential of the product rule, and note that the covariation of $e^{\alpha t}$ with $X(t)$ is zero, as it is a differentiable function $(d(e^{\alpha t})dX(t) = \alpha e^{\alpha t}dtdX(t) = 0)$, we have $dY(t) = e^{\alpha t}dX(t) + \alpha e^{\alpha t}X(t)dt$. Using the SDE for $dX(t)$ we obtain $dY(t) = \sigma e^{\alpha t}dB(t)$. This gives $Y(t) = Y(0) + \int_0^t \sigma e^{\alpha s}dB(s)$. Now the solution for $X(t)$ is

$$X(t) = e^{-\alpha t}\left(X(0) + \int_0^t \sigma e^{\alpha s}dB(s)\right). \tag{5.13}$$

The process $X(t)$ in (5.12) is known as the Ornstein-Uhlenbeck process.

We can also find the functional dependence of the solution on the Brownian motion path. Performing integration by parts, we find the function giving the strong solution

$$X(t) = F(t, (B(s), 0 \le s \le t)) = e^{-\alpha t}X(0) + \sigma B(t) - \sigma\alpha\int_0^t e^{-\alpha(t-s)}B(s)ds.$$

A more general equation is

$$dX(t) = (\beta - \alpha X(t))dt + \sigma dB(t), \qquad (5.14)$$

with the solution

$$X(t) = \frac{\beta}{\alpha} + e^{-\alpha t}\left(X(0) - \frac{\beta}{\alpha} + \int_0^t \sigma e^{\alpha s}dB(s)\right). \qquad (5.15)$$

Using Itô's formula it is easy to verify that (5.15) is indeed a solution.

Example 5.7: Consider the SDE $dX(t) = B(t)dB(t)$.
Clearly, $X(t) = X(0) + \int_0^t B(s)dB(s)$, and using integration by parts (or Itô's formula), we obtain

$$X(t) = X(0) + \frac{1}{2}(B^2(t) - t).$$

Remark 5.2: If a strong solution exists, then by definition it is adapted to the filtration of the given Brownian motion, and as such it is intuitively clear that it is a function the path $(B(s), s \leq t)$. Results of Yamada and Watanabe (1971), and Kallenberg (1996) state that provided the conditions of the existence and uniqueness theorem are satisfied, then there exists a function F such that the strong solution is given by $X(t) = F(t, (B(s), s \leq t))$. Not much is known about F in general. Often it is not easy to find this function even for Itô integrals $X(t) = \int_0^t f(B(s))dB(s)$, e.g. $X(t) = \int_0^t |B(s)|^{1/2}dB(s)$. For a representation of Itô integrals as functions of Brownian motion paths, see, for example, Rogers and Williams (1990), p.125-127.

Only some classes of SDEs admit a closed form solution. When a closed form solution is hard to find, an existence and uniqueness result is important, because without it, it is not clear what exactly the equation means. When a solution exists and is unique, then numerical methods can be employed to compute it. Similarly to ordinary differential equations, linear SDEs can be solved explicitly.

5.2   Stochastic Exponential and Logarithm

Let X have a stochastic differential, and U satisfy

$$dU(t) = U(t)dX(t), \text{ and } U(0) = 1, \text{ or } U(t) = 1 + \int_0^t U(s)dX(s). \qquad (5.16)$$

Then U is called the stochastic exponential of X, and is denoted by $\mathcal{E}(X)$. If $X(t)$ is of finite variation then the solution to (5.16) is given by $U(t) = e^{X(t)}$. For Itô processes the solution is given by

Theorem 5.2 *The only solution of (5.16) is given by*

$$U(t) = \mathcal{E}(X)(t) := e^{X(t)-X(0)-\frac{1}{2}[X,X](t)}.$$ (5.17)

PROOF: The proof of existence of a solution to (5.16) consists of verification, by using Itô's formula, of (5.17). Write $U(t) = e^{V(t)}$, with $V(t) = X(t) - X(0) - \frac{1}{2}[X,X](t)$. Then

$$d\mathcal{E}(X)(t) = dU(t) = d(e^{V(t)}) = e^{V(t)}dV(t) + \frac{1}{2}e^{V(t)}d[V,V](t).$$

Since $[X,X](t)$ is of finite variation, and $X(t)$ is continuous, $[X,[X,X]](t) = 0$, and $[V,V](t) = [X,X](t)$. Using this with the expression for $V(t)$, we obtain

$$d\mathcal{E}(X)(t) = e^{V(t)}dX(t) - \frac{1}{2}e^{V(t)}d[X,X](t) + \frac{1}{2}e^{V(t)}d[X,X](t) = e^{V(t)}dX(t),$$

and (5.16) is established. The proof of uniqueness is done by assuming that there is another process satisfying (5.16), say $U_1(t)$, and showing by integration by parts that $d(U_1(t)/U(t)) = 0$. It is left as an exercise.

□

Note that unlike in the case of the usual exponential $g(t) = \exp(f)(t) = e^{f(t)}$, the stochastic exponential $\mathcal{E}(X)$ requires the knowledge of all the values of the process up to time t, since it involves the quadratic variation term $[X,X](t)$.

Example 5.8: Stochastic exponential of Brownian motion $B(t)$ is given by $U(t) = \mathcal{E}(B)(t) = e^{B(t)-\frac{1}{2}t}$, and it satisfies for all t, $dU(t) = U(t)dB(t)$ with $U(0) = 1$.

Example 5.9: Application in Finance: Stock process and its Return process.
Let $S(t)$ denote the price of stock and assume that it is an Itô process, i.e. it has a stochastic differential. The process of the return on stock $R(t)$ is defined by the relation

$$dR(t) = \frac{dS(t)}{S(t)}.$$

In other words

$$dS(t) = S(t)dR(t)$$ (5.18)

and the stock price is the stochastic exponential of the return. Returns are usually easier to model from first principles. For example, in the Black-Scholes model it is assumed that the returns over non-overlapping time intervals are independent, and have finite variance. This assumption leads to the model for the return process $R(t) = \mu t + \sigma B(t)$. The stock price is then given by

$$\begin{aligned} S(t) &= S(0)\mathcal{E}(R)_t = S_0 e^{R(t)-R(0)-\frac{1}{2}[R,R](t)} \\ &= S(0)e^{(\mu-\frac{1}{2}\sigma^2)t+\sigma B(t)}. \end{aligned}$$ (5.19)

Stochastic Logarithm

If $U = \mathcal{E}(X)$, then the process X is called the stochastic logarithm of U, denoted $\mathcal{L}(U)$. This is the inverse operation to the stochastic exponential. For example, the stochastic exponential of Brownian motion $B(t)$ is given by $e^{B(t) - \frac{1}{2}t}$. So $B(t)$ is the stochastic logarithm of $e^{B(t) - \frac{1}{2}t}$.

Theorem 5.3 *Let U have a stochastic differential and not take value 0. Then the stochastic logarithm of U satisfies the SDE*

$$dX(t) = \frac{dU(t)}{U(t)}, \ X(0) = 0, \tag{5.20}$$

moreover

$$X(t) = \mathcal{L}(U)(t) = \ln\left(\frac{U(t)}{U(0)}\right) + \int_0^t \frac{d[U, U](t)}{2U^2(t)}. \tag{5.21}$$

PROOF: The SDE for the stochastic logarithm $\mathcal{L}(U)$ is by the definition of $\mathcal{E}(X)$. The solution (5.21) and uniqueness are obtained by Itô's formula.
                                                                                            □

Example 5.10: Let $U(t) = e^{B(t)}$. We find its stochastic logarithm $\mathcal{L}(U)$ directly and then verify (5.21). $dU(t) = e^{B(t)}dB(t) + \frac{1}{2}e^{B(t)}dt$. Hence

$$dX(t) = d\mathcal{L}(U)(t) = \frac{dU(t)}{U(t)} = dB(t) + \frac{1}{2}dt.$$

Thus
$$X(t) = \mathcal{L}(U)(t) = B(t) + \frac{1}{2}t.$$

Now, $d[U, U](t) = dU(t)dU(t) = e^{2B(t)}dt$, so that

$$\mathcal{L}(U)(t) = \ln U(t) + \int_0^t \frac{e^{2B(t)}dt}{2e^{2B(t)}} = B(t) + \int_0^t \frac{1}{2}dt = B(t) + \frac{1}{2}t,$$

which verifies (5.21).

Remark 5.3: The stochastic Logarithm is useful in financial applications (see Kallsen and Shiryaev (2002)).

5.3   Solutions to Linear SDEs

Linear SDEs form a class of SDEs that can be solved explicitly. Consider general linear SDE in one dimension

$$dX(t) = (\alpha(t) + \beta(t)X(t))\, dt + (\gamma(t) + \delta(t)X(t))\, dB(t), \tag{5.22}$$

where functions $\alpha, \beta, \gamma, \delta$ are *given* adapted processes, and are continuous functions of t. Examples considered in the previous section are particular cases of linear SDEs.

Stochastic Exponential SDEs

Consider finding solutions in the case when $\alpha(t) = 0$ and $\gamma(t) = 0$. The SDE becomes

$$dU(t) = \beta(t)U(t)dt + \delta(t)U(t)dB(t). \tag{5.23}$$

This SDE is of the form

$$dU(t) = U(t)dY(t), \tag{5.24}$$

where the Itô process $Y(t)$ is defined by

$$dY(t) = \beta(t)dt + \delta(t)dB(t).$$

The SDE (5.23) is the stochastic exponential of Y, see Section (5.2). The stochastic exponential of Y is given by

$$
\begin{aligned}
U(t) &= \mathcal{E}(Y)(t) \\
&= U(0)\exp\left(Y(t) - Y(0) - \frac{1}{2}[Y,Y](t)\right) \\
&= U(0)\exp\left(\int_0^t \beta(s)ds + \int_0^t \delta(s)dB(s) - \frac{1}{2}\int_0^t \delta^2(s)ds\right) \\
&= U(0)\exp\left(\int_0^t (\beta(s) - \frac{1}{2}\delta^2(s))ds + \int_0^t \delta(s)dB(s)\right), \tag{5.25}
\end{aligned}
$$

where $[Y,Y](t)$ is obtained from calculations $d[Y,Y](t) = dY(t)dY(t) = \delta^2(t)dt$.

General Linear SDEs

To find a solution for the Equation (5.22) in the general case, look for a solution of the form

$$X(t) = U(t)V(t), \tag{5.26}$$

where

$$dU(t) = \beta(t)U(t)dt + \delta(t)U(t)dB(t), \tag{5.27}$$

and

$$dV(t) = a(t)dt + b(t)dB(t). \tag{5.28}$$

Set $U(0) = 1$ and $V(0) = X(0)$. Note that U is given by (5.25). Taking the differential of the product it is easy to see that we can choose coefficients $a(t)$ and $b(t)$ in such a way that relation $X(t) = U(t)V(t)$ holds. The desired coefficients $a(t)$ and $b(t)$ turn out to satisfy equations

$$b(t)U(t) = \gamma(t), \text{ and } a(t)U(t) = \alpha(t) - \delta(t)\gamma(t). \tag{5.29}$$

Using the expression for $U(t)$, $a(t)$ and $b(t)$ are then determined. Thus $V(t)$ is obtained, and $X(t)$ is found to be

$$X(t) = U(t)\left(X(0) + \int_0^t \frac{\alpha(s) - \delta(s)\gamma(s)}{U(s)}ds + \int_0^t \frac{\gamma(s)}{U(s)}dB(s)\right). \tag{5.30}$$

Langevin type SDE

Let $X(t)$ satisfy
$$dX(t) = a(t)X(t)dt + dB(t), \tag{5.31}$$
where $a(t)$ is a given adapted and continuous process. When $a(t) = -\alpha$ the equation is the Langevin equation, Example 5.6.

We solve the SDE in two ways: by using the formula (5.30), and directly, similarly to Langevin's SDE.

Clearly, $\beta(t) = a(t)$, $\gamma(t) = 1$, and $\alpha(t) = \delta(t) = 0$. To find $U(t)$, we must solve $dU(t) = a(t)U(t)dt$, which gives $U(t) = e^{\int_0^t a(s)ds}$. Thus from (5.30)
$$X(t) = e^{\int_0^t a(s)ds} \left(X(0) + \int_0^t e^{-\int_0^u a(s)ds} dB(u) \right).$$

Consider the process $e^{-\int_0^t a(s)ds} X(t)$ and use integration by parts. The process $e^{-\int_0^t a(s)ds}$ is continuous and is of finite variation. Therefore it has zero covariation with $X(t)$, hence

$$
\begin{aligned}
d\left(e^{-\int_0^t a(s)ds} X(t) \right) &= e^{-\int_0^t a(s)ds} dX(t) - a(t)e^{-\int_0^t a(s)ds} X(t)dt \\
&= e^{-\int_0^t a(s)ds} dB(t).
\end{aligned}
$$

Integrating we obtain

$$e^{-\int_0^t a(s)ds} X(t) = X(0) + \int_0^t e^{-\int_0^u a(s)ds} dB(u),$$

and finally

$$X(t) = X(0)e^{\int_0^t a(s)ds} + e^{\int_0^t a(s)ds} \int_0^t e^{-\int_0^u a(s)ds} dB(u). \tag{5.32}$$

Brownian Bridge

Brownian Bridge, or pinned Brownian motion is a solution to the following SDE

$$dX(t) = \frac{b - X(t)}{T - t} dt + dB(t), \text{ for } 0 \leq t < T, \ X(0) = a. \tag{5.33}$$

This process is a transformed Brownian motion with fixed values at each end of the interval $[0, T]$, $X(0) = a$ and $X(T) = b$. The above SDE is a linear SDE, with

$$\alpha(t) = -\frac{1}{T - t}, \ \beta(t) = \frac{b}{T - t}, \ \gamma(t) = 1, \text{ and } \delta(t) = 0.$$

Identifying $U(t)$ and $V(t)$ in (5.30) we obtain

$$X(t) = a\left(1 - \frac{t}{T}\right) + b\frac{t}{T} + (T - t)\int_0^t \frac{1}{T - s}dB(s), \text{ for } 0 \le t < T. \quad (5.34)$$

Since the function under the Itô integral is deterministic, and for any $t < T$, $\int_0^t ds/(T - s)^2 < \infty$, the process $\int_0^t \frac{1}{T-s}dB(s)$ is a martingale, moreover, it is Gaussian, by Theorem 4.11. Thus $X(t)$, on $[0, T)$ is a Gaussian process with initial value $X(0) = a$. The value at T, which is $X(T) = b$, is determined by continuity, see Example 5.11 below.

Thus a Brownian Bridge is a continuous Gaussian process on $[0, T]$ with mean function $a(1 - t/T) + bt/T$ and covariance function $Cov(X(t), X(s)) = \min(s, t) - st/T$.

Example 5.11: We show that $\lim_{t \uparrow T}(T - t)\int_0^t \frac{1}{T-s}dB(s) = 0$ almost surely. Using integration by parts, (which is the same as the standard formula due to zero covariation between the deterministic term and Brownian motion), for any $t < T$

$$\int_0^t \frac{1}{T - s}dB(s) = B(t)/(T - t) - \int_0^t \frac{B(s)}{(T - s)^2}ds,$$

and

$$(T - t)\int_0^t \frac{1}{T - s}dB(s) = B(t) - (T - t)\int_0^t \frac{B(s)}{(T - s)^2}ds. \quad (5.35)$$

It is an exercise in calculus (by changing variables $u = 1/(t - s)$, or considering integrals $\int_0^{T-\delta}$ and $\int_{T-\delta}^t$) to see that for any continuous function $g(s)$,

$$\lim_{t \uparrow T}(T - t)\int_0^t \frac{g(s)}{(T - s)^2}ds = g(T).$$

Applying this with $g(s) = B(s)$ shows that the limit in (5.35) is zero.

5.4 Existence and Uniqueness of Strong Solutions

Let $X(t)$ satisfy

$$dX(t) = \mu(X(t), t)dt + \sigma(X(t), t)dB(t). \quad (5.36)$$

Theorem 5.4 (Existence and Uniqueness) *If the following conditions are satisfied*

1. *Coefficients are locally Lipschitz in x uniformly in t, that is, for every T and N, there is a constant K depending only on T and N such that for all $|x|, |y| \le N$ and all $0 \le t \le T$*

$$|\mu(x, t) - \mu(y, t)| + |\sigma(x, t) - \sigma(y, t)| < K|x - y|, \quad (5.37)$$

2. *Coefficients satisfy the linear growth condition*

$$|\mu(x,t)| + |\sigma(x,t)| \leq K(1 + |x|) \tag{5.38}$$

3. $X(0)$ *is independent of* $(B(t), 0 \leq t \leq T)$, *and* $EX^2(0) < \infty$.

Then there exists a unique strong solution $X(t)$ *of the SDE (5.36).* $X(t)$ *has continuous paths, moreover*

$$\mathrm{E}\Big(\sup_{0 \leq t \leq T} X^2(t) \Big) < C\big(1 + \mathrm{E}(X^2(0))\big), \tag{5.39}$$

where constant C *depends only on* K *and* T.

The proof of existence is carried out by successive approximations, similar to that for ordinary differential equations (Picard iterations). It can be found in Friedman (1975), p.104-107, Gihman and Skorohod (1982), Rogers and Williams (1990). It is not hard to see, by using Gronwall's lemma, that the Lipschitz condition implies uniqueness.

The Lipschitz condition (5.37) holds if, for example, partial derivatives $\frac{\partial \mu}{\partial x}(t,x)$ and $\frac{\partial \sigma}{\partial x}(t,x)$ are bounded for $|x|, |y| \leq N$ and all $0 \leq t \leq T$, which in turn is true if the derivatives are continuous (see Chapter 1).

Less Stringent Conditions for Strong Solutions

The next result is specific for one-dimensional SDEs. It is given for the case of time-independent coefficients. A similar result holds for time-dependent coefficients, see for example, Ethier and Kurtz (1986), p.298, Rogers and Williams (1990), p.265.

Theorem 5.5 (Yamada-Watanabe) *Suppose that* $\mu(x)$ *satisfies the Lipschitz condition and* $\sigma(x)$ *satisfies a Hölder condition condition of order* α, $\alpha \geq 1/2$, *that is, there is a constant* K *such that*

$$|\sigma(x) - \sigma(y)| < K|x - y|^\alpha. \tag{5.40}$$

Then the strong solution exists and is unique.

Example 5.12: (Girsanov's SDE)
$dX(t) = |X(t)|^r dB(t)$, $X(0) = 0$, $1/2 \leq r < 1$. Note that for such r, $|x|^r$ is Hölder, but not Lipschitz (see section on these conditions in Chapter 1). $X(t) \equiv 0$ is a strong solution. Since the conditions of the Theorem are satisfied, $X(t) \equiv 0$ is the only solution.

5.5 Markov Property of Solutions

The Markov property asserts that given the present state of the process, the future is independent of the past. This can be stated as follows. If $\mathcal{F}_t$ denotes the σ-field generated by the process up to time t, then for any $0 \leq s < t$

$$P(X(t) \leq y | \mathcal{F}_s) = P(X(t) \leq y | X(s)) \text{ a.s.} \tag{5.41}$$

It is intuitively clear from the heuristic Equation (5.4) that solutions to SDEs should have the Markov Property. For a small Δ, given $X(t) = x$, $X(t + \Delta)$ depends on $B(t + \Delta) - B(t)$, which is independent of the past.

We don't prove that strong solutions possess Markov property. However, by the construction of the solution on the canonical space (a weak solution), it can be seen that the Markov property holds.

Transition Function.

Markov processes are characterized by the transition probability function. Denote by

$$P(y, t, x, s) = P(X(t) \leq y | X(s) = x) \tag{5.42}$$

the conditional distribution function of the random variable $X(t)$ given that $X(s) = x$, i.e. the distribution of the values at time t given that the process was in the state x at time s.

Theorem 5.6 *Let $X(t)$ be a solution to the SDE (5.36) Then $X(t)$ has the Markov property.*

Using the law of total probability, by conditioning on all possible values z of the process at time u, for $s < u < t$, we obtain that the transition probability function $P(y, t, x, s)$ in (5.42) satisfies the Chapman-Kolmogorov equation

$$P(y, t, x, s) = \int_{-\infty}^{\infty} P(y, t, z, u) P(dz, u, x, s), \text{ for any } s < u < t. \tag{5.43}$$

In fact any function that satisfies this equation and is a distribution function in y for fixed values of the other arguments, is a transition function of some Markov process.

Example 5.13: If $P(y, t, x, s) = \int_{-\infty}^{y} \frac{1}{\sqrt{2\pi(t-s)}} e^{\frac{(u-x)^2}{2(t-s)}} du$ is the cumulative distribution function of the Normal $N(x, t-s)$ distribution, then the corresponding diffusion process is Brownian motion. Indeed, $P(B(t) \leq y | \mathcal{F}_s) = P(B(t) \leq y | B(s))$, and the conditional distribution of $B(t)$ given $B(s) = x$ is $N(x, t-s)$.

Example 5.14:   Let $X(t)$ solve SDE $dX(t) = \mu X(t)dt + \sigma X(t)dB(t)$ for some constants μ and σ. We know (see Example 5.5) that $X(t) = X(0)e^{(\mu-\sigma^2/2)t+\sigma B(t)}$. Hence $X(t) = X(s)e^{(\mu-\sigma^2/2)(t-s)+\sigma(B(t)-B(s))}$ and its transition probability function $\mathrm{P}(X(t) \leq y|X(s) = x) = \mathrm{P}(X(s)e^{(\mu-\sigma^2/2)(t-s)+\sigma(B(t)-B(s))} \leq y|X(s) = x) = \mathrm{P}(xe^{(\mu-\sigma^2/2)(t-s)+\sigma(B(t)-B(s))} \leq y|X(s) = x)$. Using independence of $B(t) - B(s)$ and $X(s)$, the conditional probability is given by $\mathrm{P}(xe^{(\mu-\sigma^2/2)(t-s)+\sigma(B(t)-B(s))} \leq y) = \mathrm{P}(e^{(\mu-\sigma^2/2)(t-s)+\sigma(B(t)-B(s))} \leq y/x)$. Thus $P(y,t,x,s) = \Phi(\frac{\ln(y/x)-(\mu-\sigma^2/2)(t-s)}{\sigma\sqrt{t-s}})$.

Remark 5.4: Introduce a useful representation, which requires us to keep track of when and where the process starts. Denote by $X_s^x(t)$ the value of the process at time t when it starts at time s from the point x. It is clear that for $0 \leq s < t$, $X_0^x(t) = X_s^{X_0^x(s)}(t)$. The Markov property states that conditionally on $X_s^{x_0}(t) = x$, the processes $X_s^{x_0}(u)$, $s \leq u \leq t$, and $X_t^x(u)$, $t \leq u$ are independent.

Definition 5.7 *A process has the strong Markov property if the relation (5.41) holds when a non-random time s is replaced by a finite stopping time τ.*

Solutions to SDEs have also the strong Markov property, meaning that given the history up to a stopping time τ, the behaviour of the process at some future time t, is independent of the past. See also Section 3.4.

If an SDE has a strong solution $X(t)$, then $X(t)$ has a transition probability function $P(y,t,x,s)$. This function can be found as a solution to the Forward or the Backward partial differential equations, see Section 5.8.

A transition probability function $P(y,t,x,s)$ may exist for SDEs without a strong solution. This function in turn determines a Markov process uniquely (all finite-dimensional distributions). This process is known as a weak solution to an SDE. In this way one can define a solution for an SDE under less stringent conditions on its coefficients. The concepts of the weak solution are considered next.

5.6   Weak Solutions to SDEs

The concept of weak solutions allows us to give a meaning to an SDE when strong solutions do not exist. Weak solutions are solutions in distribution, they can be realized (defined) on some other probability space and exist under less stringent conditions on the coefficients of the SDE.

Definition 5.8 *If there exist a probability space with a filtration, a Brownian motion $\hat{B}(t)$ and a process $\hat{X}(t)$ adapted to that filtration, such that: $\hat{X}(0)$*

*has the given distribution, for all t the integrals below are defined, and $\hat{X}(t)$
satisfies*

$$\hat{X}(t) = \hat{X}(0) + \int_0^t \mu(\hat{X}(u), u)du + \int_0^t \sigma(\hat{X}(u), u)d\hat{B}(u), \qquad (5.44)$$

then $\hat{X}(t)$ is called a weak solution to the SDE

$$dX(t) = \mu(X(t), t)dt + \sigma(X(t), t)dB(t). \qquad (5.45)$$

Definition 5.9 *A weak solution is called unique if whenever $X(t)$ and $X'(t)$
are two solutions (perhaps on different probability spaces) such that the distributions of $X(0)$ and $X'(0)$ are the same, then all finite-dimensional distributions of $X(t)$ and $X'(t)$ are the same.*

Clearly, by definition, a strong solution is also a weak solution. Uniqueness
of the strong solution (pathwise uniqueness) implies uniqueness of the weak
solution, (a result of Yamada and Watanabe (1971)). In the next example a
strong solution does not exist, but a weak solution exists and is unique.

Example 5.15: (Tanaka's SDE)

$$dX(t) = \text{sign}(X(t))dB(t), \qquad (5.46)$$

where

$$\text{sign}(x) = \begin{cases} 1 & \text{if } x \geq 0 \\ -1 & \text{if } x < 0. \end{cases}$$

Since $\sigma(x) = \text{sign}(x)$ is discontinuous, it is not Lipschitz, and conditions for the
strong existence fail. It can be shown that a strong solution to Tanaka's SDE does
not exist, for example, Gihman and Skorohod (1982), Rogers and Williams (1990)
p.151. We show that the Brownian motion is the unique weak solution of Tanaka's
SDE. Let $X(t)$ be some Brownian motion. Consider the process

$$Y(t) = \int_0^t \frac{1}{\text{sign}(X(s))}dX(s) = \int_0^t \text{sign}(X(s))dX(s).$$

$\text{sign}(X(t))$ is adapted, $\int_0^T (\text{sign}(X(t)))^2 dt = T < \infty$, and $Y(t)$ is well defined and is
a continuous martingale.

$$[Y, Y](t) = \int_0^t \text{sign}^2(X(s))d[X, X](s) = \int_0^t ds = t.$$

By Levy's theorem (which is proven later), $Y(t)$ is a Brownian motion, call it $\hat{B}(t)$,

$$\hat{B}(t) = \int_0^t \frac{dX(s)}{\text{sign}(X(s))}.$$

Rewrite the last equality in the differential notation to obtain Tanaka's SDE. Levy's
characterization theorem implies also that *any* weak solution is a Brownian motion.

Example 5.16: (Girsanov's SDE) The equation

$$dX(t) = |X(t)|^r dB(t), \tag{5.47}$$

$r > 0$, $t \geq 0$ has a strong solution $X(t) \equiv 0$. For $r \geq 1/2$, this is the only strong solution by Theorem 5.5. Therefore the are no weak solutions other than zero. For $0 < r < 1/2$ the SDE has infinitely many weak solutions (Rogers and Williams (1990) p.175). Therefore there is no strong uniqueness in this case, otherwise it would have only one weak solution. Compare this to the non-uniqueness of solution of the equation $dx(t) = 2\sqrt{|x(t)|}dt$, which has solutions $x(t) = 0$ and $x(t) = t^2$.

5.7   Construction of Weak Solutions

In this section we give results on the existence and uniqueness of weak solutions to SDEs. Construction of weak solutions requires more advanced knowledge, and this section can be skipped.

Theorem 5.10 *If for each $t > 0$, functions $\mu(x,t)$ and $\sigma(x,t)$ are bounded and continuous then the SDE (5.45) has at least one weak solution starting at time s at point x, for all s, and x. If in addition their partial derivatives with respect to x up to order two are also bounded and continuous, then the SDE (5.45) has a unique weak solution starting at time s at point x. Moreover this solution has the strong Markov property.*

These results are proved in Stroock and Varadhan (1979), ch. 6. But better conditions are available, Stroock and Varadhan (1979) Corollary 6.5.5, see also Pinsky (1995) Theorem 1.10.2.

Theorem 5.11 *If $\sigma(x,t)$ is positive and continuous and for any $T > 0$ there is K_T such that for all $x \in \mathbb{R}$*

$$|\mu(x,t)| + |\sigma(x,t)| \leq K_T(1 + |x|) \tag{5.48}$$

then there exists a unique weak solution to SDE (5.45) starting at any point $x \in \mathbb{R}$ at any time $s \geq 0$, moreover it has the strong Markov property.

Canonical Space for Diffusions

Solutions to SDEs or diffusions can be realized on the probability space of continuous functions. We indicate: how to define probability on this space by means of a transition function, how to find the transition function from the given SDE and how to verify that the constructed process indeed satisfies the given SDE.

Probability Space $(\Omega, \mathcal{F}, \mathbb{F})$

Weak solutions can be constructed on the canonical space $\Omega = C([0, \infty))$ of continuous functions from $[0, \infty)$ to $\mathbb{R}$. The Borel σ-field on Ω is the one generated by the open sets. Open sets in turn, are defined with the help of a metric, for example, an open ball of radius ϵ centered at ω is the set $D_\epsilon(\omega) = \{\omega' : d(\omega, \omega') < \epsilon\}$. The distance between two continuous functions ω_1 and ω_2 is taken as

$$d(\omega_1, \omega_2) = \sum_{n=1}^{\infty} \frac{1}{2^n} \frac{\sup_{0 \leq t \leq n} |\omega_1(t) - \omega_2(t)|}{1 + \sup_{0 \leq t \leq n} |\omega_1(t) - \omega_2(t)|}.$$

Convergence of the elements of Ω in this metric is the uniform convergence of functions on bounded closed intervals $[0, T]$. Diffusions on a finite interval $[0, T]$ can be realized on the space $C([0, T])$ with the metric

$$d(\omega_1, \omega_2) = \sup_{0 \leq t \leq T} |\omega_1(t) - \omega_2(t)|.$$

The canonical process $X(t)$ is defined by $X(t, \omega) = \omega(t)$, $0 \leq t < \infty$. It is known (for example, Dudley (1989) p.356) that the Borel σ-field $\mathcal{F}$ on $C([0, \infty))$ is given by $\sigma(X(t), 0 \leq t < \infty)$. The filtration is defined by the σ-fields $\mathcal{F}_t = \sigma(X(s), 0 \leq s \leq t)$.

Probability Measure

We outline the construction of probability measures from a given transition function $P(y, t, x, s)$. In particular, this construction gives the Wiener measure that corresponds to the Brownian motion process.

For any fixed $x \in \mathbb{R}$ and $s \geq 0$, a probability $\mathrm{P} = \mathrm{P}_{x,s}$ on $(\Omega, \mathcal{F})$ can be constructed by using properties

1. $\mathrm{P}(X(u) = x, 0 \leq u \leq s) = 1$.

2. $\mathrm{P}(X(t_2) \in B | \mathcal{F}_{t_1}) = P(B, t_2, X(t_1), t_1)$.

The second property asserts that for any Borel sets $A, B \subset \mathbb{R}$ we have

$$
\begin{aligned}
\mathrm{P}_{t_1, t_2}(A \times B) &:= \mathrm{P}(X(t_1) \in A, X(t_2) \in B) \\
&= \mathrm{E}\left(\mathrm{P}(X(t_2) \in B | \mathcal{F}_{t_1}) I(X(t_1) \in A)\right) \\
&= \mathrm{E}\left(\mathrm{P}(B, t_2, X(t_1), t_1) I((X(t_1) \in A))\right) \\
&= \int_A \int_B \mathrm{P}(dy_2, t_2, y_1, t_1) \mathrm{P}_{t_1}(dy_1),
\end{aligned}
$$

where $P_{t_1}(C) = P(X(t_1) \in C)$. This extends to the n-dimensional cylinder sets $\{\omega \in \Omega : (\omega(t_1), \ldots, \omega(t_n)) \in J_n\}$, where $J_n \subset \mathbb{R}^n$, by

$$P_{t_1,\ldots,t_{n+1}}(J_{n+1}) = \int_{J_{n+1}} P(dy_{n+1}, t_{n+1}, y_n, t_n) P_{t_1,\ldots,t_n}(dy_1 \times \ldots \times dy_n).$$

These probabilities give the finite dimensional distributions $P((\omega(t_1), \ldots, \omega(t_n)) \in J_n)$. Consistency of these probabilities is a consequence of the Chapman-Kolmogorov equation for the transition function. Thus by Kolmogorov's extension theorem P can be extended in a unique way to $\mathcal{F}$. This probability measure $P = P_{x,s}$ corresponds to the Markov process started at x at time s, denoted earlier by $X_s^x(t)$. Thus any transition function defines a probability so that the canonical process is a Markov process. We described in particular a construction of the Wiener measure, or Brownian motion.

Transition Function

Under appropriate conditions on the coefficients $\mu(x, t)$ and $\sigma(x, t)$, $P(y, t, x, s)$ is determined from a partial differential equation (PDE),

$$\frac{\partial u}{\partial s}(x, s) + L_s u(x, s) = 0, \tag{5.49}$$

called the backward PDE, involving a second order differential operator L_s,

$$L_s f(x, s) = (L_s f)(x, s) = \frac{1}{2}\sigma^2(x, s)\frac{\partial^2 f}{\partial x^2}(x, s) + \mu(x, s)\frac{\partial f}{\partial x}(x, s). \tag{5.50}$$

It follows from the key property of the transition function, that

$$f(X(t)) - \int_s^t (L_u f)(X(u))du \tag{5.51}$$

is a martingale under $P_{x,s}$ with respect to $\mathcal{F}_t$ for $t \geq s$, for any twice continuously differentiable function f vanishing outside a finite interval (with compact support), $f \in C_K^2(\mathbb{R})$.

SDE on the Canonical Space is Satisfied

Extra concepts (that of local martingales and their integrals) are needed to prove the claim rigorously. The main idea is as follows. Suppose that (5.51) holds for functions $f(x) = x$ and $f(x) = x^2$. (Although they don't have a compact support, they can be approximated by C_K^2 functions on any finite interval). Applying (5.51) to the linear function, we obtain that

$$Y(t) = X(t) - \int_s^t \mu(X(u), u)du \tag{5.52}$$

is a martingale. Applying (5.51) to the quadratic function, we obtain that

$$X^2(t) - \int_s^t \left(\sigma^2(X(u), u) + 2\mu(X(u), u)X(u) \right) du \qquad (5.53)$$

is a martingale. By the characterization property of quadratic variation for continuous martingales $Y^2(t) - [Y, Y](t)$ is a martingale, and it follows from the above relations that $[Y, Y](t) = \int_s^t \sigma^2(X(u), u)du$. One can define the Itô integral process $B(t) = \int_s^t dY(u)/\sigma(X(u), u)$. From the properties of stochastic integrals it follows that $B(t)$ is a continuous local martingale with $[B, B](t) = t$. Thus by Levy's theorem $B(t)$ is Brownian motion. Putting all of the above together and using differential notation, the required SDE is obtained. For details, see for example, Rogers and Williams (1990), p.160, also Stroock and Varadhan (1979).

Weak Solutions and the Martingale Problem

Taking the relation (5.51) as primary, Stroock and Varadhan defined a weak solution to the SDE

$$dX(t) = \mu(X(t), t)dt + \sigma(X(t), t)dB(t) \qquad (5.54)$$

as a solution to the so-called martingale problem.

Definition 5.12 *The martingale problem for the coefficients, or the operator L_s, is as follows. For each $x \in \mathbb{R}$, and $s > 0$, find a probability measure $P_{x,s}$ on $\Omega, \mathcal{F}$ such that*

1. $P_{x,s}(X(u) = x, 0 \le u \le s) = 1$,

2. For any twice continuously differentiable function f vanishing outside a finite interval the following process is a martingale under $P_{x,s}$ with respect to $\mathcal{F}_t$

$$f(X(t)) - \int_s^t (L_u f)(X(u))du. \qquad (5.55)$$

In the case when there is exactly one solution to the martingale problem, it is said that the martingale problem is *well-posed*.

Example 5.17: Brownian motion $B(t)$ is a solution to the martingale problem for the Laplace operator $L = \frac{1}{2}\frac{d^2}{dx^2}$, that is, for a twice continuously differentiable function f vanishing outside a finite interval

$$f(B(t)) - \int_0^t \frac{1}{2}f''(B(s))ds$$

is a martingale. Since Brownian motion exists and is determined by its distribution uniquely, the martingale problem for L is well-posed.

Remark 5.5: Note that if a function vanishes outside a finite interval, then its derivatives also vanish outside that interval. Thus for a twice continuously differentiable vanishing outside a finite interval function f ($f \in C_K^2$), $(L_s f)$ exists is continuous and vanishes outside that interval. This assures that the expectation of the process in (5.55) exists. If one demands only that f is twice continuously differentiable with bounded derivatives ($f \in C_b^2$), then $(L_s f)$ exists but may not be bounded, and expectation in (5.55) may not exist. If one takes ($f \in C_b^2$) then one seeks solutions to the local martingale problem, and any such solution makes the process in (5.55) into a local martingale. Local martingales are covered in Chapter 7.

As there are two definitions of weak solutions definition 5.8 and definition 5.12, we show that they are the same.

Theorem 5.13 *Weak solutions in the sense of Definition 5.8 and in the sense of Definition 5.12 are equivalent.*

PROOF: We already indicated the proof in one direction, that if the martingale problem has a solution, then the solution satisfies the SDE. The other direction is obtained by using Itô's formula. Let $X(t)$ be a weak solution in the sense of Definition 5.8. Then there is a space supporting Brownian motion $B(t)$ so that

$$X(t) = X(s) + \int_s^t \mu(X(u), u) du + \int_s^t \sigma(X(u), u) dB(u), \text{ and } X(s) = x. \quad (5.56)$$

is satisfied for all $t \geq s$. Let f be twice continuously differentiable with compact support. Applying Itô's formula to $f(X(t))$, we have

$$f(X(t)) = f(X(s)) + \int_s^t (L_u f)(X(u)) du + \int_s^t f'(X(u)) \sigma(X(u), u) dB(u). \quad (5.57)$$

Thus

$$f(X(t)) - \int_s^t (L_u f)(X(u)) du = f(X(s)) + \int_s^t f'(X(u)) \sigma(X(u), u) dB(u). \quad (5.58)$$

Since f and its derivatives vanish outside an interval, say $[-K, K]$, the functions $f'(x) \sigma(x, u)$ also vanish outside this interval, for any u. Assuming that $\sigma(x, u)$ are bounded in x on finite intervals with the same constant for all u, it follows that $|f'(x) \sigma(x, u)| < K_1$. Thus the integral $\int_s^t f'(X(u)) \sigma(X(u), u) dB(u)$ is a martingale in t, for $t \geq s$; thus the martingale problem has a solution.

□

5.8 Backward and Forward Equations

In many applications, such as Physics, Engineering and Finance, the importance of diffusions lies in their connection to PDEs, and often diffusions are specified by a PDE called the Fokker-Plank equation (introduced below (5.62), see for example, Soize (1994)). Although PDEs are hard to solve in closed form, they can be easily solved numerically. In practice it is often enough to check that conditions of the existence and uniqueness result are satisfied and then the solution can be computed by a PDE solver to the desired degree of accuracy.

In this section it is outlined how to obtain the transition function that determines the weak solution to an SDE

$$dX(t) = \mu(X(t), t)dt + \sigma(X(t), t)dB(t), \text{ for } t \geq 0. \qquad (5.59)$$

The results below are the main results from the theory of PDEs which are used for construction of diffusions (see for example Friedman (1975), Stroock and Varadhan (1979)).

Define the differential operator L_s, $0 \leq s \leq T$ by

$$L_s f(x, s) = (L_s f)(x, s) = \frac{1}{2}\sigma^2(x, s)\frac{\partial^2 f}{\partial x^2}(x, s) + \mu(x, s)\frac{\partial f}{\partial x}(x, s). \qquad (5.60)$$

The operator L_s acts on twice continuously differentiable in x functions $f(x, s)$, and the result of its action on $f(x, s)$ is another function, denoted by $(L_s f)$, the values of which at point (x, s) are given by (5.60).

Definition 5.14 *A fundamental solution of the PDE*

$$\frac{\partial u}{\partial s}(x, s) + L_s u(x, s) = 0 \qquad (5.61)$$

is a non-negative function $p(y, t, x, s)$ with following properties:

1. *it is jointly continuous in y, t, x, s, twice continuously differentiable in x, and satisfies equation (5.61) with respect to s and x.*

2. *for any bounded continuous function $g(x)$ on $\mathbb{R}$, and any $t > 0$*

$$u(x, s) = \int_{\mathbb{R}} g(y)p(y, t, x, s)dy$$

is bounded, satisfies equation (5.61) and $\lim_{s \uparrow t} u(x, s) = g(x)$, for $x \in \mathbb{R}$.

Theorem 5.15 *Suppose that $\sigma(x, t)$ and $\mu(x, t)$ are bounded and continuous functions such that*

(A1) $\sigma^2(x,t) \geq c > 0$,

(A2) $\mu(x,t)$ and $\sigma^2(x,t)$ satisfy a Hölder condition with respect to x and t, that is, for all $x, y \in \mathbb{R}$ and $s, t > 0$

$$|\mu(y,t) - \mu(x,s)| + |\sigma^2(y,t) - \sigma^2(x,s)| \leq K(|y-x|^\alpha + |t-s|^\alpha).$$

Then the PDE (5.61) has a fundamental solution $p(y,t,x,s)$, which is unique, and is strictly positive.

If in addition $\mu(x,t)$ and $\sigma(x,t)$ have two partial derivatives with respect to x, which are bounded and satisfy a Hölder condition with respect to x, then $p(y,t,x,s)$ as a function in y and t, satisfies the PDE

$$-\frac{\partial p}{\partial t} + \frac{1}{2}\frac{\partial^2}{\partial y^2}\left(\sigma^2(y,t)p\right) - \frac{\partial}{\partial y}\left(\mu(y,t)p\right) = 0. \tag{5.62}$$

Theorem 5.16 *Suppose coefficients of L_s in (5.60) satisfy conditions (A1) and (A2) of Theorem 5.15. Then PDE (5.61) has a unique fundamental solution $p(y,t,x,s)$. The function $P(y,t,x,s) = \int_{-\infty}^{y} p(u,t,x,s)du$ uniquely defines a transition probability function. Moreover, this function has the property that for any bounded function $f(x,t)$ twice continuously differentiable in x and once continuously differentiable in t ($f \in C_b^{2,1}(\mathbb{R} \times [0,t])$)*

$$\int_{\mathbb{R}} f(y,t)P(dy,t,x,s) - f(x,s) = \int_s^t \int_{\mathbb{R}} \left(\frac{\partial}{\partial u} + L_u\right)f(y,u)P(dy,u,x,s)du \tag{5.63}$$

for all $0 \leq s < t$, $x \in \mathbb{R}$.

The transition function $P(y,t,x,s)$ in the above theorem defines uniquely a Markov process $X(t)$, that is, for all x, y and $0 \leq s \leq t$

$$P(y,t,x,s) = P(X(t) \leq y|X(s) = x). \tag{5.64}$$

Equation (5.61) is a PDE in the backward variables (x,s) and is therefore called the *backward equation*, also known as Kolmogorov's backward equation. Equation (5.62) is a PDE in the forward variables (y,t) and is therefore called the *forward equation*, also known as *Fokker-Plank* equation, diffusion equation, or Kolmogorov's forward equation.

The process $X(t)$ is called a diffusion, the differential operator L_s is called its generator. The property (5.63) implies that $X(t)$ satisfies the SDE (5.59).

Remark 5.6: A weak solution exists and is unique, possesses a Strong Markov property, and has density under the conditions of Theorem 5.11, much weaker than those of Theorem 5.15.

5.9 Stratanovich Stochastic Calculus

Stochastic integrals in applications are often taken in the sense of Stratanovich calculus. This calculus is designed in such a way that its basic rules, such as the chain rule and integration by parts are the same as in the standard calculus (e.g. Rogers and Williams (1990) p.106). Although the rules of manipulations are the same, the calculi are still very different. The processes need to be adapted, just as in Itô calculus. Since Stratanovich stochastic integrals can be reduced to Itô integrals, the standard SDE theory can be used for Stratanovich stochastic differential equations. Note also that the Stratanovich Integral is more suited for generalizations of stochastic calculus on manifolds (see Rogers and Williams (1990)).

A direct definition of the Stratanovich Integral, denoted $\int_0^t Y(s)\partial X(s)$, is done as a limit in mean-square (L^2) of Stratanovich approximating sums

$$\sum_{i=0}^{n-1} \frac{1}{2}\left(Y(t_{i+1}^n) + Y(t_i^n)\right)\left(X(t_{i+1}^n) - X(t_i^n)\right), \tag{5.65}$$

when partitions $\{t_i^n\}$ become finer and finer. In Stratanovich approximating sums the average value of Y on the interval (t_i^n, t_{i+1}^n), $\frac{1}{2}\left(Y(t_{i+1}^n) + Y(t_i^n)\right)$, is taken, whereas in Itô integral the left most value of $Y(t_i^n)$ is taken. An alternative definition of the Stratanovich integral is given by using the Itô integral.

Definition 5.17 *Let X and Y be continuous adapted processes, such that the stochastic integral $\int_0^t Y(s)dX(s)$ is defined. The Stratanovich integral is defined by*

$$\int_0^t Y(s)\partial X(s) = \int_0^t Y(s)dX(s) + \frac{1}{2}[Y, X](t). \tag{5.66}$$

The Stratanovich differential is defined by

$$Y(t)\partial X(t) = Y(t)dX(t) + \frac{1}{2}d[Y, X](t). \tag{5.67}$$

Integration by Parts: Stratanovich Product rule

Theorem 5.18 *Provided all terms below are defined,*

$$X(t)Y(t) - X(0)Y(0) = \int_0^t X(s)\partial Y(s) + \int_0^t Y(s)\partial X(s), \tag{5.68}$$

$$\partial(X(t)Y(t)) = X(t)\partial Y(t) + Y(t)\partial X(t). \tag{5.69}$$

PROOF: The proof is the direct application of the stochastic product rule,

$$
\begin{aligned}
d(X(t)Y(t)) &= X(t)dY(t) + Y(t)dX(t) + d[X,Y](t) \\
&= X(t)dY(t) + \frac{1}{2}[X,Y](t) + Y(t)dX(t) + \frac{1}{2}[X,Y](t) \\
&= X(t)\partial Y(t) + Y(t)\partial X(t).
\end{aligned}
$$

$\square$

Change of Variables: Stratanovich Chain rule

Theorem 5.19 *Let X be continuous and f three times continuously differentiable (in C^3), then*

$$
\begin{aligned}
f(X(t)) - f(X(0)) &= \int_0^t f'(X(s))\partial X(s), \qquad (5.70) \\
\partial f(X(t)) &= f'(X(t))\partial X(t).
\end{aligned}
$$

PROOF: By Itô's formula $f(X(t))$ is a semimartingale, and by definition of the stochastic integral $f(X(t)) - f(X(0)) = \int_0^t df(X(s))$. By Itô's formula

$$
df(X(t)) = f'(X(t))dX(t) + \frac{1}{2}f''(X(t))d[X,X](t).
$$

Let $Y(t) = f'(X(t))$. Then according to (5.67) it is enough to show that $d[Y,X](t) = f''(X(t))d[X,X](t)$. But this follows by Itô's formula as

$$
dY(t) = df'(X(t)) = f''(X(t))dX(t) + \frac{1}{2}f'''(X(t))d[X,X](t),
$$

and

$$
d[Y,X](t) = dY(t)dX(t) = f''(X(t))dX(t)dX(t) = f''(X(t))d[X,X](t),
$$

as needed.

$\square$

Example 5.18: If $B(t)$ is Brownian motion, then its Stratanovich stochastic differential is

$$
\partial B^2(t) = 2B(t)\partial B(t),
$$

as compared to Itô differential

$$
dB^2(t) = 2B(t)dB(t) + dt.
$$

Conversion of Stratanovich SDEs into Itô SDEs

Theorem 5.20 *Suppose that $X(t)$ satisfies the following SDE in the Stratanovich sense*

$$dX(t) = \mu(X(t))dt + \sigma(X(t))\partial B(t), \qquad (5.71)$$

with $\sigma(x)$ twice continuously differentiable. Then $X(t)$ satisfies the Itô SDE

$$dX(t) = \left(\mu(X(t)) + \frac{1}{2}\sigma'(X(t))\sigma(X(t))\right)dt + \sigma(X(t))dB(t). \qquad (5.72)$$

Thus the infinitesimal drift coefficient in Itô diffusion is $\mu(x) + \frac{1}{2}\sigma'(x)\sigma(x)$ and the diffusion coefficient is the same $\sigma(x)$.

PROOF: By the definition of the Stratanovich integral $X(t)$ satisfies

$$dX(t) = \mu(X(t))dt + \sigma(X(t))dB(t) + \frac{1}{2}d[\sigma(X), B](t). \qquad (5.73)$$

Since $[\sigma(X), B](t)$ is a finite variation process, it follows that $X(t)$ solves a diffusion type SDE with the same diffusion coefficient $\sigma(X(t))$. Computing formally the bracket, we have

$$d[\sigma(X), B](t) = d\sigma(X(t))dB(t).$$

Applying Itô's formula

$$d\sigma(X(t)) = \sigma'(X(t))dX(t) + \frac{1}{2}\sigma''(X(t))d[X, X](t).$$

It follows from (5.73) that

$$d[X, B](t) = dX(t)dB(t) = \sigma(X(t))dt,$$

therefore

$$d[\sigma(X), B](t) = d\sigma(X(t))dB(t) = \sigma'(X(t))dX(t)dB(t) = \sigma'(X(t))\sigma(X(t))dt.$$

Equation (5.72) now follows from (5.73).

$\square$

Notes. Proofs and other details can be found in Dynkin (1965), Friedman (1975), Karatzas and Shreve (1988), Stroock and Varadhan (1979).

5.10 Exercises

Exercise 5.1: (Gaussian diffusions.) Show that if $X(t)$ satisfies the SDE $dX(t) = a(t)dt + b(t)dB(t)$, with deterministic bounded coefficients $a(t)$ and $b(t)$, such that $\int_0^T |a(t)|dt < \infty$, and $\int_0^T b^2(t)dt < \infty$, then $X(t)$ is a Gaussian process with independent Gaussian increments.

Exercise 5.2: Give the SDEs for $X(t) = \cos(B(t))$ and $Y(t) = \sin(B(t))$.

Exercise 5.3: Solve the SDE $dX(t) = B(t)X(t)dt + B(t)X(t)dB(t)$, $X(0) = 1$.

Exercise 5.4: Solve the SDE $dX(t) = X(t)dt + B(t)dB(t)$, $X(0) = 1$. Comment whether it is a diffusion type SDE.

Exercise 5.5: Find $d\big(\mathcal{E}(B)(t)\big)^2$.

Exercise 5.6: Let $X(t)$ satisfy $dX(t) = X^2(t)dt + X(t)dB(t)$, $X(0) = 1$. Show that $X(t)$ satisfies $X(t) = e^{\int_0^t (X(s)-1/2)ds + B(t)}$.

Exercise 5.7: By definition, the stochastic logarithm satisfies $\mathcal{L}(\mathcal{E}(X)) = X$. Show that, provided $U(t) \neq 0$ for any t, $\mathcal{E}(\mathcal{L}(U)) = U$.

Exercise 5.8: Find the stochastic logarithm of $B^2(t) + 1$.

Exercise 5.9: Let $\boldsymbol{B}(t)$ be a d-dimensional Brownian motion, and $\boldsymbol{H}(t)$ a a d-dimensional regular adapted process. Show that

$$\mathcal{E}\left(\int_0^\cdot \boldsymbol{H}(s)d\boldsymbol{B}(s)\right)(t) = \exp\left(\int_0^t \boldsymbol{H}(s)d\boldsymbol{B}(s) - \frac{1}{2}\int_0^t |\boldsymbol{H}(s)|^2\, ds\right).$$

Exercise 5.10: Find the transition probability function $P(y, t, x, s)$ for Brownian motion with drift $B(t) + t$.

Exercise 5.11: Show that under the assumptions of Theorem 5.15 the transition function $P(y, t, x, s)$ satisfies the backward equation. Give also the forward equation for $P(y, t, x, s)$ and explain why it requires extra smoothness conditions on the coefficients $\mu(x, t)$ and $\sigma(x, t)$ for it to hold.

Exercise 5.12: Let $X(t)$ satisfy the following stochastic differential equation for $0 \leq t \leq T$, $dX(t) = \sqrt{X(t) + 1}dB(t)$, and $X(0) = 0$. Assuming that Itô integrals are martingales, find $EX(t)$, and $E(X^2(t))$. Let $m(u, t) = Ee^{uX(t)}$ be the moment generating function of $X(t)$. Show that it satisfies the PDE

$$\frac{\partial m}{\partial t} = \frac{u^2}{2}\frac{\partial m}{\partial u} + \frac{u^2}{2}m.$$

Exercise 5.13: Solve the following Stratanovich stochastic differential equation $\partial U = U\partial B$, $U(0) = 1$, where $B(t)$ is Brownian motion.

Chapter 6

Diffusion Processes

In this chapter various properties of solutions of stochastic differential equations are studied. The approach taken here relies on martingales obtained by means of Itô's formula. Relationships between stochastic differential equations (SDEs) and partial differential equations (PDEs) are given, but no prior knowledge of PDEs is required. Solutions to SDEs are referred to as diffusions.

6.1 Martingales and Dynkin's Formula

Itô's formula provides a source for construction of martingales. Let $X(t)$ solve the stochastic differential equation (SDE)

$$dX(t) = \mu(X(t), t)dt + \sigma(X(t), t)dB(t), \text{ for } t \geq 0, \tag{6.1}$$

and L_t be the generator of $X(t)$, that is, the second order differential operator associated with SDE (6.1),

$$L_t f(x, t) = (L_t f)(x, t) = \frac{1}{2}\sigma^2(x, t)\frac{\partial^2 f}{\partial x^2}(x, t) + \mu(x, t)\frac{\partial f}{\partial x}(x, t). \tag{6.2}$$

Itô's formula (4.65) takes a compact form

Theorem 6.1 *For any twice continuously differentiable in x, and once in t function $f(x, t)$*

$$df(X(t), t) = \left(L_t f(X(t), t) + \frac{\partial f}{\partial t}(X(t), t)\right)dt + \frac{\partial f}{\partial x}(X(t), t)\sigma(X(t), t)dB(t). \tag{6.3}$$

Since, under appropriate conditions, the Itô integral is a martingale (see Theorem 4.7), by isolating the Itô integral martingales are obtained.

To illustrate this simple idea, let f have a bounded (by K) derivative, and use Itô's formula for $f(B(t))$. Then

$$f(B(t)) = f(0) + \int_0^t \frac{1}{2} f''(B(s))ds + \int_0^t f'(B(s))dB(s).$$

The Itô integral $\int_0^t f'(B(s))dB(s)$ is a martingale on $[0, T]$, because condition (4.10) holds, $\int_0^T (f'(B(s)))^2 ds < K^2 T < \infty$. Thus $f(B(t)) - \int_0^t \frac{1}{2} f''(B(s))ds$ is a martingale. The next result is more general.

Theorem 6.2 *Let $X(t)$ be a solution to SDE (6.1) with coefficients satisfying conditions Theorem 5.4, that is, $\mu(x, t)$ and $\sigma(x, t)$ are Lipschitz in x with the same constant for all t, and satisfy the linear growth condition (1.30), $|\mu(x, t)| + |\sigma(x, t)| \leq K(1 + |x|)$. If $f(x, t)$ is a twice continuously differentiable in x and once in t function ($C^{2,1}$) with bounded first derivative in x, then the process*

$$M_f(t) = f(X(t), t) - \int_0^t \left(L_u f + \frac{\partial f}{\partial t} \right)(X(u), u)du \qquad (6.4)$$

is a martingale.

PROOF: By Itô's formula

$$M_f(t) = \int_0^t \frac{\partial f}{\partial x}(X(u), u)\sigma(X(u), u)dB(u). \qquad (6.5)$$

By assumption $\frac{\partial f}{\partial x}(x, u)$ is bounded for all x and u, $\left(\frac{\partial f}{\partial x}(x, u) \right)^2 < K_1$. Therefore

$$\int_0^T \mathrm{E}\left(\frac{\partial f}{\partial x}(X(u), u)\sigma(X(u), u) \right)^2 du \leq K_1 \int_0^T \mathrm{E}\left(\sigma^2(X(u), u) \right)du. \qquad (6.6)$$

Using the linear growth condition,

$$\int_0^T \mathrm{E}\left(\frac{\partial f}{\partial x}(X(u), u)\sigma(X(u), u) \right)^2 du \leq 2K_1 K^2 T \left(1 + \mathrm{E}\left(\sup_{u \leq T} X^2(u) \right) \right). \qquad (6.7)$$

But $\mathrm{E}\left(\sup_{u \leq T} X^2(u) \right) < \infty$ by the existence and uniqueness result, Theorem 5.4, therefore the expression in (6.7) is finite. Thus the Itô integral is a martingale by Theorem 4.7.

□

The condition of bounded partial derivative of f can be replaced by the exponential growth condition (see for example, Pinsky (1995), Theorem 1.6.3).

Theorem 6.3 *Let $X(t)$ satisfy conditions of the previous Theorem 6.2. If $|X(0)|$ possesses moment generating function $\mathrm{E}e^{u|X(0)|} < \infty$, for all real u, then so does $|X(t)|$, $\mathrm{E}e^{u|X(t)|} < \infty$ for all $t \geq 0$. In this case*

$$M_f(t) = f(X(t), t) - \int_0^t \left(L_u f + \frac{\partial f}{\partial t} \right)(X(u), u) du \qquad (6.8)$$

is a martingale for all $f(x, t) \in C^{2,1}$, satisfying the following condition: for any t, there exist constants c_t and k_t such that for all x, all $t > 0$, and all $0 \leq u \leq t$

$$\max \left(\left| \frac{\partial f(x, u)}{\partial t} \right|, \left| \frac{\partial f(x, u)}{\partial x} \right|, \left| \frac{\partial^2 f(x, t)}{\partial x^2} \right| \right) \leq c_t e^{k_t |x|}. \qquad (6.9)$$

PROOF: The proof is given in Pinsky (1995) for bounded coefficients of the SDE, but it can be extended for this case. We give the proof when the diffusion $X(t) = B(t)$ is Brownian motion. Let $X(t) = B(t)$, then by Itô's formula $M_f(t)$ is given by

$$M_f(t) = \int_0^t \frac{\partial f(B(s), s)}{\partial x} dB(s). \qquad (6.10)$$

By the bound on $|\frac{\partial f(x,s)}{\partial x}|$, for $s \leq t$

$$\mathrm{E} \left(\frac{\partial f(B(s), s)}{\partial x} \right)^2 \leq c_t^2 \mathrm{E} \left(e^{2k_t |B(s)|} \right).$$

Writing the last expectation as an integral with respect to the density of the $N(0, t)$ distribution, it is evident that it is finite, and its integral over $[0, t]$ is finite,

$$\int_0^t \mathrm{E} \left(\frac{\partial f(B(s), s)}{\partial x} \right)^2 ds < \infty. \qquad (6.11)$$

By the martingale property of Itô integrals, (6.11) implies that the Itô integral (6.10) is a martingale. One can prove the result without use of Itô's formula, by doing calculations of integrals with respect to Normal densities (see for example, Rogers and Williams (1987) p.36).

$\square$

Corollary 6.4 *Let $f(x, t)$ solve the backward equation*

$$L_t f(x, t) + \frac{\partial f}{\partial t}(x, t) = 0, \qquad (6.12)$$

and conditions of either of the two theorems above hold. Then $f(X(t), t)$ is a martingale.

Example 6.1: Let $X(t) = B(t)$, then $(Lf)(x) = \frac{1}{2}f''(x)$. Solutions to $Lf = 0$ are linear functions $f(x) = ax + b$. Hence $f(B(t)) = aB(t) + b$ is a martingale, which is also obvious from the fact that $B(t)$ is a martingale.

Example 6.2: Let $X(t) = B(t)$. The function $f(x,t) = e^{x-t/2}$ solves the backward equation

$$\frac{1}{2}\frac{\partial^2 f}{\partial x^2}(x,t) + \frac{\partial f}{\partial t}(x,t) = 0. \tag{6.13}$$

Therefore, by the above Corollary 6.4 we recover the exponential martingale of Brownian motion $e^{B(t)-t/2}$.

Corollary 6.5 (Dynkin's Formula) *Let $X(t)$ satisfy (6.1). If the conditions of either of the above theorems hold, then for any t, $0 \leq t \leq T$,*

$$\mathrm{E}f(X(t),t) = f(X(0),0) + \mathrm{E}\int_0^t \left(L_u f + \frac{\partial f}{\partial t}\right)(X(u),u)du. \tag{6.14}$$

The result is also true if t is replaced by a bounded stopping time τ, $0 \leq \tau \leq T$.

PROOF: The bounds on the growth of the function and its partial derivatives are used to establish integrability of $f(X(t),t)$ and other terms in (6.14). Since $M_f(t)$ is a martingale, the result follows by taking expectations. For bounded stopping times the result follows by the Optional Stopping Theorem, given in Chapter 7.

$$\square$$

Example 6.3:   We show that $J = \int_0^1 s dB(s)$ has a Normal $N(0, 1/3)$ distribution, by finding its moment generating function $m(u) = \mathrm{E}(e^{uJ})$. Consider the Itô integral $X(t) = \int_0^t s dB(s)$, $t \leq 1$, and notice that $J = X_1$. As $dX(t) = t dB(t)$, $X(t)$ is an Itô process with $\mu(x,t) = 0$ and $\sigma(x,t) = t$. Take $f(x,t) = f(x) = e^{ux}$. This function satisfies conditions of Theorem 6.3. It is easy to see that $L_t f(x,t) = \frac{1}{2}t^2 u^2 e^{ux}$, note that $\frac{\partial f}{\partial t} = 0$. Therefore by Dynkin's formula

$$\mathrm{E}\left(e^{uX(t)}\right) = 1 + \frac{1}{2}u^2 \int_0^t s^2 \mathrm{E}\left(e^{uX(s)}\right)ds.$$

Denote $h(t) = \mathrm{E}\left(e^{uX(t)}\right)$, then differentiation with respect to t leads to a simple equation

$$h'(t) = \frac{1}{2}u^2 t^2 h(t), \text{ with } h(0) = 1.$$

By separating variables, $\log h(t) = \frac{1}{2}u^2 \int_0^t s^2 ds = \frac{1}{2}u^2 \frac{t^3}{3}$. Thus $h(t) = e^{\frac{1}{2}u^2 \frac{t^3}{3}}$, which corresponds to the $N(0, \frac{t^3}{3})$ distribution. Thus $X(t) = \int_0^t s dB(s)$ has $N(0, \frac{t^3}{3})$ distribution, and the result follows.

Example 6.4:   We prove that $\int_0^1 B(t)dt$ has $N(0, 1/3)$ distribution, see also Example 3.6. Using integration by parts $\int_0^1 B(t)dt = B(1) - \int_0^1 t dB(t) = \int_0^1 dB(t) - \int_0^1 t dB(t) = \int_0^1 (1-t)dB(t)$, and the result follows.

6.2 Calculation of Expectations and PDEs

Results in this section provide a method for calculation of expectations of a function or a functional of a diffusion process on the boundary. This expectation can be computed by using a solution to the corresponding partial differential equation with a given boundary condition. This connection shows that solutions to PDEs can be represented as functions (functionals) of the corresponding diffusion.

Let $X(t)$ be a diffusion satisfying the SDE for $t > s \geq 0$,

$$dX(t) = \mu(X(t), t)dt + \sigma(X(t), t)dB(t), \text{ and } X(s) = x. \qquad (6.15)$$

Backward PDE and $E\big(g(X(T))|X(t) = x\big)$

We give results on $E\big(g(X(T))|X(t) = x\big)$. Observe first that $g(X(T))$ must be integrable $(E|g(X(T))| < \infty)$ for this to make sense. Of course, if g is bounded, then this is true. Observe next that by the Markov property of $X(t)$,

$$E\left(g(X(T))|X(t)\right) = E\left(g(X(T))|\mathcal{F}_t\right).$$

The latter is a martingale, by Theorem 2.31. The last ingredient is Itô's formula, which connects this to the PDE. Again, care is taken for Itô's formula to produce a martingale term, for which assumptions on the function and its derivatives are needed, see Theorem 6.3. Apart from these requirements, the results are elegant and easy to derive.

Theorem 6.6 *Let $f(x, t)$ solve the backward equation, with L_t given by (6.2),*

$$L_t f(x, t) + \frac{\partial f}{\partial t}(x, t) = 0, \text{ with } f(x, T) = g(x). \qquad (6.16)$$

If $f(x, t)$ satisfies the conditions of Corollary 6.4, then

$$f(x, t) = E\left(g(X(T))|X(t) = x\right). \qquad (6.17)$$

PROOF: By Corollary 6.4 $f(X(t), t)$, $s \leq t \leq T$, is a martingale. The martingale property gives

$$E\left(f(X(T), T)|\mathcal{F}_t\right) = f(X(t), t).$$

On the boundary $f(x, T) = g(x)$, so that $f(X(T), T) = g(X(T))$, and

$$f(X(t), t) = E\left(g(X(T))|\mathcal{F}_t\right).$$

By the Markov property of $X(t)$, $f(X(t), t) = E\left(g(X(T))|X(t)\right)$, and the result follows.

$\square$

It is tempting to show that the expectation $\mathrm{E}\,(g(X(T))|X(t) = x) = f(x,t)$ satisfies the backward PDE, thus establishing existence of its solutions. Assume for a moment that we can use Itô's formula with $f(X(t),t)$, then

$$f(X(t),t) = f(X(0),0) + \int_0^t \left(L_s f + \frac{\partial f}{\partial s}\right)(X(s),s)ds + M(t),$$

where $M(t)$ is a martingale. As noticed earlier, $f(X(t),t) = \mathrm{E}(g(X(T))|\mathcal{F}_t)$ is a martingale. It follows that $\int_t^T (L_s f + \frac{\partial f}{\partial s})(X(s),s)ds$ is a martingale as a difference of two martingales. Since the integral with respect to ds is a function of finite variation, and a martingale is not, the latter can only be true if the integral is zero, implying the backward equation (6.16).

To make this argument precise one needs to establish the validity of Itô's formula, i.e. smoothness of the conditional expectation. This can be seen by writing the expectation as an integral with respect to the density function,

$$f(x,t) = \mathrm{E}\,(g(X(T))|X(t) = x) = \int g(y)p(y,T,x,t)dy, \qquad (6.18)$$

where $p(y,T,x,t)$ is the transition probability density. So x is now in the transition function and the smoothness in x follows. For Brownian motion $p(y,T,x,t) = \frac{1}{\sqrt{T-t}}e^{-\frac{(y-x)^2}{2\sqrt{T-t}}}$ is differentiable in x (infinitely many times) and the result follows by differentiating under the integral. For other Gaussian diffusions the argument is similar. It is harder to show this in the general case (see Theorem 5.16).

Remark 6.1: Theorem 6.6 shows that any solution of the backward equation with the boundary condition given by $g(x)$ is given by the integral of g with respect to the transition probability density, equation (6.18), affirming that the transition probability density is a fundamental solution (see Definition 5.14).

A result similar to Theorem 6.6 is obtained when zero in the rhs of the backward equation is replaced by a known function $-\phi$.

Theorem 6.7 Let $f(x,t)$ solve

$$L_t f(x,t) + \frac{\partial f}{\partial t}(x,t) = -\phi(x), \text{ with } f(x,T) = g(x). \qquad (6.19)$$

Then

$$f(x,t) = \mathrm{E}\left(\left(g(X(T)) + \int_0^T \phi(X(s))ds\right)\Big|X(t) = x\right). \qquad (6.20)$$

The proof is similar to the above Theorem 6.6 and is left as an exercise.

Feynman-Kac formula

A result more general than Theorem 6.6 is given by the Feynman-Kac formula.

Theorem 6.8 (Feynman-Kac Formula) *For given bounded functions $r(x,t)$ and $g(x)$ let*

$$C(x,t) = E\left(e^{-\int_t^T r(X(u),u)du}g(X(T))\Big|X(t) = x\right). \qquad (6.21)$$

Assume that there is a solution to

$$\frac{\partial f}{\partial t}(x,t) + L_t f(x,t) = r(x,t)f(x,t), \text{ with } f(x,T) = g(x), \qquad (6.22)$$

Then the solution is unique and $C(x,t)$ is that solution.

PROOF: We give a sketch of the proof by using Itô's formula coupled with solutions of a linear SDE. Take a solution to (6.22) and apply Itô's formula

$$df(X(t),t) = \left(\frac{\partial f}{\partial t}(X(t),t) + L_t f(X(t),t)\right)dt + \frac{\partial f}{\partial x}(X(t),t)\sigma(X(t),t)dB(t).$$

The last term is a martingale term, so write it as $dM(t)$. Now use (6.22) to obtain

$$df(X(t),t) = r(X(t),t)f(X(t),t)dt + dM(t).$$

This is a linear SDE of Langevin type for $f(X(t),t)$ where $B(t)$ is replaced by $M(t)$. Integrating this SDE between t and T, and using $T \geq t$ as a time variable and t as the origin (see (5.32) in Section 5.22) we obtain

$$f(X(T),T) = f(X(t),t)e^{\int_t^T r(X(u),u)du} + e^{\int_t^T r(X(u),u)du}\int_t^T e^{\int_t^s r(X(u),u)du}dM(s).$$

But $f(X(T),T) = g(X(T))$, and rearranging, we obtain

$$g(X(T))e^{-\int_t^T r(X(u),u)du} = f(X(t),t) + \int_t^T e^{\int_t^s r(X(u),u)du}dM(s).$$

As the last term is an integral of a bounded function with respect to martingale, it is itself a martingale with zero mean. Taking expectation given $X(t) = x$, we obtain that $C(x,t) = f(x,t)$. For other proofs see for example, Friedman (1975) and Pinsky (1995).

□

Remark 6.2: The expression $e^{-r(T-t)}E\left(g(X(T))|X(t) = x\right)$ occurs in Finance as a discounted expected payoff, where r is a constant. The discounting results in the term rf in the rhs of the backward PDE.

Example 6.5: Give a probabilistic representation of the solution $f(x, t)$ of the PDE

$$\frac{1}{2}\sigma^2 x^2 \frac{\partial^2 f}{\partial x^2} + \mu x \frac{\partial f}{\partial x} + \frac{\partial f}{\partial t} = rf, \ 0 \le t \le T, \ f(x, T) = x^2, \qquad (6.23)$$

where σ, μ and r are positive constants. Solve this PDE using the solution of the corresponding stochastic differential equation.

The SDE corresponding to L is $dX(t) = \mu X(t)dt + \sigma X(t)dB(t)$. Its solution is $X(t) = X(0)e^{(\mu - \sigma^2/2)t + \sigma B(t)}$. By the Feynman-Kac formula

$$f(x, t) = E\left(e^{-r(T-t)} X^2(T) \Big| X(t) = x\right) = e^{-r(T-t)} E(X^2(T)|X(t) = x).$$

Using $X(T) = X(t)e^{(\mu - \sigma^2/2)(T-t) + \sigma(B(T) - B(t))}$, we obtain
$E(X^2(T)|X(t) = x) = x^2 e^{(2\mu + \sigma^2)(T-t)}$, giving $f(x, t) = x^2 e^{(2\mu + \sigma^2 - r)(T-t)}$.

The following result shows that $f(x, t) = E\big(g(X(T))|X(t) = x\big)$ satisfies the backward PDE, and can be found in Gihman and Skorohod (1972), Friedman (1975). (See also Theorem 6.5.3 Friedman (1975) for $C(x, t)$)

Theorem 6.9 (Kolmogorov's Equation) *Let $X(t)$ be a diffusion with generator L_t. Assume that the coefficients $\mu(x, t)$ and $\sigma(x, t)$ of L_t are locally Lipschitz and satisfy the linear growth condition (see (1.30)). Assume in addition that they possess continuous partial derivatives with respect to x up to order two, and that they have at most polynomial growth (see (1.31)). If $g(x)$ is twice continuously differentiable and satisfies together with its derivatives a polynomial growth condition, then the function $f(x, t) = E\big(g(X(T))|X(t) = x\big)$ satisfies*

$$\frac{\partial f}{\partial t}(x, t) + L_t f(x, t) = 0, \text{in the region } 0 \le t < T, x \in \mathbb{R}, \qquad (6.24)$$

with boundary condition $f(x, T) = \lim_{t \uparrow T} f(x, t) = g(x)$.

6.3   Time Homogeneous Diffusions

The case of time-independent coefficients in SDEs corresponds to the so-called time-homogeneous diffusions,

$$dX(t) = \mu(X(t))dt + \sigma(X(t))dB(t). \qquad (6.25)$$

Theorem 6.10 *Assume that there is a unique weak solution to (6.25). Then the transition probability function of the solution $P(y, t, x, s) = P(y, t - s, x, 0)$ depends only on $t - s$.*

PROOF: Denote by (X, B) the weak solution to 6.25. By the definition of the transition function $P(y, t, x, s) = P(X(t) \leq y | X(s) = x) = P(X_s^x(t) \leq y)$, where the process $X_s^x(t)$ satisfies $X_s^x(s) = x$ and for $t > 0$

$$X_s^x(s + t) = x + \int_s^{s+t} \mu(X_s^x(u))du + \int_s^{s+t} \sigma(X_s^x(u))dB(u). \qquad (6.26)$$

Let $Y(t) = X_s^x(s + t)$, and $B_1(t) = B(s + t) - B(s)$, $t \geq 0$. Then $B_1(t)$ is a Brownian motion and from the above equation, $Y(t)$ satisfies for $t \geq 0$

$$Y(t) = x + \int_0^t \mu(Y(v))dv + \int_0^t \sigma(Y(v))dB_1(v), \text{ and } Y(0) = x. \qquad (6.27)$$

Put $s = 0$ in (6.26) to obtain

$$X_0^x(t) = x + \int_0^t \mu(X_0^x(v))dv + \int_0^t \sigma(X_0^x(v))dB(v), \text{ and } X_0^x(0) = x. \quad (6.28)$$

Thus $Y(t)$ and $X_0^x(t)$ satisfy the same SDE. Hence $Y(t)$ and $X_0^x(t)$ have the same distribution. Therefore for $t > s$

$$
\begin{aligned}
P(y, t, x, s) &= P(X_s^x(t) \leq y) = P(Y(t - s) \leq y) \\
&= P(X_0^x(t - s) \leq y) = P(y, t - s, x, 0).
\end{aligned}
$$

$\square$

Since the transition function of a homogeneous diffusion depends on t and s only through $t - s$, it is denoted as

$$P(t, x, y) = P(y, t + s, x, s) = P(y, t, x, 0) = P(X(t) \leq y | X(0) = x), \quad (6.29)$$

and it gives the probability for the process to go from x to $(-\infty, y]$ during time t. Its density $p(t, x, y)$, when it exists, is the density of the conditional distribution of $X(t)$ given $X(0) = x$.

The generator L of a time-homogeneous diffusion is given by

$$Lf(x) = \frac{1}{2}\sigma^2(x)f''(x) + \mu(x)f'(x). \qquad (6.30)$$

Under appropriate conditions (conditions (A1) and (A2) of Theorem 5.15) $p(t, x, y)$ is the fundamental solution of the backward equation (5.61), which becomes in this case

$$\frac{\partial p}{\partial t} = Lp = \frac{1}{2}\sigma^2(x)\frac{\partial^2 p}{\partial x^2} + \mu(x)\frac{\partial p}{\partial x}. \qquad (6.31)$$

If moreover, $\sigma(x)$ and $\mu(x)$ have derivatives, $\sigma'(x)$, $\mu'(x)$, and $\sigma''(x)$, which are bounded and satisfy a Hölder condition, then $p(t, x, y)$ satisfies the forward equation in t and y for any fixed x, which becomes

$$\frac{\partial p}{\partial t}(t, x, y) = \frac{1}{2}\frac{\partial^2}{\partial y^2}\Big(\sigma^2(y)p(t, x, y)\Big) - \frac{\partial}{\partial y}\Big(\mu(y)p(t, x, y)\Big). \qquad (6.32)$$

In terms of the generator, the backward and the forward equations are written as

$$\frac{\partial p}{\partial t} = Lp, \quad \frac{\partial p}{\partial t} = L^*p, \qquad (6.33)$$

where

$$(L^*f)(y) = \frac{1}{2}\frac{\partial^2}{\partial y^2}\Big(\sigma^2(y)f(y)\Big) - \frac{\partial}{\partial y}\Big(\mu(y)f(y)\Big),$$

denotes the operator appearing in equation (6.32), and is known as the adjoint operator to L. (The adjoint operator is defined by the requirement that whenever the following integrals exist, $\int g(x)Lf(x)dx = \int f(x)L^*g(x)dx$.)

Example 6.6: The generator of Brownian motion $L = \frac{1}{2}\frac{d^2}{dx^2}$ is called the Laplacian. The backward equation for the transition probability density is

$$\frac{\partial p}{\partial t} = Lp = \frac{1}{2}\frac{\partial^2 p}{\partial x^2}. \qquad (6.34)$$

Since the distribution of $B(t)$ when $B(0) = x$ is $N(x, t)$, the transition probability density is given by the density of $N(x, t)$, and is the fundamental solution of PDE (6.34)

$$p(t, x, y) = \frac{1}{\sqrt{2\pi t}}e^{-\frac{(y-x)^2}{2t}}.$$

The adjoint operator L^* is the same as L, so that L is self-adjoint. A stronger result than Theorem 6.9 holds. It is possible to show, see for example, Karatzas and Shreve (1988) p.255, that if $\int e^{-ay^2}|g(y)|dy < \infty$ for some $a > 0$, then $f(x, t) = \mathrm{E}_x g(B(t))$ for $t < 1/2a$ satisfies the heat equation with initial condition $f(0, x) = g(x)$, $x \in \mathbf{R}$. A result of Widder 1944, states that any non-negative solution to the heat equation can be represented as $\int p(t, x, y)dF(y)$ for some non-decreasing function F.

Example 6.7: The Black-Scholes SDE

$$dX(t) = \mu X(t)dt + \sigma X(t)dB(t)$$

for constants μ and σ. The generator of this diffusion is

$$Lf(x) = \frac{1}{2}\sigma^2 x^2 f''(x) + \mu x f'(x). \qquad (6.35)$$

Its density is the fundamental solution of the PDE

$$\frac{\partial p}{\partial t} = \frac{1}{2}\sigma^2 x^2 \frac{\partial^2 p}{\partial x^2} + \mu x \frac{\partial p}{\partial x}.$$

The transition probability function of $X(t)$ was found in Example 5.14. Its density is $p(t, x, y) = \frac{\partial}{\partial y}\Phi\big(\frac{\ln(y/x) - (\mu - \sigma^2/2)(t-s)}{\sigma\sqrt{t-s}}\big)$.

Itô's Formula and Martingales

If $X(t)$ is a solution of (6.25) then Itô's formula takes the form: for any twice continuously differentiable $f(x)$

$$df(X(t)) = Lf(X(t))dt + f'(X(t))\sigma(X(t))dB(t). \qquad (6.36)$$

Theorem 6.2 and Theorem 6.3 for time homogeneous diffusions becomes

Theorem 6.11 *Let X be a solution to SDE (6.25) with coefficients satisfying conditions of Theorem 5.4, that is, $\mu(x)$ and $\sigma(x)$ are Lipschitz and satisfy the linear growth condition $|\mu(x)| + |\sigma(x)| \leq K(1 + |x|)$. If $f(x)$ is twice continuously differentiable in x with derivatives growing not faster than exponential satisfying condition (6.9), then the following process is a martingale.*

$$M_f(t) = f(X(t)) - \int_0^t Lf(X(u))du. \qquad (6.37)$$

Weak solutions to (6.25) are defined as solution to the martingale problem, by requiring existence of a filtered probability space, with an adapted process $X(t)$, so that

$$f(X(t)) - \int_0^t Lf(X(u))du \qquad (6.38)$$

is a martingale for any twice continuously differentiable f vanishing outside a finite interval, see Section 5.8). Equation (6.38) also allows us to identify generators.

Remark 6.3: The concept of generator is a central concept in studies of Markov processes. The generator of a time-homogeneous Markov process (not necessarily a diffusion process) is a linear operator defined by:

$$Lf(x) = \lim_{t \to 0} \frac{E\Big(f(X(t))|X_0 = x\Big) - f(x)}{t}. \qquad (6.39)$$

If the above limit exists we say that f is in the domain of the generator. If $X(t)$ solves (6.25) and f is bounded and twice continuously differentiable, then from (6.39) the generator for a diffusion is obtained. This can be seen by interchanging the limit and the expectation (dominated convergence), using Taylor's formula. Generators of pure jump processes, such as birth-death processes, are given later. For the theory of construction of Markov processes from their generators and their studies see for example, Dynkin (1965), Ethier and Kurtz (1986), Stroock and Varadhan (1979), Rogers and Williams (1990).

The result on existence and uniqueness of weak solutions (Theorem 5.11) becomes

Theorem 6.12 *If $\sigma(x)$ is positive and continuous and for any $T > 0$ there is K_T such that for all $x \in \mathbb{R}$*

$$|\mu(x)| + |\sigma(x)| \le K_T(1 + |x|) \tag{6.40}$$

then there exists a unique weak solution to SDE (6.25) starting at any point $x \in \mathbb{R}$, moreover the solution has the strong Markov property.

The following result is specific for one-dimensional homogeneous diffusions and does not carry over to higher dimensions.

Theorem 6.13 (Engelbert-Schmidt) *The SDE*

$$dX(t) = \sigma(X(t))dB(t)$$

has a weak solution for every initial value $X(0)$ if and only if for all $x \in \mathbb{R}$ the condition

$$\int_{-a}^{a} \frac{dy}{\sigma^2(x+y)} = \infty \quad \text{for all } a > 0$$

implies $\sigma(x) = 0$. The weak solution is unique if the above condition is equivalent to $\sigma(x) = 0$.

Corollary 6.14 *If $\sigma(x)$ is continuous (on $\mathbb{R}$) or bounded away from zero, then the above SDE has a unique weak solution.*

Example 6.8: By the above corollary Tanaka's SDE, Example 5.15,

$$dX(t) = \text{sign}(X(t))dB(t), \quad X(0) = 0,$$

has a unique weak solution.

6.4　Exit Times from an Interval

The main tool for studying various properties of diffusions is the result on exit times from an interval. Define $\tau_{(a,b)}$ to be the first time the diffusion exits (a, b), $\tau_{(a,b)} = \inf\{t > 0 : X(t) \notin (a, b)\}$. Since $X(t)$ is continuous, $X(\tau_{(a,b)}) = a$ or b. It was shown in Theorem 2.35 that τ is a stopping time, moreover, since the filtration is right-continuous, $\{\tau < t\}$ and $\{\tau \ge t\}$ are in $\mathcal{F}_t$ for all t. In this section results on $\tau_{(a,b)}$ are given. As the interval (a, b) is fixed, denote in this section $\tau_{(a,b)} = \tau$.

The fact that the process started in (a, b) remains in (a, b) for all $t < \tau$ allows to construct martingales, without additional assumptions on functions and coefficients. The following important result for analyzing diffusions, which is also known as Dynkin's formula, is established first. Introduce T_a and T_b as the hitting times of a and b, $T_a = \inf\{t > 0 : X(t) = a\}$, with the convention that the infimum of an empty set is infinity. Clearly, $\tau = \min(T_a, T_b) = T_a \wedge T_b$. The next result is instrumental for obtaining properties of τ.

Theorem 6.15 *Let $X(t)$ be a diffusion with a continuous $\sigma(x) > 0$ on $[a,b]$ and $X(0) = x$, $a < x < b$. Then for any twice continuously differentiable function $f(x)$ on $\mathbb{R}$ the following process is a martingale*

$$f(X(t \wedge \tau)) - \int_0^{t \wedge \tau} Lf(X(s))ds. \tag{6.41}$$

Consequently,

$$\mathbb{E}_x \left(f(X(t \wedge \tau)) - \int_0^{t \wedge \tau} Lf(X(s))ds \right) = f(x). \tag{6.42}$$

PROOF: Using Itô's formula and replacing t by $t \wedge \tau$

$$f(X(t \wedge \tau)) - \int_0^{t \wedge \tau} Lf(X(s))ds = f(x) + \int_0^{t \wedge \tau} f'(X(s))\sigma(X(s))dB(s). \tag{6.43}$$

Write the Itô integral as $\int_0^t I(s \leq \tau)f'(X(s))\sigma(X(s))dB(s)$. By Theorem 2.35 $\{\tau \geq s\}$ are in $\mathcal{F}_s$ for all s. Thus $I(s \leq \tau)$ is adapted. Now, for any $s \leq \tau$, $X(s) \in [a,b]$. Since $f'(x)\sigma(x)$ is continuous on $[a,b]$, it is bounded on $[a,b]$, say by K. Thus for any $s \leq t$, $|I(s \leq \tau)f'(X(s))\sigma(X(s))| \leq K$, and expectation $\mathbb{E}\int_0^t I(s \leq \tau)(f'(X(s))\sigma(X(s)))^2 ds < K^2 t$ is finite. Therefore the Itô integral $\int_0^t I(s \leq \tau)f'(X(s))\sigma(X(s))dB(s)$ is a martingale for $t \leq T$, and any T. Since it is a martingale it has a constant mean, and taking expectations in (6.43) formula (6.42) is obtained.

□

The next result establishes in particular that τ has a finite expectation, consequently it is finite with probability one.

Theorem 6.16 *Let $X(t)$ be a diffusion with generator L with continuous $\sigma(x) > 0$ on $[a,b]$ and $X(0) = x$, $a < x < b$. Then $\mathbb{E}_x(\tau) = v(x)$ satisfies the following differential equation*

$$Lv = -1, \tag{6.44}$$

with $v(a) = v(b) = 0$.

PROOF: Take $v(x)$ satisfying (6.44). By the previous Theorem 6.15

$$\mathbb{E}_x \left(v(X(t \wedge \tau)) - \int_0^{t \wedge \tau} Lv(X(s))ds \right) = v(x). \tag{6.45}$$

But $Lv = -1$, therefore

$$\mathbb{E}_x \left(v(X(t \wedge \tau)) \right) + \mathbb{E}_x(t \wedge \tau) = v(x), \tag{6.46}$$

and
$$E_x(t \wedge \tau) = v(x) - E_x\Big(v(X(t \wedge \tau))\Big). \tag{6.47}$$

$(t \wedge \tau)$ increases to τ as $t \to \infty$. Since the functions v and X are continuous, $v(X(t \wedge \tau)) \to v(X(\tau))$ as $t \to \infty$. $X(t \wedge \tau) \in (a, b)$ for any t and $v(x)$ is bounded on $[a, b]$, say by K, therefore $E_x(v(X(t \wedge \tau))) \le K$. It follows from the above equation (6.47) that $E_x(\tau) < \infty$, hence τ is almost surely finite, moreover by dominated convergence $E_x(v(X(t \wedge \tau))) \to E_x(v(X(\tau)))$. But $X(\tau) = a$ or b, so that $v(X(\tau)) = 0$. Thus from (6.47) $E_x(\tau) = v(x)$.

$\square$

The probability that the process reaches b before it reaches a, that is, $P_x(T_b < T_a)$ is used to obtain further properties. This probability is calculated with the help of the function $S(x)$, which is a solution to the equation

$$\frac{1}{2}\sigma^2(x)S''(x) + \mu(x)S'(x) = 0, \text{ or } LS = 0. \tag{6.48}$$

Any solution to (6.48) is called a *harmonic function* for L. Only positive harmonic functions are of interest, and ruling out constant solutions, it is easy to see (see Example 6.9 below) that for any such function

$$S'(x) = C \exp\left(-\int_{x_0}^{x} \frac{2\mu(s)}{\sigma^2(s)} ds\right). \tag{6.49}$$

$S'(x)$ is either positive for all x, if $C > 0$, or negative for all x if $C < 0$. Consequently, if S is not identically constant, $S(x)$ is monotone. Assume that $\sigma(x)$ is continuous and positive, and $\mu(x)$ is continuous. Then any L-harmonic function comes from the general solution to (6.48), which is given by

$$S(x) = \int^{x} \exp\left(-\int^{u} \frac{2\mu(y)}{\sigma^2(y)} dy\right) du, \tag{6.50}$$

and involves two undetermined constants.

Example 6.9: We show that harmonic functions for L are given by (6.50). S must solve

$$\frac{1}{2}\sigma^2(x)S''(x) + \mu(x)S'(x) = 0. \tag{6.51}$$

This equation leads to (with $h = S'$)

$$h'/h = -2\mu(x)/\sigma^2(x),$$

and provided $\mu(x)/\sigma^2(x)$ is integrable,

$$S'(x) = e^{-\int^{x} \frac{2\mu(y)}{\sigma^2(y)} dy}.$$

Integrating again we find $S(x)$.

Theorem 6.17 *Let $X(t)$ be a diffusion with generator L with continuous $\sigma(x) > 0$ on $[a, b]$. Let $X(0) = x$, $a < x < b$. Then*

$$P_x(T_b < T_a) = \frac{S(x) - S(a)}{S(b) - S(a)}, \qquad (6.52)$$

where $S(x)$ is given by (6.50).

PROOF: By Theorem 6.15

$$E_x\Big(S(X(t \wedge \tau)) - \int_0^{t \wedge \tau} LS(X(s))ds \Big) = S(x). \qquad (6.53)$$

But $LS = 0$, therefore

$$E_x\Big(S(X(t \wedge \tau)) \Big) = S(x). \qquad (6.54)$$

Since τ is finite it takes values T_b with probability $P_x(T_b < T_a)$ and T_a with the complimentary probability. It is not a bounded stopping time, but by taking limit as $t \to \infty$, we can assert by dominated convergence, that

$$ES(X(\tau)) = ES(X(0)) = S(x).$$

Expanding the expectation on the left and rearranging gives the result.

□

Remark 6.4: Note that the ratio in (6.52) remains the same no matter what non-constant solution $S(x)$ to the equation (6.50) is used.

Note that although the proof is given under the assumption of continuous drift $\mu(x)$, the result holds true for $\mu(x)$ bounded on finite intervals (see for example, Pinsky (1995)).

The above theorem has a number of far reaching corollaries.

Corollary 6.18 *Let $X(t)$ be a diffusion with zero drift on (a, b), $X(0) = x$, $a < x < b$. Then*

$$P_x(T_b < T_a) = \frac{x - a}{b - a}. \qquad (6.55)$$

PROOF: If $\mu(x) = 0$ on (a, b), then $S(x)$ is a linear function on (a, b) and the result follows from (6.52).

□

Diffusion $S(X(t))$ has zero drift by Itô's formula and (6.48). The exit probabilities from an interval are proportional to the distances from the end points. This explains why the function $S(x)$ is called the *scale* function. The diffusion $S(X(t))$ is said to be on the *natural scale*.

Example 6.10: We specify $P_x(T_b < T_a)$ for Brownian motion and Ornstein-Uhlenbeck processes. Brownian motion is in natural scale, since $\mu(x) = 0$. Thus $S(x) = x$, and $P_x(T_b < T_a) = (x - a)/(b - a)$.
For Ornstein-Uhlenbeck process with parameters $\mu(x) = -\alpha x$, $\sigma^2(x) = \sigma^2$,

$$S(x) = \int^x \exp\left(\frac{\alpha}{\sigma^2} y^2\right) dy.$$

Consequently

$$P_x(T_b < T_a) = \frac{\int_a^x \exp\left(\frac{\alpha}{\sigma^2} y^2\right) dy}{\int_a^b \exp\left(\frac{\alpha}{\sigma^2} y^2\right) dy}.$$

Under standard assumptions, there is a positive probability for the diffusion process to reach any point from any starting point.

Corollary 6.19 *Let $X(t)$ be a diffusion satisfying the assumptions of Theorem 6.17. Then for any $x, y \in (a, b)$*

$$P_x(T_y < \infty) > 0. \tag{6.56}$$

Indeed, for $x < y$, $T_y < T_b$, and $P_x(T_y < \infty) > P_x(T_b < \infty) > P_x(T_b < T_a) > 0$, and similarly for $y < x$.

As an application of the properties of the scale function, a better result for the existence and uniqueness of strong solutions is obtained. The transformation $Y(t) = S(X(t))$ results in a diffusion with no drift which allows us to waive assumptions on the drift.

Theorem 6.20 (Zvonkin) *Suppose that $\mu(x)$ is bounded and $\sigma(x)$ is Lipschitz and is bounded away from zero. Then the strong solution exists and is unique. In particular any SDE of the form*

$$dX(t) = \mu(X(t))dt + \sigma dB(t), \text{ and } X(0) = x_0, \tag{6.57}$$

with any bounded $\mu(x)$ and constant σ has a unique strong solution.

PROOF: If $Y(t) = S(X(t))$, then $dY(t) = \sigma(X(t))S'(X(t))dB(t)$. Notice that $S(x)$ is strictly increasing, therefore $X(t) = h(Y(t))$, where h is the inverse to S. Thus SDE for $Y(t)$ is $dY(t) = \sigma(h(Y(t)))S'(h(Y(t)))dB(t) = \sigma_Y(Y(t))dB(t)$. The rest of the proof consists in verification that $\sigma_Y(x) = \sigma(h(x))S'(h(x))$ is locally Lipschitz and, under the stated assumptions, satisfies conditions of the existence and uniqueness result.

□

6.5 Representation of Solutions of ODEs

Solutions to some PDEs have stochastic representations. Such representations are given by Theorems 6.6 and 6.8. Here we show that if a solution to an ODE satisfying a given boundary conditions exists, then it has a representation as the expectation of a diffusion process stopped at the boundary.

Theorem 6.21 *Let $X(t)$ be a diffusion with generator L with time-independent coefficients, $L = \frac{1}{2}\sigma^2(x)\frac{d^2}{dx^2} + \mu(x)\frac{d}{dx}$, continuous $\sigma(x) > 0$ on $[a, b]$, and $X(0) = x$, $a < x < b$. If f is twice continuously differentiable in (a, b) and continuous on $[a, b]$ and solves*

$$Lf = -\phi \text{ in } (a, b), \quad f(a) = g(a), f(b) = g(b) \tag{6.58}$$

for some bounded functions g and ϕ, then f has the representation

$$f(x) = \mathrm{E}_x\big(g(X(\tau))\big) + \mathrm{E}_x\Big(\int_0^\tau \phi(X(s))ds\Big), \tag{6.59}$$

where τ is the exit time from (a, b). In particular if $\phi \equiv 0$, the solution of (6.58) is given by

$$f(x) = \mathrm{E}_x\big(g(X(\tau))\big). \tag{6.60}$$

PROOF: The proof is immediate from Theorem 6.15. Indeed, by (6.42)

$$\mathrm{E}_x\Big(f(X(t \wedge \tau)) - \int_0^{t \wedge \tau} Lf(X(u))du\Big) = f(x).$$

Since τ is finite, by taking limits as $t \to \infty$ by dominated convergence

$$\mathrm{E}_x\big(f(X(\tau))\big) = f(x) + \mathrm{E}_x\Big(\int_0^\tau Lf(X(u))du\Big).$$

But $Lf(X(u)) = -\phi(X(u))$ for any $u < \tau$, and $X(\tau)$ is in the boundary $\{a, b\}$, where $f(x) = g(x)$, and the result follows.

$\square$

Example 6.11: Let $X = B$ be Brownian motion. Consider the solution of the problem

$$\frac{1}{2}f''(x) = 0 \text{ in } (a, b), f(a) = 0, f(b) = 1.$$

Here we solve this problem directly and verify the result of Theorem 6.21. Clearly, the solution is a linear function, and from boundary conditions it follows that it must be $f(x) = (x - a)/(b - a)$. By (6.60) this solution has the representation

$$
\begin{aligned}
f(x) &= \mathrm{E}_x\big(g(B(\tau))\big) \\
&= g(a)\mathrm{P}_x(T_a < T_b) + g(b)(1 - \mathrm{P}_x(T_a < T_b)) = \mathrm{P}_x(T_b < T_a).
\end{aligned}
$$

As we know, for Brownian motion, $\mathrm{P}_x(T_b < T_a) = (x - a)/(b - a)$, and the result is verified.

6.6 Explosion

Explosion refers to the situation when the process reaches infinite values in finite time. For example, the function $1/(1-t)$, $t < 1$, explodes at $t = 1$. Similarly, the solution $x(t)$ to the ordinary differential equation

$$dx(t) = (1 + x^2(t))dt, \quad x(0) = 0$$

explodes. Indeed, consider $x(t) = \tan(t)$, which approaches infinity as t approaches $\pi/2$. The time of explosion is $\pi/2$. Similar situation can occur with solutions to SDEs, except the time of explosion will be random. Solutions can be considered until the time of explosion.

Let diffusion $X(t)$ satisfy SDE on $\mathbb{R}$

$$dX(t) = \mu(X(t))dt + \sigma(X(t))dB(t), \text{ and } X(0) = x. \qquad (6.61)$$

Let $D_n = (-n, n)$ for $n = 1, 2, \ldots$. $\tau_n = \tau_{D_n}$ is the first time the process has absolute value n. Since a diffusion process is continuous, it must reach level n before it reaches level $n + 1$. Therefore τ_n are non-decreasing, hence they converge to a limit $\tau_\infty = \lim_{n \to \infty} \tau_n$. Explosion occurs on the set $\{\tau_\infty < \infty\}$, because on this set, by continuity of $X(t)$, $X(\tau_\infty) = \lim_{n \to \infty} X(\tau_n)$. Thus $|X(\tau_\infty)| = \lim_{n \to \infty} |X(\tau_n)| = \lim_{n \to \infty} n = \infty$, and infinity is reached in finite time on this set.

Definition 6.22 *Diffusion started from x explodes if* $\mathrm{P}_x(\tau_\infty < \infty) > 0$.

Note that under appropriate conditions on the coefficients, if diffusion explodes when started at some x_0, then it explodes when started at *any* $x \in \mathbb{R}$. Indeed, if for any x, y, $\mathrm{P}_x(T_y < \infty) > 0$ (see Corollary 6.19), then $\mathrm{P}_y(\tau_\infty < \infty) \geq \mathrm{P}_y(T_x < \infty)\mathrm{P}_x(\tau_\infty < \infty) > 0$.

The result below gives necessary and sufficient conditions for explosions. It is known as Feller's test for explosions.

Theorem 6.23 *Suppose $\mu(x)$, $\sigma(x)$ are bounded on finite intervals, and $\sigma(x) > 0$ and is continuous. Then the diffusion process explodes if and only if one of the two following conditions holds. There exists x_0 such that*

1. $\int_{-\infty}^{x_0} \exp\left(-\int_{x_0}^{x} \frac{2\mu(s)}{\sigma^2(s)}ds\right)\left(\int_{x}^{x_0} \frac{\exp\left(\int_{x_0}^{y} \frac{2\mu(s)}{\sigma^2(s)}ds\right)}{\sigma^2(y)}dy\right)dx < \infty.$

2. $\int_{x_0}^{\infty} \exp\left(-\int_{x_0}^{x} \frac{2\mu(s)}{\sigma^2(s)}ds\right)\left(\int_{x_0}^{x} \frac{\exp\left(\int_{x_0}^{y} \frac{2\mu(s)}{\sigma^2(s)}ds\right)}{\sigma^2(y)}dy\right)dx < \infty.$

The proof relies on the analysis of exit times τ_n with the aid of Theorem 6.16 and the Feynman-Kac formula (see for example, Gihman and Skorohod (1972), p.163, Pinsky (1995) p.213-214).

If the drift coefficient $\mu(x) \equiv 0$, then both conditions in the above theorem fail, since $\int_x^{x_0} \sigma^{-2}(y)dy \not\to 0$ as $x \to -\infty$, hence the following result.

Corollary 6.24 *SDEs of the form $dX(t) = \sigma(X(t))dB(t)$ do not explode.*

Example 6.12: Consider the SDE $dX(t) = cX^r(t)dt + dB(t)$, $c > 0$. Solutions of $dx(t) = cx^r(t)dt$ explode if and only if $r > 1$, see Exercise 6.11. Here $\sigma(x) = 1$ and $\mu(x) = cx^r$, $c > 0$, and $D = (\alpha, \beta) = (0, \infty)$. It is clear that this diffusion drifts to $+\infty$ due to the positive drift for any $r > 0$. However, explosion occurs only in the case of $r > 1$, that is, $P_x(\tau_D < \infty) > 0$ if $r > 1$, and $P_x(\tau_D < \infty) = 0$ if $r \leq 1$. The integral in part 2 of the above theorem is

$$\int_{x_0}^\infty \frac{\int_{x_0}^x \exp\left(\frac{2c}{r+1} y^{r+1}\right)dy}{\exp\left(\frac{2c}{r+1} x^{r+1}\right)} dx.$$

Using l'Hôpital rule, it can be seen that the function under the integral is of order x^{-r} as $x \to \infty$. Since $\int_{x_0}^\infty x^{-r}dx < \infty$ if and only if $r > 1$, the result is established (see Pinsky (1995)).

Example 6.13: Consider the SDE $dX(t) = X^2(t)dt + X^r(t)dB(t)$. Using the integral test, it can be seen that if $r < 3/2$, there is no explosion, and if $r > 3/2$ there is an explosion.

6.7 Recurrence and Transience

Let $X(t)$ be a diffusion on $\mathbb{R}$. There are various definitions of recurrence and transience in the literature, however, under the imposed assumptions on the coefficients, they are all equivalent.

Definition 6.25 *A point x is called recurrent for diffusion $X(t)$ if the probability of the process coming back to x infinitely often is one, that is,*

$$P_x(X(t) = x \text{ for a sequence of } t\text{'s increasing to infinity }) = 1.$$

Definition 6.26 *A point x is called transient for diffusion $X(t)$ if*

$$P_x(\lim_{t\to\infty} |X(t)| = \infty) = 1.$$

If all points of a diffusion are recurrent, the diffusion itself is called recurrent. If all points of a diffusion are transient, the diffusion itself is called transient.

Theorem 6.27 *Let $X(t)$ be a diffusion on $\mathbb{R}$ satisfying assumptions of the existence and uniqueness result Theorem 5.11, that is, $\mu(x)$ and $\sigma(x)$ are bounded on finite intervals, $\sigma(x)$ is continuous and positive and $\mu(x)$, $\sigma(x)$ satisfy the linear growth condition. Then*

1. *If there is one recurrent point then all points are recurrent.*

2. *If there are no recurrent points, then the diffusion is transient.*

To prove this result two fundamental properties of diffusions are used. The first is the strong Markov property, and the second is the *strong Feller* property, which states that for any bounded function $f(x)$, $\mathrm{E}_x f(X(t))$ is a continuous function in x for any $t > 0$. Both these properties hold under the stated conditions. It also can be seen that the recurrence is equivalent to the property: for any x, y $\mathrm{P}_x(T_y < \infty) = 1$, where T_y is the hitting time of y. By the above Theorem, transience is equivalent to the property: for any x, y $\mathrm{P}_x(T_y < \infty) <$ 1, see for example, Pinsky (1995). To decide whether $\mathrm{P}_x(T_y < \infty) < 1$, the formula (6.52) for the probability of exit from one end of an interval in terms of the scale function is used. If a diffusion does not explode, then the hitting time of infinity is defined as $T_\infty = \lim_{b\to\infty} T_b = \infty$ and $T_{-\infty} = \lim_{a\to-\infty} T_a = \infty$. Recall that $S(x) = \int_{x_0}^{x} \exp\left(-\int_{x_0}^{u} \frac{2\mu(s)}{\sigma^2(s)} ds\right) du$, and that by (6.52) $\mathrm{P}_x(T_b < T_a) = (S(x) - S(a))/(S(b) - S(a))$. Take any $y > x$, then

$$\mathrm{P}_x(T_y < \infty) = \lim_{a\to-\infty} \mathrm{P}_x(T_y < T_a) = \lim_{a\to-\infty} \frac{S(x) - S(a)}{S(y) - S(a)}. \tag{6.62}$$

Thus if $S(-\infty) = \lim_{a\to-\infty} S(a) = \infty$, then $\mathrm{P}_x(T_y < \infty) = 1$.
    Similarly, for $y < x$,

$$\mathrm{P}_x(T_y < \infty) = \lim_{b\to\infty} \mathrm{P}_x(T_y < T_b) = \lim_{b\to\infty} \frac{S(x) - S(b)}{S(y) - S(b)}.$$

Thus if $S(\infty) = \lim_{b\to\infty} S(b) = \infty$, then $\mathrm{P}_x(T_y < \infty) = 1$. Thus for any y, $\mathrm{P}_x(T_y < \infty) = 1$, which is the recurrence property.
    If one of the values $S(-\infty)$ or $S(\infty)$ is finite, then for some y, $\mathrm{P}_x(T_y < \infty) < 1$, which is the transience property. Thus the necessary and sufficient conditions for recurrence and transience are given by

Theorem 6.28 *Let operator $L = \frac{1}{2}\sigma^2(x)\frac{d^2}{dx^2} + \mu(x)\frac{d}{dx}$ have coefficients that satisfy assumptions of the above Theorem 6.27. Denote for a fixed x_0,*

$$I_1 = \int_{-\infty}^{x_0} \exp\left(-\int_{x_0}^{u} \frac{2\mu(s)}{\sigma^2(s)} ds\right) du \text{ and } I_2 = \int_{x_0}^{\infty} \exp\left(-\int_{x_0}^{u} \frac{2\mu(s)}{\sigma^2(s)} ds\right) du.$$

The diffusion corresponding to L is recurrent if and only if both I_1 and I_2 are infinite, and transient otherwise, that is, when one of I_1 or I_2 is finite.

6.8 Diffusion on an Interval

Consider diffusion on an interval (α, β), where one of the ends is finite. The main difference between this case and diffusion on the whole line, is that the finite end of the interval may be attained in finite time, the situation analogous to explosion in the former case. Take α finite, that is $-\infty < \alpha < \beta \le \infty$. The case of finite β is similar. Write the scale function in the form

$$S(x) = \int_{x_1}^{x} \exp(-\int_{x_0}^{u} \frac{2\mu(s)}{\sigma^2(s)} ds) du, \qquad (6.63)$$

where $x_0, x_1 \in (\alpha, \beta)$. By using stopping of $S(X(t))$, a martingale is obtained, and by the same argument as before the probabilities of exit from an interval $(a, b) \subset (\alpha, \beta)$ are given by (6.52)

$$P_x(T_a < T_b) = (S(b) - S(x))/(S(b) - S(a)).$$

If $S(\alpha) = -\infty$, then the above probability can be made arbitrarily small by taking $a \to \alpha$. This means that $P_x(T_\alpha < T_b) = 0$, and the boundary α is not attained before b for any b. If $S(\alpha) > -\infty$, then $P_x(T_\alpha < T_b) > 0$. A result similar to Theorem 6.28 holds.

Theorem 6.29 *Let $L_1 = \int_{\alpha}^{b} \exp\left(-\int_{b}^{u} \frac{2\mu(s)}{\sigma^2(s)} ds\right) du$. If $L_1 = \infty$ then the diffusion attains the point b before α, for any initial point $x \in (\alpha, b)$. If $L_1 < \infty$, then let $L_2 = \int_{\alpha}^{b} \frac{1}{\sigma^2(y)} \int_{\alpha}^{y} \exp\left(-\int_{b}^{x} \frac{2\mu(s)}{\sigma^2(s)} ds\right) \exp\left(\int_{b}^{y} \frac{2\mu(s)}{\sigma^2(s)} ds\right) dy$.*

1. *If $L_2 < \infty$ then for all $x \in (\alpha, b)$ the diffusion exits (α, b) in finite time, moreover $P_x(T_\alpha < \infty) > 0$.*

2. *If $L_2 = \infty$ then either, the exit time of (α, b) is infinite and $\lim_{t\to\infty} X(t) = \alpha$, or the exit time of (α, b) is finite and $P_x(T_b < T_\alpha) = 1$.*

Example 6.14: Consider a diffusion given by the SDE

$$dX(t) = n dt + 2\sqrt{X(t)} dB(t), \qquad (6.64)$$

where n is a positive integer. It will be seen later that $X(t)$ is the squared distance from the origin of the Brownian motion in n dimensions, see (6.82) Section 6.10. If T_0 is the first visit of zero, we show that if $n \ge 2$, then $P_x(T_0 = \infty) = 1$. This means that $X(t)$ never visits zero, that is $P(X(t) > 0$, for all $t \ge 0) = 1$. But for $n = 1$ $P_x(T_0 < \infty) = 1$. For $n = 2$ the scale function $S(x)$ is given by $S(x) = \ln x$, so that for any $b > 0$ $P_1(T_0 < T_b) = (S(1) - S(b))/(S(0) - S(b)) = 0$, hence $P_1(T_0 < \infty) = 0$. For $n \ge 3$, the scale function $S(x) = (1 - x^{-n/2+1})/(1 - n/2)$. Therefore $P_1(T_0 < \infty) = (S(1) - S(\infty))/(S(0) - S(\infty)) = 0$. Thus for any $n \ge 2$,

$P_1(T_0 = \infty) = 1$. Directly, $\alpha = 0$, by the above theorem $L_1 = \infty$, and the result follows. When $n = 1$ calculations show that $L_1 < \infty$ and also $L_2 < \infty$, thus $P_1(T_0 < \infty) > 0$.

Remark 6.5:   There is a classification of boundary points depending on the constants L_1, L_2 and L_3, where $L_3 = \int_\alpha^b \frac{1}{\sigma^2(y)} \exp\left(\int_{x_0}^y \frac{2\mu(s)}{\sigma^2(s)} ds \right) dy$. The boundary α is called

1. natural, if $L_1 = \infty$;

2. attracting, if $L_1 < \infty$, $L_2 = \infty$;

3. absorbing, if $L_1 < \infty$, $L_2 < \infty$, $L_3 = \infty$;

4. regular, if $L_1 < \infty$, $L_2 < \infty$, $L_3 < \infty$.

See for example, Gihman and Skorohod (1972), p.165.

6.9 Stationary Distributions

Consider the diffusion process given by the SDE

$$dX(t) = \mu(X(t))dt + \sigma(X(t))dB(t),$$

with $X(0)$ having a distribution $\nu(x) = P(X(0) \leq x)$. The distribution $\nu(x)$ is called *stationary* or *invariant* for the diffusion process $X(t)$ if for any t the distribution of $X(t)$ is the same as $\nu(x)$. If $P(t, x, y)$ denotes the transition probability function of the process $X(t)$, that is, $P(t, x, y) = P(X(t) \leq y | X(0) = x)$, then then an invariant $\nu(x)$ satisfies

$$\nu(y) = \int P(t, x, y) d\nu(x). \qquad (6.65)$$

To justify (6.65) use the total probability formula and the fact that the stationary distribution is the distribution of $X(t)$ for all t,

$$P(X_0 \leq y) = P(X_t \leq y) = \int P(X_t \leq y | X_0 = x) d\nu(x). \qquad (6.66)$$

If the stationary distribution has a density, $\pi(x) = d\nu(x)/dx$, then $\pi(x)$ is called a stationary or invariant density. If $p(t, x, y) = \partial P(t, x, y)/\partial y$ denotes the density of $P(t, x, y)$, then a stationary π satisfies

$$\pi(y) = \int p(t, x, y) \pi(x) dx. \qquad (6.67)$$

Under appropriate conditions on the coefficients (μ and σ are twice continuously differentiable with second derivatives satisfying a Hölder condition) an invariant density exists if and only if the following two conditions hold

1. $\int_{-\infty}^{x_0} \exp\left(-\int_{x_0}^{x} \frac{2\mu(s)}{\sigma^2(s)} ds\right) dx = \int_{x_0}^{\infty} \exp\left(-\int_{x_0}^{x} \frac{2\mu(s)}{\sigma^2(s)} ds\right) dx = \infty,$

2. $\int_{-\infty}^{\infty} \frac{1}{\sigma^2(x)} \exp\left(\int_{x_0}^{x} \frac{2\mu(s)}{\sigma^2(s)} ds\right) dx < \infty.$

Furthermore, if an invariant density is twice continuously differentiable, then it satisfies the ordinary differential equation

$$L^*\pi = 0, \text{ that is, } \frac{1}{2}\frac{\partial^2}{\partial y^2}\left(\sigma^2(y)\pi\right) - \frac{\partial}{\partial y}\left(\mu(y)\pi\right) = 0. \qquad (6.68)$$

Moreover, any solution of this equation with finite integral defines an invariant probability density. For rigorous proof see for example, Pinsky (1995), p.219 and p.181. To justify equation (6.68) heuristically, recall that under appropriate conditions the density of $X(t)$ satisfies the forward (Fokker-Plank) equation (5.62). If the system is in a stationary regime, its distribution does not change with time, which means that the derivative of the density with respect to t is zero, resulting in equation (6.68).

Equation (6.68) can be solved, as it can be reduced to a first order differential equation (see (1.34)). Using the integrating factor $\exp\left(-\int_a^x \frac{2\mu(y)}{\sigma^2(y)} dy\right)$, we find that the solution is given by

$$\pi(x) = \frac{C}{\sigma^2(x)} \exp\left(\int_{x_0}^{x} \frac{2\mu(y)}{\sigma^2(y)} dy\right), \qquad (6.69)$$

where C is found from $\int \pi(x) dx = 1$.

Example 6.15: For Brownian motion condition 1. above is true, but condition 2. fails. Thus no stationary distribution exists. The forward equation for the invariant distribution is

$$\frac{1}{2}\frac{\partial^2 p}{\partial x^2} = 0,$$

which has for its solutions linear functions of x and none of these has a finite integral.

Example 6.16: The forward equation for the Ornstein-Uhlenbeck process is

$$\frac{\partial p}{\partial t} = L^* p = \frac{1}{2}\sigma^2\frac{\partial^2 p}{\partial x^2} - \alpha\frac{\partial}{\partial x}(xp),$$

and the equation for the invariant density is

$$\frac{1}{2}\sigma^2\frac{\partial^2 p}{\partial x^2} - \alpha\frac{\partial}{\partial x}(xp) = 0.$$

The solution is given by

$$\pi(x) = \frac{C}{\sigma^2}\exp\left(\int_0^x \frac{-2\alpha}{\sigma^2} dy\right) = \frac{C}{\sigma^2}\exp\left(-\frac{\alpha}{\sigma^2}x^2\right). \qquad (6.70)$$

This shows that if α is negative, no stationary distribution exists, and if α is positive then the stationary density is Normal $N(0, \sigma^2/(2\alpha))$. The fact that $N(0, \sigma^2/(2\alpha))$ is a stationary distribution can be easily verified directly from representation (5.13).

Remark 6.6: The Ornstein-Uhlenbeck process has the following properties: it is a Gaussian process with continuous paths, it is Markov, and it is stationary, provided the initial distribution is the stationary distribution $N(0, \frac{\sigma^2}{2\alpha})$. Stationarity means that finite-dimensional distributions do not change with shift in time. For Gaussian processes stationarity is equivalent to the covariance function to be a function of $|t - s|$ only, i.e. $Cov(X(t), X(s)) = h(|t - s|)$ (see Exercise (6.3)). The Ornstein-Uhlenbeck process is *the only* process that is simultaneously Gaussian, Markov and stationary (see for example Breiman (1968), p.350).

Invariant Measures

A measure ν is called invariant for $X(t)$ if it satisfies the equation
$\nu(B) = \int_{-\infty}^{\infty} P(t, x, B) d\nu(x)$ for all intervals B. In equation (6.65) intervals of the form $B = (-\infty, y]$ were used. The general equation reduces to (6.65) if $C = \nu(\mathbb{R}) < \infty$. In this case ν can be normalized to a probability distribution. If $C = \infty$ this is impossible. Densities of invariant measures, when they exist and are smooth enough, satisfy equation (6.67). Conversely, any positive solution to (6.67) is a density of an invariant measure.

For Brownian motion $\pi(x) = 1$ is a solution of the equation (6.67). This is seen as follows. Since $p(t, x, y)$ is the density of the $N(x, t)$ distribution, $p(t, x, y) = \frac{1}{\sqrt{2\pi t}} \exp((y - x)^2/(2t))$. Note that for a fixed y, as a function of x, it is also the density of the $N(y, t)$ distribution. Therefore it integrates to unity, $\int_{\mathbb{R}} p(t, x, y) dx = 1$. Thus $\pi(x) = 1$ is a positive solution of the equation (6.67). In this case the density 1 corresponds to the Lebesgue measure, which is an invariant measure for Brownian motion. Since $\int_{\mathbb{R}} 1 dx = \infty$, it can not be normalized to a probability density. Note also that since the mean of the $N(y, t)$ distribution is y, we have

$$\int_{-\infty}^{\infty} x p(t, x, y) dx = y.$$

So that $\pi(x) = x$ is also a solution of the equation (6.67), but it is not a positive solution, and therefore is not a density of an invariant measure.

An interpretation of the invariant measure which is not a probability measure may be given by the density of a large number (infinite number) of particles with locations corresponding to the invariant measure, all diffusing according to the diffusion equation. Then at any time the density of the particles at any location will be preserved.

6.10 Multi-dimensional SDEs

We cover the concepts very briefly, relying on analogy with the one-dimensional case, but state the differences arising due to the increase in dimension. Let $\boldsymbol{X}(t)$ be a diffusion in n dimensions, described by the multi-dimensional SDE

$$d\boldsymbol{X}(t) = \boldsymbol{b}(\boldsymbol{X}(t), t)dt + \sigma(\boldsymbol{X}(t), t)d\boldsymbol{B}(t), \qquad (6.71)$$

where σ is $n \times d$ matrix valued function, $\boldsymbol{B}$ is d-dimensional Brownian motion, see section 4.7, $\boldsymbol{X}, \boldsymbol{b}$ are n-dimensional vector valued functions. In coordinate form this reads

$$dX_i(t) = b_i(\boldsymbol{X}(t), t)dt + \sum_{j=1}^{d} \sigma_{ij}(\boldsymbol{X}(t), t)dB_j(t), \quad i = 1, \ldots, n, \qquad (6.72)$$

and it means that for all $t > 0$ and $i = 1, \ldots, n$

$$X_i(t) = X_i(0) + \int_0^t b_i(\boldsymbol{X}(u), u)du + \sum_{j=1}^{d} \int_0^t \sigma_{ij}(\boldsymbol{X}(u), u)dB_j(u). \qquad (6.73)$$

The coefficients of the SDE are: the vector $\boldsymbol{b}(\boldsymbol{x}, t)$ and the matrix $\sigma(\boldsymbol{x}, t)$.

An existence and uniqueness result for strong solutions, under the assumption of locally Lipschitz coefficients holds in the same form, see Theorem 5.4, except for absolute values that should be replaced by the norms. The norm of the vector is its length, $|\boldsymbol{b}| = \sqrt{\sum_{i=1}^{n} b_i^2}$. The norm of the matrix σ is defined by $|\sigma|^2 = \text{trace}(\sigma \sigma^{Tr})$, with σ^{Tr} being the transposed of σ. The trace$(a) = \sum_{i=1}^{n} a_{ii}$. The matrix $a = \sigma \sigma^{Tr}$ is called the diffusion matrix.

Theorem 6.30 *If the coefficients are locally Lipschitz in $\boldsymbol{x}$ with a constant independent of t, that is, for every N, there is a constant K depending only on T and N such that for all $|\boldsymbol{x}|, |\boldsymbol{y}| \leq N$ and all $0 \leq t \leq T$*

$$|\boldsymbol{b}(\boldsymbol{x}, t) - \boldsymbol{b}(\boldsymbol{y}, t)| + |\sigma(\boldsymbol{x}, t) - \sigma(\boldsymbol{y}, t)| < K|\boldsymbol{x} - \boldsymbol{y}|, \qquad (6.74)$$

then for any given $\boldsymbol{X}(0)$ the strong solution to SDE (6.71) is unique. If in addition to condition (6.74) the linear growth condition holds

$$|\boldsymbol{b}(\boldsymbol{x}, t)| + |\sigma(\boldsymbol{x}, t)| \leq K_T(1 + |\boldsymbol{x}|),$$

$\boldsymbol{X}(0)$ is independent of $\boldsymbol{B}$, and $E|\boldsymbol{X}(0)|^2 < \infty$, then the strong solution exists and is unique on $[0, T]$, moreover

$$E\Big(\sup_{0 \leq t \leq T} |\boldsymbol{X}(t)|^2 \Big) < C\big(1 + E|\boldsymbol{X}(0)|^2\big), \qquad (6.75)$$

where constant C depends only on K and T.

Note that unlike in the one-dimensional case, the Lipschitz condition on σ can not be weakened in general to a Hölder condition, i.e. there is no Yamada-Watanabe-type result for multi-dimensional SDEs.

The quadratic covariation is easy to work out from (6.72), by taking into account that independent Brownian motions have zero quadratic covariation.

$$d[X_i, X_j](t) = dX_i(t)dX_j(t) = a_{ij}(\boldsymbol{X}(t), t)dt. \tag{6.76}$$

It can be shown that if $\boldsymbol{X}(t)$ is a solution to (6.71) then

$$\mathrm{E}\Big(X_i(t + \Delta) - x_i | \boldsymbol{X}(t) = \boldsymbol{x}\Big) = b_i(\boldsymbol{x}, t)\Delta + o(\Delta)$$

$$\mathrm{E}\Big((X_i(t + \Delta) - x_i)(X_j(t + \Delta) - x_j) | \boldsymbol{X}(t) = \boldsymbol{x}\Big) = a_{ij}(\boldsymbol{x}, t)\Delta + o(\Delta),$$

as $\Delta \to 0$. Thus $\boldsymbol{b}(\boldsymbol{x}, t)$ is the coefficient in the infinitesimal mean of the displacement from point $\boldsymbol{x}$ at time t, and $a(\boldsymbol{x}, t)$ is approximately the coefficient in the infinitesimal covariance of the displacement.

Weak solutions can be defined as solutions to the martingale problem. Let the operator L_t, acting on twice continuously differentiable functions from $\mathbb{R}^n$ to $\mathbb{R}$, be

$$L_t = \sum_{i=1}^{n} b_i(\boldsymbol{x}, t)\frac{\partial}{\partial x_i} + \frac{1}{2}\sum_{i=1}^{n}\sum_{j=1}^{n} a_{ij}(\boldsymbol{x}, t)\frac{\partial^2}{\partial x_i \partial x_j}. \tag{6.77}$$

Note that L_t depends on σ only through a. Then $\boldsymbol{X}(t)$ is a weak solution started at $\boldsymbol{x}$ at time s, if

$$f(\boldsymbol{X}(t)) - \int_s^t (L_u f)(\boldsymbol{X}(u))du \tag{6.78}$$

is a martingale for any twice continuously differentiable function f vanishing outside a compact set in $\mathbb{R}^n$. This process is called a diffusion with generator L_t. In the case of time-independent coefficients, the process is a time-homogeneous diffusion with generator L.

Theorem 6.31 *Assume that $a(\boldsymbol{x}, t)$ is continuous and satisfies condition (A)*

$$\text{(A)} \quad \sum_{i,j=1}^{n} a_{ij}(\boldsymbol{x}, t)v_i v_j > 0, \text{for all } \boldsymbol{x} \in \mathbb{R}^n \text{ and } \boldsymbol{v} \neq \boldsymbol{0}$$

and $\boldsymbol{b}(\boldsymbol{x}, t)$ is bounded on bounded sets. Then there exists a unique weak solution up to the time of explosion. If, in addition, the linear growth condition is satisfied, that is, for any $T > 0$ there is K_T such that for all $x \in \mathbb{R}$

$$|\boldsymbol{b}(\boldsymbol{x}, t)| + |a(\boldsymbol{x}, t)| \leq K_T(1 + |\boldsymbol{x}|), \tag{6.79}$$

then there exists a unique weak solution to the martingale problem (6.78) start-
ing at any point $x \in \mathbb{R}$ *at any time* $s \geq 0$, *moreover this solution has the strong*
Markov property.

Since the weak solution is defined in terms of the generator, which itself depends on σ only through a, the weak solution to (6.71) can be constructed
using a single Brownian motion provided the matrix a remains the same. If
a single SDE is equivalent to a number of SDEs, heuristically, it means that
there is as much randomness in a d-dimensional Brownian motion as there is
in a single Brownian motion. Replacement of a system of SDEs by a single
one is shown in detail for the Bessel process.

Note that the equation $\sigma\sigma^{Tr} = a$ has many solutions for σ, the matrix
square root is non-unique. However, if $a(\boldsymbol{x}, t)$ is non-negative definite for all $\boldsymbol{x}$
and t, and has for entries twice continuously differentiable functions of $\boldsymbol{x}$ and
t, then it has a locally Lipschitz square root $\sigma(\boldsymbol{x}, t)$ of the same dimension as
$a(\boldsymbol{x}, t)$ (see for example Friedman (1975) Theorem 6.1.2).

Bessel Process

Let $\boldsymbol{B}(t) = (B_1(t), B_2(t), \ldots, B_d(t))$ be the d-dimensional Brownian motion,
$d \geq 2$. Denote by $R(t)$ its squared distance from the origin, that is,

$$R(t) = \sum_{i=1}^{d} B_i^2(t). \tag{6.80}$$

The SDE for $R(t)$ is given by (using $d(B^2(t)) = 2B(t)dB(t) + dt$)

$$dR(t) = d\,dt + 2\sum_{i=1}^{d} B_i(t)dB_i(t). \tag{6.81}$$

In this case we have one equation driven by d independent Brownian motions.
Clearly, $b(x) = d$, $\sigma(x)$ is $(1 \times d)$ matrix $2(B_1(t), B_2(t), \ldots, B_d(t))$, so that
$a(X(t)) = \sigma(X(t))\sigma^{Tr}(X(t)) = 4\sum_{i=1}^{d} B_i^2(t) = 4R(t)$ is a scalar. Thus the
generator of $X(t)$ is given by $L = d\frac{d}{dx} + \frac{1}{2}(4x)\frac{d^2}{dx^2}$. But the same generator
corresponds to the process $X(t)$ satisfying the SDE below driven by a single
Brownian motion

$$dX(t) = d\,dt + 2\sqrt{X(t)}dB(t). \tag{6.82}$$

Therefore the squared distance process $R(t)$ in (6.80) satisfies SDE (6.82).
This SDE was considered in Example 6.14. The Bessel process is defined as
the distance from the origin, $Z(t) = \sqrt{\sum_{i=1}^{d} B_i^2(t)} = \sqrt{R(t)}$. Since $R(t)$ has
the same distribution as $X(t)$ given by (6.82), by Itô's formula, $Z(t)$ satisfies

$$dZ(t) = \frac{d-1}{2Z(t)}dt + dB(t). \tag{6.83}$$

Using the one-dimensional SDE (6.82) for $R(t)$ we can decide on the recurrence, transience and attainability of $\mathbf{0}$ of Brownian motion in dimensions 2 and higher. It follows from Example 6.14 that in one and two dimensions Brownian motion is recurrent, but in dimensions three and higher it is transient. It was also shown there that in dimension two and above Brownian motion never visits zero. See also Karatzas and Shreve (1988), p.161-163.

Itô's Formula, Dynkin's Formula

Let $\mathbf{X}(t) = (X_1(t), \ldots, X_n(t))$ be a diffusion in $\mathbb{R}^n$ with generator L, (the general case is similar, but in what follows time-homogeneous case will be considered). Let $f : \mathbb{R}^n \to \mathbb{R}$ be a twice continuously differentiable (C^2) function. Then Itô's formula states that

$$df(\mathbf{X}(t)) = \sum_{i=1}^{n} \frac{\partial f}{\partial x_i}(\mathbf{X}(t))dX_i(t) + \frac{1}{2}\sum_{i=1}^{n}\sum_{j=1}^{n} a_{ij}(\mathbf{X}(t))\frac{\partial^2 f}{\partial x_i \partial x_j}(\mathbf{X}(t))dt.$$

(6.84)

It can be regarded as a Taylor's formula expansion where

$$dX_i dX_j \equiv d[X_i, X_j] \quad \text{and} \quad dX_i dX_j dX_k \equiv 0.$$

Itô's formula can be written with the help of the generator as

$$df(\mathbf{X}(t)) = (Lf)(\mathbf{X}(t)) + \sum_{i=1}^{n}\sum_{j=1}^{d} \frac{\partial f}{\partial x_i}(\mathbf{X}(t))\sigma_{ij}(\mathbf{X}(t))dB_j(t). \quad (6.85)$$

The analogues of Theorems 6.3 and 6.15 hold. It is clear from the above (6.85) that if partial derivatives of f are bounded, and $\sigma(\mathbf{x})$ is bounded, then

$$f(\mathbf{X}(t)) - \int_0^t Lf(\mathbf{X}(u))du \quad (6.86)$$

is a martingale. (Without the assumption of functions being bounded , it is a local martingale).

Theorem 6.32 *Suppose that the assumptions of Theorem 6.31 hold. Let $D \subset \mathbb{R}^n$ be a bounded domain (an open and simply connected set) in $\mathbb{R}^n$. Let $\mathbf{X}(0) = \mathbf{x}$ and denote by τ the exit time from D, $\tau_D = \inf\{t > 0 : \mathbf{X}(t) \in \partial D\}$. Then for any twice continuously differentiable f,*

$$f(\mathbf{X}(t \wedge \tau_D)) - \int_0^{t \wedge \tau_D} Lf(\mathbf{X}(u))du \quad (6.87)$$

is a martingale.

It can be shown that under conditions of the above theorem

$$\sup_{x \in D} E_x(\tau_D) < \infty. \tag{6.88}$$

As a corollary the following is obtained

Theorem 6.33 *Suppose that the assumptions of Theorem 6.31 hold. If f is twice continuously differentiable in D, continuous on ∂D, and solves*

$$Lf = -\phi \text{ in } D \text{ and } f = g \text{ on } \partial D. \tag{6.89}$$

for some bounded functions g and ϕ. Then $f(x)$, $x \in D$, has representation

$$f(x) = E_x\big(g(X(\tau_D))\big) + E_x\Big(\int_0^{\tau_D} \phi(X(s))ds\Big). \tag{6.90}$$

In particular if $\phi \equiv 0$, solution has representation as

$$f(x) = E_x\big(g(X(\tau_D))\big).$$

The proof is exactly the same as for Theorem 6.21 in one dimension.

Definition 6.34 *A function $f(x)$ is said to be L-harmonic on D if it is twice continuously differentiable on D and $Lf(x) = 0$ for $x \in D$.*

The following result follows from (6.86).

Corollary 6.35 *For any bounded L-harmonic function on D with bounded derivatives, $f(X(t \wedge \tau_D))$ is a martingale.*

Example 6.17: Denote by $\Delta = \frac{1}{2}\sum_{i=1}^{3} \frac{\partial^2}{\partial x_i^2}$ the three-dimensional Laplacian. This operator is the generator of three-dimensional Brownian motion $B(t)$, $L = \Delta$. Let $D = \{x : |x| > r\}$. Then $f(x) = 1/|x|$ is harmonic on D, that is $Lf(x) = 0$ for all $x \in D$. To see this perform differentiation and verify that $\sum_{i=1}^{3} \frac{\partial^2}{\partial x_i^2}\Big(\frac{1}{\sqrt{x_1^2+x_2^2+x_3^2}}\Big) = 0$ at any point $x \neq 0$. Note that in one dimension all harmonic functions for the Laplacian are linear functions, whereas in higher dimensions there are many more. It is easy to see that $1/|x|$ and its derivatives are bounded on D, consequently if $B(0) = x \neq 0$, then $1/|B(t \wedge \tau_D)|$ is a martingale.

The Backward (Kolmogorov's) equation in higher dimensions is the same as in one dimension, with the obvious replacement of the state variable $x \in \mathbb{R}^n$. We have seen that solutions to the backward equation can be expressed by means of diffusion, Theorem 6.33. However, it is a formidable task to prove that if $X(t)$ is a diffusion, and $g(x)$ is a smooth function on D then $f(x) = E_x\big(gX(t)\big)$ solves the backward equation (see for example, Friedman (1975)).

Theorem 6.36 *Let $g(x)$ be a function with two continuous derivatives satisfying a polynomial growth condition, that is, the function and its derivatives in absolute value do not exceed $K(1+|x|^m)$ for some constants $K, m > 0$. Let $X(t)$ satisfy (6.71). Assume that coefficients $b(x,t)$, $\sigma(x,t)$ are Lipschitz in x uniformly in t, satisfy the linear growth condition, and their two derivatives satisfy a polynomial growth condition. Let*

$$f(x,t) = \mathrm{E}_x\big(g(X(T))|X(t) = x\big) \tag{6.91}$$

Then f has continuous derivatives in x, which can be computed by differentiating (6.91) under the expectation sign. Moreover f has a continuous derivative in t, and solves the backward PDE

$$
\begin{aligned}
L_t f + \frac{\partial f}{\partial t} &= 0, \text{ in } \mathbb{R}^n \times [0, T) \\
f(x, T) &\to g(x), \text{ as } t \uparrow T.
\end{aligned}
\tag{6.92}
$$

The fundamental solution of (6.92) gives the transition probability function of the diffusion (6.71).

Remark 6.7: (Diffusions on manifolds)
The PDEs above can also be considered when the state variable x belongs to a manifold, rather than $\mathbb{R}^n$. The fundamental solution then corresponds to the diffusion on the manifold and represents the way heat propagates on that manifold. It turns out that various geometric properties of the manifold can be obtained from the properties of the fundamental solution, Molchanov (1975).

Remark 6.8: The Feynman-Kac formula holds also in the multi-dimensional case in the same way as in one dimension. If 0 in the right hand side of the PDE (6.92) is replaced by rf, for a bounded function r, then the solution satisfying a given boundary condition $f(x, T) = g(x)$ has a representation

$$f(x,t) = \mathrm{E}\left(e^{-\int_t^T r(X(u),u)du} g(X(T))|X(t) = x\right).$$

See Karatzas and Shreve (1988).

Recurrence, Transience and Stationary Distributions

Properties of recurrence and transience of multi-dimensional diffusions, solutions to (6.71), are defined similarly to the one-dimensional case. However, in higher dimensions a diffusion $X(t)$ is recurrent if for any starting point $x \in \mathbb{R}^n$ the process will visit a ball around any point $y \in \mathbb{R}$ of radius ϵ, $D_\epsilon(y)$, however small, with probability one.

$$\mathrm{P}_x(X(t) \in D_\epsilon(y) \text{ for a sequence of } t\text{'s increasing to infinity }) = 1.$$

A diffusion $X(t)$ on $\mathbb{R}^n$ is transient if for any starting point $x \in \mathbb{R}^n$ the process will leave any ball, however large, never to return. It follows by a diffusion analysis of the squared lengths of the multi-dimensional Brownian motion (see Example 6.14) that in dimensions one and two Brownian motion is recurrent, but it is transient in dimensions three and higher.

For time-homogeneous diffusions under conditions of Theorem 6.31 on the coefficients, recurrence is equivalent to the property that the process started at any point x hits the closed ball $\bar{D}_\epsilon(y)$ around any point y in finite time. Under these conditions, there is a dichotomy, a diffusion is either transient or recurrent. Invariant measures are defined in exactly the same way as in one dimension. Stationary distributions are finite invariant measures; they may exist only if a diffusion is recurrent. A diffusion is recurrent and admits a stationary distribution if and only if the expected hitting time of $\bar{D}_\epsilon(y)$ from x is finite. When this property holds diffusion is also called *ergodic* or *positive recurrent*.

In general there are no necessary and sufficient conditions for recurrence and ergodicity for multi-dimensional diffusions, however there are various tests for these properties. The method of Lyapunov functions, developed by R.Z. Khasminskii, consists of finding a suitable function f, such that $Lf \leq 0$ outside a ball around zero. If $\lim_{|x| \to \infty} f(x) = \infty$, then the process is transient. If f is ultimately decreasing, then the process is recurrent. If $Lf \leq -\epsilon$ for some $\epsilon > 0$ outside a ball around zero, with $f(x)$ bounded from below in that domain, then a diffusion is positive recurrent. Proofs consist of an application of Itô's formula coupled with the martingale theory (convergence property of supermartingales). See for details Bhattacharya (1978), Hasminskii (1980), Pinsky (1995).

Higher Order Random Differential Equations

Similarly to ODEs higher order random differential equations have interpretations as multi-dimensional SDEs. For example, a second order random differential equation of the form

$$\ddot{x} + h(x, \dot{x}) = \dot{B}, \tag{6.93}$$

where $\dot{x}(t) = dx(t)/dt$, $\ddot{x}(t) = d^2x(t)/dt^2$, and $\dot{B}$ denotes the White noise, has interpretation as the following two-dimensional SDE by letting

$$
\begin{aligned}
x_1(t) &= x(t), & (6.94) \\
x_2(t) &= dx_1(t)/dt. & (6.95)
\end{aligned}
$$

$$
\begin{aligned}
dX_1(t) &= X_2(t)dt, & (6.96) \\
dX_2(t) &= -h(X_1(t), X_2(t)) + dB(t). & (6.97)
\end{aligned}
$$

Such equations are considered in Section 14.2 of Chapter 14.

Higher n-th order random equations are interpreted in a similar way: by letting $X_1(t) = X(t)$ and $dX_i(t) = X_{i+1}(t)dt$, $i = 1, \ldots, n - 1$, an n-dimensional SDE is obtained.

Notes. Most of the material can be found in Friedman (1975), Gihman and Skorohod (1982), Stroock and Varadhan (1979), Karatzas and Shreve (1988), Rogers and Williams (1990), Pinsky (1995).

6.11   Exercises

Exercise 6.1: Show that for any u, $f(x, t) = \exp(ux - u^2t/2)$ solves the backward equation for Brownian motion. Take derivatives, first, second, etc., of $\exp(ux - u^2t/2)$ with respect to u, and set $u = 0$, to obtain that functions x, $x^2 - t$, $x^3 - 3tx$, $x^4 - 6tx^2 + 3t^2$, etc. also solve the backward equation (6.13). Deduce that $B^2(t) - t$, $B(t)^3 - 3tB(t)$, $B^4(t) - 6tB^2(t) + 3t^2$ are martingales.

Exercise 6.2: Find the generator for the Ornstein-Uhlenbeck process, write the backward equation and give its fundamental solution. Verify that it satisfies the forward equation.

Exercise 6.3: Let $X(t)$ be a stationary process. Show that the covariance function $\gamma(s, t) = Cov(X(s), X(t))$ is a function of $|t - s|$ only. Hint: take $k = 2$. Deduce that for Gaussian processes stationarity is equivalent to the requirements that the mean function is a constant and the covariance function is a function of $|t - s|$.

Exercise 6.4: $X(t)$ is a diffusion with coefficients $\mu(x) = cx$ and $\sigma(x) = 1$. Give its generator and show that $X^2(t) - 2c \int_0^t X^2(s)ds - t$ is a martingale.

Exercise 6.5: $X(t)$ is a diffusion with $\mu(x) = 2x$ and $\sigma^2(x) = 4x$. Give its generator L. Solve $Lf = 0$, and give a martingale M_f. Find the SDE for the process $Y(t) = \sqrt{X(t)}$, and give the generator of $Y(t)$.

Exercise 6.6: Find $f(x)$ such that $f(B(t) + t)$ is a martingale.

Exercise 6.7: $X(t)$ is a diffusion with coefficients $\mu(x, t), \sigma(x, t)$. Find a differential equation for $f(x, t)$ such that $Y(t) = f(X(t), t)$ has infinitesimal diffusion coefficient equal to 1.

Exercise 6.8: Show that the mean exit time of a diffusion from an interval, which (by Theorem 6.16) satisfies the ODE (6.44) is given by

$$v(x) = - \int_a^x 2G(y) \int_a^y \frac{ds}{\sigma^2(s)G(s)}dy + \int_a^b 2G(y) \int_a^y \frac{ds}{\sigma^2(s)G(s)}dy \frac{\int_a^x G(s)ds}{\int_a^b G(s)ds},$$
$$(6.98)$$

where $G(x) = \exp\left(-\int_a^x \frac{2\mu(s)}{\sigma^2(s)}ds\right)$.

Exercise 6.9: Find $P_x(T_b < T_a)$ for Brownian motion with drift when $\mu(x) = \mu$ and $\sigma^2(x) = \sigma^2$.

Exercise 6.10: Give a probabilistic representation of the solution $f(x,t)$ of the PDE

$$\frac{1}{2}\frac{\partial^2 f}{\partial x^2} + \frac{\partial f}{\partial t} = 0, \ \ 0 \le t \le T, \ \ f(x,T) = x^2.$$

Solve this PDE using the solution of the corresponding stochastic differential equation.

Exercise 6.11: Show that the solution of the following ordinary differential equation $dx(t) = cx^r(t)dt$, $c > 0$, $x(0) = x_0 > 0$, explodes if and only if $r > 1$.

Exercise 6.12: Investigate for explosions the following process

$$dX(t) = X^2(t)dt + \sigma X^\alpha(t)dB(t).$$

Exercise 6.13: Show that Brownian motion $B(t)$ is recurrent. Show that $B(t) + t$ is transient.

Exercise 6.14: Show that the Ornstein-Uhlenbeck process is positively recurrent. Show that the limiting distribution for the Ornstein-Uhlenbeck process (5.6) exists, and is given by its stationary distribution. Hint: the distribution of $\sigma e^{-\alpha t}\int_0^t e^{\alpha s}dB_s$ is Normal, find its mean and variance, and take limits.

Exercise 6.15: Show that the square of the Bessel process $X(t)$ in (6.64) comes arbitrarily close to zero when $n = 2$, that is, $P(T_y < \infty) = 1$ for any small $y > 0$, but when $n \ge 3$, $P(T_y < \infty) < 1$.

Exercise 6.16: Let diffusion $X(t)$ have $\sigma(x) = 1$, $\mu(x) = -1$ for $x < 0$, $\mu(x) = 1$ for $x > 0$ and $\mu(0) = 0$. Show that $\pi(x) = e^{-|x|}$ is a stationary distribution for X.

Exercise 6.17: Let diffusion on (α, β) be such that the transition probability density $p(t, x, y)$ is symmetric in x and y, $p(t, x, y) = p(t, y, x)$ for all x, y and t. Show that if (α, β) is a finite interval, then the uniform distribution is invariant for the process $X(t)$.

Exercise 6.18: Investigate for absorption at zero the following process (used as a model for interst rates, the square root model of Cox, Ingersoll and Ross). $dX(t) = b(a - X(t))dt + \sigma\sqrt{X(t)}dB(t)$, where parameters b, a and σ are constants.

Chapter 7

Martingales

Martingales play a central role in the modern theory of stochastic processes and stochastic calculus. Martingales constructed from a Brownian motion were considered in Section 3.3 and martingales arising in diffusions in Section 6.1. Martingales have a constant expectation, which remains the same under random stopping. Martingales converge almost surely. Stochastic integrals are martingales. These are the most important properties of martingales, which hold under some conditions.

7.1 Definitions

The main ingredient in the definition of a martingale is the concept of conditional expectation, consult Chapter 2 for its definition and properties.

Definition 7.1 *A stochastic process $M(t)$, where time t is continuous $0 \leq t \leq T$, or discrete $t = 0, 1, \ldots, T$, adapted to a filtration $\mathbb{F} = (\mathcal{F}_t)$ is a martingale if for any t, $M(t)$ is integrable, that is, $\mathrm{E}|M(t)| < \infty$ and for any t and s with $0 \leq s < t \leq T$,*

$$\mathrm{E}(M(t)|\mathcal{F}_s) = M(s) \quad a.s. \tag{7.1}$$

$M(t)$ is a martingale on $[0, \infty)$ if it is integrable and the martingale property (7.1) holds for any $0 \leq s < t < \infty$.

Definition 7.2 *A stochastic process $X(t)$, $t \geq 0$ adapted to a filtration $\mathbb{F}$ is a supermartingale (submartingale) if it is integrable, and for any t and s, $0 \leq s < t \leq T$*

$$\mathrm{E}(X(t)|\mathcal{F}_s) \leq X(s), \quad \left(\mathrm{E}(X(t)|\mathcal{F}_s) \geq X(s)\right) \quad a.s.$$

If $X(t)$ is a supermartingale, then $-X(t)$ is a submartingale. The mean of a supermartingale is non-increasing with t, the mean of a submartingale is non-decreasing in t, and the mean of a martingale is constant in t. This property is used in a test for a super(sub)martingale to be a true martingale.

Theorem 7.3 *A supermartingale $M(t)$, $0 \leq t \leq T$, is a martingale if and only if* $\mathrm{E}M(T) = \mathrm{E}M(0)$.

PROOF: If M is a martingale, then $\mathrm{E}M(T) = \mathrm{E}M(0)$ follows by the martingale property with $s = 0$ and $t = T$. Conversely, suppose $M(t)$ is a supermartingale and $\mathrm{E}M(T) = \mathrm{E}M(0)$. If for some t and s we have a strict inequality, $\mathrm{E}(M(t)|\mathcal{F}_s) < M(s)$ on a set of positive probability, then by taking expectations, we obtain $\mathrm{E}M(t) < \mathrm{E}M(s)$. Since the expectation of a supermartingale is non-increasing, $\mathrm{E}M(T) \leq \mathrm{E}M(t) < \mathrm{E}M(s) \leq \mathrm{E}M(0)$. But this contradicts the condition of the theorem $\mathrm{E}M(T) = \mathrm{E}M(0)$. Thus for all t and s the inequality $\mathrm{E}(M(t)|\mathcal{F}_s) \leq M(s)$ must be an equality almost surely.
□

We refer to Theorem 2.32 on the existence of the regular right-continuous version for supermartingales. Regular right-continuous versions of processes will be taken.

Square Integrable Martingales

A special role in the theory of integration is played by square integrable martingales.

Definition 7.4 *A random variable X is square integrable if $\mathrm{E}(X^2) < \infty$. A process $X(t)$ on the time interval $[0,T]$, where T can be infinite, is square integrable if $\sup_{t \in [0,T]} \mathrm{E}X^2(t) < \infty$ ($\sup_{t \geq 0} \mathrm{E}X^2(t) < \infty$), i.e. second moments are bounded.*

Example 7.1:

1. Brownian motion $B(t)$ on a finite time interval $0 \leq t \leq T$ is a square integrable martingale, since $\mathrm{E}B^2(t) = t < T < \infty$. Similarly, $B^2(t) - t$ is a square integrable martingale. They are not square integrable when $T = \infty$.

2. If $f(x)$ is bounded and continuous function on $\mathbf{R}$, then Itô integrals $\int_0^t f(B(s))dB(s)$ and $\int_0^t f(s)dB(s)$ are square integrable martingales on any finite time interval $0 \leq t \leq T$. Indeed, by (4.7), an Itô integral is a martingale, and since $|f(x)| \leq K$,

$$\mathrm{E}\left(\int_0^t f(B(s))dB(s)\right)^2 = \mathrm{E}\left(\int_0^t f^2(B(s))ds\right) \leq K^2 t \leq K^2 T < \infty.$$

If moreover, $\int_0^\infty f^2(s)ds < \infty$ then $\int_0^t f(s)dB(s)$ is a square integrable martingale on $[0,\infty)$.

7.2 Uniform Integrability

To appreciate the definition of uniform integrability of a process, recall what is meant by integrability of a random variable X. It is called integrable if $E|X| < \infty$. It is easy to see that this holds if and only if

$$\lim_{n \to \infty} E\big(|X|I(|X| > n)\big) = 0. \tag{7.2}$$

Indeed, if X is integrable then (7.2) holds by the dominated convergence, since $\lim_{n \to \infty} |X|I(|X| > n) = 0$ and $|X|I(|X| > n) \leq |X|$. Conversely, let n be large enough for the rhs in (7.2) to be finite. Then

$$E|X| = E\big(|X|I(|X| > n)\big) + E\big(|X|I(|X| \leq n)\big) < \infty,$$

since the first term is finite by (7.2) and the second is bounded by n.

Definition 7.5 *A process $X(t)$, $0 \leq t \leq T$ is called uniformly integrable if $E\big(|X(t)|I(|X(t)| > n)\big)$ converges to zero as $n \to \infty$ uniformly in t, that is,*

$$\lim_{n \to \infty} \sup_t E\big(|X(t)|I(|X(t)| > n)\big) = 0, \tag{7.3}$$

where the supremum is over $[0, T]$ in the case of a finite time interval and $[0, \infty)$ if the process is considered on $0 \leq t < \infty$.

Example 7.2: We show that if $X(t)$, $0 \leq t \leq T$ is uniformly integrable, then it is integrable, that is, $\sup_t E|X(t)| < \infty$. Indeed,

$$\sup_t E|X(t)| < \sup_t E\big(|X(t)|I(|X(t)| > n)\big) + n.$$

Since $X(t)$ is uniformly integrable, the first term converges to zero as $n \to \infty$, in particular it is bounded, and the result follows.

Sufficient conditions for uniform integrability are given next.

Theorem 7.6 *If the process X is dominated by an integrable random variable, $|X(t)| \leq Y$ and $E(Y) < \infty$, then it is uniformly integrable. In particular, if $E(\sup_t |X(t)|) < \infty$, then it is uniformly integrable.*

PROOF: $E\big(|X(t)|I(|X(t)| > n)\big) < E\big(|Y|I(|Y| > n)\big) \to 0$, as $n \to \infty$.

$\square$

Note that there are uniformly integrable processes (martingales) which are not dominated by an integrable random variable, so that the sufficient condition for uniform integrability $E(\sup_t |X(t)|) < \infty$ is not necessary for uniform integrability. Another sufficient condition for uniform integrability is given by the following result, see for example Protter (1992), p.9, Liptser and Shiryaev (2001), p. 17.

Theorem 7.7 *If for some positive, increasing, convex function $G(x)$ on $[0, \infty)$ such that $\lim_{x \to \infty} G(x)/x = \infty$,*

$$\sup_{t \leq T} \mathrm{E}\big(G(|X(t)|)\big) < \infty, \tag{7.4}$$

then $X(t)$, $t \leq T$ is uniformly integrable.

We omit the proof. In practice the above result is used with $G(x) = x^r$ for $r > 1$, and uniform integrability is checked by using moments. For second moments $r = 2$, we have: square integrability implies uniform integrability.

Corollary 7.8 *If $X(t)$ is square integrable, that is, $\sup_t \mathrm{E} X^2(t) < \infty$, then it is uniformly integrable.*

In view of this, examples of uniformly integrable martingales are provided by square integrable martingales given in Example 7.1. The following result provides a construction of uniformly integrable martingales.

Theorem 7.9 (Doob's, Levy's martingale) *Let Y be an integrable random variable, that is, $\mathrm{E}|Y| < \infty$ and define*

$$M(t) = \mathrm{E}(Y|\mathcal{F}_t). \tag{7.5}$$

Then $M(t)$ is a uniformly integrable martingale.

PROOF: It is easy to see that $M(t)$ is a martingale. Indeed, by the law of double expectation, $\mathrm{E}(M(t)|\mathcal{F}_s) = \mathrm{E}\big(\mathrm{E}(Y|\mathcal{F}_t)|\mathcal{F}_s\big) = \mathrm{E}(Y|\mathcal{F}_s) = M(s)$. The proof of uniform integrability is more involved. It is enough to establish the result for $Y \geq 0$ as the general case will follow by considering Y^+ and Y^-. If $Y \geq 0$ then $M(t) \geq 0$ for all t. We show next that $M^* = \sup_{t \leq T} M(t) < \infty$. If not, there is a sequence of $t_n \uparrow \infty$ such that $M(t_n) \uparrow \infty$. By monotone convergence, $\mathrm{E} M(t_n) \uparrow \infty$, which is a contradiction, as $\mathrm{E} M(t_n) = \mathrm{E} Y < \infty$. Now, by the general definition of conditional expectation, see (2.16), $\mathrm{E}\big(M(t)I(M(t) > n)\big) = \mathrm{E}\big(YI(M(t) > n)\big)$. Since $\{M(t) > n\} \subseteq \{M^* > n\}$, $\mathrm{E}\big(YI(M(t) > n)\big) \leq \mathrm{E}\big(YI(M^* > n)\big)$. Thus $\mathrm{E}\big(M(t)I(M(t) > n)\big) \leq \mathrm{E}\big(YI(M^* > n)\big)$. Since the right-hand side does not depend on t, $\sup_{t \leq T} \mathrm{E}\big(M(t)I(M(t) > n)\big) \leq \mathrm{E}\big(YI(M^* > n)\big)$. But this converges to zero as $n \to \infty$, because M^* is finite and Y is integrable.
□

The martingale in (7.5) is said to be *closed* by Y. An immediate corollary is

Corollary 7.10 *Any martingale $M(t)$ on a finite time interval $0 \leq t \leq T < \infty$ is uniformly integrable and is closed by $M(T)$.*

It will be seen in the next section that a uniformly integrable martingale on $[0, \infty)$ is also of the form (7.5). That is, there exists a random variable, called $M(\infty)$ such that the martingale property holds for all $0 \leq s < t$, including $t = \infty$.

7.3 Martingale Convergence

In this section martingales on the infinite time interval $[0, \infty)$ are considered.

Theorem 7.11 (Martingale Convergence Theorem) *If $M(t)$, $0 \leq t < \infty$, is an integrable martingale (supermartingale or submartingale), that is, if $\sup_{t \geq 0} E|M(t)| < \infty$, then there exists an almost sure limit $\lim_{t \to \infty} M(t) = Y$ and Y is an integrable random variable.*

The proof of this result is due to Doob and it is too involved to be given here. If $M(t)$ is a martingale, then the condition $\sup_{t \geq 0} E|M(t)| < \infty$ is equivalent to any of the following conditions:

- $\lim_{t \to \infty} E|M(t)| < \infty$. This is because $|x|$ is a convex function, implying that $|M(t)|$ is a submartingale, and expectation of a submartingale is an increasing function of t. Hence the supremum is the same as the limit.

- $\lim_{t \to \infty} EM^+(t) < \infty$. This is because $E|M(t)| = EM^+(t) + EM^-(t)$. If $EM(t) = c$, then $EM(t) = EM^+(t) - EM^-(t) = c$ and $EM^+(t) = EM^-(t) + c$.

- $\lim_{t \to \infty} EM^-(t) < \infty$.

If $M(t)$ is a submartingale, it is enough to demand $\sup_t EM^+(t) < \infty$, and if it is a supermartingale it is enough to demand $\sup_t EM^-(t) < \infty$, for the existence of a finite limit.

Corollary 7.12

1. *Uniformly integrable martingales converge almost surely.*

2. *Square integrable martingales converge almost surely.*

3. *Positive martingales converge almost surely.*

4. *Submartingales bounded from above (negative) converge almost surely.*

5. *Supermartingales bounded from below (positive) converge almost surely.*

PROOF: Since uniformly integrable martingales are integrable, they converge. Since square integrable martingales are uniformly integrable, they converge. If $M(t)$ is positive then $|M(t)| = M(t)$, and $E|M(t)| = EM(t) = EM(0) < \infty$.
□

Note that expectations $EM(t)$ may or may not converge to the expectation of the limit EY (see Example 7.3). The case when $EY = \lim_{t \to \infty} EM(t)$, is precisely when $M(t)$ is uniformly integrable. The next result, given without proof, shows that uniformly integrable martingales have form (7.5).

Theorem 7.13 *If $M(t)$ is a uniformly integrable martingale then it converges to a random variable Y almost surely and in L^1. Conversely, if $M(t)$ is a martingale that converges in L^1 to a random variable Y, then $M(t)$ is uniformly integrable, and it converges almost surely to Y. In any case $M(t) = \mathrm{E}(Y|\mathcal{F}_t)$.*

Example 7.3: (Exponential martingale of Brownian motion.)
Let $M(t) = e^{B(t)-t/2}$. Then $M(t)$, $t \geq 0$ is a martingale. Since it is positive, it converges by Corollary 7.12 almost surely to a limit Y. By the Law of Large Numbers for Brownian motion $B(t)/t$ converges almost surely to zero. Thus $M(t) = e^{t(B(t)/t-1/2)} \to 0$ as $t \to \infty$. Thus $Y = 0$ almost surely. Therefore $M(t)$ is not uniformly integrable, as $\mathrm{E}Y = 0 \neq 1 = \mathrm{E}M(t)$.

Example 7.4: Let $f(s)$ be non-random, such that $\int_0^\infty f^2(s)ds < \infty$. We show that $M(t) = \int_0^t f(s)dB(s)$ is a uniformly integrable martingale and find a representation for the closing random variable.
Since $\int_0^\infty f^2(s)ds < \infty$, $\int_0^t f(s)dB(s)$ is defined for all $t > 0$ and is a martingale. Since $\sup_{t>0} \mathrm{E}(M^2(t)) = \sup_{t>0} \int_0^t f^2(s)ds = \int_0^\infty f^2(s)ds < \infty$, $M(t)$ is uniformly integrable. Thus it converges almost surely to Y. Convergence is also in L^1, that is $\mathrm{E}|M(t) - Y| \to 0$ as $t \to \infty$. Denote $Y = M(\infty) = \int_0^\infty f(s)dB(s)$, then we have shown that $Y - M(t) = \int_t^\infty f(s)dB(s)$ converges to zero almost surely and in L^1. Y is the closing variable. Indeed,
$\mathrm{E}(Y|\mathcal{F}_t) = \mathrm{E}(M(\infty)|\mathcal{F}_t) = \mathrm{E}(\int_0^\infty f(s)dB(s)|\mathcal{F}_t) = \int_0^t f(s)dB(s) = M(t)$.

Example 7.5: A bounded positive martingale $M(t) = \mathrm{E}(I(Y > 0)|\mathcal{F}_t)$ with $Y = \int_0^\infty f(s)dB(s)$, where $f(s)$ is non-random and $\int_0^\infty f^2(s)ds < \infty$, from the previous example.

$$
\begin{aligned}
M(t) &= \mathrm{E}(I(Y > 0)|\mathcal{F}_t) = P(Y > 0|\mathcal{F}_t) \\
&= \mathrm{P}\left(\int_t^\infty f(s)dB(s) > -\int_0^t f(s)dB(s) \middle| \mathcal{F}_t\right) \\
&= \Phi\left(\frac{\int_0^t f(s)dB(s)}{\sqrt{\int_t^\infty f^2(s)ds}}\right),
\end{aligned}
\tag{7.6}
$$

where the last equality is due to normality of the Itô integral for a non-random f.
By taking f to be zero on (T, ∞), a result is obtained for martingales of the form $\mathrm{E}(I(\int_0^T f(s)dB(s) > 0)|\mathcal{F}_t)$. In particular, by taking $f(s) = 1_{[0,T]}(s)$, we obtain that

$$\Phi(B(t)/\sqrt{T-t})$$

is a positive bounded martingale on $[0, T]$. Its distribution for $t \leq T$ is left as an exercise.

7.4 Optional Stopping

In this section we consider results on stopping martingales at random times. Recall that a random time τ is called a stopping time if for any $t > 0$ the sets $\{\tau \leq t\} \in \mathcal{F}_t$. For filtrations generated by a process X, τ is a stopping time if it is possible to decide whether τ has occurred or not by observing the process up to time t. A martingale stopped at a random time τ is the process $M(t \wedge \tau)$. A Basic Stopping result, given here without proof, states that a martingale stopped at a stopping time is a martingale, in particular $EM(\tau \wedge t) = EM(0)$. This equation is used most frequently.

Theorem 7.14 *If $M(t)$ is a martingale and τ is a stopping time, then the stopped process $M(\tau \wedge t)$ is a martingale. Moreover,*

$$EM(\tau \wedge t) = EM(0). \tag{7.7}$$

This result was proved in discrete time (see Theorem 3.39). We refer to (7.7) as the Basic Stopping equation.

Remark 7.1: We stress that in this theorem $M(\tau \wedge t)$ is a martingale with respect to the original filtration $\mathcal{F}_t$. Since it is adapted to $\mathcal{F}_{\tau \wedge t}$, it is also an $\mathcal{F}_{\tau \wedge t}$-martingale (see Exercise 7.1).

Example 7.6: (Exit of Brownian Motion from an Interval)
Let $B(t)$ be Brownian motion started at x and τ be the first time when $B(t)$ exits the interval (a, b), $a < x < b$, that is, $\tau = \inf\{t : B(t) = a \text{ or } b\}$. Clearly, τ is a stopping time. By the basic stopping result (7.7), $EB(t \wedge \tau) = B(0) = x$. By definition of τ, $|B(t \wedge \tau)| \leq \max(|a|, |b|)$. Thus one can take $t \to \infty$, and use dominated convergence to obtain $EB(\tau) = x$. But $B(\tau) = b$ with probability p and $B_\tau = a$ with probability $1 - p$. From these equations we obtain that $p = (x - a)/(b - a)$ is the probability that Brownian motion reaches b before it reaches a.

If M is a martingale then $EM(t) = EM(0)$. If τ is a stopping time, then $EM(\tau)$ may be different to $EM(0)$, as the next example shows.

Example 7.7: Let $B(t)$ be Brownian motion started at 0 and τ is the hitting time of 1. Then by definition $B(\tau) = 1$ and $EB(\tau) = 1 \neq 0 = EB(0)$.

However, under some additional assumptions on the martingale or on the stopping time, the random stopping does not alter the expected value. The following result gives sufficient conditions for optional stopping to hold.

Theorem 7.15 (Optional Stopping) *Let $M(t)$ be a martingale.*

1. *If $\tau \leq K < \infty$ is a bounded stopping time then $EM(\tau) = EM(0)$.*

2. *If $M(t)$ is uniformly integrable, then for any stopping time τ, $EM(\tau) = EM(0)$.*

The first statement follows from the Basic Stopping Result by taking $t > K$ and applying (7.7), $EM(0) = EM(t \wedge \tau) = EM(\tau)$.

Applied to gambling this shows that when betting on a martingale, on average no loss or gain is made, even if a clever stopping rule is used, provided it is bounded.

We don't give a proof of the second statement, the difficult point is in showing that $M(\tau)$ is integrable.

Theorem 7.16 *Let $M(t)$ be a martingale and τ a finite stopping time. If $E|M(\tau)| < \infty$, and*

$$\lim_{t \to \infty} E\big(M(t)I(\tau > t)\big) = 0, \qquad (7.8)$$

then $EM(\tau) = EM(0)$.

PROOF: Write $M(\tau \wedge t)$ as

$$M(\tau \wedge t) = M(t)I(t < \tau) + M(\tau)I(t \geq \tau). \qquad (7.9)$$

Using Basic Stopping Result (7.7), $EM(\tau \wedge t) = EM(0)$. Taking expectations in (7.9), we have

$$EM(0) = E\big(M(t)I(t < \tau)\big) + E\big(M(\tau)I(t \geq \tau)\big). \qquad (7.10)$$

Now take the limit in (7.10) as $t \to \infty$. Since τ is finite, $I(t \geq \tau) \to I(\tau < \infty) = 1$. $|M(\tau)|I(t \geq \tau) \leq |M(\tau)|$, integrable. Hence $E\big(M(\tau)I(t \geq \tau)\big) \to EM(\tau)$ by dominated convergence. It is assumed that $E\big(M(t)I(t < \tau)\big) \to 0$ as $t \to \infty$, and the result follows.
$$\square$$

The Basic Stopping result or Optional Stopping are used to find the distribution of stopping times for Brownian motion and Random Walks.

Example 7.8: (Hitting times of Brownian Motion)
We derive the Laplace transform of hitting times, from which it also follows that they are finite. Let $B(t)$ be a Brownian motion starting at 0, and $T_b = \inf\{t : B(t) = b\}$, $b > 0$. Consider the exponential martingale of Brownian motion $e^{uB(t)-u^2t/2}$, $u > 0$, stopped at T_b, $e^{uB(t \wedge T_b)-(t \wedge T_b)u^2/2}$. Using the Basic Stopping result (7.7)

$$Ee^{uB(t \wedge T_b)-(t \wedge T_b)u^2/2} = 1.$$

The martingale is bounded from above by e^{ub} and it is positive. If we take it as already proven that T_b is finite, $P(T_b < \infty) = 1$, then we obtain by taking $t \to \infty$ that $Ee^{ub-(t \wedge T_b)u^2/2} = 1$. Replacing u by $\sqrt{2u}$, we obtain the Laplace transform of T_b

$$\psi_{T_b}(u) = E\left(e^{-uT_b}\right) = e^{-b\sqrt{2u}}. \qquad (7.11)$$

We now show the finiteness of T_b. Write the expectation of the stopped martingale

$$\mathrm{E}\left(e^{ub-T_b u^2/2}I(T_b \leq t)\right) + \mathrm{E}\left(e^{uB(t)-tu^2/2}I(T_b > t)\right) = 1. \tag{7.12}$$

The term $\mathrm{E}\left(e^{uB(t)-tu^2/2}I(T_b > t)\right) \leq \mathrm{E}\left(e^{ub-tu^2/2}I(T_b > t)\right) \leq e^{ub-tu^2/2} \to 0$, as $t \to \infty$. Thus taking limits in the above equation (7.12),
$\mathrm{E}\left(e^{ub-T_b u^2/2}I(T_b \leq t)\right) \to \mathrm{E}\left(e^{ub-T_b u^2/2}I(T_b < \infty)\right) = 1$. Therefore
$\mathrm{E}\left(e^{-T_b u^2/2}I(T_b < \infty)\right) = e^{-ub}$. But $e^{-T_b u^2}I(T_b = \infty) = 0$, therefore by adding this term, we can write $\mathrm{E}\left(e^{-T_b u^2/2}\right) = e^{-ub}$. It follows in particular that
$\mathrm{P}(T_b < \infty) = \lim_{u\downarrow 0} \psi(u) = 1$, and T_b is finite. Hence (7.11) is proved. The distribution of T_b corresponding to the transform (7.11) is given in Theorem 3.18.

The following result is in some sense the converse to the Optional Stopping Theorem.

Theorem 7.17 *Let $X(t)$, $t \geq 0$, be such that for any bounded stopping time τ, $X(\tau)$ is integrable and $\mathrm{E}X(\tau) = \mathrm{E}X(0)$. Then $X(t)$, $t \geq 0$ is a martingale.*

PROOF: The proof consists of checking the martingale property by using appropriate stopping times. Since a deterministic time t is a stopping time, $X(t)$ is integrable. Without loss of generality take $X(0) = 0$. Next we show that for $t > s$, $\mathrm{E}(X(t)|\mathcal{F}_s) = X(s)$. In other words, we need to show that for any $s < t$ and any set $B \in \mathcal{F}_s$

$$\mathrm{E}\big(X(t)I(B)\big) = \mathrm{E}\big(X(s)I(B)\big). \tag{7.13}$$

Fix a set $B \in \mathcal{F}_s$ and for any $t > s$, define a stopping time $\tau = sI(B) + tI(B^c)$. We have $\mathrm{E}(X(\tau)) = \mathrm{E}\big(X(s)I(B)\big) + \mathrm{E}\big(X(t)I(B^c)\big)$. Since $\mathrm{E}X(\tau) = 0$,
$\mathrm{E}\big(X(s)I(B)\big) = \mathrm{E}X(\tau) - \mathrm{E}\big(X(t)I(B^c)\big) = -\mathrm{E}\big(X(t)I(B^c)\big)$.
As the right hand side of the above equality does not depend on s, it follows that (7.13) holds.

$\square$

The following result is sometimes known as the Optional Sampling Theorem (see for example, Rogers and Williams (1990)).

Theorem 7.18 (Optional Sampling) *Let $M(t)$ be a uniformly integrable martingale, and $\tau_1 \leq \tau_2 \leq \infty$ two stopping times. Then*

$$\mathrm{E}\big(M(\tau_2)|\mathcal{F}_{\tau_1}\big) = M(\tau_1), \quad a.s. \tag{7.14}$$

Optional Stopping of Discrete Time Martingales

We consider next the case of discrete time $t = 0, 1, 2\ldots$, and martingales arising in a Random Walk.

Gambler's Ruin

Consider a game played by two people by betting on the outcomes of tosses of a coin. You win \$1 if Heads come up and lose \$1 if Tails come up. The game stops when one party has no money left. You start with x, and your opponent with b dollars. Then S_n, the amount of money you have at time n is a Random Walk (see Section 3.12). The Gambler's ruin problem is to find the probabilities of ruin of the players.

In this game the loss of one person is the gain of the other (a zero sum game). Assuming that the game will end in a finite time τ (this fact will be shown later), it follows that the ruin probabilities of the players add up to one.

Consider first the case of the fair coin. Then

$$S_n = x + \sum_{i=1}^{n} \xi_i, \quad P(\xi_i = 1) = \frac{1}{2}, \quad P(\xi_i = -1) = \frac{1}{2},$$

is a martingale (see Theorem 3.33). Let τ be the time when the game stops, the first time the amount of money you have is equal to 0 (your ruin) or $x + b$ (your opponent's ruin). Then τ is a stopping time. Denote by u the probability of your ruin. It is the probability of you losing your initial capital x before winning b dollars. Thus

$$P(S_\tau = 0) = u \text{ and } P(S_\tau = x + b) = 1 - u. \tag{7.15}$$

Formally applying the Optional Stopping Theorem

$$E(S_\tau) = S_0 = x. \tag{7.16}$$

But

$$E(S_\tau) = (x + b) \times (1 - u) + 0 \times u = (x + b)u.$$

These equations give

$$u = \frac{b}{x + b}. \tag{7.17}$$

So that the ruin probabilities are given by a simple calculation using martingale stopping.

We now justify the steps. S_n is a martingale, and τ is a stopping time. By Theorem 7.14 the stopped process $S_{n \wedge \tau}$ is a martingale. It is non-negative and bounded by $x + b$, by the definition of τ. Thus $S_{n \wedge \tau}$ is a uniformly integrable martingale. Hence it converges almost surely to a finite limit Y, (with $EY = x$), $\lim_{n \to \infty} S_{n \wedge \tau} = Y$. By Theorem 3.33 $S_n^2 - n$ is a martingale, and so is $S_{n \wedge \tau}^2 - n \wedge \tau$. Thus for all n by taking expectation

$$E(S_{n \wedge \tau}^2) = E(n \wedge \tau) + E(S_0^2). \tag{7.18}$$

By dominated convergence, the lhs has a finite limit, therefore there is a finite limit $\lim_{n\to\infty} E(n \wedge \tau)$. Expanding this, $E(n \wedge \tau) \geq nP(\tau > n)$, we can see that for a limit to exist it must be

$$\lim_{n\to\infty} P(\tau > n) = 0, \qquad (7.19)$$

so that $P(\tau < \infty) = 1$, and τ is finite. (Note that a standard proof of finiteness of τ is done by using Markov Chain Theory, the property of recurrence of states in a Random Walk.) Writing $E(S_{n\wedge\tau}) = x$ and taking limits as $n \to \infty$, the equation (7.16) is obtained (alternatively, the conditions of the Optional Stopping Theorem hold). This concludes a rigorous derivation of the ruin probability in an unbiased Random Walk.

We now consider the case when the Random Walk is biased, $p \neq q$.

$$S_n = x + \sum_{i=1}^{n} \xi_i, \quad P(\xi_i = 1) = p, \quad P(\xi_i = -1) = q = 1 - p.$$

In this case the exponential martingale of the Random Walk $M_n = (q/p)^{S_n}$ is used (see Theorem 3.33). Stopping this martingale, we obtain the ruin probability

$$u = \frac{(q/p)^{b+x} - (q/p)^x}{(q/p)^{b+x} - 1}. \qquad (7.20)$$

Justification of the equation $E(M_\tau) = M_0$ is similar to the previous case.

Hitting Times in Random Walks

Let S_n denote a Random Walk on the integers started at $S_0 = x$, $S_n = S_0 + \sum_{i=1}^{n} \xi_i$, $P(\xi_i = 1) = p$, $P(\xi_i = -1) = q = 1 - p$, with arbitrary p, and T_b the first hitting time of b, $T_b = \inf\{n : S_n = b\}$ (infimum of an empty set is infinity). Without loss of generality take the starting state $x = 0$, otherwise consider the process $S_n - x$. Consider hitting the level $b > 0$, for $b < 0$ consider the process $-S_n$.

We find the Laplace transform of T_b, $\psi(\lambda) = E(e^{-\lambda T_b})$, $\lambda > 0$, by stopping the exponential martingale of the Random Walk $M_n = e^{uS_n - nh(u)}$, where $h(u) = \ln E(e^{u\xi_1})$, and u is arbitrary (see Section 3.12).

$$E(M_{n\wedge T_b}) = E\left(e^{uS_{n\wedge T_b} - (n\wedge T_b)h(u)}\right) = 1. \qquad (7.21)$$

Take u, so that $h(u) = \lambda > 0$. Write the expectation in (7.21) as

$$E\left(e^{uS_{n\wedge T_b} - (n\wedge T_b)h(u)}\right) = E\left(e^{uS_{T_b} - T_b h(u)}I(T_b \leq n)\right) + E\left(e^{uS_n - nh(u)}I(T_b > n)\right).$$

The first term equals to $E\left(e^{ub-T_b h(u)} I(T_b \le n)\right)$. The second term converges to zero, because by definition of T_b,

$$E\left(e^{uS_n - nh(u)} I(T_b > n)\right) \le E\left(e^{ub - nh(u)} I(T_b > n)\right) \le e^{ub - nh(u)} \to 0.$$

Now taking limits in (7.21), using dominated convergence we obtain

$$E\left(e^{-h(u)T_b} I(T_b < \infty)\right) = e^{-ub}.$$

Note that $e^{-h(u)T_b} I(T_b = \infty) = 0$, therefore by adding this term, we can write

$$E\left(e^{-h(u)T_b}\right) = e^{-ub}. \tag{7.22}$$

This is practically the Laplace transform of T_b, it remains to replace $h(u)$ by λ, by taking $u = h^{(-1)}(\lambda)$, with $h^{(-1)}$ being the inverse of h. Thus the Laplace transform of T_b is given by

$$\psi(\lambda) = E\left(e^{-\lambda T_b}\right) = e^{-h^{(-1)}(\lambda)b}. \tag{7.23}$$

To find $h^{(-1)}(\lambda)$, solve $h(u) = \lambda$, which is equivalent to $E(e^{u\xi_1}) = e^{\lambda}$, or $pe^u + (1-p)e^{-u} = e^{\lambda}$. There are two values for $e^u = (e^{\lambda} \pm \sqrt{e^{2\lambda} - 4p(1-p)})/(2p)$, but only one corresponds to a Laplace transform, (7.22). Thus we have

$$\psi(\lambda) = E\left(e^{-\lambda T_b}\right) = \left(\frac{2p}{e^{\lambda} + \sqrt{e^{2\lambda} - 4p(1-p)}}\right)^b. \tag{7.24}$$

Using a general result on Laplace transform of a random variable,

$$P(T_b < \infty) = \lim_{\lambda \downarrow 0} \psi(\lambda) = \left(\frac{2p}{1 + |1 - 2p|}\right)^b. \tag{7.25}$$

It now follows that the hitting time T_b of b is finite if and only if $p \ge 1/2$. For $p < 1/2$, there is a positive probability that level b is never reached, $P(T_b = \infty) = 1 - (\frac{p}{1-p})^b$.

When the hitting time of level b is finite, it may or may not have a finite expectation. If $p \ge 1/2$ we have

$$E(T_b) = -\psi'(0) = \begin{cases} \frac{l}{2p-1} & \text{if } p > 1/2 \\ \infty & \text{if } p = 1/2. \end{cases}$$

Thus we have shown that when $p \ge 1/2$ any positive state will be reached from 0 in a finite time, but when $p = 1/2$ the average time for it to happen is infinite.

The results obtained above are known as transience ($p \ne 1/2$) and recurrence ($p = 1/2$) of the Random Walk, and are usually obtained by Markov Chains Theory.

Example 7.9: (Optional stopping of discrete time martingales)
Let $M(t)$ be a discrete time martingale and τ be a stopping time such that $E|M(\tau)| < \infty$.

1. If $E\tau < \infty$ and $|M(t+1) - M(t)| \leq K$, then $EM(\tau) = EM(0)$.
2. If $E\tau < \infty$ and $E\big(|M(t+1) - M(t)||\mathcal{F}_t\big) \leq K$, then $EM(\tau) = EM(0)$.

PROOF: We prove the first statement.
$M(t) = M(0) + \sum_{i=0}^{t-1} \big(M(i+1) - M(i)\big)$. This together with the bound on increments gives

$$M(t) \leq |M(0)| + \sum_{i=0}^{t-1} |M(i+1) - M(i)| \leq |M(0)| + Kt.$$

Take for simplicity non-random $M(0)$ Then

$$EM(t)I(\tau > t) \leq |M(0)|P(\tau > t) + KtP(\tau > t).$$

The last term converges to zero, $tP(\tau > t) \leq E(\tau I(\tau > t)) \to 0$, by dominated convergence due to $E(\tau) < \infty$. Thus condition (7.8) holds, and the result follows. The proof of the second statement is similar and is left as an exercise. □

7.5 Localization and Local Martingales

As it was seen earlier in Chapter 4, Itô integrals $\int_0^t X(s)dB(s)$ are martingales under the additional condition $\int_0^t X^2(s)ds < \infty$. In general, stochastic integrals with respect to martingales are only local martingales rather than true martingales. This is the main reason for introducing local martingales. We have also seen that for the calculation of expectations stopping and truncations are often used. These ideas give rise to the following

Definition 7.19 *A property of a stochastic process $X(t)$ is said to hold locally if there exists a sequence of stopping times τ_n, called the localizing sequence, such that $\tau_n \uparrow \infty$ as $n \to \infty$ and for each n the stopped processes $X(t \wedge \tau_n)$ has this property.*

For example, the uniform integrability property holds locally for any martingale. By Theorem 7.13 a martingale convergent in L^1 is uniformly integrable. Here $M(t \wedge n) = M(n)$ for $t > n$, and therefore $\tau_n = n$ is a localizing sequence.
Local martingales are defined by localizing the martingale property.

Definition 7.20 *An adapted process $M(t)$ is called a local martingale if there exists a sequence of stopping times τ_n, such that $\tau_n \uparrow \infty$ and for each n the stopped processes $M(t \wedge \tau_n)$ is a uniformly integrable martingale in t.*

As we have just seen, any martingale is a local martingale. Examples of local martingales which are not martingales are given below.

Example 7.10: $M(t) = 1/|\boldsymbol{B}(t)|$, where $\boldsymbol{B}(t)$ is the three-dimensional Brownian motion, $\boldsymbol{B}(0) = \mathbf{x} \neq \mathbf{0}$. We have seen in Example 6.17 that if D_r is the complementary set to the ball of radius r centered at the origin, $D_r = \{\mathbf{z} : |\mathbf{z}| > r\}$, then $f(\mathbf{z}) = 1/|\mathbf{z}|$ is a harmonic function for the Laplacian on D_r. Consequently $1/|\boldsymbol{B}(t \wedge \tau_{D_r})|$ is a martingale, where τ_{D_r} is the time of exit from D_r. Take now τ_n be the exit time from $D_{1/n}$, that is, $\tau_n = \inf\{t > 0 : |\boldsymbol{B}(t)| = 1/n\}$. Then for any fixed n, $1/|\boldsymbol{B}(t \wedge \tau_n)|$ is a martingale. τ_n increase to, say, τ and by continuity, $\boldsymbol{B}(\tau) = \mathbf{0}$. As Brownian motion in three dimensions never visits the origin (see Example 6.14), it follows by continuity that τ is infinite. Thus $M(t)$ is a local martingale. To see that it is not a true martingale, recall that in three dimensions Brownian motion is transient and $|\boldsymbol{B}(t)| \to \infty$ as $t \to \infty$. Therefore $\mathrm{E}M(t) \to 0$, whereas $\mathrm{E}M(0) = 1/|\mathbf{x}|$. Since the expectation of a martingale is constant, $M(t)$ is not a martingale.

Example 7.11: (Itô integrals.)
Let $M(t) = \int_0^t e^{B^2(s)} dB(s)$, $t > 1/4$, where B is Brownian motion in one dimension with $B(0) = 0$. Let $\tau_n = \inf\{t > 0 : e^{B^2(t)} = n\}$. Then for $t \leq \tau_n$, the integrand is bounded by n. By the martingale property of Itô integrals, $M(t \wedge \tau_n)$ is a martingale in t for any n. By continuity, $\exp(B^2(\tau)) = \infty$, thus $\tau_n \to \tau = \infty$. Therefore $M(t)$ is a local martingale. To see that it is not a martingale notice that for $t > 1/4$, $\mathrm{E}\left(e^{B^2(t)}\right) = \infty$, implying that $M(t)$ is not integrable.

Remark 7.2: Note that it is not enough for a local martingale to be integrable in order to be a true martingale. For example, positive local martingales are integrable, but in general they are not martingales, but only supermartingales (see Theorem 7.23) below. Even uniformly integrable local martingales may not be martingales. However, if a local martingale is dominated by an integrable random variable then it is a martingale.

Theorem 7.21 *Let $M(t)$, $0 \leq t < \infty$, be a local martingale such that $|M(t)| \leq Y$, with $\mathrm{E}Y < \infty$. Then M is a uniformly integrable martingale.*

PROOF: Let τ_n be a localizing sequence. Then for any n and $s < t$

$$\mathrm{E}(M(t \wedge \tau_n)|\mathcal{F}_s) = M(s \wedge \tau_n). \tag{7.26}$$

M is clearly integrable, since $\mathrm{E}|M(t)| \leq \mathrm{E}Y < \infty$. Since $\lim_{n \to \infty} M(t \wedge \tau_n) = M(t)$, by dominated convergence of conditional expectations $\lim_{n \to \infty} \mathrm{E}(M(t \wedge \tau_n)|\mathcal{F}_s) = \mathrm{E}(M(t)|\mathcal{F}_s)$. Since $\lim_{n \to \infty} M(s \wedge \tau_n) = M(s)$, the martingale property is established by taking limits in (7.26). If a martingale is dominated by an integrable random variable then it is uniformly integrable (see Theorem 7.6).

$\square$

Corollary 7.22 *Let $M(t)$, $0 \leq t < \infty$, be a local martingale such that for all t, $\mathrm{E}(\sup_{s \leq t} |M(s)|) < \infty$. Then it is a martingale, and as such it is uniformly* *integrable on any finite interval $[0, T]$. If in addition $\mathrm{E}(\sup_{t \geq 0} |M(t)|) < \infty$, then $M(t)$, $t \geq 0$, is uniformly integrable on $[0, \infty)$.*

In financial applications we meet positive local martingales.

Theorem 7.23 *A non-negative local martingale $M(t)$, $0 \leq t \leq T$, is a super-martingale, that is, $\mathrm{E}M(t) < \infty$, and for any $s < t$, $\mathrm{E}(M(t)|\mathcal{F}_s) \leq M(s)$.*

PROOF: Let τ_n be a localizing sequence Then since $M(t \wedge \tau_n) \geq 0$, by Fatou's lemma

$$\mathrm{E}\big(\liminf_{n \to \infty} M(t \wedge \tau_n)\big) \leq \liminf_{n \to \infty} \mathrm{E}\big(M(t \wedge \tau_n)\big). \tag{7.27}$$

Since the limit exists, the lower limit is the same, that is, $\lim_{n \to \infty} M(t \wedge \tau_n) = M(t)$ implies $\liminf_{n \to \infty} M(t \wedge \tau_n) = M(t)$. But $\mathrm{E}M(t \wedge \tau_n) = \mathrm{E}M(0 \wedge \tau_n) = \mathrm{E}M(0)$ by the martingale property of $M(t \wedge \tau_n)$. Therefore by taking limits, $\mathrm{E}M(t) \leq \mathrm{E}M(0)$, so that M is integrable. The supermartingale property is established similarly. Using Fatou's lemma for conditional expectations,

$$\mathrm{E}\big(\liminf_{n \to \infty} M(t \wedge \tau_n)|\mathcal{F}_s\big) \leq \liminf_{n \to \infty} \mathrm{E}\big(M(t \wedge \tau_n)|\mathcal{F}_s\big) = M(s \wedge \tau_n), \tag{7.28}$$

Taking limits as $n \to \infty$ we obtain $\mathrm{E}\big(M(t)|\mathcal{F}_s\big) \leq M(s)$ almost surely.

$\square$

From this result and Theorem 7.3 we obtain

Theorem 7.24 *A non-negative local martingale $M(t)$, $0 \leq t \leq T$, is a martingale if and only if $\mathrm{E}M(T) = M(0)$.*

For a general local martingale a necessary and sufficient condition to be a uniformly integrable martingale is described in terms of the property of Dirichlet class (D). This class of processes also arises in other areas of calculus and is given in the next section.

Dirichlet Class (D)

Definition 7.25 *A process X is of Dirichlet class, (D), if the family $\{X(\tau) : \tau$ a finite stopping time$\}$ is uniformly integrable.*

Any uniformly integrable martingale M is of class (D). Indeed, by Theorem 7.13, M is closed by $Y = M(\infty)$, $M(\tau) = \mathrm{E}(Y|\mathcal{F}_\tau)$, and the last family is uniformly integrable.

Using localization one can show the other direction, and we have a theorem

Theorem 7.26 *A local martingale M is a uniformly integrable martingale if and only if it is of class (D).*

PROOF: Suppose that M is a local martingale of class (D).

Let τ_n be a localizing sequence, so that $M(t \wedge \tau_n)$ is a uniformly integrable martingale in t. Then for $s < t$,

$$M(s \wedge \tau_n) = \mathrm{E}(M(t \wedge \tau_n)|\mathcal{F}_s). \tag{7.29}$$

The martingale property of M is obtained by taking $n \to \infty$ in both sides of the equation.

Since $\tau_n \to \infty$, $M(s \wedge \tau_n) \to M(s)$ almost surely. $s \wedge \tau_n$ is a finite stopping time, and because M is in (D), the sequence of random variables $\{M(s \wedge \tau_n)\}_n$ is uniformly integrable. Thus $M(s \wedge \tau_n) \to M(s)$ also in L^1, that is,

$$\mathrm{E}|M(s \wedge \tau_n) - M(s)| \to 0. \tag{7.30}$$

Using the properties of conditional expectation,

$$\begin{aligned}
\mathrm{E}\big|\mathrm{E}(M(t \wedge \tau_n)|\mathcal{F}_s) - \mathrm{E}(M(t)|\mathcal{F}_s)\big| &= \mathrm{E}\big|\mathrm{E}(M(t \wedge \tau_n) - M(t) \,|\mathcal{F}_s)\big| \\
&\leq \mathrm{E}\left(\mathrm{E}\big|M(t \wedge \tau_n) - M(t)\big| \,|\mathcal{F}_s\right) \\
&= \mathrm{E}\big|M(t \wedge \tau_n) - M(t)\big|.
\end{aligned}$$

The latter converges to zero by (7.30). This implies $\mathrm{E}(M(t \wedge \tau_n)|\mathcal{F}_s) \to \mathrm{E}(M(t)|\mathcal{F}_s)$ as $n \to \infty$. Taking limits in (7.29) as $n \to \infty$ establishes the martingale property of M. Since it is in (D), by taking $\tau = t$, it is uniformly integrable.

$\square$

7.6   Quadratic Variation of Martingales

Quadratic variation of a process $X(t)$ is defined as a limit in probability

$$[X, X](t) = \lim \sum_{i=1}^{n} (X(t_i^n) - X(t_{i-1}^n))^2, \tag{7.31}$$

where the limit is taken over partitions:

$$0 = t_0^n < t_1^n < \ldots < t_n^n = t,$$

with $\delta_n = \max_{0 \leq i \leq n}(t_i^n - t_{i-1}^n) \to 0$. If $M(t)$ is a martingale, then $M^2(t)$ is is a submartingale, and its mean increases (unless $M(t)$ is a constant). By compensating $M^2(t)$ by some increasing process, it is possible to make it into a martingale. The process which compensates $M^2(t)$ to a martingale turns out to be the quadratic variation process of M. It can be shown that quadratic variation of martingales exists and is characterized by the above property.

Theorem 7.27

1. *Let $M(t)$ be a martingale with finite second moments, $E(M^2(t)) < \infty$ for all t. Then its quadratic variation process $[M, M](t)$ defined in (7.31) exists, moreover $M^2(t) - [M, M](t)$ is a martingale.*

2. *If M is a local martingale, then $[M, M](t)$ exists, moreover $M^2(t) - [M, M](t)$ is a local martingale.*

PROOF: We outline the proof of the first statement only. The second follows for locally square integrable martingales by localization. For local martingales the result follows from representation of quadratic variation by means of stochastic integrals. For a full proof see for example, Liptser and Shiryaev (1989), p.56-59.

$$E\big(M(t)M(s)\big) = EE\big(M(t)M(s)|\mathcal{F}_s\big) = E\big(M(s)E(M(t)|\mathcal{F}_s)\big) = E(M^2(s)).$$
$$(7.32)$$

Using this it is easy to obtain

$$E(M(t) - M(s))^2 = E(M^2(t)) - E(M^2(s)). \qquad (7.33)$$

It is easy to see that the sums in the definition of quadratic variation $[M, M](t)$ have constant mean, that of $EM^2(t)$. It is possible, but is not easy, to prove that these sums converge in probability to the limit $[M, M](t)$. Now using property (7.33), we can write

$$
\begin{aligned}
E\big(M^2(t) - M^2(s)|\mathcal{F}_s\big) &= E\big((M(t) - M(s))^2|\mathcal{F}_s\big) \\
&= E\Big(\sum_{i=0}^{n-1}(M(t_{i+1}) - M(t_i))^2|\mathcal{F}_s\Big), \quad (7.34)
\end{aligned}
$$

where $\{t_i\}$ is a partition of $[s, t]$. Taking the limit as the size of the partition goes to zero, we obtain

$$E\big(M^2(t) - M^2(s)|\mathcal{F}_s\big) = E\big([M, M](t) - [M, M](s)|\mathcal{F}_s\big). \qquad (7.35)$$

Rearranging, we obtain the martingale property of $M^2(t) - [M, M](t)$.

$\square$

For the next result note that if M is a martingale, then for any t

$$E(M(t) - M(0))^2 = E(M^2(t)) - E(M^2(0)),$$

which shows that $E(M^2(t)) > E(M^2(0))$ unless $M(t) = M(0)$ a.s. Thus M^2 can not be a martingale on $[0, t]$, unless $M(t) = M(0)$. If $M(t) = M(0)$, then for all $s < t$, $M(s) = E(M(t)|\mathcal{F}_s) = M(0)$, and M is a constant on $[0, t]$.

Theorem 7.28 *Let M be a martingale with $M(0) = 0$. If for some t, $M(t)$ is not identically zero, then $[M, M](t) > 0$. Conversely, if $[M, M](t) = 0$, then $M(s) = 0$ a.s. for all $s \leq t$. The result also holds for local martingales.*

PROOF: We prove the result for square integrable martingales; for local martingales it can be shown by localization. Suppose that $[M, M](t) = 0$ for some $t > 0$. Then, since $[M, M]$ is non-decreasing, $[M, M](s) = 0$ for all $s \leq t$. By Theorem 7.27 $M^2(s)$, $s \leq t$, is a martingale. In particular, $E(M^2(t)) = 0$. This implies that $M(t) = 0$ a.s., which is a contradiction. Therefore $[M, M](t) > 0$.

Conversely, if $[M, M](t) = 0$, the same argument shows that $M(t) = 0$ a.s., and by the martingale property $M(s) = 0$ for all $s \leq t$.

$\square$

It also follows from the proof that M and $[M, M]$ have same intervals of constancy. This theorem implies remarkably that a continuous martingale which is not a constant has infinite variation on any interval.

Theorem 7.29 *Let M be a continuous local martingale, and fix any t. If $M(t)$ is not identically equal to $M(0)$, then M has infinite variation over $[0, t]$.*

PROOF: $M(t) - M(0)$ is a martingale, null at zero, with its value at time t not equal identically to zero. By the above theorem M has a positive quadratic variation on $[0, t]$, $[M, M](t) > 0$. By Theorem 1.10 a continuous process of finite variation on $[0, t]$ has zero quadratic variation over this interval. Therefore M must have infinite variation over $[0, t]$.

$\square$

Corollary 7.30 *If a continuous local martingale has finite variation over an interval, then it must be a constant over that interval.*

Remark 7.3: Note that there are martingales with finite variation, but by the previous result they can not be continuous. An example of such a martingale is the Poisson process martingale $N(t) - t$.

7.7   Martingale Inequalities

$M(t)$ denotes a martingale or a local martingale on the interval $[0, T]$ with possibly $T = \infty$.

Theorem 7.31 *If $M(t)$ is a martingale (or a positive submartingale) then for $p \geq 1$*

$$P(\sup_{s \leq t} |M(s)| \geq a) \leq a^{-p} \sup_{s \leq t} E(|M(s)|^p). \qquad (7.36)$$

If $p > 1$, then

$$E\big(\sup_{s \leq t} |M(s)|^p\big) \leq \Big(\frac{p}{p-1}\Big)^p E\big(|M(t)|^p\big). \tag{7.37}$$

The case of $p = 2$ is called Doob's inequality for martingales.

$$E\big(\sup_{s \leq T} M(s)^2\big) \leq 4E\big(M^2(T)\big). \tag{7.38}$$

As a consequence, if for $p > 1$, $\sup_{t \leq T} E\big(|M(t)|^p\big) < \infty$, then $M(t)$ is uniformly integrable (This is a particular case of Theorem 7.7).

Theorem 7.32 *If M is locally square integrable martingale with $M(0) = 0$, then*

$$P(\sup_{t \leq T} |M(t)| > a) \leq a^{-2} E\left([M, M](T)\right). \tag{7.39}$$

Theorem 7.33 (Davis' Inequality) *There are constants $c > 0$ and $C < \infty$ such that for any local martingale $M(t)$, null at zero,*

$$cE\left(\sqrt{[M, M](T)}\right) \leq E\left(\sup_{t \leq T} |M(t)|\right) \leq CE\left(\sqrt{[M, M](T)}\right). \tag{7.40}$$

Theorem 7.34 (Burkholder-Gundy Inequality) *There are constants c_p and C_p depending only on p, such that for any local martingale $M(t)$, null at zero,*

$$c_p E\left([M, M](T)^{p/2}\right) \leq E\left((\sup_{t \leq T} |M(t)|)^p\right) \leq C_p E\left([M, M](T)^{p/2}\right), \tag{7.41}$$

for $1 < p < \infty$. If moreover, $M(t)$ is continuous, then the result holds also for $0 < p \leq 1$.

The above inequalities hold when T is a stopping time.

Proofs of these inequalities involve concepts of stochastic calculus for general processes and can be found, for example, in Protter (1992), Rogers and Williams (1990), Liptser and Shiryayev (1989).

We use the above inequalities to give sufficient conditions for a local martingale to be a true martingale.

Theorem 7.35 *Let $M(t)$ be a local martingale, null at zero, such that $E\left(\sqrt{[M, M](t)}\right) < \infty$ for all t. Then $M(t)$ is a uniformly integrable martingale on $[0, T]$, for any finite T.*

If moreover, $E[M, M](t) < \infty$, then $M(t)$ is a martingale with $EM^2(t) = E[M, M](t) < \infty$ for all t.

If $\sup_{t < \infty} E[M, M](t) < \infty$, then $M(t)$ is a square integrable martingale.

PROOF: By the Davis inequality, $\sup_{t \leq T} |M(t)|$ is an integrable random variable, $\mathrm{E}\left(\sup_{t \leq T} |M(t)|\right) \leq C\mathrm{E}\left(\sqrt{[M,M](T)}\right) < \infty$. Thus $M(t)$ is dominated by an integrable random variable on any finite time interval. Therefore it is a uniformly martingale by Theorem 7.21, and the first claim is proved.

The condition $\mathrm{E}[M,M](t) < \infty$ implies the previous condition $\mathrm{E}\left(\sqrt{[M,M](t)}\right) < \infty$, as for $X \geq 0$, $\mathrm{E}(X) \geq (\mathrm{E}(\sqrt{X}))^2$, due to $Var(\sqrt{X}) \geq 0$. Thus M is a martingale. Alternatively, use the Burkholder-Gundy inequality with $p = 2$. Next, recall that by Theorem 7.27, if $M(t)$ is a martingale with $\mathrm{E}\left(M^2(t)\right) < \infty$, then $M^2(t) - [M,M](t)$ is a martingale. In particular, for any finite t

$$\mathrm{E}\left(M^2(t)\right) = \mathrm{E}[M,M](t),$$

and the second statement is proved. To prove the third, notice that since both sides in the above equation are non-decreasing, they have a limit. Since by assumption $\lim_{t \to \infty} \mathrm{E}[M,M](t) < \infty$, $\sup_{t < \infty} \mathrm{E}M^2(t) < \infty$, and $M(t)$, $0 \leq t < \infty$ is a square integrable martingale.

□

Application to Itô integrals

Let $X(t) = \int_0^t H(s)dB(s)$. Being an Itô integral, X is a local martingale. Its quadratic variation is given by $[X,X](t) = \int_0^t H^2(s)ds$. The Burkholder-Gundy inequality with $p = 2$ gives $\mathrm{E}\left(\sup_{t \leq T} X^2(t)\right) \leq C\mathrm{E}\left([X,X](T)\right) = \mathrm{E}\int_0^T H^2(s)ds$. If $\mathrm{E}\left(\int_0^T H^2(s)ds\right) < \infty$, then $X(t)$ is a square integrable martingale. Thus we recover the known fact that $\mathrm{E}(X^2(t)) = \mathrm{E}\left(\int_0^t H^2(s)ds\right)$.

The Davis inequality gives $\mathrm{E}(\sup_{t \leq T} \int_0^t H(s)dB(s)) \leq C\mathrm{E}\left(\sqrt{\int_0^t H^2(s)ds}\right)$. Thus the condition

$$\mathrm{E}\left(\sqrt{\int_0^t H^2(s)ds}\right) < \infty \tag{7.42}$$

is a sufficient condition for the Itô integral to be a martingale and, in particular, to have zero mean. This condition, however, does not assure second moments.

7.8 Continuous Martingales. Change of Time

Brownian motion is the basic continuous martingale from which all continuous martingales can be constructed, either by random change of time, given in this section, or by stochastic integration, as will be seen in the next chapter. The starting point is a result that characterizes a Brownian motion.

Levy's Characterization of Brownian Motion

Theorem 7.36 (Levy) *A process M with $M(0) = 0$ is a Brownian motion if and only of it is a continuous local martingale with quadratic variation process $[M, M](t) = t$.*

PROOF: If M is a Brownian motion, then it is a continuous martingale with $[M, M](t) = t$.

Let $M(t)$ be a continuous local martingale with $[M, M](t) = t$. Then $uM(t)$ is a continuous local martingale with $[uM, uM](t) = u^2 t$. We show that

$$U(t) = e^{uM(t) - u^2 t/2} = e^{uM(t) - [uM, uM](t)/2} \tag{7.43}$$

is a martingale. Once this is established, the rest of the proof follows by an application of the martingale property.

The general theory of integration with respect to martingales is required to show the martingale property of $U(t)$. It is an an easy corollary of a general result on stochastic exponential martingales (Theorem 8.17, Corollary 8.18). Writing the martingale property, we have

$$E\left(e^{uM(t) - u^2 t/2} | \mathcal{F}_s\right) = e^{uM(s) - u^2 s/2}, \tag{7.44}$$

from which it follows that

$$E\left(e^{u(M(t) - M(s))} | \mathcal{F}_s\right) = e^{u^2(t-s)/2}. \tag{7.45}$$

Since the right hand side of (7.45) is non-random, it follows that $M(t)$ has independent increments. Taking expectation in (7.45), we obtain

$$E\left(e^{u(M(t) - M(s))}\right) = e^{u^2(t-s)/2}, \tag{7.46}$$

which shows that the increment of the martingale $M(t) - M(s)$ has Normal distribution with mean zero and variance $(t - s)$. Therefore M is a continuous process with independent Gaussian increments, hence it is Brownian motion. □

Example 7.12: Any solution of Tanaka's SDE in Example 5.15 is a Brownian motion (weak uniqueness).

$dX(t) = \text{sign}(X(t))dB(t)$, where $\text{sign}(x) = 1$ if $x \geq 0$ and -1 if $x < 0$. $X(0) = 0$. $X(t) = \int_0^t \text{sign}(X(s))dB(s)$. Since it is an Itô integral, it is a local martingale (even a martingale, as the condition for it to be a martingale holds). It is continuous, and its quadratic variation is given by $[X, X](t) = \int_0^t \text{sign}^2(X(s))ds = t$. Therefore it is a Brownian motion.

Change of Time for Martingales

The main result below states that a continuous martingale M is a Brownian motion with a change of time, where time is measured by the quadratic variation $[M,M](t)$, namely, there is a Brownian motion $B(t)$, such that $M(t) = B([M,M](t))$. This $B(t)$ is constructed from $M(t)$. Define

$$\tau_t = \inf\{s : [M,M](s) > t\}. \tag{7.47}$$

If $[M,M](t)$ is strictly increasing, then τ_t is its inverse.

Theorem 7.37 (Dambis, Dubins-Schwarz) *Let $M(t)$ be a continuous martingale, null at zero, such that $[M,M](t)$ is non-decreasing to ∞, and τ_t defined by (7.47). Then the process $B(t) = M(\tau_t)$ is a Brownian motion with respect to the filtration $\mathcal{F}_{\tau_t}$. Moreover, $[M,M](t)$ is a stopping time with respect to this filtration, and the martingale M can be obtained from the Brownian motion B by the change of time $M(t) = B([M,M](t))$. The result also holds when M is a continuous local martingale.*

We outline the idea of the proof, for details see, for example, Rogers and Williams (1990), p.64, Karatzas and Shreve (1988) p.174, Protter (1992) p.81, Revuz and Yor (1998) p. 181.

PROOF: Let $M(t)$ be a local martingale. τ_t defined by (7.47) are finite stopping times, since $[M,M](t) \to \infty$. Thus $\mathcal{F}_{\tau_t}$ are well defined, (see Chapter 2 for the definition of $\mathcal{F}_\tau$). Note that $\{[M,M](s) \le t\} = \{\tau_t \ge s\}$. This implies that $[M,M](s)$ are stopping times for $\mathcal{F}_{\tau_t}$. Since $[M,M](s)$ is continuous $[M,M](\tau_t) = t$. Let $X(t) = M(\tau_t)$. Then it is a continuous local martingale, since M and $[M,M]$ have the same intervals of constancy (see the comment following Theorem 7.28). Using Theorem 7.27 we obtain $EX^2(t) = E[X,X](t) = E[M,M](\tau_t) = t$. Thus X is a Brownian motion by Levy's characterization Theorem 7.36. The second part is proven as follows. Recall that M and $[M,M]$ have the same intervals of constancy. Thus $X([M,M](t)) = M(\tau_{[M,M](t)}) = M(t)$.

$\square$

Example 7.13:   Let $M(t) = \int_0^t f(s)dB(s)$, with f continuous and non-random. Then M is a Gaussian martingale. Its quadratic variation is given by $[M,M](t) = \int_0^t f^2(s)ds$. For example, with $f(s) = s$, $M(t) = \int_0^t sdB(s)$ and $[M,M](t) = \int_0^t s^2 ds = t^3/3$. In this example $[M,M](t)$ is non-random and increasing. τ_t is given by its inverse, $\tau_t = (3t)^{1/3}$. Let $X(t) = M(\tau_t) = \int_0^{\sqrt[3]{3t}} sdB(s)$. Then, clearly, X is continuous, as a composition of continuous functions. It is also a martingale with quadratic variation $\tau_t^3/3 = t$. Hence, by the Levy's theorem, it is a Brownian motion, $X(t) = \hat{B}(t)$. By the above theorem, $M(t) = \hat{B}(t^3/3)$.

Example 7.14: If $M(t) = \int_0^t H(s)dB(s)$ is an Itô integral, then it is a local martingale with quadratic variation $[M, M](t) = \int_0^t H^2(s)ds$. If $\int_0^\infty H^2(s)ds = \infty$, then $M(t) = \hat{B}\left(\int_0^t H^2(s)ds\right)$, where $\hat{B}(t)$ is Brownian motion and can be recovered from $M(t)$ with the appropriate change of time.

Example 7.15: (Brownian Bridge as Time Changed Brownian motion)
The SDE for Brownian Bridge (5.34) contains as its only stochastic term $\int_0^t \frac{1}{T-s}dB(s)$. Since for any $t < T$, it is a continuous martingale with quadratic variation $[Y, Y](t) = \int_0^t \frac{1}{(T-s)^2}ds = \frac{t}{T(T-t)}$, it follows by the DDS Theorem

$$Y(t) = \hat{B}\left(\frac{t}{T(T-t)}\right),$$

for some Brownian motion $\hat{B}$. Therefore SDE (5.34) has the following representation

$$X(t) = a\left(1 - \frac{t}{T}\right) + b\frac{t}{T} + (T-t)\hat{B}\left(\frac{t}{T(T-t)}\right), \text{ for } 0 \leq t \leq T. \tag{7.48}$$

In this representation $t = T$ is allowed and understood by continuity, since the limit of $tB(1/t)$ as $t \to 0$ is zero by the Law of Large Numbers for Brownian motion.

Change of Time in SDEs

We use the Change of Time (DDS) Theorem for constructing weak solutions of some SDEs. Let

$$X(t) = \int_0^t \sqrt{f'(t)}dB(t), \tag{7.49}$$

where $f(t)$ is an adapted, positive, increasing, differentiable process, null at zero. It is a local martingale with quadratic variation $[X, X](t) = \int_0^t f'(s)ds = f(t)$. Thus $\tau_t = f^{(-1)}(t)$, the inverse of f, and according to the Change of Time Theorem, the process $X(f^{(-1)}(t)) = \hat{B}(t)$ is a Brownian motion (with respect to $\mathcal{F}_{\tau_t}$), and

$$X(t) = \hat{B}(f(t)). \tag{7.50}$$

Thus from equations (7.49) and (7.50) we have

Theorem 7.38 *Let $f(t)$ be an adapted, positive, increasing, differentiable process, and*

$$dX(t) = \sqrt{f'(t)}dB(t). \tag{7.51}$$

Then the process $B(f(t))$ is a weak solution.

We can write equation (7.51) as follows: for a Brownian motion B, and a function f, there is a Brownian motion $\hat{B}$, such that

$$d\hat{B}(f(t)) = \sqrt{f'(t)}dB(t). \tag{7.52}$$

In the case of non-random change of time in Brownian motion $B(f(t))$, it is easy to check directly that $M(t) = B(f(t))$ is a martingale (with respect to the filtration $\mathcal{F}_{f(t)}$). The quadratic variation of $B(f(t))$ is $[M, M](t) = [B(f), B(f)](t) = f(t)$, was calculated directly in the Example 4.24, equation (4.63).

Example 7.16: (Ornstein-Uhlenbeck process as Time Changed Brownian motion) With $f(t) = \sigma^2(e^{2\alpha t} - 1)/(2\alpha))$, the process $B(\sigma^2(e^{2\alpha t} - 1)/(2\alpha)$ is a weak solution to the SDE

$$dX(t) = \sigma e^{\alpha t} dB(t).$$

Consider $U(t) = e^{-\alpha t} X(t)$. Integrating by parts, $U(t)$ satisfies

$$dU(t) = -\alpha U(t)dt + \sigma dB(t). \tag{7.53}$$

Recall that the solution to this SDE is given by (5.13). Thus $U(t)$ is an Ornstein-Uhlenbeck process (see Example 5.6). Thus an Ornstein-Uhlenbeck process has representation

$$U(t) = e^{-\alpha t} B(\sigma^2(e^{2\alpha t} - 1)/(2\alpha)). \tag{7.54}$$

To have $U(0) = x$, take $B(t)$ to be a Brownian motion started at x. Note that in equations (7.53) and (7.54), $B(t)$ denotes different Brownian motions.

Next we construct a weak solution to the SDEs of the form

$$dX(t) = \sigma(X(t))dB(t)$$

with $\sigma(x) > 0$ such that

$$G(t) = \int_0^t \frac{ds}{\sigma^2(B(s))}$$

is finite for finite t, and increases to infinity, $\int_0^\infty \frac{ds}{\sigma^2(B(s))} = \infty$ almost surely. Then $G(t)$ is adapted, continuous and strictly increasing to $G(\infty) = \infty$. Therefore it has inverse

$$\tau_t = G^{(-1)}(t). \tag{7.55}$$

Note that for each fixed t, τ_t is a stopping time, as it is the first time the process $G(s)$ hits t, and that τ_t is increasing.

Theorem 7.39 *The process $X(t) = B(\tau_t)$ is a weak solution to the SDE*

$$dX(t) = \sigma(X(t))dB(t). \tag{7.56}$$

PROOF: $X(t) = B(\tau_t) = B(G^{(-1)}(t))$. Using equation (7.52) with $f = G^{(-1)}$, we obtain

$$dB(G^{(-1)}(t)) = \sqrt{(G^{(-1)})'(t)}d\hat{B}(t).$$

$$(G^{(-1)})'(t) = \frac{1}{G'(G^{(-1)}(t))} = \frac{1}{1/\sigma^2(B(G^{(-1)}(t)))} = \sigma^2(B(\tau_t)). \qquad (7.57)$$

Thus we obtain

$$dB(\tau_t) = \sigma^2(B(\tau_t))d\hat{B}(t),$$

and the result is proved.

$\square$

Another proof, by considering the martingale problem, is given next. Note that (7.57) gives $d\tau_t = \sigma^2(B(\tau_t))dt$.

PROOF: The diffusion operator for the SDE (7.56) is given by $Lf(x) = \frac{1}{2}\sigma^2(x)f''(x)$. We show that $X(t) = B(\tau_t)$ is a solution to the martingale problem for L. Indeed, we know (see Example 5.17) that for any twice continuously differentiable function f vanishing outside a compact interval, the process $M(t) = f(B(t)) - \int_0^t \frac{1}{2}f''(B(s))ds$ is a martingale. Since τ_t are increasing stopping times it can be shown (by using Optional Stopping Theorem 7.18) that the process $M(\tau_t)$ is also a martingale, that is, $f(B(\tau_t)) - \int_0^{\tau_t} \frac{1}{2}f''(B(s))ds$ is a martingale. Now perform the change of variable $s = \tau_u$, and observe from (7.57) that $d\tau_t = \sigma^2(B(\tau_t))dt$, to obtain that the process $f(B(\tau_t)) - \int_0^t \frac{1}{2}\sigma^2(B(\tau_u))f''(B(\tau_u))du$ is a martingale. But since $X(t) = B(\tau_t)$, this being the same as $f(X(t)) - \int_0^t \frac{1}{2}\sigma^2(X(u))f''(X(u))du$ is a martingale, and $X(t)$ solves the martingale problem for L.

$\square$

An application of Theorem 7.37 gives a result on uniqueness of the solution of SDE (7.56). This result is weaker than Theorem 6.13 of Engelbert-Schmidt.

Theorem 7.40 *Let $\sigma(x)$ be a positive function bounded away from zero, $\sigma(x) \geq \delta > 0$. Then the stochastic differential equation (7.56) has a unique weak solution.*

PROOF: Let $X(t)$ be a weak solution to (7.56). Then $X(t)$ is a local martingale and there is a Brownian motion $\beta(t)$, such that $X(t) = \beta([X,X](t))$. Now,

$$[X,X](t) = \int_0^t \sigma^2(X(s))ds = \int_0^t \sigma^2(\beta([X,X](s)))ds.$$

Thus $[X,X](t)$ is a solution to the ordinary differential equation (ODE) $da(t) = \sigma^2(\beta(a(t))dt$. Since the solution to this ODE is unique, the solution to (7.56) is unique.

$\square$

A more general change of time is done for the stochastic differential equation

$$dX(t) = \mu(X(t))dt + \sigma(X(t))dB(t). \qquad (7.58)$$

Let $g(x)$ be a positive function for which $G(t) = \int_0^t g(X(s))ds$ is finite for finite t and increases to infinity almost surely. Define $\tau_t = G^{(-1)}(t)$.

Theorem 7.41 *Let $X(t)$ be a solution to (7.58) and define $Y(t) = X(\tau_t)$. Then $Y(t)$ is a weak solution to the stochastic differential equation*

$$dY(t) = \frac{\mu(Y(t))}{g(Y(t))}dt + \frac{\sigma(Y(t))}{\sqrt{g(Y(t))}}dB(t), \quad with \ Y(0) = X(0). \tag{7.59}$$

One can use the change of time on an interval $[0, T]$, for a stopping time T.

Example 7.17: (Lamperti's Change of Time)
    Let $X(t)$ satisfy the SDE (Feller's branching diffusion)

$$dX(t) = \mu X(t)dt + \sigma\sqrt{X(t)}dB(t), \quad X(0) = x > 0, \tag{7.60}$$

with positive constants μ and σ. Lamperti's change of time is $G(t) = \int_0^t X(s)ds$. Here $g(x) = x$. Then $Y(t) = X(\tau_t)$ satisfies the SDE

$$\begin{aligned} dY(t) &= \frac{\mu Y(t)}{Y(t)}dt + \frac{\sigma\sqrt{Y(t)}}{\sqrt{Y(t)}}dB(t), \\ &= \mu dt + \sigma dB(t) \ \text{with} \ Y(0) = x, \end{aligned}$$

and

$$Y(t) = x + \mu t + \sigma B(t). \tag{7.61}$$

In other words, with a random change of time, the Branching diffusion is a Brownian motion with drift. At the (random) point where $G(t)$ stops increasing its inverse τ_t, defined as the right-inverse $\tau_t = \inf\{s : G(s) = t\}$, also remains the same. This happens at the point of time when $X(t) = 0$. It can be seen that once the process is at zero, it stays at 0 forever. Let $T = \inf\{t : X(t) = 0\}$. T is a stopping time, and $Y(t)$ is the Brownian motion stopped at that time.

The other direction is also true, a Branching diffusion can be obtained from a Brownian motion with drift. Let $Y(t)$ satisfy (7.61), and and let $T = \inf\{t : Y(t) = 0\}$. $Y(t) > 0$ for $t \leq T$. Define

$$G(t) = \int_0^{t \wedge T} \frac{1}{Y(s)}ds,$$

and let τ_t be the inverse of G, which is well defined on $[0, T)$. Then $X(t) = Y(\tau_t)$ satisfies the SDE (7.60) stopped when it hits zero.

Remark 7.4: Any solution to an SDE with time independent coefficients can be obtained from Brownian motion by using change of variables and random time change (Gihman and Skorohod (1972), p.113).

There are three main methods used for solving SDEs: change of state space, that is, change of variable (Itô's formula), change of time and change of measure. We have seen examples of SDEs solved by using change of variables, and change of time. The change of measure approach will be covered later.

Notes. Material for this chapter is based on Protter (1992), Rogers and Williams (1990), Gihman and Skorohod (1972), Liptser and Shiryayev (1977), (1989), Revuz and Yor (1998).

7.9 Exercises

Exercise 7.1: Let $M(t)$ be an $\mathcal{F}_t$-martingale and denote its natural filtration by $\mathcal{G}_t$. Show that $M(t)$ is a $\mathcal{G}_t$-martingale.

Exercise 7.2: Show that an increasing integrable process is a submartingale.

Exercise 7.3: Show that if $X(t)$ is a submartingale and g is a non-decreasing convex function such that $\mathrm{E}|g(X(t))| < \infty$, then $g(X(t))$ is a submartingale.

Exercise 7.4: Show that $M(t)$ is a square integrable martingale if and only if $M(t) = \mathrm{E}(Y|\mathcal{F}_t)$, where Y is square integrable, $\mathrm{E}(Y^2) < \infty$.

Exercise 7.5: (Expected exit time of Brownian motion from (a, b).)
Let $B(t)$ be a Brownian motion started at $x \in (a, b)$, and $\tau = \inf\{t : B(t) = a \text{ or } b\}$. By stopping the martingale $M(t) = B(t)^2 - t$, show that $\mathrm{E}_x(\tau) = (x - a)(b - x)$.

Exercise 7.6: Find the probability of $B(t) - t/2$ reaching a before it reaches b when started at x, $a < x < b$. Hint: use the exponential martingale $M(t) = e^{B(t)-t/2}$.

Exercise 7.7: Find the expected length of the game in Gambler's ruin, when

- betting is done on a fair coin

- betting is done on a biased coin

Exercise 7.8: Give the probability of ruin when playing a game of chance against an infinitely rich opponent (with initial capital $b \to \infty$).

Exercise 7.9: (Ruin Probability in Insurance) A Discrete Time Risk Model for the surplus U_n of an insurance company at the end of year n, $n = 1, 2, \ldots$ is given by $U_n = U_0 + cn - \sum_{k=1}^{n} X_k$, where c is the total annual premium, X_k is the total (aggregate) claim in year k. The time of ruin T is the first time when the surplus becomes negative, $T = \min\{n : U_n < 0\}$, with $T = \infty$ if $U_n \geq 0$ for all n. Assume that $\{X_k, k = 1, 2, \cdots\}$ are i.i.d. random variables, and there exists a constant $R > 0$ such that $\mathrm{E}\left(e^{-R(c-X_1)}\right) = 1$. Show that for all n, $\mathrm{P}_x(T \leq n) \leq e^{-Rx}$, where $U_0 = x$ the initial funds, and the ruin probability $\mathrm{P}_x(T < \infty) \leq e^{-Rx}$. Hint: show that $M_n = e^{-RU_n}$ is a martingale, and use the Optional Stopping Theorem.

Exercise 7.10: (Ruin Probability in Insurance continued) Find the bound on the ruin probability when the aggregate claims have $N(\mu, \sigma^2)$ distribution. Give the initial amount x required to keep the ruin probability below level α.

Exercise 7.11: Let $B(t)$ be a Brownian motion starting at zero and T be the first exit time from $(-1, 1)$, that is, the first time when $|B|$ takes value 1. Use Davis' inequality to show that $E(\sqrt{T}) < \infty$.

Exercise 7.12: Let $B(t)$ be a Brownian motion, $X(t) = \int_0^t \text{sign}(B(s))dB(s)$. Show that X is also a Brownian motion.

Exercise 7.13: Let $M(t) = \int_0^t e^s dB(s)$. Find $g(t)$ such that $M(g(t))$ is a Brownian motion.

Exercise 7.14: Let $B(t)$ be a Brownian motion. Give an SDE for $e^{-\alpha t}B(e^{2\alpha t})$.

Exercise 7.15: Prove the change of time result in SDEs, Theorem 7.41.

Exercise 7.16: Let $X(t)$ satisfy SDE $dX(t) = \mu(t)dt + \sigma(t)dB(t)$ on $[0, T]$. Show that $X(t)$ is a local martingale if and only if $\mu(t) = 0$ a.e.

Exercise 7.17: $f(x, t)$ is differentiable in t and twice in x. It is known that $X(t) = f(B(t), t)$ is of finite variation. Show that f is a function of t alone.

Exercise 7.18: Let $Y(t) = \int_0^t B(s)dB(s)$ and $W(t) = \int_0^t \text{sign}(B(s))dB(s)$. Show that $dY(t) = \sqrt{t + 2Y(t)}dW(t)$. Show uniqueness of the weak solution of the above SDE.

Chapter 8

Calculus For Semimartingales

In this chapter rules of calculus are given for the most general processes for which stochastic calculus is developed, called semimartingales. A semimartingale is process consisting of a sum of a local martingale and a finite variation process. Integration with respect to semimartingales involves integration with respect to local martingales, and these integrals generalize the Itô integral where integration is done with respect to a Brownian motion. Important concepts, such as compensators and the sharp bracket processes are introduced, and Itô's formula in its general form is given.

8.1 Semimartingales

In stochastic calculus only regular processes are considered. These are either continuous processes, or right-continuous with left limits, or left-continuous with right limits. The regularity of the process implies that it can have at most countably many discontinuities, and all of them are jumps (Chapter 1). The definition of a semimartingale presumes a given filtration and processes which we consider are adapted to it. Following the classical approach, see for example, Metivier (1982), Liptser and Shiryayev (1989) p.85, a semimartingale, is a local martingale plus a process of finite variation. More precisely,

Definition 8.1 *A regular right-continuous with left limits (càdlàg) adapted process is a semimartingale if it can be represented as a sum of two processes: a local martingale $M(t)$ and a process of finite variation $A(t)$, with $M(0) = A(0) = 0$, and*

$$S(t) = S(0) + M(t) + A(t). \tag{8.1}$$

Example 8.1: (Semimartingales)

1. $S(t) = B^2(t)$, where $B(t)$ is a Brownian motion is a semimartingale. $S(t) = M(t) + t$, where $M(t) = B^2(t) - t$ is a martingale and $A(t) = t$ is a finite variation process.

2. $S(t) = N(t)$, where $N(t)$ is a Poisson process with rate λ, is a semimartingale, as it is a finite variation process.

3. One way to obtain semimartingales from known semimartingales is by applying a twice continuously differentiable (C^2) transformation. If $S(t)$ is a semimartingale and f is a C^2 function, then $f(S(t))$ is also a semimartingale. The decomposition of $f(S(t))$ into martingale part and finite variation part is given by Itô's formula, given later. In this way we can assert that, for example, the geometric Brownian motion $e^{\sigma B(t)+\mu t}$ is a semimartingale.

4. A right-continuous with left limits (càdlàg) deterministic function $f(t)$ is a semimartingale if and only if it is of finite variation. Thus $f(t) = t\sin(1/t)$, $t \in (0,1]$, $f(0) = 0$ is continuous, but not a semimartingale (see Example 1.7).

5. A diffusion, that is, a solution to a stochastic differential equation with respect to Brownian motion, is a semimartingale. Indeed, the Itô integral with respect to $dB(t)$ is a local martingale and the integral with respect to dt is a process of finite variation.

6. Although the class of semimartingales is rather large, there are processes which are not semimartingales. Examples are: $|B(t)|^\alpha$, $0 < \alpha < 1$, where $B(t)$ is the one-dimensional Brownian motion; $\int_0^t (t-s)^{-\alpha} dB(s)$, $0 < \alpha < 1/2$. It requires analysis to show that the above processes are not semimartingales.

For a semimartingale X, the process of jumps ΔX is defined by

$$\Delta X(t) = X(t) - X(t-), \tag{8.2}$$

and represents the jump at point t. If X is continuous, then of course, $\Delta X = 0$.

8.2   Predictable Processes

In this section we describe the class of predictable processes. This class of processes has a central role in the theory. In particular, only predictable processes can be integrated with respect to a semimartingale. Recall that in discrete time a process H is predictable if H_n is $\mathcal{F}_{n-1}$ measurable, that is, H is known with certainty at time n on the basis of information up to time $n-1$. Predictability in continuous time is harder to define. We recall some general definitions of processes starting with the class of adapted processes.

Definition 8.2 *A process X is called adapted to filtration $\mathbb{F} = (\mathcal{F}_t)$, if for all t, $X(t)$ is $\mathcal{F}_t$-measurable.*

In construction of the stochastic integral $\int_0^t H(u)dS(u)$, processes H and S are taken to be adapted to $\mathbb{F}$. For a general semimartingale S, the requirement that H is adapted is too weak, it fails to assure measurability of some basic constructions. H must be *predictable*. The exact definition of predictable processes involves σ-fields generated on $\mathbb{R}^+ \times \Omega$ and is given later in Section 8.13. Note that left-continuous processes are predictable, in the sense that $H(t) = \lim_{s \uparrow t} H(s) = H(s-)$. So that if the values of the process before t are known, then the value at t is determined by the limit. For our purposes it is enough to describe a subclass of predictable processes which can be defined constructively.

Definition 8.3 *H is predictable if it is one of the following:*

a) *a left-continuous adapted process, in particular, a continuous adapted process.*

b) *a limit (almost sure, in probability) of left-continuous adapted processes.*

c) *a regular right-continuous process such that, for any stopping time τ, H_τ is $\mathcal{F}_{\tau-}$-measurable, the σ-field generated by the sets $A \cap \{T < t\}$, where $A \in \mathcal{F}_t$.*

d) *a Borel-measurable function of a predictable process.*

Example 8.2: Poisson process $N(t)$ is right-continuous and is obviously adapted to its natural filtration. It can be shown, see Example 8.31, that it is not predictable. Its left-continuous modification $N(t-) = \lim_{s \uparrow t} N(s)$ is predictable, because it is adapted and left-continuous by a). Any measurable function (even right-continuous) of $N(t-)$ is also predictable by d).

Example 8.3: Right-continuous adapted processes may not be predictable, even though they can be approached by left-continuous processes, for example, $X_\epsilon(t) = \lim_{\epsilon \to 0} X((t + \epsilon)-)$.

Example 8.4: Let T be a stopping time. This means that for any t, the set $\{T > t\} \in \mathcal{F}_t$. Consider the process $X(t) = I_{[0,T]}(t)$. It is adapted, because its values are determined by the set $\{T \leq t\}$ ($X(t) = 1$ if and only if $\omega \in \{T \leq t\}$), and $\{T \leq t\} = \{T > t\}^c \in \mathcal{F}_t$. $X(t)$ is also left-continuous. Thus it is a predictable process by a). We also see that T is a stopping time if and only if the process $X(t) = I_{[0,T]}(t)$ is adapted.

Example 8.5: It will be seen later that when filtration is generated by Brownian motion, then any right-continuous adapted process is predictable. This is why in the definition of the Itô integral right-continuous functions are allowed as integrands.

8.3   Doob-Meyer Decomposition

Recall that a process is a submartingale if for all $s < t$, $\mathrm{E}(X(t)|\mathcal{F}_s) \geq X(s)$ almost surely

Theorem 8.4 *If X is a submartingale or a local submartingale, then there exists a local martingale $M(t)$ and a unique increasing predictable process $A(t)$, locally integrable, such that*

$$X(t) = X(0) + M(t) + A(t). \tag{8.3}$$

If $X(t)$ is a submartingale of Dirichlet class (D) (see Definition 7.25), then the process A is integrable, that is, $\sup_t \mathrm{E}A(t) < \infty$, and $M(t)$ is a uniformly integrable martingale.

Example 8.6:

1. Let $X(t) = B^2(t)$ on a finite interval $t \leq T$. $X(t)$ is a submartingale. Decomposition (8.3) holds with $M(t) = B^2(t) - t$ and $A(t) = t$. Since the interval is finite, M is uniformly integrable and A is integrable.

2. Let $X(t) = B^2(t)$ on the infinite interval $t \geq 0$. Then (8.3) holds with $M(t) = B^2(t) - t$ and $A(t) = t$. Since the interval is infinite M is a martingale, and A is locally integrable; for example take the localizing sequence $\tau_n = n$.

3. Let $X(t) = N(t)$ be a Poisson process with intensity λ. Then X is a submartingale. Decomposition (8.3) holds with $M(t) = N(t) - \lambda t$ and $A(t) = \lambda t$.

Proof of Theorem 8.4 can be found in Rogers and Williams (1990), p.372-375, Dellacherie (1972), Meyer (1966).

Doob's Decomposition

For processes in discrete time, decomposition (8.3) is due to Doob, and it is simple to obtain. Indeed, if X_n is a submartingale, then clearly, $X_{n+1} = X_0 + \sum_{i=0}^{n} (X_{i+1} - X_i)$. By adding and subtracting $\mathrm{E}(X_{i+1}|\mathcal{F}_i)$, we obtain the Doob decomposition

$$X_{n+1} = X_0 + \sum_{i=0}^{n} \left((X_{i+1} - \mathrm{E}(X_{i+1}|\mathcal{F}_i)) \right) + \sum_{i=0}^{n} \left(\mathrm{E}(X_{i+1}|\mathcal{F}_i) - X_i \right), \tag{8.4}$$

where martingale and an increasing process are given by

$$M_{n+1} = \sum_{i=0}^{n} (X_{i+1} - \mathrm{E}(X_{i+1}|\mathcal{F}_i)) \text{ and } A_{n+1} = \sum_{i=0}^{n} \left(\mathrm{E}(X_{i+1}|\mathcal{F}_i) - X_i \right) \tag{8.5}$$

A_n is increasing due to the submartingale property, $\mathrm{E}(X_{i+1}|\mathcal{F}_i) - X_i \geq 0$ for all i. It is also predictable, because $\mathrm{E}(X_{n+1}|\mathcal{F}_n)$ and all other terms are $\mathcal{F}_n$ measurable.

It is much harder to prove decomposition (8.3) in continuous time and this was done by Meyer.

8.4 Integrals with respect to Semimartingales

In this section the stochastic integral $\int_0^T H(t)dS(t)$ is defined, where $S(t)$ is a semimartingale. Due to representation $S(t) = S(0) + M(t) + A(t)$ the integral with respect to $S(t)$ is the sum of two integrals one with respect to a local martingale $M(t)$ and the other with respect to a finite variation process $A(t)$. The integral with respect to $A(t)$ can be done path by path as the Stieltjes integral, since $A(t)$, although random, is of finite variation.

The integral with respect to the martingale $M(t)$ is new, it is the stochastic integral $\int_0^T H(t)dM(t)$. When $M(t)$ is Brownian motion $B(t)$, it is the Itô integral, defined in Chapter 4. But now martingales are allowed to have jumps and this makes the theory more complicated. The key property used in the definition of the Itô integral is that on finite intervals Brownian motion is a square integrable martingale. This property in its local form plays an important role in the general case. Conditions for the existence of the integral with respect to a martingale involves the martingale's quadratic variation, which was introduced in Section 7.6.

Stochastic Integral with respect to Martingales

For a simple predictable process $H(t)$, given by

$$H(t) = H(0)I_0 + \sum_{i=0}^{n-1} H_i I_{(T_i, T_{i+1}]}(t), \tag{8.6}$$

where $0 = T_0 \leq T_1 \leq \ldots \leq T_n \leq T$ are stopping times and H_i's are $\mathcal{F}_{T_i}$-measurable, the stochastic integral is defined as the sum

$$\int_0^T H(t)dS(t) = \sum_{i=0}^{n-1} H_i \big(M(T_{i+1}) - M(T_i) \big). \tag{8.7}$$

If $M(t)$ is a locally square integrable martingale, then by the L^2 theory (Hilbert space theory) one can extend the stochastic integral from simple predictable processes to the class of predictable processes H such that

$$\sqrt{\int_0^T H^2(t)d[M, M](t)} \quad \text{is locally integrable.} \tag{8.8}$$

If $M(t)$ is a *continuous* local martingale, then the stochastic integral is defined for a wider class of predictable processes H satisfying

$$\int_0^T H^2(t)d[M, M](t) < \infty \quad \text{a.s.} \tag{8.9}$$

Properties of Stochastic Integrals with respect to Martingales

1. Local martingale property. If $M(t)$ is a local martingale, the integral $\int_0^t H(s)dM(s)$ is a local martingale.

2. Isometry property. If $M(t)$ is a square integrable martingale, and H satisfies

$$\mathrm{E}\left(\int_0^T H^2(s)d[M,M](s) \right) < \infty, \tag{8.10}$$

then $\int_0^t H(s)dM(s)$ is a square integrable martingale with zero mean and variance

$$\mathrm{E}\left(\int_0^t H(s)dM(s) \right)^2 = \mathrm{E}\left(\int_0^t H^2(s)d[M,M](s) \right). \tag{8.11}$$

3. If a local martingale $M(t)$ is of finite variation, then the stochastic integral is indistinguishable from the Stieltjes integral.

Example 8.7: Consider Itô integrals with respect to Brownian motion. Since $B(t)$ is a square integrable martingale on $[0,T]$, with $[B,B](t) = t$, we recover that for a predictable H, such that $\mathrm{E}\int_0^T H^2(s)ds < \infty$, $\int_0^t H(s)dB(s)$ is a square integrable martingale with zero mean and variance $\int_0^t \mathrm{E}H^2(s)ds$.

Stochastic Integrals with respect to Semimartingales

Let S be a semimartingale with representation

$$S(t) = S(0) + M(t) + A(t), \tag{8.12}$$

where M is a local martingale and A is a finite variation process. Let H be a predictable process such that conditions (8.13) and (8.8) hold.

$$\int_0^T |H(t)|dV_A(t) < \infty, \tag{8.13}$$

where $V_A(t)$ is the variation process of A. Then the stochastic integral is defined as the sum of integrals,

$$\int_0^t H(t)dS(t) = \int_0^T H(t)dM(t) + \int_0^T H(t)dA(t). \tag{8.14}$$

Since a representation of a semimartingale (8.12) is not unique, one should check that the the stochastic integral does not depend on the representation used. Indeed, if $S(t) = S(0) + M_1(t) + A_1(t)$ is another representation, then

$(M - M_1)(t) = -(A - A_1)(t)$. So that $M - M_1$ is a local martingale of finite variation. But for such martingales stochastic and Stieltjes integrals are the same, and it follows that $\int H(t)dM_1(t) + \int H(t)dA_1(t) = \int H(t)dM(t) + \int H(t)dA(t) = \int H(t)dS(t)$.

Since the integral with respect to a local martingale is a local martingale, and the integral with respect to a finite variation process is a process of finite variation, it follows that a stochastic integral with respect to a semimartingale is a semimartingale.

For details see Liptser and Shiryayev (1989), p.90-116.

Example 8.8: Let $N(t)$ be a Poisson process. $N(t)$ is of finite variation and the integral $\int_0^T N(t)dN(t)$ is well defined as a Stieltjes integral, $\int_0^T N(t)dN(t) = \sum_{\tau_i \leq T} N(\tau_i)$, where τ_i's are the jumps of $N(t)$. However, $\int_0^T N(t)dN(t)$ is not the stochastic integral, since $N(t)$ is not predictable, but $\int_0^T N(t-)dN(t)$ is. It is indistinguishable from the integral in the sense of Stieltjes $\int_0^T N(t-)dN(t) = \sum_{\tau_i \leq T} N(\tau_{i-1})$.

Properties of Stochastic Integrals with respect to Semimartingales

Let X be a semimartingale and H a predictable processes, such that the stochastic integral exists for $0 \leq t \leq T$, and denote

$$(H \cdot X)(t) := \int_0^t H(s)dX(s).$$

Then the stochastic integral $H \cdot X$ has the following properties

1. The jumps of the integral occur at the points of jumps of X, and $\Delta(H \cdot X)(t) = H(t)\Delta X(t)$. In particular, a stochastic integral with respect to a continuous semimartingale is continuous.

2. If τ is a stopping time, then the stopped integral is the integral with respect to the stopped semimartingale,
$$\int_0^{t \wedge \tau} H(s)dX(s) = \int_0^t H(s)I(s \leq \tau)dX(s) = \int_0^t H(s)dX(s \wedge \tau).$$

3. If X is of finite variation, then $\int_0^t H(s)dX(s)$ is indistinguishable from the Stieltjes integral computed path by path.

4. Associativity. If $Y(t) = \int_0^t H(s)dX(s)$ is a semimartingale, and if K is a predictable process, such that $(K \cdot Y)(t) = \int_0^t K(s)dY(s)$ is defined, then $K \cdot Y = K \cdot (H \cdot X) = (KH) \cdot X$, that is,
$$\int_0^t K(s)dY(s) = \int_0^t K(s)H(s)dX(s).$$

8.5 Quadratic Variation and Covariation

If X, Y are semimartingales on the common space, then the quadratic co-variation process, also known as the *the square bracket* process and denoted $[X, Y](t)$, is defined, as usual, by

$$[X, Y](t) = \lim \sum_{i=0}^{n-1} (X(t_{i+1}^n) - X(t_i^n))(Y(t_{i+1}^n) - Y(t_i^n)), \qquad (8.15)$$

where the limit is taken over shrinking partitions $\{t_i^n\}_{i=0}^n$ of the interval $[0, t]$ when $\delta_n = \max_i(t_{i+1}^n - t_i^n) \to 0$ and is in probability. Taking $Y = X$ we obtain the quadratic variation process of X.

Example 8.9: We have seen that quadratic variation of Brownian motion $B(t)$ is $[B, B](t) = t$ and of Poisson process $N(t)$ is $[N, N](t) = N(t)$.

Properties of Quadratic Variation

We give the fundamental properties of the quadratic variation process with some explanations, but omit the proofs.

1. If X is a semimartingale, then $[X, X]$ exists and is an adapted process.

2. It is clear from the definition that quadratic variation over non-overlapping intervals is the sum of the quadratic variation over each interval. As such, $[X, X](t)$ is non-decreasing function of t. Consequently $[X, X](t)$ is a function of finite variation.

3. It follows from the definition (8.15) that $[X, Y]$ is bilinear and symmetric, that is, $[X, Y] = [Y, X]$ and

$$[\alpha X + Y, \beta U + V] = \alpha\beta[X, U] + \alpha[X, V] + \beta[Y, U] + [Y, V]. \qquad (8.16)$$

4. Polarization identity.

$$[X, Y] = \frac{1}{2}\Big([X + Y, X + Y] - [X, X] - [Y, Y]\Big). \qquad (8.17)$$

This property follows directly from the previous one.

5. $[X, Y](t)$ is a regular right-continuous (càdlàg) function of finite variation. This follows from the polarization identity, as $[X, Y]$ is the difference of two increasing functions.

6. The jumps of the quadratic covariation process occur only at points where *both* processes have jumps,

$$\Delta[X, Y](t) = \Delta X(t)\Delta Y(t). \qquad (8.18)$$

7. If one of the processes X or Y is of finite variation, then

$$[X, Y](t) = \sum_{s \leq t} \Delta X(s) \Delta Y(s). \tag{8.19}$$

Notice that although the summation is taken over all s not exceeding t, there are at most countably many terms different from zero.

The following property is frequently used, it follows directly from (8.19).

Corollary 8.5 *If $X(t)$ is a continuous semimartingale of finite variation, then it has zero quadratic covariation with any other semimartingale $Y(t)$.*

Quadratic Variation of Stochastic Integrals

The quadratic covariation of stochastic integrals has the following property

$$\left[\int_0^{\cdot} H(s)dX(s), \int_0^{\cdot} K(s)dY(s) \right](t) = \int_0^t H(s)K(s)d[X, Y](s). \tag{8.20}$$

In particular the quadratic variation of a stochastic integral is given by

$$\left[\int_0^{\cdot} H(s)dX(s), \int_0^{\cdot} H(s)dX(s) \right](t) = \int_0^t H^2(s)d[X, X](s), \tag{8.21}$$

and

$$\begin{aligned}
\left[\int_0^{\cdot} H(s)dX(s), Y \right](t) &= \left[\int_0^{\cdot} H(s)dX(s), \int_0^{\cdot} 1dY(s) \right](t) \\
&= \int_0^t H(s)d[X, Y](s). \tag{8.22}
\end{aligned}$$

Quadratic variation has a representation in terms of the stochastic integral.

Theorem 8.6 *Let X be a semimartingale null at zero. Then*

$$[X, X](t) = X^2(t) - 2\int_0^t X(s-)dX(s). \tag{8.23}$$

For a partition $\{t_i^n\}$ of $[0, t]$, consider

$$v_n(t) = \sum_{i=0}^{n-1} (X(t_{i+1}^n) - X(t_i^n))^2.$$

For any fixed t the sequence $v_n(t)$ converges in probability to a limit $[X, X](t)$ as $\max_{i \leq n}(t_{i+1}^n - t_i^n) \to 0$. Moreover, there is a subsequence n_k, such that the processes $v_{n_k}(t)$, converge uniformly on any bounded time interval to the process $X^2(t) - 2\int_0^t X(s-)dX(s)$.

We justify (8.23) heuristically. By opening the brackets in the sum for $v_n(t)$, adding and subtracting $X^2(t_i)$, we obtain

$$v_n(t) = \sum_{i=0}^{n-1} \left(X^2(t_{i+1}^n) - X^2(t_i^n) \right) - 2 \sum_{i=0}^{n-1} X(t_i^n) \left(X(t_{i+1}^n) - X(t_i^n) \right).$$

The first sum is $X^2(t)$. The second has for its limit in probability $\int_0^t X(s-)dX(s)$. This argument can be made into a rigorous proof, see for example, Metivier (1982), p.175. Alternatively (8.23) can be established by using Itô's formula.

Using the polarization identity, it is easy to see

Corollary 8.7 *For semimartingales X, Y the quadratic covariation process is given by*

$$[X, Y](t) = X(t)Y(t) - X(0)Y(0) - \int_0^t X(s-)dY(s) - \int_0^t Y(s-)dX(s). \quad (8.24)$$

This is also known as the integration by parts or product rule formula.

8.6   Itô's Formula for Continuous Semimartingales

If $X(t)$ is a *continuous* semimartingale and f is a twice continuously differentiable function, then $Y(t) = f(X(t))$ is a semimartingale and admits the following representation

$$f(X(t)) - f(X(0)) = \int_0^t f'(X(s))dX(s) + \frac{1}{2} \int_0^t f''(X(s))d[X, X](s). \quad (8.25)$$

In differential form this is written as

$$df(X(t)) = f'(X(t))dX(t) + \frac{1}{2}f''(X(t))d[X, X](t). \quad (8.26)$$

It follows, in particular, that $f(X(t))$ is also a semimartingale, and its decomposition into the martingale part and the finite variation part can be obtained from Itô's formula by splitting the stochastic integral with respect to $X(t)$ into the integral with respect to a local martingale $M(t)$ and a finite variation process $A(t)$.

We have given a justification of Itô's formula and examples of its use in Chapters 4 and 6.

Remark 8.1:

1. The differentiability properties of f may be relaxed. If, for example, X is of finite variation, then f needs to be only once continuously differentiable. f can be defined only on an open set, rather than a whole line, but then X must take its values almost surely in this set. For example, if X is a positive semimartingale, then Itô's formula can be used with $f = \ln$.

2. Itô's formula holds for convex functions (Protter (1992) p.163), and more generally, for functions which are the difference of two convex functions. This is the Meyer-Itô (Itô-Tanaka) formula, see for example, Protter (1992) p.167, Rogers and Williams (1990), p.105, Revuz and Yor p.208. In particular, if f is a convex function on $\mathbb{R}$ and $X(t)$ is a semimartingale, then $f(X(t))$ is also a semimartingale. See also Section 8.7.

3. It follows from Itô's formula that if a semimartingale X is continuous with nil quadratic variation $[X, X](t) = 0$, then the differentiation rule is the same as in the ordinary calculus. If $X(t)$ is a Brownian motion, then $d[X, X](t) = dt$ and we recover formulae (4.39) and (4.53). If X has jumps, then the formula has an extra term (see Section 8.10).

The following result is a direct corollary to Itô's formula.

Corollary 8.8 *Let $X(t)$ be a continuous semimartingale and f be twice continuously differentiable. Then*

$$[f(X), f(X)](t) = \int_0^t \Big(f'(X(s)) \Big)^2 d[X, X](s). \tag{8.27}$$

PROOF: Since $[X, X]$ is of finite variation it follows from (8.25)

$$[f(X), f(X)](t) = \left[\int_0^{\cdot} f'(X(s)) dX(s), \int_0^{\cdot} f'(X(s)) dX(s) \right](t).$$

$\square$

Itô's Formula for Functions of Several Variables

Let $f : \mathbb{R}^n \to \mathbb{R}$ be C^2, and let $\boldsymbol{X}(t) = (X_1(t), \ldots, X_n(t))$ be a *continuous* semimartingale in $\mathbb{R}^n$, that is, each X_i is a continuous semimartingale. Then $f(\boldsymbol{X})$ is a semimartingale, and has the following representation.

$$
\begin{aligned}
f(\boldsymbol{X}(t)) - f(\boldsymbol{X}(0)) &= \int_0^t \sum_{i=1}^n \frac{\partial f}{\partial x_i}(\boldsymbol{X}(s)) dX_i(s) \\
&+ \frac{1}{2} \int_0^t \sum_{i=1}^n \sum_{j=1}^n \frac{\partial^2 f}{\partial x_i \partial x_j}(\boldsymbol{X}(s)) d[X_i, X_j](s).
\end{aligned} \tag{8.28}
$$

8.7   Local Times

Let $X(t)$ be a continuous semimartingale. Consider $|X(t) - a|$, $a \in \mathbb{R}$. The function $|x - a|$ is not differentiable at a, but at any other point its derivative is given by $\text{sign}(x - a)$, where $\text{sign}(x) = 1$ for $x > 0$ and $\text{sign}(x) = -1$ for $x \leq 0$. It is possible to extend Itô's formula for this case and prove (see Rogers and Williams (1990), p.95-102, Protter (1992) p.165-167)

Theorem 8.9 (Tanaka's Formula) *Let $X(t)$ be a continuous semimartingale. Then for any $a \in \mathbb{R}$ there exists a continuous non-decreasing adapted process $L^a(t)$, called the local time at a of X, such that*

$$|X(t) - a| = |X(0) - a| + \int_0^t \text{sign}(X(s) - a)dX(s) + L^a(t). \qquad (8.29)$$

As a function in a, $L^a(t)$ is right-continuous with left limits. For any fixed a as a function in t $L^a(t)$ increases only when $X(t) = a$, that is, $L^a(t) = \int_0^t I(X(s) = a)dL^a(s)$. Moreover, if $X(t)$ is a continuous local martingale, then $L^a(t)$ is jointly continuous in a and t.

Remark 8.2: Heuristically Tanaka's formula can be justified by a formal application of Itô's formula to the function $\text{sign}(x)$. The derivative of $\text{sign}(x)$ is zero everywhere but at zero, where it is not defined. However, it is possible to define the derivative as a generalized function or a Schwartz distribution, in which case it is equal to 2δ. Thus the second derivative of $|x - a|$ is $\delta(x - a)$ in the generalized function sense. The local time at a of X is defined as $L^a(t) = \int_0^t \delta(X(s) - a)ds$. Formal use of Itô's formula gives 8.29.

Theorem 8.10 (Occupation Times Formula) *Let $X(t)$ be a continuous semimartingale with local time $L^a(t)$. Then for any bounded measurable function $g(x)$*

$$\int_0^t g(X(s))d[X, X](s) = \int_{-\infty}^{\infty} g(a)L^a(t)da. \qquad (8.30)$$

In particular

$$[X, X](t) = \int_{-\infty}^{\infty} L^a(t)da. \qquad (8.31)$$

Example 8.10: Let $X(t) = B(t)$ be Brownian motion. Then its local time at zero process, $L^0(t)$ satisfies (Tanaka's formula)

$$L^0(t) = |B(t)| - \int_0^t \text{sign}(B(s))dB(s). \qquad (8.32)$$

The occupation times formula (8.30) becomes

$$\int_0^t g(B(s))ds = \int_{-\infty}^\infty g(a)L^a(t)da. \qquad (8.33)$$

The time Brownian motion spends in a set $A \subset \mathbf{R}$ up to time t is given by (with $g(x) = I_A(x)$)

$$\int_0^t I_A(B(s))ds = \int_{-\infty}^\infty I_A(a)L^a(t)da = \int_A L^a(t)da. \qquad (8.34)$$

Remark 8.3: Taking $A = (a, a + da)$ and $g(x) = I_{(a,a+da)}(x)$ its indicator in (8.33), $L^a(t)da$ is the time Brownian motion spends in $(a, a + da)$ up to time t, which explains the name "local time". The time Brownian motion spends in a set A is $\int_A L^a(t)da$, therefore the name "occupation times density" formula (8.34). For a continuous semimartingale the formula (8.30) is the "occupation times density" formula relative to the random "clock" $d[X, X](s)$.

Example 8.11: $X(t) = |B(t)|$ is a semimartingale, since $|x|$ is a convex function. Its decomposition into the martingale and finite variation parts is given by Tanaka's formula (8.32). $\sqrt{|B(t)|}$ is not a semimartingale, see Protter (1992), p.169-170.

Example 8.12: The function $(x - a)^+$ is important in financial application, as it gives the payoff of a financial stock option. The Meyer-Tanaka's formula for $(x - a)^+$

$$(X(t) - a)^+ = (X(0) - a)^+ + \int_0^t I(X(s) > a)dX(s) + \frac{1}{2}L_t^a. \qquad (8.35)$$

Theorem 8.11 *Let $L^a(t)$ be the local time of Brownian motion at a, and $f_t(a)$ the density of $N(0, t)$ at a. Then*

$$\mathrm{E}(L^a(t)) = \int_0^t f_s(a)da \quad hence \quad \frac{d\mathrm{E}(L^a(t))}{dt} = f_t(a). \qquad (8.36)$$

PROOF: Taking expectation in both sides of equation (8.33) and changing the order of integration, we obtain for any positive and bounded g

$$\int_{-\infty}^\infty g(a)f_s(a)dsda = \int_{-\infty}^\infty g(a)\mathrm{E}(L^a(t))da.$$

The result follows, since g is arbitrary.

□

A similar result can be established for continuous semimartingales by using equation (8.30) (e.g. Klebaner (2002)).

Remark 8.4: Local times can also be defined for discontinuous semimartingales. For any fixed a, $L^a(t)$ is a continuous non-decreasing function in t, and it increases only at points of continuity of X where it is equal to a, that is, $X(t-) = X(t) = a$. The formula (8.30) holds with quadratic variation $[X, X]$ replaced by its continuous part $[X, X]^c$, see for example, Protter (1992), p.168.

8.8 Stochastic Exponential

The stochastic exponential (also known as the semimartingale, or Doléans-Dade exponential) is a stochastic analogue of the exponential function. Recall that if $f(t)$ is a smooth function then $g(t) = e^{f(t)}$ is the solution to the differential equation $dg(t) = g(t)df(t)$. The stochastic exponential is defined as a solution to a similar stochastic equation. The stochastic exponential of Itô processes was introduced in Section 5.2. For a semimartingale X, its stochastic exponential $\mathcal{E}(X)(t) = U(t)$ is defined as the unique solution to the equation

$$U(t) = 1 + \int_0^t U(s-)dX(s) \quad \text{or} \quad dU(t) = U(t-)dX(t); \quad \text{with } U(0) = 1. \quad (8.37)$$

As an application of Itô's formula and the rules of stochastic calculus we prove

Theorem 8.12 *Let X be a continuous semimartingale. Then its stochastic exponential is given by*

$$U(t) = \mathcal{E}(X)(t) = e^{X(t)-X(0)-\frac{1}{2}[X,X](t)}. \quad (8.38)$$

PROOF: Write $U(t) = e^{V(t)}$, with $V(t) = X(t) - X(0) - \frac{1}{2}[X,X](t)$. Then

$$dU(t) = d(e^{V(t)}) = e^{V(t)}dV(t) + \frac{1}{2}e^{V(t)}d[V,V](t).$$

Using the fact that $[X,X](t)$ is a continuous process of finite variation, we obtain $[X,[X,X]](t) = 0$, and $[V,V](t) = [X,X](t)$. Using this, we obtain

$$dU(t) = e^{V(t)}dX(t) - \frac{1}{2}e^{V(t)}d[X,X](t) + \frac{1}{2}e^{V(t)}d[X,X](t) = e^{V(t)}dX(t),$$

or $dU(t) = U(t)dX(t)$. Thus $U(t)$ defined by (8.38) satisfies (8.37). To show uniqueness, let $V(t)$ be another solution to (8.37), and consider $V(t)/U(t)$. By integration by parts

$$d(\frac{V(t)}{U(t)}) = V(t)d(\frac{1}{U(t)}) + \frac{1}{U(t)}dV(t) + d[V, \frac{1}{U}](t).$$

By Itô's formula, using that $U(t)$ is continuous and satisfies (8.37)

$$d(\frac{1}{U(t)}) = -\frac{1}{U(t)}dX(t) + \frac{1}{U(t)}d[X,X](t),$$

which leads to

$$d(\frac{V(t)}{U(t)}) = -\frac{V(t)}{U(t)}dX(t) + \frac{V(t)}{U(t)}dX(t) + \frac{V(t)}{U(t)}d[X,X](t) - \frac{V(t)}{U(t)}d[X,X](t) = 0.$$

Thus $V(t)/U(t) = const. = V(0)/U(0) = 1$.

$\square$

Properties of the stochastic exponential are given by the following result.

Theorem 8.13 *Let X and Y be semimartingales on the same space. Then*

    1. $\mathcal{E}(X)(t)\mathcal{E}(Y)(t) = \mathcal{E}(X + Y + [X, Y])(t)$

    2. If X is continuous, $X(0) = 0$, then $(\mathcal{E}(X)(t))^{-1} = \mathcal{E}(-X + [X, X])(t)$.

The proof uses the integration by parts formula and is left as an exercise.

Example 8.13: (Stock process and its Return process.)
An application in finance is provided by the relation between the stock process and its return process. The return is defined by $dR(t) = dS(t)/S(t-)$. Hence the stock price is the stochastic exponential of the return, $dS(t) = S(t-)dR(t)$, and $S(t) = S(0)\mathcal{E}(R)(t)$.

Stochastic Exponential of Martingales

Stochastic exponential $U = \mathcal{E}(M)$ of a martingale, or a local martingale, $M(t)$ is a stochastic integral with respect to $M(t)$. Since stochastic integrals with respect to martingales or local martingales are local martingales, $\mathcal{E}(M)$ is a local martingale. In applications it is important to have conditions for $\mathcal{E}(M)$ to be a true martingale.

Theorem 8.14 (Martingale exponential) *Let $M(t), 0 \leq t \leq T < \infty$ be a continuous local martingale null at zero. Then its stochastic exponential $\mathcal{E}(M)$ is given by $e^{M(t)-\frac{1}{2}[M,M](t)}$ and it is a continuous positive local martingale. Consequently, it is a supermartingale, it is integrable and has a finite non-increasing expectation. It is a martingale if any of the following conditions hold.*

    1. $\mathrm{E}\big(e^{M(T)-\frac{1}{2}[M,M](T)}\big) = 1.$

    2. For all $t \geq 0$, $\mathrm{E}\left(\int_0^t e^{2M(s)-[M,M](s)}d[M, M](s)\right) < \infty.$

    3. For all $t \geq 0$, $\mathrm{E}\left(\int_0^t e^{2M(s)}d[M, M](s)\right) < \infty.$

Moreover, if the expectations above are bounded by $K < \infty$, then $\mathcal{E}(M)$ is a square integrable martingale.

PROOF: By Theorem 8.12, $\mathcal{E}(M)(t) = e^{M(t)-\frac{1}{2}[M,M](t)}$, therefore it is positive. Being a stochastic integral with respect to a martingale, it is a local martingale. Thus $\mathcal{E}(M)$ is a supermartingale, as a positive local martingale, see Theorem 7.23. A supermartingale has a non-increasing expectation, and is a martingale if and only if its expectation at T is the same as at 0 (see Theorem 7.3). This gives the first condition. The second condition follows from Theorem 7.35, which states that if a local martingale has finite expectation of its quadratic variation, then it is a martingale. So that if $E[\mathcal{E}(M), \mathcal{E}(M)](t) < \infty$, then $\mathcal{E}(M)$ is a martingale. By the quadratic variation of an integral

$$[\mathcal{E}(M), \mathcal{E}(M)](t) = \int_0^t e^{2M(s) - [M,M](s)} d[M, M](s). \qquad (8.39)$$

The third condition follows from the second, since $[M, M]$ is positive and increasing. The last statement follows by Theorem 7.35, since the bound implies $\sup_{0 \le t} E[\mathcal{E}(M), \mathcal{E}(M)](t) < \infty$.

$\square$

Theorem 8.15 (Kazamaki's condition) *Let M be a continuous local martingale with $M(0) = 0$. If $e^{\frac{1}{2}M(t)}$ is a submartingale, then $\mathcal{E}(M)$ is a martingale.*

The result is proven in Revuz and Yor (1999), p. 331.

Theorem 8.16 *Let M be a continuous martingale with $M(0) = 0$. If*

$$E\left(e^{\frac{1}{2}M(T)}\right) < \infty, \qquad (8.40)$$

then $\mathcal{E}(M)$ is a martingale on $[0, T]$.

PROOF: By Jensen's inequality (see Exercise (7.3)) if g is a convex function, and $E|g(M(t)| < \infty$ for $t \le T$, then $Eg(M(t)) \le Eg(M(T))$ and $g(M(t))$ is a submartingale. Since $e^{x/2}$ is convex, the result follows by Kazamaki's condition.

$\square$

Theorem 8.17 (Novikov's condition) *Let M be a continuous local martingale with $M(0) = 0$. Suppose that for each $t \le T$*

$$E\left(e^{\frac{1}{2}[M,M](t)}\right) < \infty. \qquad (8.41)$$

Then $\mathcal{E}(M)$ is a martingale with mean one. In particular, if for each t there is a constant K_t such that $[M, M](t) < K_t$, then $\mathcal{E}(M)(t)$, $t \le T$ is a martingale.

PROOF: The condition (8.41) implies that $[M, M](t)$ has moments. By the BDG inequality (7.41), $\sup_{t \le T} M(t)$ is integrable, therefore $M(t)$ is a martingale. Next, using the formula for $\mathcal{E}(M)$ and Jensen's inequality, $E\sqrt{X} \le \sqrt{E(X)}$,

$$E(e^{\frac{1}{2}M(T)}) = E\sqrt{\mathcal{E}(M)(T)e^{\frac{1}{2}[M,M](T)}} \le \sqrt{E\mathcal{E}(M)(T)E\left(e^{\frac{1}{2}[M,M](T)}\right)} < \infty. \qquad (8.42)$$

The last expression is finite since $E\mathcal{E}(M)(T) < \infty$, as $\mathcal{E}(M)$ is a supermartingale, and by the condition of the theorem. This shows that the condition (8.40) holds and the result follows.

□

Another proof can be found in Karatzas and Shreve (1988), p.198.

Now we can complete the missing step in the proof of Levy's theorem on characterization of Brownian motion (Theorem 7.36).

Corollary 8.18 *If $M(t)$ is a continuous local martingale with $[M, M](t) = t$, then $U(t) = \mathcal{E}(uM)(t) = e^{uM(t)-u^2t/2}$ is a martingale.*

PROOF: Clearly $uX(t)$ is a continuous local martingale with quadratic variation $u^2 t$. The result follows by Novikov's condition.

However, it is possible to give a proof from first principles, by using the BDG inequality, and the the fact that under the condition (8.10) stochastic integrals with respect to martingales are martingales. Since U is a stochastic exponential it satisfies the SDE

$$U(t) = 1 + \int_0^t U(s)d(uM(s)).$$

A sufficient condition for a stochastic integral to be a martingale is finiteness of the expectation of its quadratic variation. Thus it is enough to check the condition

$$E \int_0^T U^2(t)d[uM, uM](t) = u^2 E \int_0^T U^2(t)dt < \infty.$$

To see this use the BDG inequality with $p = 2$ (for $U(t) - 1$ to have 0 at $t = 0$).

$$E(U^2(T) - 1) \le E(\sup_{t \le T} U(t))^2 - 1 \le CE[U, U](T) = C \int_0^T E(U^2(t))dt.$$

Let $h(t) = E(U^2(t))$. Then

$$h(T) \le 1 + C \int_0^T h(t)dt.$$

Gronwall's inequality (Theorem 1.20) implies that $h(T) \le e^{CT} < \infty$. Thus $E(U^2(T)) < \infty$, and $U(t)$ is a martingale.

□

The next section gives more information and tools for processes with jumps.

8.9   Compensators and Sharp Bracket Process

A process N is called increasing if all of its realizations $N(t)$ are non-decreasing functions of t. A process N is of finite variation if all of its realizations $N(t)$ are functions of finite variation, $V_N(t) < \infty$ for all t, where $V_N(t)$ is the variation of N on $[0, t]$.

Definition 8.19 *An increasing process N, $t \geq 0$, is called integrable if $\sup_{t \geq 0} \mathrm{E}N(t) < \infty$.*
A finite variation process N is of integrable variation if its variation process is integrable, $\sup_{t \geq 0} \mathrm{E}V_N(t) < \infty$.
A finite variation process N is of locally integrable variation if there is a sequence of stopping times τ_n such that $\tau_n \uparrow \infty$ so that $N(t \wedge \tau_n)$ is of integrable variation, that is, $\sup_{t \geq 0} \mathrm{E}V_N(t \wedge \tau_n) < \infty$.

Example 8.14: A Poisson process $N(t)$ with parameter λ is of finite but not integrable variation, since for any t, $V_N(t) = N(t) < \infty$, but $\sup_{t \geq 0} \mathrm{E}V_N(t) = \infty$. It is of locally integrable variation, since $\sup_{t \geq 0} \mathrm{E}V_N(t \wedge n) = \lambda n < \infty$. Here $\tau_n = n$.

Example 8.15: It can be seen that a finite variation process $N(t)$ with bounded jumps $|\Delta N(t)| \leq c$ is of locally integrable variation. If $\tau_n = \inf\{t : V_N(t) \geq n\}$, then $N(t \wedge \tau_n)$ has variation bounded by $n + c$. τ_n are stopping times, as first times of boundary crossing.

Definition 8.20 *Let $N(t)$ be an adapted process of integrable or locally integrable variation. Its compensator $A(t)$ is the unique predictable process such that $M(t) = N(t) - A(t)$ is a local martingale.*

Existence of compensators is assured by the Doob-Meyer decomposition.

Theorem 8.21 *Let $N(t)$ be an adapted process of integrable or locally integrable variation. Then its compensator exists. Moreover, it is locally integrable.*

PROOF: As a finite variation process is a difference of two increasing processes, it is enough to establish the result for increasing processes. By localization it is possible to assume that it is integrable. But an increasing integrable process is a submartingale, and the result follows by the Doob-Meyer decomposition Theorem 8.4.

$\square$

Remark 8.5: The condition $M = N - A$ is a local martingale is equivalent to the condition (see Liptser and Shiryayev (1989), p.33)

$$\mathrm{E} \int_0^\infty H(s)dN(s) = \mathrm{E} \int_0^\infty H(s)dA(s), \tag{8.43}$$

for any positive predictable process H. Sometimes this integral condition (8.43) is taken as the definition of the compensator, e.g. Karr (1986).

The compensator of N is also called the dual predictable projection of N (Rogers and Williams (1990), p.350, Liptser and Shiryayev (1989), p.33).

Note that the compensator is unique with respect to the given filtration and probability. If the filtration or probability are changed, then the compensator will also change.

Recall that the quadratic variation process $[X, X](t)$ of a semimartingale X exists and is non-decreasing. Consider now semimartingales with integrable ($\sup_{t \geq 0} \mathrm{E}[X, X](t) < \infty$) or locally integrable quadratic variation.

Definition 8.22 *The sharp bracket (or angle bracket, or predictable quadratic variation) $\langle X, X \rangle (t)$ process of a semimartingale X is the compensator of $[X, X](t)$. That is, it is the unique predictable process that makes $[X, X](t) - \langle X, X \rangle (t)$ into a local martingale.*

Example 8.16: Let N be a Poisson process. It is of finite variation and changes only by jumps (pure jump process), which are of size 1, $\Delta N(t) = 0$ or 1, and $(\Delta N(t))^2 = \Delta N(t)$. Its quadratic variation is the process $N(t)$ itself,

$$[N, N](t) = \sum_{0 \leq s \leq t} (\Delta N(s))^2 = \sum_{0 \leq s \leq t} \Delta N(s) = N(t).$$

Clearly $\sup_{0 \leq t \leq T} \mathrm{E}[N, N](t) = T$. Thus N is of integrable variation on $[0, T]$. t is non-random, therefore predictable. Since $[N, N](t) - t = N(t) - t$ is a martingale,

$$\langle N, N \rangle (t) = t.$$

Example 8.17: Let B be a Brownian motion. Its quadratic variation is $[B, B](t) = t$, and since it is non-random, it is predictable. Hence $\langle B, B \rangle (t) = t$, and the martingale part in the Doob-Meyer decomposition of $[B, B](t)$ is trivial, $M(t) \equiv 0$.

The last example generalizes to any continuous semimartingale.

Theorem 8.23 *If $X(t)$ is a continuous semimartingale with integrable quadratic variation, then $\langle X, X \rangle (t) = [X, X](t)$, and there is no difference between the sharp and the square bracket processes.*

PROOF: The quadratic variation jumps at the points of jumps of X and $\Delta[X, X](s) = (\Delta X(s))^2$. Since X has no jumps, $[X, X](t)$ is continuous. $[X, X](t)$ is predictable as a continuous and adapted process, the martingale part in the Doob-Meyer decomposition of $[X, X](t)$ is trivial, $M(t) \equiv 0$, and $\langle X, X \rangle (t) = [X, X](t)$.

□

Example 8.18: Let X be a diffusion solving the SDE $dX(t) = \mu(X(t))dt + \sigma(X(t))dB(t)$. Then $[X, X](t) = \int_0^t \sigma^2(X(s))ds = \langle X, X \rangle (t)$.

Sharp Bracket for Square Integrable Martingales

Let M be a square integrable martingale, that is, $\sup_t \mathrm{E} M^2(t) < \infty$. Recall that the quadratic variation of M has the property

$$M^2(t) - [M, M](t) \text{ is a martingale.} \qquad (8.44)$$

M^2 is a submartingale, since x^2 is a convex function. Using the Doob-Meyer decomposition for submartingales, Theorem 8.4, we can prove the following

Theorem 8.24 *Let M be a square integrable martingale. Then the sharp bracket process $\langle M, M \rangle (t)$ is the unique predictable increasing process for which*

$$M^2(t) - \langle M, M \rangle (t) \text{ is a martingale.} \qquad (8.45)$$

PROOF: By the definition of the sharp bracket process, $[M, M](t) - \langle M, M \rangle (t)$ is a martingale. As a difference of two martingales, $M^2(t) - \langle M, M \rangle (t)$ is also a martingale. Since $\langle M, M \rangle (t)$ is predictable and $M^2(t)$ is a submartingale, uniqueness follows by the Doob-Meyer decomposition.

$\square$

By taking expectations in (8.45) we obtain a useful corollary.

Corollary 8.25 *Let $M(t)$ be a square integrable martingale. Then*

$$\mathrm{E}\big(M^2(T)\big) = \mathrm{E}[M, M](t) = \mathrm{E}\langle M, M \rangle (t). \qquad (8.46)$$

This result allows us to use Doob's martingale inequality with the sharp bracket

$$\mathrm{E}\big((\sup_{s \leq T} M(s))^2\big) \leq 4\mathrm{E}\big(M^2(T)\big) = 4\mathrm{E}\big(\langle M, M \rangle (T)\big). \qquad (8.47)$$

By using localization in the proof of Theorem 8.24 one can show

Theorem 8.26 *Let M be a locally square integrable martingale, then the predictable quadratic variation $\langle M, M \rangle (t)$ is the unique predictable process for which $M^2(t) - \langle M, M \rangle (t)$ is a local martingale.*

Next result allows to decide when a local martingale is a martingale by using the predictable quadratic variation.

Theorem 8.27 *Let $M(t)$, $0 \leq t \leq T < \infty$, be a local martingale such that for all t, $\mathrm{E}\langle M, M \rangle (t) < \infty$. Then M is a square integrable martingale, moreover $\mathrm{E} M^2(t) = \mathrm{E}[M, M](t) = \mathrm{E}\langle M, M \rangle (t)$. If $T = \infty$, and $\sup_{t < \infty} \mathrm{E}\langle M, M \rangle (t) < \infty$, then $M(t)$ is a square integrable martingale on $[0, \infty)$.*

PROOF: $[M,M](t) - \langle M,M \rangle (t)$ is a local martingale. Let τ_n be a localizing sequence. Then $E[M,M](t \wedge \tau_n) = E \langle M,M \rangle (t \wedge \tau_n)$. Since both sides are non-decreasing, they converge to the same limit as $n \to \infty$. But $\lim_{n \to \infty} E \langle M,M \rangle (t \wedge \tau_n) = E \langle M,M \rangle (t) < \infty$. Therefore $E[M,M](t) = E \langle M,M \rangle (t) < \infty$. Thus the conditions of Theorem 7.35 are satisfied and the result follows.

□

Since a continuous local martingale is locally square integrable, we obtain

Corollary 8.28 *The sharp bracket process (predictable quadratic variation) for a continuous local martingale exists.*

Continuous Martingale Component of a Semimartingale

A function of finite variation has a decomposition into continuous and discrete parts. A semimartingale is a sum of a process of finite variation and a local martingale. It turns out that one can decompose any local martingale into a continuous local martingale and a purely discontinuous one. Such decomposition requires a different approach to the case of finite variation processes.

Definition 8.29 *A local martingale is purely discontinuous if it is orthogonal to any continuous local martingale. Local martingales M and N are orthogonal if MN is a local martingale.*

Example 8.19: A compensated Poisson process $\bar{N}(t) = N(t) - t$ is a purely discontinuous martingale. Let $M(t)$ be any continuous local martingale. Then by the integration by parts formula (8.24)

$$M(t)\bar{N}(t) = \int_0^t M(s-)d\bar{N}(s) + \int_0^t \bar{N}(s-)dM(s) + [M,\bar{N}](t).$$

Since $\bar{N}$ is of finite variation, by the property (8.19) of quadratic covariation $[M,\bar{N}](t) = \sum_{s \leq t} \Delta M(s) \Delta \bar{N}(s)$. But M is continuous, $\Delta M(s) = 0$, and $[M,\bar{N}](t) = 0$. Therefore, $M(t)\bar{N}(t)$ is a sum of two stochastic integrals with respect to local martingales, and itself is a local martingale.

It is possible to show that any local martingale M has a unique decomposition

$$M = M^c + M^d,$$

where M^c is a continuous and M^d a purely discontinuous local martingale (see for example Liptser and Shiryayev (1989), p.39, Protter (1992), Jacod and Shiryaev (1987)).

If X is a semimartingale with representation

$$X(t) = X(0) + M(t) + A(t), \tag{8.48}$$

with a local martingale M, then M^c is called the continuous martingale component of X and is denoted by X^{cm}. Even if the above representation of X is not unique, the continuous martingale component of X is the same for all representations. Indeed, if $X(t) = X(0) + M_1(t) + A_1(t)$, is another representation, then $(M - M_1)(t) = -(A - A_1)(t)$. Hence $(M - M_1)$ is a martingale of finite variation. Hence its continuous component is also a martingale of finite variation. But a continuous martingale of finite variation is a constant. This implies that $M^c - M_1^c = 0$. Thus $X^{cm} = M^c$ is the same for all representations. If X is of finite variation then the martingale part is zero and by the uniqueness of X^{cm} we have

Corollary 8.30 *If X is a semimartingale of finite variation, then $X^{cm} \equiv 0$.*

For example, the compensated Poisson process $N(t) - t$ has zero continuous martingale component.

It can be shown that

$$\langle X^{cm}, X^{cm} \rangle = [X, X]^c, \tag{8.49}$$

where $[X, X]^c$ is the continuous part of the finite variation process $[X, X]$. Of course, because X^{cm} is continuous $\langle X^{cm}, X^{cm} \rangle = [X^{cm}, X^{cm}]$.

Let $\Delta X(s) = X(s) - X(s-)$ and put $X(0-) = 0$ and $[X, X]^c(0) = 0$. Since the jumps of quadratic variation satisfy (see (8.18))

$$\Delta[X, X](s) = (\Delta X(s))^2,$$

$$
\begin{aligned}
[X, X](t) &= [X, X]^c(t) + \sum_{0 < s \le t} \Delta[X, X](s) = [X, X]^c(t) + \sum_{0 < s \le t} (\Delta X(s))^2, \\
&= \langle X^{cm}, X^{cm} \rangle + \sum_{0 < s \le t} (\Delta X(s))^2. \tag{8.50}
\end{aligned}
$$

Since the quadratic variation $[X, X]$ for a semimartingale exists, and the predictable the quadratic variation $\langle X^{cm}, X^{cm} \rangle$ exists (Corollary 8.28) we obtain

Corollary 8.31 *If X is a semimartingale then for each t*

$$\sum_{s \le t} (\Delta X(s))^2 < \infty. \tag{8.51}$$

Conditions for Existence of a Stochastic Integral

The class of processes H for which the stochastic integral with respect to a martingale M can be defined depends in an essential way on the properties of the predictable quadratic variation $\langle M, M \rangle$ of M. Consider integrals with respect to a locally square integrable martingale M, possibly discontinuous. The stochastic integral $\int_0^T H(s)dM(s)$ can be defined for predictable processes H such that

$$\int_0^T H^2(t)d\langle M, M \rangle(t) < \infty, \tag{8.52}$$

and in this case the integral $\int_0^t H(s)dM(s)$, $0 \le t \le T$, is a local martingale.

The class of processes H that can be integrated against M is wider when $\langle M, M \rangle(t)$ is continuous, and even wider when $\langle M, M \rangle(t)$ is absolutely continuous (can be represented as an integral with respect to dt). These classes and conditions are given, for example, in Liptser and Shiryaev (2001), p. 191.

Example 8.20: Let filtration $\mathbb{F}$ be generated by a Brownian motion $B(t)$ and a Poisson process $N(t)$. The process $N(t-)$ is a left-continuous modification of $N(t)$. By definition, $N(t-) = \lim_{s \uparrow t} N(s)$. Being left-continuous it is predictable. Condition (8.52) holds. The integral $\int_0^t N(s-)dB(s)$ is a well defined a stochastic integral of a predictable process with respect to a martingale B.

Properties of the Predictable Quadratic Variation

The predictable quadratic variation (the sharp bracket process) has similar properties to the quadratic variation (the square bracket) process. We list them without proof. All the processes below are assumed to be semimartingales with locally integrable quadratic variation.

1. $\langle X, X \rangle(t)$ is increasing in t.

2. $\langle X, Y \rangle$ is bilinear and symmetric,

$$\langle \alpha X + Y, \beta U + V \rangle = \alpha\beta \langle X, U \rangle + \alpha \langle X, V \rangle + \beta \langle Y, U \rangle + \langle Y, V \rangle. \tag{8.53}$$

3. Polarization identity. $\langle X, Y \rangle = \frac{1}{4}(\langle X + Y, X + Y \rangle - \langle X - Y, X - Y \rangle)$.

4. $\langle X, Y \rangle$ is a predictable process of finite variation.

5. $\langle X, Y \rangle = 0$ if X or Y is of finite variation and one of them is continuous.

6. The sharp bracket process of stochastic integrals $\langle H \cdot X, K \cdot Y \rangle(t)$.

$$\left\langle \int_0^{\cdot} H(s)dX(s), \int_0^{\cdot} K(s)dY(s) \right\rangle(t) = \int_0^t H(s)K(s)d\langle X, Y \rangle(s), \tag{8.54}$$

in particular $\left\langle \int_0^{\cdot} H(s)dX(s), \int_0^{\cdot} H(s)dX(s) \right\rangle(t) = \int_0^t H^2(s)d\langle X, X \rangle(s)$,

$\left\langle \int_0^{\cdot} H(s)dX(s), Y \right\rangle(t) = \left\langle \int_0^{\cdot} H(s)dX, \int_0^{\cdot} dY \right\rangle(t) = \int_0^t H(s)d\langle X, Y \rangle(s)$.

Recall that stochastic integrals with respect to local martingales are again local martingales. Using the sharp bracket of the integral together with Theorem 8.27 we obtain

Theorem 8.32 *Let $M(t)$, $0 \leq t \leq T$ be a local martingale and $H(t)$ be a predictable process such that* $E\left(\int_0^T H^2(s)d\langle M, M\rangle(s)\right) < \infty$. *Then* $\int_0^t H(s)dM(s)$ *is a square integrable martingale, moreover*

$$\left\langle \int_0^{\cdot} H(s)dM(s), \int_0^{\cdot} H(s)dM(s) \right\rangle (t) = \int_0^t H^2(s)d\langle M, M\rangle(s). \qquad (8.55)$$

Using this result, we obtain the isometry property for stochastic integrals in terms of the sharp bracket process.

$$E\left(\int_0^t H(s)dM(s)\right)^2 = E\left(\int_0^t H^2(s)d\langle M, M\rangle(s)\right) \qquad (8.56)$$

Example 8.21: Let $M(t)$ be the compensated Poisson process, $M(t) = N(t) - t$, and H be predictable, satisfying $E\int_0^T H^2(t)dt < \infty$. Then $\int_0^t H(s)dM(s)$ is a martingale, moreover

$$E\int_0^t H(s)dM(s) = 0, \text{ and } E\left(\int_0^t H(s)dM(s)\right)^2 = E\int_0^t H^2(s)ds.$$

8.10   Itô's Formula for Semimartingales

Let $X(t)$ be a semimartingale and f be a C^2 function. Then $f(X(t))$ is a semimartingale, and Itô's formula holds

$$f(X(t)) - f(X(0)) = \int_0^t f'(X(s-))dX(s) + \frac{1}{2}\int_0^t f''(X(s-))d[X, X](s) \quad (8.57)$$

$$+ \sum_{s \leq t} \left(f(X(s)) - f(X(s-)) - f'(X(s-))\Delta X(s) - \frac{1}{2}f''(X(s-))(\Delta X(s))^2 \right).$$

The quadratic variation $[X, X]$ jumps at the points of jumps of X and its jumps $\Delta[X, X](s) = (\Delta X(s))^2$. Thus the jump part of the integral $\int_0^t f''(X(s-))d[X, X](s)$ is given by $\sum_{s \leq t} f''(X(s-))(\Delta X(s))^2$, leading to an equivalent form of the formula

$$f(X(t)) - f(X(0)) = \int_0^t f'(X(s-))dX(s) + \frac{1}{2}\int_0^t f''(X(s-))d[X, X]^c(s)$$

$$+ \sum_{s \leq t} \left(f(X(s)) - f(X(s-)) - f'(X(s-))\Delta X(s) \right), \qquad (8.58)$$

where $[X, X]^c$ is the continuous component of the finite variation function $[X, X]$. Using the relationship between the square and the sharp brackets (8.50), we can write Itô's formula with the sharp bracket process of X, provided the sharp bracket exists,

$$f(X(t)) - f(X(0)) = \int_0^t f'(X(s-))dX(s) + \frac{1}{2} \int_0^t f''(X(s-))d\langle X^{cm}, X^{cm}\rangle(s)$$

$$+ \sum_{s \le t} \Big(f(X(s)) - f(X(s-)) - f'(X(s-))\Delta X(s)\Big), \quad (8.59)$$

where X^{cm} denotes the continuous martingale part of X.

Example 8.22: Let $N(t)$ be a Poisson process. We calculate $\int_0^t N(s-)dN(s)$. The answer can be derived from the integration by parts formula (8.24), but now we use (8.59). Since $(N(t) - t)^{cm} = 0$ (by Corollary 8.30)

$$N^2(t) = 2 \int_0^t N(s-)dN(s) + \sum_{s \le t} \Big(N^2(s) - N^2(s-) - 2N(s-)\Delta N(s)\Big).$$

Since $N(s) = N(s-) + \Delta N(s)$,
$(N(s-) + \Delta N(s))^2 - N^2(s-) - 2N(s-)\Delta N(s) = (\Delta N(s))^2 = \Delta N(s)$, and the sum simplifies to $\sum_{s \le t} \Delta N(s) = N(t)$. Thus we obtain

$$\int_0^t N(s-)dN(s) = \frac{1}{2} \Big(N^2(t) - N(t)\Big).$$

A formula (8.59) for a function of n variables reads: $\boldsymbol{X}(t) = (X^1(t), \ldots, X^n(t))$ is a semimartingale and f is a C^2 function of n variables,

$$f(\boldsymbol{X}(t)) - f(\boldsymbol{X}(0)) = \sum_{i=1}^n \int_0^t \frac{\partial f}{\partial x_i}(\boldsymbol{X}(s-))dX^i(s)$$

$$+ \frac{1}{2} \sum_{i,j=1}^n \int_0^t \frac{\partial^2 f}{\partial x_i \partial x_j}(\boldsymbol{X}(s-))d\langle X^{i,cm}, X^{j,cm}\rangle(s)$$

$$+ \sum_{s \le t} \Big(f(\boldsymbol{X}(s)) - f(\boldsymbol{X}(s-)) - \sum_{i=1}^n \frac{\partial}{\partial x_i} f(\boldsymbol{X}(s-))\Delta X^i(s)\Big). \quad (8.60)$$

Itô's formula can be found in many texts, Protter (1992), p. 71, Rogers and Williams (1990), p. 394, Liptser and Shiryayev (1989), Métivier (1982), Dellacherie and Meyer (1982).

8.11   Stochastic Exponential and Logarithm

As an application of Itô's formula and the rules of stochastic calculus we outline a proof of the following result.

Theorem 8.33 *Let X be a semimartingale. Then the stochastic equation*

$$U(t) = 1 + \int_0^t U(s-)dX(s). \tag{8.61}$$

has a unique solution, called the stochastic exponential of X, and this solution is given by

$$U(t) = \mathcal{E}(X)(t) = e^{X(t)-X(0)-\frac{1}{2}\langle X,X\rangle^c(t)} \prod_{s\leq t}\left(1 + \Delta X(s)\right)e^{-\Delta X(s)}. \tag{8.62}$$

Formula (8.62) can be written by using quadratic variation as follows

$$\mathcal{E}(X)(t) = e^{X(t)-X(0)-\frac{1}{2}[X,X](t)} \prod_{s\leq t}(1 + \Delta X(s))e^{(-\Delta X(s))+\frac{1}{2}(\Delta X(s))^2}. \tag{8.63}$$

PROOF: Let $Y(t) = X(t) - X(0) - \frac{1}{2}\langle X, X\rangle^c(t)$ and
$V(t) = \prod_{s\leq t}\left(1+\Delta X(s)\right)e^{-\Delta X(s)}$. Note that although the product is taken for all $s \leq t$, there are at most countably many points at which $\Delta X(s) \neq 0$, (by the regularity property of the process), hence there are at most countably many elements different from 1 in the product. We show that the product converges. Since by (8.51) $\sum_{s\leq t}(\Delta X(s))^2 < \infty$, there are only finitely many points s at which $|\Delta X(s)| > 0.5$, which give a finite non-zero contribution to the product. Taking the product with over s at which $|\Delta X(s)| \leq 1/2$, and taking logarithm, it is enough to show that $\sum_{s\leq t}|\ln(1 + \Delta X(s)) - \Delta X(s)|$ converges. But this follows from the inequality $|\ln(1 + \Delta X(s)) - \Delta X(s)| \leq (\Delta X(s))^2$ by (8.51). To see that $U(t)$ defined by (8.62) satisfies (8.61) use Itô's formula applied to the function $f(Y(t), V(t))$ with $f(x_1, x_2) = e^{x_1}x_2$. For the uniqueness of the solution of (8.61) and other details see Liptser and Shiryayev (1989), p.123.
□

Example 8.23:   The stochastic exponential (8.62) of a Poisson process is easily seen to be $\mathcal{E}(N)(t) = 2^{N(t)}$.

If $U = \mathcal{E}(X)$ is the stochastic exponential of X, then $X = \mathcal{L}(U)$ is the stochastic logarithm of U, satisfying equation (8.61)

$$dX(t) = \frac{dU(t)}{U(t-)}, \text{ or } \mathcal{L}(\mathcal{E}(X)) = X.$$

For Itô processes an expression for $X(t)$ is given in Theorem 5.3, for general case see Exercise 8.17.

8.12   Martingale (Predictable) Representations

In this section we give results on representation of martingales by stochastic integrals of predictable processes, also called *predictable* representations. Let $M(t)$ be a martingale, $0 \leq t \leq T$, adapted to the filtration $\mathbb{F} = (\mathcal{F}_t)$, and $H(t)$ be a predictable process satisfying $\int_0^T H^2(s) d \langle M, M \rangle (s) < \infty$ with probability one. Then $\int_0^t H(s) dM(s)$ is a local martingale. The predictable representation property means that the converse is also true. Let $\mathbb{F}^M = (\mathcal{F}_t^M)$ denote the natural filtration of M.

Definition 8.34 *A local martingale M has the predictable representation property if for any $\mathbb{F}^M$-local martingale X there is a predictable process H such that*

$$X(t) = X(0) + \int_0^t H(s) dM(s). \qquad (8.64)$$

This definition is different to the classical one for martingales with jumps, see Remark 8.7 below, but is the same for continuous martingales.

Brownian motion has the predictable representation property (see, for example Revuz and Yor (1999) p. 209, Liptser and Shiryayev (2001) I p. 170).

Theorem 8.35 (Brownian Martingale Representation)
Let $X(t)$, $0 \leq t \leq T$, be a local martingale adapted to the Brownian filtration $\mathbb{F}^B = (\mathcal{F}_t)$. Then there exists a predictable process $H(t)$ such that $\int_0^T H^2(s) ds < \infty$ with probability one, and equation (8.65) holds

$$X(t) = X(0) + \int_0^t H(s) dB(s). \qquad (8.65)$$

Moreover, if Y is an integrable $\mathcal{F}_T$-measurable random variable, $\mathbb{E}|Y| < \infty$, then

$$Y = \mathbb{E}Y + \int_0^T H(t) dB(t). \qquad (8.66)$$

If in addition, Y and B have jointly a Gaussian distribution, then the process $H(t)$ in (8.66) is deterministic.

PROOF: We don't prove the representation of a martingale, but only the representation for a random variable based on it.
Take $X(t) = \mathbb{E}(Y|\mathcal{F}_t)$. Then $X(t)$, $0 \leq t \leq T$, is a martingale (see Theorem 7.9). Hence by the martingale representation there exists H, such that $X(t) = X(0) + \int_0^t H(s) dB(s)$. Taking $t = T$ gives the result.

$\square$

Remark 8.6: A functional of the path of the Brownian motion $B_{[0,T]}$ is a random variable Y, $\mathcal{F}_T$-measurable. Theorem 8.35 states that under the above assumptions, any functional of Brownian motion has the form (8.66).

Since Itô integrals are continuous, and any local martingale of a Brownian filtration is an Itô integral, it follows that all local martingales of a Brownian filtration are continuous. In fact we have the following result (the second statement is not straightforward)

Corollary 8.36

1. *All local martingales of the Brownian filtration are continuous.*

2. *All right-continuous adapted processes are predictable.*

Corollary 8.37
Let $X(t)$, $0 \leq t \leq T$, be a square integrable martingale adapted to the Brownian filtration $\mathbb{F}$. Then there exists a predictable process $H(t)$ such that $\mathrm{E} \int_0^T H^2(s)ds < \infty$ and representation (8.65) holds. Moreover,

$$\langle X, B \rangle (t) = \int_0^t H(s)ds, \quad and \quad H(t) = \frac{d \langle X, B \rangle (t)}{dt}. \tag{8.67}$$

The equation (8.67) follows from (8.65) by the rule of the sharp bracket for integrals.

Example 8.24: (Representation of martingales)
1. $X(t) = B^2(t) - t$. Then $X(t) = \int_0^t 2B(s)dB(s)$. Here $H(t) = 2B(t)$, which can also be found by using (8.67).
2. Let $X(t) = f(B(t), t)$ be a martingale. By Itô's formula $dX(t) = \frac{\partial f}{\partial x}(B(t), t)dB(t)$. Thus $H(t) = \frac{\partial f}{\partial x}(B(t), t)$. This also shows that

$$\frac{\langle f(B, t), B \rangle (t)}{dt} = \frac{\partial f}{\partial x}(B(t), t).$$

Example 8.25: (Representation of random variables)
1. If $Y = \int_0^T B(s)ds$, then $Y = \int_0^T (T - s)dB(s)$.
2. $Y = B^2(1)$. Then $M(t) = \mathrm{E}(B^2(1)|\mathcal{F}_t) = B^2(t) + (1 - t)$. Using Itô's formula for $M(t)$ we obtain $B^2(1) = 1 + 2 \int_0^1 B(t)dB(t)$.

Similar results hold for the Poisson process filtration.

Theorem 8.38 (Poisson Martingale Representation)
Let $M(t)$, $0 \leq t \leq T$, be a local martingale adapted to the Poisson filtration. Then there exists a predictable process $H(t)$ such that

$$M(t) = M(0) + \int_0^t H(s)d\bar{N}(s), \tag{8.68}$$

where $\bar{N}(t) = N(t) - t$ is the compensated Poisson process.

When a filtration is larger than the natural filtration of a martingale, then there is the following result (Revuz and Yor (1999) p. 209, Liptser and Shiryaev (2001) p. 170).

Theorem 8.39 *If $M(t)$, $0 \leq t \leq T$, is any continuous local martingale, and X a continuous $\mathbb{F}^M$-local martingale. Then X has a representation*

$$X(t) = X(0) + \int_0^t H(s)dM(s) + Z(t), \tag{8.69}$$

where H is predictable and $\langle M, Z \rangle = 0$; (consequently $\langle X - Z, Z \rangle = 0$).

Example 8.26: Let $\mathbb{F}$ be generated by two independent Brownian motions B and W, and let $M(t) = \int_0^t W(s)dB(s)$. It is a martingale, as a stochastic integral satisfying $\mathrm{E} \int_0^t W^2(s)ds < \infty$. We show that M does not have the predictable representation property.
$\langle M, M \rangle (t) = \int_0^t W^2(s)ds$. Hence $W^2(t) = \frac{d\langle S,S \rangle(t)}{dt}$, which shows that $W^2(t)$ is $\mathcal{F}_t^M$-measurable. Hence the martingale $X(t) = W^2(t) - t$ is adapted to $\mathcal{F}_t^M$, but is not an integral of a predictable process with respect to M. By Itô's formula $X(t) = 2 \int_0^t W(s)dW(s)$. Hence $\langle X, M \rangle (t) = \int_0^t W^2(s)d\langle W, B \rangle (s) = 0$. Suppose there is H, such that $X(t) = \int_0^t H(u)dM(u)$. Then by (8.67) $H(t) = \frac{d\langle X,M \rangle(t)}{dt} = 0$, implying that $X(t) = 0$, which is a contradiction.
    This example has an application in Finance, it shows non-completeness of a stochastic volatility model.

Example 8.27: Let $\mathbb{F}$ be generated by a Brownian motions B and a Poisson process N, and let $M(t) = B(t) + N(t) - t = B(t) + \bar{N}(t)$, where $\bar{N}(t) = N(t) - t$. M is a martingale, as a sum of two martingales. We show that M does not have the predictable representation property.
$[M, M](t) = [B, B](t) + [N, N](t) = t + N(t)$. This shows that $N(t) = [M, M](t) - t$ is $\mathcal{F}_t^M$-measurable. Hence $B(t) = M(t) - N(t) + t$ is $\mathcal{F}_t^M$-measurable. Thus the martingale $X(t) = \int_0^t N(s-)dB(s)$ is $\mathcal{F}_t^M$-measurable, but it does not have a predictable representation. If it did, $\int_0^t N(s-)dB(s) = \int_0^t H(s)dB(s) + \int_0^t H(s)d\bar{N}(s)$, and $\int_0^t (N(s-) - H(s)) dB(s) = \int_0^t H(s)d\bar{N}(s)$. Since the integral on the rhs is of finite variation, $H(s) = N(s-)$, for almost all s. Thus $\int_0^t H(s)d\bar{N}(s) = 0$. This is the same as $\int_0^t N(s-)dN(s) = \int_0^t N(s-)ds$. But this is impossible. To see the contradiction, let $t = T_2$ the time of the second jump of N. Then $\int_0^{T_2} N(s-)dN(s) = 1$ and $\int_0^{T_2} N(s-)ds = T_2 - T_1$.
    This example shows non-completeness of models of stock prices with jumps. It can be generalized to a model that supports a martingale with a jump component.

Remark 8.7: The Definition 8.34 of the predictable representation property given here agrees with the standard definition given for continuous martingales

but is different to the definition for predictable representation with respect
to semimartingales, given in Liptser and Shiryaev (1989) p.250, Jacod and
Shiryaev (1987) p.172, Protter (1992) p.150. The general definition allows for
different predictable functions h and H to be used in the integrals with respect
to the continuous martingale part M^c and the discrete martingale part M^d of
M,

$$X(t) = X(0) + \int_0^t h(s)dM^c(s) + \int_0^t H(s)dM^d(s).$$

In this definition, the martingale M in Example 8.27 has the predictable representation property.

The definition given here is more suitable for financial applications. According to the financial mathematics theory an option can be priced if it can
be replicated, which means that it is an integral of a predictable process H
with respect to the discounted stock price process M, which is a martingale.
The process H represents the number of shares bought/sold so it does not
make sense to have H consist of two different components.

8.13   Elements of the General Theory

The basic setup consists of the probability space $(\Omega, \mathcal{F}, P)$, where $\mathcal{F}$ is a σ-field
on Ω and P is a probability on $\mathcal{F}$. A stochastic process is a map from $\mathbb{R}^+ \times \Omega$
to $\mathbb{R}$, namely $(t, \omega) \to X(t, \omega)$. $\mathbb{R}^+$ has the Borel σ-field of measurable sets,
and $\mathcal{F}$ is the σ-field of measurable sets on Ω. Only measurable processes are
considered, that is for any $A \in \mathcal{B}$

$$\{(t, \omega) : X(t, \omega) \in A\} \in \mathcal{B}(\mathbb{R}^+) \times \mathcal{F}.$$

Theorem 8.40 (Fubini) *Let $X(t)$ be a measurable stochastic process. Then*

1. *P-a.s. the functions $X(t, \omega)$ (trajectories) are (Borel) measurable.*

2. *If $EX(t)$ exists for all t, then it is measurable as a function of t.*

3. *If $\int_a^b E|X(t)|dt < \infty$ P-a.s. then almost all trajectories $X(t)$ are inte-
   grable and $\int_a^b EX(t)dt = E \int_a^b X(t)dt$.*

Let $\mathbb{F}$ be a filtration of increasing σ-fields on Ω. Important classes of processes
are introduced via measurability with respect to various σ-fields of subsets of
$\mathbb{R}^+ \times \Omega$: adapted, progressively measurable processes, optional processes and
predictable processes, given in the order of inclusion.

X is adapted if, for all t, $X(t)$ is $\mathcal{F}_t$ measurable. X is progressively measurable if, for any t, $\{(s \leq t, \omega) : X(s, \omega) \in A\} \in \mathcal{B}([0, t]) \times \mathcal{F}_t$. Any progressively measurable process is clearly adapted. It can be shown that any right or left continuous process is progressively measurable.

Definition 8.41

1. *The σ-field generated by the adapted left-continuous processes is called the predictable σ-field $\mathcal{P}$.*

2. *The σ-field generated by the adapted right-continuous processes is called the optional σ-field $\mathcal{O}$.*

3. *A process is called predictable if it is measurable with respect to the predictable σ-field $\mathcal{P}$; it is called optional if it is measurable with respect to the optional σ-field $\mathcal{O}$.*

Remarks

1. The predictable σ-field $\mathcal{P}$ is also generated by the adapted continuous processes.

2. Define $\mathcal{F}_{t-} = \sigma\left(\cup_{s<t} \mathcal{F}_s\right)$ the smallest σ-field containing $\mathcal{F}_s$ for all $s < t$, and $\mathcal{F}_{0-} = \mathcal{F}_0$. $\mathcal{F}_{t-}$ represents the information available prior to t. Then the predictable σ-field $\mathcal{P}$ is generated by the sets $[s, t) \times A$ with $s < t$ and $A \in \mathcal{F}_{t-}$.

3. Predictable and optional σ-fields are also generated respectively by simple adapted left-continuous processes and simple adapted right-continuous processes.

4. Since a left-continuous adapted process $H(t)$ can be approximated by right-continuous adapted processes, $(H(t) = \lim_{\epsilon \to 0} H((t - \epsilon)+))$, any predictable process is also optional. Therefore $\mathcal{P} \subseteq \mathcal{O}$.

5. The Poisson process is right-continuous and it can be shown that it cannot be approximated by left-continuous adapted processes. Therefore there are optional processes which are not predictable, $\mathcal{P} \subset \mathcal{O}$.

6. In discrete time optional is the same as adapted.

Stochastic Sets

Subset of $\mathbb{R}^+ \times \Omega$ are called stochastic sets. If A is a stochastic set, then "its projection on Ω" $\pi_A = \{\omega : \exists t \text{ such that } (t, \omega) \in A\}$. A is called *evanescent* if $P(\pi_A) = 0$.

Two processes $X(t)$ and $Y(t)$ are called indistinguishable if the stochastic set $A = \{(t, \omega) : X(t, \omega) \neq Y(t, \omega)\}$ is an evanescent set. A process indistinguishable from zero is called P-negligible.

If τ_1 and τ_2 are stopping times, a closed stochastic interval is defined as $[[\tau_1, \tau_2]] = \{(t, \omega) : \tau_1(\omega) \leq t \leq \tau_2(\omega)\}$. Similarly half-closed $[[\tau_1, \tau_2[[,]]\tau_1, \tau_2]]$, and open $]]\tau_1, \tau_2[[$ stochastic intervals are defined. Double brackets are used to emphasize that the intervals are subsets of $\mathbb{R}^+ \times \Omega$ and to distinguish them from intervals on $\mathbb{R}^+$.

The stochastic interval $[[\tau, \tau]] = \{(t, \omega) : \tau(\omega) = t\}$ is called the graph of the stopping time τ.

A stochastic set A is called *thin* if there are stopping times τ_n, such that $A = \cup_n [[\tau_n]]$.

Example 8.28: Let $N(t)$ be a Poisson process with rate λ, and $A = \{\Delta N \neq 0\}$. Then $A = \cup_n [[\tau_n]]$, where τ_n is the time of the n-th jump. Hence A is a thin set.

It can be shown that for any regular right-continuous process the stochastic set of jumps $\{\Delta X \neq 0\}$ is a thin set (Liptser and Shiryayev (1989), p.4).

Classification of Stopping Times

Recall that τ is a stopping time with respect to filtration $\mathbb{F}$ if for all $t \geq 0$ the event $\{\tau \leq t\} \in \mathcal{F}_t$. If $\mathbb{F}$ is right-continuous, then also $\{\tau < t\} \in \mathcal{F}_t$. Events observed before or at time τ are described by the σ-field $\mathcal{F}_\tau$, defined as the collection of sets $\{A \in \mathcal{F} : \text{for any } t \, A \cap \{\tau \leq t\} \in \mathcal{F}_t\}$. Events observed before time τ are described by the σ-field $\mathcal{F}_{\tau-}$, the σ-field generated by $\mathcal{F}_0$ and the sets $A \cap \{\tau > t\}$, where $A \in \mathcal{F}_t$, $t > 0$.

There are three types of stopping times that are used in stochastic calculus:

1. predictable stopping times,

2. accessible stopping times,

3. totally inaccessible stopping times.

τ is a predictable stopping time if there exists a sequence of stopping times τ_n, $\tau_n < \tau$, and $\lim_n \tau_n = \tau$. In this case it is said that the sequence τ_n announces τ.

Example 8.29: If τ is a stopping time then for any constant $a > 0$, $\tau + a$ is a predictable stopping time. Indeed, it can be approached by $\tau_n = \tau + a - 1/n$.

Example 8.30: Let $B(t)$ be Brownian motion started at zero, $\mathbb{F}$ its natural filtration and τ the first hitting time of 1, that is, $\tau = \inf\{t : B(t) = 1\}$. τ is a predictable stopping time, since $\tau_n = \inf\{t : B(t) = 1 - 1/n\}$ converge to τ.

τ is an accessible stopping time if it is possible to announce τ, but with different sequences on different parts of Ω, that is, $[[\tau]] \subset \cup_n [[\tau_n]]$, where τ_n are predictable stopping times. All other types of stopping times are called totally inaccessible.

Example 8.31: Let $N(t)$ be Poisson process, $\mathbf{F}$ its natural filtration and τ the time of the first jump, $\tau = \inf\{t : N(t) = 1\}$. τ is a totally inaccessible stopping time. Any predictable stopping time $\tau_n < \tau$ is a constant, since $\mathcal{F}_t \cap \{t < \tau\}$ is trivial. But τ has a continuous distribution (exponential), thus it cannot be approached by constants.

The optional σ-field is generated by the stochastic intervals $[[0, \tau[[$, where τ is a stopping time. The predictable σ-field is generated by the stochastic intervals $[[0, \tau]]$.

A set A is called predictable if its indicator is a predictable process, $I_A \in \mathcal{P}$. The following results allow us to decide on predictability.

Theorem 8.42 *Let $X(t)$ be a predictable process and τ be a stopping time. Then*

1. *$X(\tau)I(\tau < \infty)$ is $\mathcal{F}_{\tau-}$ measurable,*

2. *the stopped process $X(t \wedge \tau)$ is predictable.*

For a proof see, for example, Liptser and Shiryayev (1989) p. 13.

Theorem 8.43 *An adapted regular process is predictable if and only if for any predictable stopping time τ the random variable $X(\tau)I(\tau < \infty)$ is $\mathcal{F}_{\tau-}$ measurable and for each totally inaccessible stopping time τ one of the following two conditions hold*

1. *$X(\tau) = X(\tau-)$ on $\tau < \infty$*

2. *the set $\{\Delta X \neq 0\} \cap [[\tau]]$ is P-evanescent.*

For a proof see, for example, Liptser and Shiryayev (1989) p. 16.

Theorem 8.44 *A stopping time τ is predictable if and only if for any bounded martingale M, $\mathrm{E}(M(\tau)I(\tau < \infty)) = \mathrm{E}(M(\tau-)I(\tau < \infty))$.*

Theorem 8.45 *The compensator $A(t)$ is continuous if and only if the jump times of the process $X(t)$ are totally inaccessible.*

See, for example, Liptser and Shiryayev (2001) for the proof.

Example 8.32: The compensator of the Poisson process is t, which is continuous. By the above result the jump times of the Poisson process are totally inaccessible. This was shown in Example 8.31.

It can be shown that for Brownian motion and diffusions any stopping time is predictable. This implies that the class of optional processes is the same as the class of predictable processes.

Theorem 8.46 *For Brownian motion filtration any martingale (local martingale) is continuous and any positive stopping time is predictable. Any optional process is also predictable,* $\mathcal{O} = \mathcal{P}$.

Similar result holds for diffusions (see for example, Rogers and Williams (1990), p.338).

Remark 8.8: It can be shown that $\langle X, X \rangle$ is the conditional quadratic variation of $[X, X]$ conditioned on the predictable events $\mathcal{P}$.

8.14   Random Measures and Canonical Decomposition

The canonical decomposition of semimartingales with jumps uses the concepts of a random measure and its compensator, as well as integrals with respect to random measures. We do not use this material elsewhere in the book. However, the canonical decomposition is often met in research papers.

Random Measure for a Single Jump

Let ξ be a random variable. For a Borel set $A \subset \mathbb{R}$ define

$$\mu(\omega, A) = I_A(\xi(\omega)) = I(\xi(\omega) \in A). \tag{8.70}$$

Then μ is a random measure, meaning that for each $\omega \in \Omega$, $\mu(\omega, A)$ is a measure when A varies, $A \in \mathcal{B}(\mathbb{R})$. Its (random) distribution function has a single jump of size 1 at ξ. The following random Stieltjes integrals consist of a single term:

$$\int_{\mathbb{R}} x\mu(\omega, dx) = \xi(\omega), \text{ and for a function } h, \int_{\mathbb{R}} h(x)\mu(\omega, dx) = h(\xi(\omega)).$$
$$\tag{8.71}$$

There is a special notation for this random integral

$$h * \mu := \int_{\mathbb{R}} h(x)\mu(dx). \tag{8.72}$$

Random Measure of Jumps and its Compensator in Discrete Time

Let $X_0, \ldots X_n, \ldots$ be a sequence of random variables, adapted to $\mathcal{F}_n$, and let $\xi_n = \Delta X_n = X_n - X_{n-1}$. Let $\mu_n = I_A(\xi_n)$ be jump measures, and let

$$\nu_n(A) = E(\mu_n(A)|\mathcal{F}_{n-1}) = E(I_A(\xi_n)|\mathcal{F}_{n-1}) = P(\xi_n \in A|\mathcal{F}_{n-1})$$

be the conditional distributions, $n = 0, 1, \ldots$ Define

$$\mu((0,n], A) = \sum_{i=1}^{n} \mu_i(A), \text{ and } \nu((0,n], A) = \sum_{i=1}^{n} \nu_i(A). \tag{8.73}$$

Then for each A the sequence $\mu((0,n], A) - \nu((0,n], A)$ is a martingale. The measure $\mu((0,n], A)$ is called the measure of jumps of the sequence X_n, and $\nu((0,n], A)$ its compensator (A does not include 0). Clearly, the measure $\mu = \{\mu((0,n])\}_{n \geq 1}$ admits representation

$$\mu = \nu + (\mu - \nu), \tag{8.74}$$

where $\nu = \{\nu((0,n])\}_{n \geq 1}$ is predictable, and $\mu - \nu = \{\mu((0,n]) - \nu((0,n])\}_{n \geq 1}$ is a martingale. This is Doob's decomposition for random measures. With notation (8.72)

$$X_n = (x * \mu)_n. \tag{8.75}$$

Regular conditional distributions exist and for a function $h(x)$ the conditional expectations can be written as integrals with respect to these

$$E(h(\xi_n)|\mathcal{F}_{n-1}) = \int_{\mathbb{R}} h(x)\nu_n(dx),$$

provided $h(\xi_n)$ is integrable, $E|h(\xi_n)| < \infty$.

Assume now that ξ_n are integrable, then its Doob's decomposition (8.4)

$$X_n = X_0 + \sum_{i=1}^{n} E(\xi_i|\mathcal{F}_{i-1}) + \sum_{i=1}^{n} (\xi_i - E(\xi_i|\mathcal{F}_{i-1})) = X_0 + A_n + M_n. \tag{8.76}$$

Using random measures and their integrals, we can express A_n and M_n as

$$A_n = \sum_{i=1}^{n} E(\xi_i|\mathcal{F}_{i-1}) = \sum_{i=1}^{n} \int_{\mathbb{R}} x\nu_i(dx) = (x * \nu)_n, \tag{8.77}$$

$$M_n = \sum_{i=1}^{n} (\xi_i - E(\xi_i|\mathcal{F}_{i-1})) = \sum_{i=1}^{n} \int_{\mathbb{R}} x(\mu_i(dx) - \nu_i(dx)) = (x * (\mu - \nu))_n.$$

Thus the semimartingale decomposition of X is given by using the random measure and its compensator

$$X_n = X_0 + (x * \nu)_n + (x * (\mu - \nu))_n. \tag{8.78}$$

However, the jumps of X, $\xi_n = \Delta X_n$, may not be integrable. Then the term $(x * \nu)_n$ is not defined. In this case a truncation function is used, such as $h(x) = xI(|x| \leq 1)$, and a similar decomposition is achieved, called the canonical decomposition,

$$
\begin{aligned}
X_n &= X_0 + \sum_{i=1}^{n} \mathrm{E}(h(\xi_i)|\mathcal{F}_{i-1}) + \sum_{i=1}^{n}(h(\xi_i) - \mathrm{E}(\xi_i|\mathcal{F}_{i-1})) + \sum_{i=1}^{n}(\xi_i - h(\xi_i)) \\
&= X_0 + (h * \nu)_n + (h * (\mu - \nu))_n + ((x - h(x)) * \mu)_n.
\end{aligned}
\tag{8.79}
$$

The above representation has well-defined terms and in addition it has another advantage that carries over to the continuous time case.

Random Measure of Jumps and its Compensator

Let X be a semimartingale. For a fixed t consider the jump $\Delta X(t)$. Taking $\xi = \Delta X(t)$, we obtain the measure of the jump at t

$$
\mu(\{t\}, A) = I_A(\Delta X(t)), \quad \text{with} \quad \int_{\mathbb{R}} x\mu(\{t\}, dx) = \Delta X(t).
\tag{8.80}
$$

Now consider the measure of jumps of X (in $\mathbb{R}^+ \times \mathbb{R}^0$, with $\mathbb{R}^0 = \mathbb{R} \setminus 0$)

$$
\mu((0, t] \times A) = \sum_{0 < s \leq t} I_A(\Delta X(s))
\tag{8.81}
$$

for a Borel set A that does not include zero (there are no jumps of size 0, if $\Delta X(t) = 0$ then t is a point of continuity of X). It is possible to define the compensator of μ, such that $\mu((0, t] \times A) - \nu((0, t] \times A)$ is a local martingale.

For the canonical decomposition of semimartingales, similar to (8.79), firstly large jumps are taken out, then the small jumps are compensated as follows. Consider

$$
(x - h(x)) * \mu(t) = \int_0^t \int_{\mathbb{R} \setminus 0} (x - h(x))\mu(ds, dx) = \sum_{s \leq t}(\Delta X_s - h(\Delta X_s)),
$$

where $h(x)$ is a truncation function. This is a sum over "large" jumps with $|\Delta X(s)| > 1$ (since $x - h(x) = 0$ for $|x| \leq 1$). Since the sum of squares of jumps is finite, (Corollary 8.31, (8.51)), the above sum has only finitely many terms, hence it is finite. Thus the following canonical decomposition of a semimartingale is obtained

$$
\begin{aligned}
X(t) &= X(0) + A(t) + X^{cm}(t) + (h(x) * (\mu - \nu))(t) + ((x - h(x)) * \mu)(t), \\
&= X(0) + A(t) + X^{cm}(t) + \int_0^t \int_{|x| \leq 1} x d(\mu - \nu) + \int_0^t \int_{|x| > 1} x d\mu,
\end{aligned}
$$

where A is a predictable process of finite variation, X^{cm} is a continuous martingale component of X, μ is the measure of jumps of X, and ν its compensator. For a proof see Liptser and Shiryayev (1989), p.188, Shiryaev (1999), p. 663.

Let $C = \langle X^{cm}, X^{cm} \rangle$ (it always exists for continuous processes). The following three processes appearing in the canonical decomposition (A, C, ν) are called the *triplet of predictable characteristics* of the semimartingale X.

Notes. Material for this chapter is based on Protter (1992), Rogers and Williams (1990), Metivier (1982), Liptser and Shiryayev (1989), Shiryaev (1999).

8.15 Exercises

Exercise 8.1: Let $\tau_1 < \tau_2$ be stopping times. Show that $I_{(\tau_1, \tau_2]}(t)$ is a simple predictable process.

Exercise 8.2: Let $H(t)$ be a regular adapted process, not necessarily left-continuous. Show that for any $\delta > 0$, $H(t - \delta)$ is predictable.

Exercise 8.3: Show that a continuous process is locally integrable. Show that a continuous local martingale is locally square integrable.

Exercise 8.4: M is a local martingale and $E \int_0^T H^2(s) d[M, M](s) < \infty$. Show that $\int_0^t H(s) dM(s)$ is a square integrable martingale.

Exercise 8.5: Find the variance of $\int_0^1 N(t-) dM(t)$, where M is the compensated Poisson process $M(t) - t$.

Exercise 8.6: If S and T are stopping times, show that

1. $S \wedge T$ and $S \vee T$ are stopping times.

2. The events $\{S = T\}$, $\{S \leq T\}$ and $\{S < T\}$ are in $\mathcal{F}_S$.

3. $\mathcal{F}_S \cap \{S \leq T\} \subset \mathcal{F}_T \cap \{S \leq T\}$.

Exercise 8.7: Let U be a positive random variable on a probability space $(\Omega, \mathcal{F}, P)$, and $\mathcal{G}$ be a sub-σ-field of $\mathcal{F}$.

1. Let $t \geq 0$. Show that $\mathcal{F}_t := \{A \in \mathcal{F} : \exists B \in \mathcal{G} \text{ such that } A \cap \{U > t\} = B \cap \{U > t\}\}$ is a σ-field.

2. Show that $\mathcal{F}_t$ is a right-continuous filtration on $(\Omega, \mathcal{F}, P)$.

3. Show that U is a stopping time for $\mathcal{F}_t$.

4. What are $\mathcal{F}_0$, $\mathcal{F}_{U-}$ and $\mathcal{F}_U$ equal to?

Exercise 8.8: Let $U_1, U_2, \ldots$ be (strictly) positive random variables on a probability space $(\Omega, \mathcal{F}, P)$, and $\mathcal{G}_1, \mathcal{G}_2, \ldots$ be sub-σ-fields of $\mathcal{F}$. Suppose that for all n, $U_1, U_2, \ldots, U_n$ are $\mathcal{G}_n$-measurable and denote by T_n the random variable $\sum_{i=1}^n U_i$. Set $\mathcal{F}_t = \bigcap_n \{A \in \mathcal{F} : \exists B_n \in \mathcal{G}_n$ such that $A \cap \{T_n > t\} = B_n \cap \{T_n > t\}\}$.

1. Show that $\mathcal{F}_t$ is a right-continuous filtration on $(\Omega, \mathcal{F}, P)$.

2. Show that for all n, T_n is a stopping time for $\mathcal{F}_t$.

3. Suppose that $\lim_n T_n = \infty$ a.s. Show that $\mathcal{F}_{T_n} = \mathcal{G}_{n+1}$ and $\mathcal{F}_{T_n-} = \mathcal{G}_n$.

Exercise 8.9: Let $B(t)$ be a Brownian motion and $H(t)$ be a predictable process. Show that $M(t) = \int_0^t H(s)dB(s)$ is a Brownian motion if and only if $\text{Leb}(\{t : |H(t)| \neq 1\}) = 0$ a.s.

Exercise 8.10: Let T be a stopping time. Show that the process $M(t) = 2B(t \wedge T) - B(t)$ obtained by reflecting $B(t)$ at time T, is a Brownian motion.

Exercise 8.11: Let $B(t)$ and $N(t)$ be respectively a Brownian motion and a Poisson process on the same space. Denote by $\bar{N}(t) = N(t) - t$ the compensated Poisson process. Show that the following processes are martingales: $B(t)\bar{N}(t)$, $\mathcal{E}(B)(t)\bar{N}(t)$ and $\mathcal{E}(\bar{N})(t)B(t)$.

Exercise 8.12: $X(t)$ solves the SDE $dX(t) = \mu X(t)dt + aX(t-)dN(t) + \sigma X(t)dB(t)$. Find the condition for $X(t)$ to be a martingale.

Exercise 8.13: Find the predictable representation for $Y = B^5(1)$.

Exercise 8.14: Find the predictable representation for the martingale $e^{B(t)-t/2}$.

Exercise 8.15: Let $Y = \int_0^1 \text{sign}(B(s))dB(s)$. Show that Y has a Normal distribution. Show that there is no deterministic function $H(s)$, such that $Y = \int_0^1 H(s)dB(s)$. This shows that the assumption that Y, B are jointly Gaussian in Theorem 8.35 is indispensable.

Exercise 8.16: Find the quadratic variation of $|B(t)|$.

Exercise 8.17: (Stochastic Logarithm)
Let U be a semimartingale such that $U(t)$ and $U(t-)$ are never zero. Show that there exists a unique semimartingale X with $X(0) = 0$, $(X = \mathcal{L}(U))$ such that $dX(t) = \frac{dU(t)}{U(t-)}$, and

$$X(t) = \ln\left|\frac{U(t)}{U(0)}\right| + \frac{1}{2}\int_0^t \frac{d<U^c, U^c>(s)}{U^2(s-)} - \sum_{s \leq t}\left(\ln\left|\frac{U(s)}{U(s-)}\right| + 1 - \frac{U(s)}{U(s-)}\right),$$

(8.82)

see Kallsen and Shiryaev (2002).

Chapter 9

Pure Jump Processes

In this Chapter we consider pure jump processes, that is, processes that change only by jumps. Counting processes and Markov Jump processes are defined and their semimartingale representation is given. This representation allows us to see the process as a solution to a stochastic equation driven by discontinuous martingales.

9.1 Definitions

A counting process is determined by a sequence of non-negative random variables T_n, satisfying $T_n < T_{n+1}$ if $T_n < \infty$ and $T_n = T_{n+1}$ if $T_n = \infty$. T_n can be considered as the time of the n-th occurrence of an event, and they are often referred to as arrival times. $N(t)$ counts the number of events that occurred up to time t, that is,

$$N(t) = \sum_{n=1}^{\infty} I(T_n \le t), \ N(0) = 0. \tag{9.1}$$

$N(t)$ is piece-wise constant and has jumps of size one at the points T_n. Such processes are also known as simple point processes to distinguish them from more general *marked point processes*, which are described by a sequence (T_n, Z_n) for some random variables Z_n. Z_n, for example, may describe the size of jump at T_n.

The pure jump process X is defined as follows.

$$X(t) = X(0) + \sum_{n=1}^{\infty} I(T_n \le t) Z_n. \tag{9.2}$$

Note that X in (9.2) is right-continuous, piece-wise constant with the time of the n-th jump at T_n, and $Z_n = X(T_n) - X(T_n-) = X(T_n) - X(T_{n-1})$ is the size of the jump at T_n.

9.2 Pure Jump Process Filtration

The filtration $\mathbb{F}$ considered here is the natural filtration of the process. We recall related definitions. For a process X its natural filtration is defined by (the augmentation of) σ-fields $\mathcal{F}_t = \sigma(X(s), 0 \leq s \leq t)$ and represents the information obtained by observing the process on $[0, t]$. The strict past is the information obtained by observing the process on $[0, t)$ and is denoted by $\mathcal{F}_{t-} = \sigma(X(s), 0 \leq s < t)$.

A non-negative random variable τ, which is allowed to be infinity, is a stopping time if $\{\tau \leq t\} \in \mathcal{F}_t$ for every t. Thus by observing the process on $[0, t]$ it is possible to deduce whether τ has occurred.

Information obtained from observing the process up to a stopping time τ is $\mathcal{F}_\tau$, defined by $\mathcal{F}_\tau = \{A \in \mathcal{F} : \text{ for any } t, \ A \cap \{\tau \leq t\} \in \mathcal{F}_t\}$. The strict past of X at τ is described by the σ-field $\mathcal{F}_{\tau-} = \sigma(A \cap \{t < \tau\} : t \geq 0, \ A \in \mathcal{F}_t) \bigvee \mathcal{F}_0$. Note that $\tau \in \mathcal{F}_{\tau-}$ (take $A = \{\tau > t\} \in \mathcal{F}_t$). Clearly, $\mathcal{F}_{\tau-} \subset \mathcal{F}_\tau$.

Clearly, the arrival times T_n are stopping times for $\mathbb{F}$. They are usually taken as a localizing sequence. Note that $\mathcal{F}_{T_n} = \sigma((T_i, Z_i), i \leq n)$ and that $X(T_n-) = X(T_{n-1})$, since for t satisfying $T_{n-1} \leq t < T_n$, $X(t) = X(T_{n-1})$ and this value is kept until the next jump at time T_n. As $T_n \in \mathcal{F}_{T_n-}$, $\mathcal{F}_{T_n-} = \sigma((T_i, Z_i), i \leq n-1, T_n)$. Thus Z_n, the jump size at T_n, is the only information in $\mathcal{F}_{T_n}$ not available in $\mathcal{F}_{T_n-}$.

It can be shown that $\mathcal{F}_t$ is right-continuous, that is, $\mathcal{F}_t = \mathcal{F}_{t+}$, as well as the following result, which is essential in finding compensators.

Theorem 9.1 *If τ is a stopping time, then for each n there is a random variable ζ_n which is $\mathcal{F}_{T_n}$ measurable such that*

$$\tau \wedge T_{n+1} = (T_n + \zeta_n) \wedge T_{n+1} \text{ on } \{T_n \leq \tau\}. \tag{9.3}$$

Proofs of the above results can be found for example, in Liptser and Shiryayev (1974), Karr (1986).

Since compensators are predictable processes, it is important to have some criteria to decide on predictability. By definition, any adapted left continuous or continuous process is predictable. The following construction, which is often met in calculations, results in a predictable process.

Theorem 9.2 *Let T_n be the arrival times in a pure jump process, and for all $n = 0, 1, \ldots,$ $Y_n(t)$ be an adapted process such that for any $t \in (T_n, T_{n+1}]$ it is $\mathcal{F}_{T_n}$ measurable. Then the process $X(t) = \sum_{n=0}^\infty Y_n(t) I(T_n < t \leq T_{n+1})$ is predictable.*

PROOF: We outline the proof. The process $I_n(t) = I(T_n < t \leq T_{n+1})$ is predictable. Indeed, $I_n(t) = I(T_n < t \leq T_{n+1}) = I(t \leq T_{n+1}) - I(t \leq T_n)$. Since for each n, T_n is a stopping time, $\{T_n \geq t\} \in \mathcal{F}_t$ and therefore $I_n(t)$ is adapted. But it is left-continuous, hence it is predictable. Since $Y_n(t)$ is "known" when $T_n < t \leq T_{n+1}$, $X(t)$ is predictable.

□

Assumptions

$T_\infty = \lim_{n \to \infty} T_n$ exists, since T_n's are non-decreasing. The results given below hold for $t < T_\infty$, and in order to avoid repetitions we assume throughout, that $T_\infty = \infty$, unless stated otherwise. In the case of $T_\infty < \infty$ there are infinitely many jumps on the finite time interval $[0, T_\infty)$, and it is said that explosion occurs. We assume that there are no explosions. Sufficient conditions for Markov Jump processes are given later.

Assume that the jumps are integrable, $\mathrm{E}|Z_n| < \infty$ for all n. Under this assumption X is locally integrable, since $\mathrm{E}|X(t \wedge T_n)| \leq \sum_{i=1}^{n} \mathrm{E}|Z_i| < \infty$, and therefore it has a uniquely defined compensator A.

$M(t) = X(t) - A(t)$ is the local martingale associated with X, also called the innovation martingale.

9.3 Itô's Formula for Processes of Finite Variation

If a semimartingale X is of finite variation, then its continuous martingale part $X^{cm} = 0$, consequently $\langle X, X \rangle^c(t) = \langle X^{cm}, X^{cm} \rangle(t) = 0$. Therefore the term containing the second derivative f'' disappears in (8.59). Moreover, since X is of finite variation its continuous part X^c satisfies $dX^c(t) = dX(t) - \Delta X(t)$, and Itô's formula takes form: for any C^1 function f

$$f(X(t)) - f(X(0)) = \int_0^t f'(X(s-))dX^c(s) + \sum_{s \leq t} \Big(f(X(s)) - f(X(s-)) \Big). \quad (9.4)$$

If the continuous part X^c is zero, then the formula is an identity, representing a function as the sum of its jumps. A similar formula holds for a function of n variables.

Stochastic Exponential

The stochastic exponential (8.62) of finite variation processes simplifies to

$$\mathcal{E}(X)(t) = e^{X(t) - X(0)} \prod_{s \leq t} (1 + \Delta X(s)) e^{-\Delta X(s)} = e^{X^c(t)} \prod_{s \leq t} (1 + \Delta X(s)), \quad (9.5)$$

where X^c is the continuous part of X. The last equality is due to
$X^c(t) = X(t) - X(0) - \sum_{s \leq t} \Delta X(s)$.

Example 9.1: Let $X(t)$ be a process with jumps of size one (counting process), so
that $\Delta X(s) = 0$ or 1. Its stochastic exponential is given by

$$\mathcal{E}(X)(t) = \prod_{s \leq t} \left(1 + \Delta X(s) \right) = 2^{X(t) - X(0)}. \tag{9.6}$$

Integration by Parts for Processes of Finite Variation

The integration by parts formula is obtained directly from the integral representation of the quadratic covariation (8.24) and (1.20). Recall that if X and Y are of finite variation, then their quadratic covariation is given by

$$[X, Y](t) = \sum_{0 < s \leq t} \Delta X(s) \Delta Y(s) \quad \text{and} \quad [X](t) = [X, X](t) = \sum_{0 < s \leq t} (\Delta X(s))^2. \tag{9.7}$$

Using (8.24) we obtain

$$X(t)Y(t) - X(0)Y(0) = \int_0^t X(s-)dY(s) + \int_0^t Y(s-)dX(s) + \sum_{0 < s \leq t} \Delta X(s) \Delta Y(s). \tag{9.8}$$

Remark 9.1: The following formula holds for finite variation processes.

$$\sum_{s \leq t} \Delta X(s) \Delta Y(s) = \int_0^t \Delta X(s) dY(s). \tag{9.9}$$

Indeed, by letting $Y^c(t) = Y(t) - \sum_{s \leq t} \Delta Y(s)$ be the continuous part of Y, we have

$$\int_0^t \Delta X(s) dY(s) - \sum_{s \leq t} \Delta X(s) \Delta Y(s) = \int_0^t \Delta X(s) dY^c(s) = 0, \tag{9.10}$$

since $Y^c(s)$ is continuous and $\Delta X(s)$ is different from zero at mostly countably many points.

9.4   Counting Processes

Let N be a counting process, then it is a pure jump process with jumps of size one. Thus

$$[N, N](t) = \sum_{s \leq t} (\Delta N(s))^2 = N(t). \tag{9.11}$$

Theorem 9.3 *The compensator A and the sharp bracket process of N are the same, $A = \langle N, N \rangle$.*

PROOF: N is of locally integrable variation, since $N(t \wedge T_n) \leq n$ and T_n is a localizing sequence. A is the unique predictable process such that $N(t) - A(t)$ is a local martingale. $\langle N, N \rangle$ is the unique predictable process such that $[N, N](t) - \langle N, N \rangle (t)$ is a local martingale. The result follows from (9.11) and the uniqueness of the compensator.

□

Heuristically the compensator for a counting process is given by

$$dA(t) = \mathrm{E}\big(dN(t)|\mathcal{F}_{t-}\big), \tag{9.12}$$

where $\mathcal{F}_{t-}$ denotes the information available prior to time t by observing the process over $[0, t)$, and $t + dt > t$. $dM(t) = d(N - A)(t) = dN(t) - dA(t)$ is that part of $dN(t)$ that cannot be foreseen from the observations of N over $[0, t)$.

The next result shows the relation between the sharp bracket of the martingale and the compensator in a counting process. Since the proof is a straight application of stochastic calculus rules, it is given below. It is useful for calculation of the variance of M, indeed by Theorem 8.24 $\mathrm{E}M^2(t) = \mathrm{E} \langle M, M \rangle (t)$.

Theorem 9.4 *Let $M = N - A$. Then $\langle M, M \rangle (t) = \int_0^t (1 - \Delta A(s)) dA(s)$. In particular, if A is continuous, then $\langle M, M \rangle (t) = A(t)$.*

PROOF: By integration by parts (9.8), we have

$$M^2(t) = 2 \int_0^t M(s-) dM(s) + \sum_{s \leq t} (\Delta M(s))^2. \tag{9.13}$$

Use $\Delta M(s) = \Delta N(s) - \Delta A(s)$ and expand the sum to obtain

$$\sum_{s \leq t} (\Delta M(s))^2 = \sum_{s \leq t} \Delta N(s) - 2 \sum_{s \leq t} \Delta N(s) \Delta A(s) + \sum_{s \leq t} (\Delta A(s))^2,$$

where we used that since N is a simple process $(\Delta N(s))^2 = \Delta N(s)$. Thus we obtain by using formula (9.9) and $N = M + A$,

$$
\begin{aligned}
M^2(t) &= 2 \int_0^t M(s-) dM(s) + N(t) - 2 \int_0^t \Delta A(s) dN(s) + \int_0^t \Delta A(s) dA(s) \\
&= \int_0^t \Big(2M(s-) + 1 - 2\Delta A(s) \Big) dM(s) + \int_0^t (1 - \Delta A(s)) dA(s).
\end{aligned}
$$

The process in the first integral is predictable, because $M(s-)$ is adapted and left-continuous, $A(s)$ is predictable, so that $\Delta A(s)$ is also predictable. Therefore the first integral is a local martingale. The second integral is predictable, since A is predictable. Thus the above equation is the Doob-Meyer decomposition of M^2. The result now follows by the uniqueness of the Doob-Meyer decomposition.

□

We give examples of processes and their compensators next.

Point Process of a Single Jump

Let T be a random variable and the process N have a single jump at the random time t, that is, $N(t) = I(T \leq t)$. Let F be the distribution function of T.

Theorem 9.5 *The compensator $A(t)$ of $N(t) = I(T \leq t)$ is given by*

$$A(t) = \int_0^{t \wedge T} \frac{dF(s)}{1 - F(s-)}. \tag{9.14}$$

PROOF: $A(t)$ is clearly predictable. To show that $N(t) - A(t)$ is a martingale, by Theorem 7.17 it is enough to show that $EN(S) = EA(S)$ for any stopping time S. By Theorem 9.1 there exists an $\mathcal{F}_0$-measurable (i.e. almost surely constant) random variable ζ such that

$$\{S \geq T\} = \{S \wedge T = T\} = \{\zeta \wedge T = T\} = \{T \leq \zeta\}.$$

Therefore

$$\begin{aligned}
EN(S) &= P(S \geq T) = P(T \leq \zeta) \\
&= \int_0^\zeta dF(t) = \int_0^\zeta \frac{P(T \geq t)}{1 - F(t-)} dF(t) \\
&= E\left(\int_0^\zeta \frac{I(T \geq t)}{1 - F(t-)} dF(t)\right) = E\left(\int_0^{\zeta \wedge T} \frac{dF(t)}{1 - F(t-)}\right) \\
&= E\left(\int_0^{S \wedge T} \frac{dF(t)}{1 - F(t-)}\right).
\end{aligned}$$

Thus (9.14) is obtained.

□

If F has a density f, then $A(t) = \int_0^{t \wedge T} h(s) ds$, where

$$h(t) = \frac{f(t)}{1 - F(t)}, \tag{9.15}$$

is called the hazard function and it gives the likelihood of the occurrence of the jump at t given that the jump has not occurred before t.

Compensators of Counting Processes

The next result gives an explicit form of the compensator of a general counting process. Since the proof uses the same ideas as above, it is omitted.

Theorem 9.6 *Let N be a counting process generated by the sequence T_n. Denote by $U_{n+1} = T_{n+1} - T_n$ the inter-arrival times, $T_0 = 0$. Let $F_n(t) = P(U_{n+1} \leq t | T_1, \ldots, T_n)$ denote the regular conditional distributions, and $F_0(t) = P(T_1 \leq t)$. Then the compensator $A(t)$ is given by*

$$A(t) = \sum_{i=0}^{\infty} \int_0^{t \wedge T_{i+1} - t \wedge T_i} \frac{dF_i(s)}{1 - F_i(s-)}. \tag{9.16}$$

Note that if the conditional distributions F_n in the above theorem are continuous with $F_n(0) = 0$, then by changing variables we have

$$\int_0^a \frac{dF_n(s)}{1 - F_n(s-)} = \int_0^a \frac{dF_n(s)}{1 - F_n(s)} = -\log(1 - F_n(a)), \tag{9.17}$$

and equation (9.16) can be simplified accordingly.

Renewal Process

A renewal process N is a point process in which all inter-arrival times are independent and identically distributed, that is, $T_1, T_2 - T_1, \ldots T_{n+1} - T_n$ are i.i.d. with distribution function $F(x)$. In this case all the conditional distributions F_n in the previous theorem are given by F. As a result of Theorem 9.6 we obtain the following corollary.

Corollary 9.7 *Assume that the inter-arrival distribution is continuous and $F(0) = 0$. Then the compensator of the renewal process is given by*

$$A(t) = -\sum_{n=1}^{\infty} \log \left(1 - F(t \wedge T_n - t \wedge T_{n-1}) \right). \tag{9.18}$$

Stochastic Intensity

Definition 9.8 *If $A(t) = \int_0^t \lambda(s)ds$, where $\lambda(t)$ is a positive predictable process, then $\lambda(t)$ is called the stochastic intensity of N.*

Note that if $A(t)$ is deterministic and differentiable with derivative $A'(t)$, and $A(t) = \int_0^t A'(s)ds$, then $\lambda(t) = A'(t)$ is predictable and is the stochastic intensity. If $A(t)$ is random and differentiable with derivative $A'(t)$, satisfying $A(t) = \int_0^t A'(s)ds$, then $\lambda(t) = A'(t-)$. Indeed, $A(t) = \int_0^t A'(s-)ds$.

If the stochastic intensity exists, then by definition of the compensator, $N(t) - \int_0^t \lambda(s)ds$ is a local martingale. For counting processes a heuristic interpretation of the intensity is given by

$$\lambda(t)dt = dA(t) = \mathrm{E}(dN(t)|\mathcal{F}_{t-}) = \mathrm{P}(dN(t) = 1|\mathcal{F}_{t-}). \qquad (9.19)$$

If the stochastic intensity exists, then the compensator is continuous and by Theorem 9.4 the sharp bracket of the martingale $M = N - A$ is given by

$$\langle M, M \rangle\,(t) = \int_0^t \lambda(s)ds. \qquad (9.20)$$

Example 9.2:

1. A deterministic point process is its own compensator, so it does not have stochastic intensity.

2. Stochastic intensity for the renewal process with continuous inter-arrival distribution F is given by $h(V(t-))$, where h is the hazard function and $V(t) = t - T_{N(t)}$, called the age process. This can be seen by differentiating the compensator in (9.18).

3. Stochastic intensity for the renewal process with a discrete inter-arrival distribution F does not exist.

Non-homogeneous Poisson Processes

Theorem 9.9 *Let $N(t)$ be point process with a continuous deterministic compensator $A(t)$. Then it has independent Poisson distributed increments, that is, the distribution of $N(t) - N(s)$ is Poisson with parameter $A(t) - A(s)$, $0 \leq s < t$.*

If $A(t)$ has a density $\lambda(t)$, that is, $A(t) = \int_0^t \lambda(s)ds$, then $N(t)$ is called the non-homogeneous Poisson process with rate $\lambda(s)$.

PROOF: We prove the result by an application of the stochastic exponential. $M(t) = N(t) - A(t)$ is a martingale. For a fixed $0 < u < 1$, $-uM(t)$ is also a martingale. Consider $\mathcal{E}(uM)(t)$. By (9.5)

$$
\begin{aligned}
\mathcal{E}(-uM)(t) &= e^{uA(t)} \prod_{s \leq t}(1 - u\Delta M(s)) \\
&= e^{uA(t)} \prod_{s \leq t}(1 - u\Delta N(s)) = e^{uA(t)}(1 - u)^{N(t)} \\
&= e^{uA(t)+N(t)\log(1-u)}, \qquad (9.21)
\end{aligned}
$$

where we have used that $\Delta N(s)$ is zero or one. Stochastic exponential of a martingale is always a local martingale, but in this case it is also a true martingale by Theorem 7.21. Indeed, since $A(t)$ is deterministic and non-decreasing,

$$\mathrm{E}\sup_{t\leq T} e^{uA(t)+N(t)\log(1-u)} \leq e^{uA(T)}\mathrm{E}\sup_{t\leq T} e^{N(t)\log(1-u)} \leq e^{uA(T)} < \infty,$$

and the condition of Theorem 7.21 is satisfied. Taking expectations in (9.21) we obtain the moment generating function of $N(t)$,

$$\mathrm{E}\big((1-u)^{N(t)}\big) = e^{-uA(t)}$$

or with $v = 1 - u$,

$$\mathrm{E}(v^{N(t)}) = e^{(1-v)A(t)}. \tag{9.22}$$

This shows that $N(t)$ is a Poisson random variable with parameter $A(t)$. If we take conditional expectation in (9.21) and use the martingale property, we obtain in the same way that for all $s < t$

$$\mathrm{E}\big(v^{N(t)-N(s)}\big|\mathcal{F}_s\big) = e^{(1-v)(A(t)-A(s))}, \tag{9.23}$$

which shows that the distribution of $N(t) - N(s)$ does not depend on the past and is Poisson.

$\square$

A similar result holds if the compensator is deterministic but discontinuous. The proof is more involved and can be found for example, in Liptser and Shiryayev (1974) p. 279, where the form of the distribution of the increments is also given.

Theorem 9.10 *Let $N(t)$ be a point process with a deterministic compensator $A(t)$. Then it has independent increments.*

The following result states that a point process with a continuous, but possibly random, compensator can be transformed into Poisson process by a random change of time. Compare this result to change of time for continuous martingales, Dambis, Dubins-Schwarz Theorem 7.37.

Theorem 9.11 *Let a counting process $N(t)$ have a continuous compensator $A(t)$ and $\lim_{t\to\infty} A(t) = \infty$. Define $\rho(t) = \inf\{s \geq 0 : A(s) = t\}$. Let $K(t) = N(\rho(t))$ and $\mathcal{G}_t = \mathcal{F}_{\rho(t)}$. Then the process $K(t)$ with respect to filtration $\mathcal{G}_t$ is Poisson with rate 1.*

The proof can be found in Liptser and Shiryayev (1974) p 280, (2001) and is not given here. To convince ourselves of the result, consider the case when $A(t)$ is strictly increasing, then $\rho(t)$ is the usual inverse, that is $A(\rho(t)) = t$. Then $\mathrm{E}K(t) = \mathrm{E}N(\rho(t)) = \mathrm{E}A(\rho(t)) = t$, so that K has the right mean.

For more information on point processes see for example, Liptser and Shiryayev (2001), (1989) and Karr (1986).

Compensators of Pure Jump Processes

Let for all $t \geq 0$

$$X(t) = X(0) + \sum_{n=1}^{\infty} Z_n I(T_n \leq t), \qquad (9.24)$$

be a pure jump point process generated by the sequence (T_n, Z_n). By using the same arguments as in the proof of Theorem 9.6 we can show the following result.

Theorem 9.12 Let $F_n(t) = \mathrm{P}(T_{n+1} - T_n \leq t | \mathcal{F}_{T_n})$ denote the regular conditional distributions of inter-arrival times, $F_0(t) = \mathrm{P}(T_1 \leq t)$, and $m_n = \mathrm{E}(Z_{n+1} | \mathcal{F}_{T_n}) = \mathrm{E}(X(T_{n+1}) - X(T_n) | \mathcal{F}_{T_n})$ denote the conditional expectation of the jump sizes. Then the compensator $A(t)$ is given by

$$A(t) = \sum_{n=0}^{\infty} m_n \int_0^{t \wedge T_{n+1} - t \wedge T_n} \frac{dF_n(s)}{1 - F_n(s-)}. \qquad (9.25)$$

The following observation is frequently used in the calculus of pure jump processes.

Theorem 9.13 If $X(t)$ is a pure jump process, then for a function f, $f(X(t))$ is also a pure jump process with the same jump times T_n. The size of the jump at T_n is given by $Z_n' = f(X(T_n)) - f(X(T_n-))$. Consequently if f is such that $\mathrm{E}|Z_n'| < \infty$, then the compensator of $f(X)$ is given by (9.25) with m_n replaced by $m_n' = \mathrm{E}(Z_{n+1}' | \mathcal{F}_{T_n})$.

Theorem 9.14 Let $X(t)$ be a pure jump point process generated by the sequence (T_n, Z_n). Assume conditions and notations of Theorem 9.12 and that the compensator $A(t)$ is continuous. Suppose in addition that $\mathrm{E}Z_n^2 < \infty$ and $v_n = \mathrm{E}(Z_{n+1}^2 | \mathcal{F}_{T_n})$. Let $M(t) = X(t) - A(t)$, then $\langle M, M \rangle(t)$ is given by (9.25) with m_n replaced by v_n.

PROOF: Since A is assumed to be continuous, (and it is always of finite variation)

$$[M, M](t) = [X - A, X - A](t) = [X, X](t) = \sum_{0 < s \leq t} (\Delta X(s))^2. \qquad (9.26)$$

Thus $[M, M]$ is a pure jump process with jump times T_n and jump sizes $(\Delta X(T_n))^2 = Z_n^2$. Thus $[M, M](t)$ is a pure jump process generated by the sequence (T_n, Z_n^2). $\langle M, M \rangle(t)$ is its compensator. The result follows by Theorem 9.12.

$\square$

A particular case of pure jump processes is the class of processes with exponentially distributed inter-arrival times and independent jump sizes. This is the class of Markov Jump processes.

9.5 Markov Jump Processes

Definitions

Let for all $t \geq 0$

$$X(t) = X(0) + \sum_{n=1}^{\infty} Z_n I(T_n \leq t), \qquad (9.27)$$

where T_n and Z_n have the following conditional distributions. Given that $X(T_n) = x$, $T_{n+1} - T_n$ is exponentially distributed with mean $\lambda^{-1}(x)$, and independent of the past; and the jump $Z_{n+1} = X(T_{n+1}) - X(T_n)$ is independent of the past and has a distribution that depends only on x.

$$F_n(t) = P(T_{n+1} - T_n \leq t | \mathcal{F}_{T_n}) = 1 - e^{-\lambda(X(T_n))t}, \qquad (9.28)$$

and for some family of distribution functions $K(x, \cdot)$

$$
\begin{aligned}
P(X(T_{n+1}) - X(T_n) \leq t | \mathcal{F}_{T_n}) &= K(X(T_n), t), \\
E(X(T_{n+1}) - X(T_n) | \mathcal{F}_{T_n}) &= m(X(T_n)) = m_n.
\end{aligned} \qquad (9.29)
$$

It is intuitively clear that X defined in this way possesses the Markov property, due to the lack of memory of the exponential distribution. We omit the proof.

Heuristically, Markov Jump processes can be described as follows. If the process is in state x then it stays there for an exponential length of time with mean $\lambda^{-1}(x)$ (parameter $\lambda(x)$) after which it jumps from x to a new state $x + \xi(x)$, where $P(\xi(x) \leq t) = K(x, t)$ denotes the distribution of the jump from x. The parameters of the process are: the function $\lambda(x)$ (the holding time parameter) and distributions $K(x, \cdot)$ (distributions of jump sizes).

$\lambda(x)$ is always non-negative. If for some x, $\lambda(x) = 0$ then once the process gets into x it stays in x forever, in this case the state x is called absorbing. We shall assume that there are no absorbing states. If $\lambda(x) = \infty$ then the process leaves x instantaneously. We assume that $\lambda(x)$ is finite on finite intervals, so that there are no instantaneous states.

For construction, classification and properties of Markov Jump processes see for example, Breiman (1968), Chung (1967), Ethier and Kurtz (1986).

The Compensator and the Martingale

We derive the compensator heuristically first and then give the precise result. Suppose that $X(t) = x$. By the lack of memory property of the exponential distribution, it does not matter how long the process has already spent in x, the jump from x will still occur after exponentially distributed (with parameter $\lambda(x)$) random time. Therefore the conditional probability that a jump occurs in $(t, t+dt)$, given that it has not occurred before, is (with U having exponential $\exp(\lambda(x))$ distribution)

$$P(U \leq dt) = 1 - \exp(-\lambda(x)dt) \approx \lambda(x)dt.$$

When the jump occurs its size is $\xi(x)$, with mean $m(x) = E(\xi(x))$. Therefore

$$dA(t) = E(dX(t)|\mathcal{F}_{t-}) = \lambda(X(t))m(X(t))dt. \tag{9.30}$$

Assume that $\lim_{n\to\infty} T_n = T_\infty = \infty$ (such processes are called regular, and sufficient conditions for this are given in a later section). If $T_\infty < \infty$ then the compensator is given for times $t < T_\infty$.

Theorem 9.15 *Let X be a Markov jump process such that for all x, the holding time parameter is positive, $\lambda(x) > 0$, and the size of the jump from x is integrable with mean $m(x)$.*

1. *The compensator of X is given by*

$$A(t) = \int_0^t \lambda(X(s))m(X(s))ds. \tag{9.31}$$

2. *Suppose the second moments of the jumps are finite, $v(x) = E\xi^2(x) < \infty$. Then the sharp bracket of the local martingale $M(t) = X(t) - A(t)$ is given by*

$$\langle M, M \rangle (t) = \int_0^t \lambda(X(s))v(X(s))ds. \tag{9.32}$$

PROOF: The formula (9.31) follows from (9.25). Indeed, using the exponential form of F_n (9.28), we have

$$A(t) = \sum_{i=0}^{\infty} m(X(T_i))\lambda(X(T_i))(t \wedge T_{i+1} - t \wedge T_i). \tag{9.33}$$

Note that since the process X has a constant value $X(T_i)$ on the time interval $[T_i, T_{i+1})$, we have for a function f

$$\int_0^t f(X(s))ds = \sum_{i=0}^{\infty} f(X(T_i))(t \wedge T_{i+1} - t \wedge T_i), \tag{9.34}$$

and taking $f(x) = \lambda(x)m(x)$ gives the result. The formula (9.32) follows from Theorem 9.14.

$\square$

9.6 Stochastic Equation for Jump Processes

It is clear from Theorem 9.15 that a Markov Jump process X has a semi-martingale representation

$$X(t) = X(0) + A(t) + M(t) = X(0) + \int_0^t \lambda(X(s))m(X(s))ds + M(t). \quad (9.35)$$

This is the integral form of the stochastic differential equation for X

$$dX(t) = \lambda(X(t))m(X(t))dt + dM(t), \quad (9.36)$$

and is driven by the purely discontinuous, finite variation martingale M. By analogy with diffusions the infinitesimal mean is $\lambda(x)m(x)$ and the infinitesimal variance is $\lambda(x)v(x)$.

It is useful to have conditions assuring that the local martingale M in the representation (9.35) is a martingale. Such conditions are given in the next result.

Theorem 9.16 *Suppose that for all x*

$$\lambda(x)\mathrm{E}(|\xi(x)|) \leq C(1 + |x|), \quad (9.37)$$

then representation (9.35) holds with a zero mean martingale M. If in addition

$$\lambda(x)v(x) \leq C(1 + x^2), \quad (9.38)$$

then M is square integrable with

$$\langle M, M \rangle (t) = \int_0^t \lambda(X(s))v(X(s))ds. \quad (9.39)$$

In particular, we have the following corollary.

Corollary 9.17 *Suppose that the conditions of the above Theorem 9.16 hold. Then*

$$\mathrm{E}X(t) = \mathrm{E}X(0) + \mathrm{E}\int_0^t \lambda(X(s))m(X(s))ds, \quad (9.40)$$

$$Var(M(t) - M(0)) = \mathrm{E}\int_0^t \lambda(X(s))v(X(s))ds. \quad (9.41)$$

The proof of the result can be found in Hamza and Klebaner (1995).

An application of the stochastic equation approach to the model of Birth-Death processes is given in Chapter 13.

Remark 9.2: Markov Jump processes with countably or finitely many states are called Markov Chains. The jump variables have discrete distributions in a Markov Chain, whereas they can have a continuous distribution in a Markov process. Markov Jump processes are also known as Markov Chains with general state space.

Remark 9.3: A model for a randomly evolving population can be served by a Markov Chain on non-negative integers. In this case the states of the process are the possible values for the population size. The traditional way of defining a Markov Chain is by the infinitesimal probabilities: for integer i and j and small δ

$$P(X(t+\delta) = j | X(t) = i) = \lambda_{ij}\delta + o(\delta),$$

where $\lim_{\delta \to 0} o(\delta)/\delta = 0$. In this section we presented an almost sure, path by path representation of a Markov Jump process. Another representation related to the Poisson process can be found in Ethier and Kurtz (1986).

Generators and Dynkin's Formula

Let X be the Markov Jump process described by (9.27). Suppose that $\lambda(x)$ is bounded, that is $\sup_x \lambda(x) < \infty$. Define the following linear operator L acting on bounded functions by

$$Lf(x) = \lambda(x)\mathrm{E}\big(f(x + \xi(x)) - f(x)\big) = \lambda(x)m_f(x). \qquad (9.42)$$

Theorem 9.18 (Dynkin's Formula) *Let L be as above, and define $M^f(t)$ by*

$$M^f(t) = f(X(t)) - f(X(0)) - \int_0^t Lf(X(s))ds. \qquad (9.43)$$

Then M^f is a martingale, and consequently

$$\mathrm{E}f(X(t)) = f(X(0)) + \mathrm{E}\int_0^t Lf(X(s))ds. \qquad (9.44)$$

PROOF: The proof of these formulae follows from Theorem 9.13 and Theorem 9.15. The process $f(X(t))$ is a pure jump process with the same arrival times T_n, and jumps of size $Z'_n = f(X(T_n)) - f(X(T_n-))$ (in this case it is also Markov). The jumps are bounded, since f is bounded. The mean of the jump from x is given by $\mathrm{E}(Z'_{n+1}|\mathcal{F}_{T_n}, X(T_n) = x) = \mathrm{E}\big(f(x + \xi(x)) - f(x)\big) = m_f(x)$. Therefore the compensator of $f(X(t))$ is, from Theorem 9.15, $\int_0^t \lambda(X(s))m_f(X(s))ds$. This implies that M^f is a local martingale. If $\lambda(x)$ is bounded, then $M(t)$ is bounded by a constant Ct on any finite time interval $[0, t]$. Since a bounded local martingale is a martingale, M^f is a martingale. Moreover it is square integrable on any finite time interval.   □

The linear operator L in (9.42) is called the generator of X. It can be shown that with its help one can define probabilities on the space of of right-continuous functions, so that the coordinate process is the Markov process, cf. solution to the martingale problem for diffusions.

Theorem 9.16 above can be seen as a generalization of Dynkin's formula for the identity function $f(x) = x$ and the quadratic function $f(x) = x^2$. These are particular cases of a more general theorem

Theorem 9.19 *Assume that there are no explosions, that is, $T_n \uparrow \infty$. Suppose that for an unbounded function f there is a constant C such that for all x*

$$\lambda(x)\mathrm{E}|f(x + \xi(x)) - f(x)| \le C(1 + |f(x)|). \tag{9.45}$$

Suppose also that $\mathrm{E}|f(X(0))| < \infty$. Then for all t, $0 \le t < \infty$, $\mathrm{E}|f(X(t))| < \infty$, moreover M^f in (9.43) is a martingale and (9.44) holds.

For the proof see Hamza and Klebaner (1995).

9.7 Explosions in Markov Jump Processes

Let $X(t)$ be a jump Markov process taking values in $\mathbb{R}$. If the process is in states x then it stays there for an exponential length of time with mean $\lambda^{-1}(x)$ after which it jumps from x. If $\lambda(x) \to \infty$ for a set of values x, then the expected duration of stay in state x, $\lambda^{-1}(x) \to 0$, and the time spent in x becomes shorter and shorter. It can happen that the process jumps infinitely many times in a finite time interval, that is, $\lim_{n\to\infty} T_n = T_\infty < \infty$. This phenomenon is called explosion. The terminology comes from the case when the process takes only integer values and $\lambda(x)$ can tend to infinity only when $x \to \infty$. In this case there are infinitely many jumps only if the process reaches infinity in finite time.

When the process does not explode it is called *regular*. In other words, the regularity assumption is

$$T_\infty = \lim_{n\to\infty} T_n = \infty. \ a.s. \tag{9.46}$$

If $\mathrm{P}(T_\infty < \infty) > 0$, then it said that the process explodes on the set $\{T_\infty < \infty\}$.

A necessary and sufficient condition for non-explosion is given by

Theorem 9.20

$$\sum_{0}^{\infty} \frac{1}{\lambda(X(T_n))} = \infty. \ a.s. \tag{9.47}$$

PROOF: We sketch the proof, see Breiman (1968) for details. Let ν_n be a sequence of independent exponentially distributed random variables with parameter λ_n. Then it is easy to see (by taking a Laplace transform) that $\sum_0^n \nu_i < \infty$ converges in distribution if and only if $\sum_0^\infty \frac{1}{\lambda_n} < \infty$. Using the result that a series of *independent* random variables converges almost surely if and only if it converges in distribution, we have $\sum_0^\infty \nu_n < \infty$ almost surely if and only if $\sum_0^\infty \frac{1}{\lambda_n} < \infty$. Condition (9.47) now follows, since the conditional distribution of $T_{n+1} - T_n$, given the information up to the n-th jump, $\mathcal{F}_{T_n}$, is exponential with parameter $\lambda(X(T_n))$.

$\square$

Condition (9.47) is hard to check in general, since it involves the variables $X(T_n)$. A simple condition, which is essentially in terms of the function $\lambda(x)m(x)$ (the drift), is given below.

Theorem 9.21 *Assume that $X \geq 0$, and there exists a monotone function $f(x)$ such that $\lambda(x)m(x) \leq f(x)$ and $\int_0^\infty \frac{dx}{f(x)} = \infty$. Then the process X does not explode, that is, $\mathrm{P}(X(t) < \infty$ for all $t,\ 0 \leq t < \infty) = 1$.*

According to this, the first condition of the result on integrability, Theorem 9.16, guarantees that the process does not explode. Proof of Theorem 9.21 and other sharp sufficient conditions for regularity in terms of the parameters of the process are given in Kersting and Klebaner (1995), (1996).

Notes. Material for this chapter is based on Liptser and Shiryayev (1974), (1989), Karr (1986), and the research papers quoted in the text.

9.8 Exercises

Exercise 9.1: Let $U \geq 0$ and denote $h(x) = \int_0^x \frac{dF(s)}{1-F(s-)}$, where $F(x)$ is the distribution function of U. Show that for $a > 0$ $\mathrm{E}h(U \wedge a) = \int_0^a dF(x)$.

Exercise 9.2: Show that when the distribution of inter-arrival times is exponential the formula (9.18) gives the compensator λt, hence $N(t)$ is a Poisson process.

Exercise 9.3: Show that when the distribution of inter-arrival times is Geometric then $N(t)$ has a Binomial distribution.

Exercise 9.4: Prove (9.25).

Exercise 9.5: Let $X(t)$ be a pure jump process with first k moments. Let $m_k(x) = \mathrm{E}\xi^k(x)$ and assume for all $i = 1, \ldots, k$, $\lambda(x)\mathrm{E}|\xi(x)|^k \leq C|x|^i$. Show that

$$\mathrm{E}X^k(t) = \mathrm{E}X^k(0) + \sum_{i=0}^{k-1} \binom{k}{i} \int_0^t \mathrm{E}\Big(\lambda(X(s))X^i(s)m_{k-i}(X(s))\Big)ds$$

Chapter 10

Change of Probability Measure

In this chapter we describe what happens to random variables and processes when the original probability measure is changed to an equivalent one. Change of measure for processes is done by using Girsanov's theorem.

10.1 Change of Measure for Random Variables

Change of Measure on a Discrete Probability Space

We start with a simple example. Let $\Omega = \{\omega_1, \omega_2\}$ with probability measure P given by $p(\omega_1) = p$, $p(\omega_2) = 1 - p$.

Definition 10.1 Q *is equivalent to* P *(Q $\sim$ P) if they have same null sets, i e.* $Q(A) = 0$ *if and only if* $P(A) = 0$.

Let Q be a new probability measure equivalent to P. In this example, this means that $Q(\omega_1) > 0$ and $Q(\omega_2) > 0$ (or $0 < Q(\omega_1) < 1$). Put $Q(\omega_1) = q$, $0 < q < 1$.

Let now $\Lambda(\omega) = \frac{Q(\omega)}{P(\omega)}$, that is, $\Lambda(\omega_1) = \frac{Q(\omega_1)}{P(\omega_1)} = \frac{q}{p}$, and $\Lambda(\omega_2) = \frac{1-q}{1-p}$.
By definition of Λ, for all ω

$$Q(\omega) = \Lambda(\omega)P(\omega). \tag{10.1}$$

Let now X be a random variable. The expectation of X under the probability P is given by

$$E_P(X) = X(\omega_1)P(\omega_1) + X(\omega_2)P(\omega_2) = pX(\omega_1) + (1 - p)X(\omega_2)$$

and under the probability Q

$$E_Q(X) = X(\omega_1)Q(\omega_1) + X(\omega_2)Q(\omega_2).$$

From (10.1)

$$E_Q(X) = X(\omega_1)\Lambda(\omega_1)P(\omega_1) + X(\omega_2)\Lambda(\omega_2)P(\omega_2) = E_P(\Lambda X). \qquad (10.2)$$

Take $X = 1$, then

$$E_Q(X) = 1 = E_P(\Lambda). \qquad (10.3)$$

On the other hand, take any random variable $\Lambda > 0$, such that $E_P(\Lambda) = 1$, and define Q by (10.1).

Then Q is a probability, because $Q(\omega_i) = \Lambda(\omega_i)P(\omega_i) > 0$, $i = 1, 2$, and

$$\begin{aligned}
Q(\Omega) &= Q(\omega_1) + Q(\omega_2) = \Lambda(\omega_1)P(\omega_1) + \Lambda(\omega_2)P(\omega_2) \\
&= E_P(\Lambda) = 1.
\end{aligned}$$

Q is equivalent to P, since Λ is strictly positive.

Thus we have shown that for any equivalent change of measure $Q \sim P$ there is a positive random variable Λ, such that $E_P(\Lambda) = 1$, and $Q(\omega) = \Lambda(\omega)P(\omega)$. This is the simplest version of the general result, the Radon-Nikodym Theorem. The expectation under Q of a random variable X is given by,

$$E_Q(X) = E_P(\Lambda X). \qquad (10.4)$$

By taking indicators $I(X \in A)$ we obtain the distribution of X under Q

$$Q(X \in A) = E_P(\Lambda I(X \in A)).$$

These formulae, obtained here for a simple example, hold also in general.

Change of Measure for Normal Random Variables

Consider first a change of a Normal probability measure on $\mathbb{R}$. Let μ be any real number, $f_\mu(x)$ denote the probability density of $N(\mu, 1)$ distribution, and P_μ the $N(\mu, 1)$ probability measure on $\mathbb{R}$ $(\mathcal{B}(\mathbb{R}))$. Then, it is easy to see that

$$f_\mu(x) = \frac{1}{\sqrt{2\pi}} e^{-\frac{1}{2}(x-\mu)^2} = f_0(x)e^{\mu x - \mu^2/2}. \qquad (10.5)$$

Put

$$\Lambda(x) = e^{\mu x - \mu^2/2}. \qquad (10.6)$$

Then the above equation reads

$$f_\mu(x) = f_0(x)\Lambda(x). \qquad (10.7)$$

By the definition of a density function, the probability of a set A on the line, is the integral of the density over this set,

$$P(A) = \int_A f(x)dx = \int_A dP. \qquad (10.8)$$

In infinitesimal notations this relation is written as

$$dP = P(dx) = f(x)dx. \qquad (10.9)$$

Hence the relation between the densities (10.7) can be written as a relation between the corresponding probability measures

$$f_\mu(x)dx = f_0(x)\Lambda(x)dx, \quad \text{and} \quad P_\mu(dx) = \Lambda(x)P_0(dx). \qquad (10.10)$$

By a property of the expectation (integral) (2.3), if a random variable $X \geq 0$ then $EX = 0$ if and only if $P(X = 0) = 1$,

$$P_\mu(A) = \int_A \Lambda(x)P_0(dx) = E_{P_0}(I_A\Lambda) = 0$$

implies

$$P_0(I_A\Lambda = 0) = 1.$$

Since $\Lambda(x) > 0$ for all x, this implies $P_0(A) = 0$. Using that $\Lambda < \infty$, the other direction follows: if $P_0(A) = 0$ then $P_\mu(A) = E_{P_0}(I_A\Lambda) = 0$, which proves that these measures have same null sets and are equivalent.

Another notation for Λ is

$$\Lambda = \frac{dP_\mu}{dP_0}, \quad \frac{dP_\mu}{dP_0}(x) = e^{\mu x - \mu^2/2}. \qquad (10.11)$$

This shows that any $N(\mu, 1)$ probability is obtained by an equivalent change of probability measure from the $N(0, 1)$ distribution.

We now give the same result in terms of changing the distribution of a random variable.

Theorem 10.2 *Let X have $N(0, 1)$ under* P, *and* $\Lambda(X) = e^{\mu X - \mu^2/2}$. *Define measure* Q *by*

$$Q(A) = \int_A \Lambda(X)dP = E_P(I_A\Lambda(X)) \quad or \quad \frac{dQ}{dP}(X) = \Lambda(X). \qquad (10.12)$$

Then Q *is an equivalent probability measure, and* X *has* $N(\mu, 1)$ *distribution under* Q.

PROOF: We show first that Q defined by (10.12) is indeed a probability. $Q(\Omega) = E_Q(1) = E_P(1\Lambda(X)) = E_P(e^{\mu X - \mu^2/2}) = e^{-\mu^2/2}E_P(e^{\mu X}) = 1$. The last equality is because $E_P(e^{\mu X}) = e^{\mu^2/2}$ for the $N(0,1)$ random variable. Other properties of probability Q follow from the corresponding properties of the integral. Consider the moment generating function of X under Q. $E_Q(e^{uX}) = E_P(e^{uX}\Lambda(X)) = E_P(e^{(u+\mu)X - \mu^2/2}) = e^{-\mu^2/2}E_P(e^{(u+\mu)X}) = e^{u\mu + u^2/2}$, which corresponds to the $N(\mu, 1)$ distribution.

$\square$

Remark 10.1: If P is a probability such that X is $N(0,1)$, then $X + \mu$ has $N(\mu, 1)$ distribution under the same P. This is an operation on the outcomes x. When we change the probability measure P to $Q = P_\mu$, we leave the outcomes as they are, but assign different probabilities to them (more precisely to sets of outcomes). Under the new measure the same X has $N(\mu, 1)$ distribution.

Example 10.1: (Simulations of rare events)
Change of probability measure is useful in simulation and estimation of probabilities of rare events. Consider estimation of $\Pr(N(6,1) < 0)$ by simulations. This probability is about 10^{-10}. Direct simulation is done by the expression

$$\Pr(N(6,1) < 0) \approx \frac{1}{n}\sum_{i=1}^{n} I(x_i < 0), \qquad (10.13)$$

where x_i's are the observed values from the $N(6,1)$ distribution. Note that in a million runs, $n = 10^6$, we should expect no values below zero, and the estimate is 0.

Let Q be the $N(6,1)$ distribution on **R**. Consider Q as changed from $N(0,1) = P$, as follows

$$\frac{dQ}{dP}(x) = \frac{dN(6,1)}{dN(0,1)}(x) = \Lambda(x) = e^{\mu x - \mu^2/2} = e^{6x-18}.$$

So we have

$$Q(A) = N(6,1)(A) = E_P\big(\Lambda(x)I(A)\big) = E_{N(0,1)}\big(e^{6x-18}I(A)\big).$$

In our case $A = (-\infty, 0)$, and $\Pr(N(6,1) < 0) = Q(A)$.
Thus we are led to the following estimate

$$\Pr(N(6,1) < 0) = Q(A) \approx \frac{1}{n}\sum_{i=1}^{n} e^{6x_i - 18}I(x_i < 0) = e^{-12}\frac{1}{n}\sum_{i=1}^{n} e^{6(x_i-1)}I(x_i < 0),$$
$$(10.14)$$

with x_i generated from $N(0,1)$ distribution. Of course, about half of the observations x_i's will be negative, resulting in a more precise estimate, even for a small number of runs n.

Next we give the result for random variables, set up in a way similar to Girsanov's theorem for processes, changing measure to remove the mean.

Theorem 10.3 (Removal of the mean) *Let X have $N(0,1)$ distribution under* P, *and* $Y = X + \mu$. *Then there is an equivalent probability* Q $\sim$ P, *such that* Y *has* $N(0,1)$ *under* Q. $(dQ/dP)(X) = \Lambda(X) = e^{-\mu X - \mu^2/2}$.

PROOF: Similarly to the previous proof, Q is a probability measure, with the same null sets as P.

$$E_Q(e^{uY}) = E_P(e^{u(X+\mu)}\Lambda(X)) = E_P(e^{(u-\mu)X+u\mu-\frac{1}{2}\mu^2}) = e^{u\mu-\frac{1}{2}\mu^2}E_P(e^{(u-\mu)X}).$$

Using the moment generating function of $N(0,1)$, $E_P(e^{(u-\mu)X}) = e^{(u-\mu)^2/2}$, which gives $E_Q(e^{uY}) = e^{u^2/2}$, establishing $Y \sim N(0,1)$ under Q.

□

Finally, we give the change of one Normal probability to another.

Theorem 10.4 *Let X have $N(\mu_1, \sigma_1^2)$ distribution, call it* P. *Define* Q *by*

$$(dQ/dP)(X) = \Lambda(X) = \frac{\sigma_1}{\sigma_2}e^{\frac{(X-\mu_1)^2}{2\sigma_1^2} - \frac{(X-\mu_2)^2}{2\sigma_2^2}}. \qquad (10.15)$$

Then X has $N(\mu_2, \sigma_2^2)$ distribution under Q.

PROOF: The form of Λ is easily verified as the ratio of Normal densities. This proves the statement for P and Q on $\mathbb{R}$. On a general space, the proof is similar to the above, by working out the moment generating function of X under Q.

□

10.2 Change of Measure on a General Space

Let two probability measures P and Q be defined on the same space.

Definition 10.5 Q *is called absolutely continuous with respect to* P, *written as* Q $\ll$ P, *if* $Q(A) = 0$ *whenever* $P(A) = 0$. P *and* Q *are called equivalent if* Q $\ll$ P *and* P $\ll$ Q.

Theorem 10.6 (Radon-Nikodym) *If* Q $\ll$ P, *then there exists a random variable Λ, such that $\Lambda \geq 0$, $E_P\Lambda = 1$, and*

$$Q(A) = E_P(\Lambda I(A)) = \int_A \Lambda dP \qquad (10.16)$$

for any measurable set A. Λ is P-almost surely unique. Conversely, if there exists a random variable Λ with the above properties and Q is defined by (10.16), then it is a probability measure and Q $\ll$ P.

The random variable Λ in the above theorem is called the Radon-Nikodym derivative or the density of Q with respect to P, and is denoted by dQ/dP. It follows from (10.16) that if $Q \ll P$, then expectations under P and Q are related by

$$E_Q X = E_P(\Lambda X), \qquad (10.17)$$

for any random variable X integrable with respect to Q.

Calculations of expectations are sometimes made easier by using a change of measure.

Example 10.2: (A Lognormal calculation for a financial option) $E(e^X I(X > a))$, where X has $N(\mu, 1)$ distribution (P). Take $\Lambda(X) = e^X/Ee^X = e^{X-\mu-\frac{1}{2}}$, and $dQ = \Lambda(X)dP$. Then $E\Lambda(X) = 1$, $0 < \Lambda(X) < \infty$, so $Q \sim P$.

$$
\begin{aligned}
E_P(e^X I(X > a)) &= e^{\mu+\frac{1}{2}} E_P(\Lambda(X)I(X > a)) \\
&= E_Q(I(X > a)) = Q(X > a).
\end{aligned}
$$

$$E_Q(e^{uX}) = E_P(e^{uX}\Lambda(X)) = E_P(e^{(u+1)X-\mu-\frac{1}{2}}) = e^{u\mu+u+u^2/2},$$

which is the transform of $N(\mu + 1, 1)$ distribution. Thus the Q distribution of X is $N(\mu+1, 1)$, and $E(e^X I(X > a)) = Q(X > a) = \Pr(N(\mu+1, 1) > a) = 1-\Phi(a-\mu-1)$.

We give the definition of the "opposite" concept to the absolute continuity of two measures.

Definition 10.7 *Two probability measures* P *and* Q *defined on the same space are called singular if there exist a set* A *such that* $P(A) = 0$ *and* $Q(A) = 1$.

Singularity means that by observing an outcome we can decide with certainty on the probability model.

Example 10.3: 1. Let $\Omega = \mathbf{R}^+$, P is the exponential distribution with parameter 1, and Q is the Poisson distribution with parameter 1. Then P and Q are singular. Indeed the set of non-negative integers has Q probability 1, and P probability 0.

2. We shall see later that the probability measures induced by processes $\sigma B(t)$, where B is a Brownian motion, are singular for different values of σ.

Expectations for an absolutely continuous change of probability measure are related by the formula (10.17). Conditional expectations are related by an equation, similar to Bayes formula.

Theorem 10.8 (General Bayes formula) *Let* $\mathcal{G}$ *be a sub-σ-field of* $\mathcal{F}$ *on which two probability measures* Q *and* P *are defined. If* $Q \ll P$ *with* $dQ = \Lambda dP$ *and* X *is* Q*-integrable, then* ΛX *is* P*-integrable and* Q*-a.s.*

$$E_Q(X|\mathcal{G}) = \frac{E_P(X\Lambda|\mathcal{G})}{E_P(\Lambda|\mathcal{G})}. \qquad (10.18)$$

PROOF: We check the definition of conditional expectation given $\mathcal{G}$, (2.15) for any bounded $\mathcal{G}$-measurable random variable ξ

$$E_Q(\xi X) = E_Q(\xi E_Q(X|\mathcal{G})).$$

Clearly the rhs of (10.18) is $\mathcal{G}$-measurable.

$$
\begin{aligned}
E_Q\left(\frac{E_P(X\Lambda|\mathcal{G})}{E_P(\Lambda|\mathcal{G})}\xi\right) &= E_P\left(\Lambda\frac{E_P(X\Lambda|\mathcal{G})}{E_P(\Lambda|\mathcal{G})}\xi\right) \\
&= E_P\left(E_P(\Lambda|\mathcal{G})\frac{E_P(X\Lambda|\mathcal{G})}{E_P(\Lambda|\mathcal{G})}\xi\right) \\
&= E_P\left(E_P(X\Lambda|\mathcal{G})\xi\right) \\
&= E_P(X\Lambda\xi) = E_Q(X\xi).
\end{aligned}
$$

$\square$

In the rest of this section we address the general case of two measures, not necessarily equivalent. Let P and Q be two measures on the same probability space. Consider the measure $\nu = \frac{P+Q}{2}$. Then, clearly, $P \ll \nu$ and $Q \ll \nu$. Therefore, by the Radon-Nikodym Theorem, there exist Λ_1 and Λ_2, such that $dP/d\nu = \Lambda_1$ and $dQ/d\nu = \Lambda_2$. Introduce the notation $x^{(+)} = 1/x$, if $x \neq 0$, and 0 otherwise, so that $x^{(+)}x = I(x \neq 0)$. We have

$$
\begin{aligned}
Q(A) &= \int_A \Lambda_2 d\nu = \int_A \Lambda_2(\Lambda_1^{(+)}\Lambda_1 + 1 - \Lambda_1^{(+)}\Lambda_1)d\nu \\
&= \int_A \Lambda_2\Lambda_1^{(+)}\Lambda_1 d\nu + \int_A (1 - \Lambda_1^{(+)}\Lambda_1)\Lambda_2 d\nu \\
&= \int_A \Lambda_2\Lambda_1^{(+)} dP + \int_A (1 - \Lambda_1^{(+)}\Lambda_1)dQ \\
&:= Q^c(A) + Q^s(A),
\end{aligned}
$$

where $Q^c(A) = \int_A \Lambda_2\Lambda_1^{(+)} dP$ is absolutely continuous with respect to P, and $Q^s(A) = \int_A (1 - \Lambda_1^{(+)}\Lambda_1)dQ$. The set $A = \{\Lambda_1 > 0\}$ has P probability one,

$$P(A) = \int_A \Lambda_1 d\nu = \int_\Omega \Lambda_1 d\nu = P(\Omega) = 1,$$

but has measure zero under Q^s, because on this set $\Lambda_1^{(+)}\Lambda_1 = 1$ and $Q^s(A) = 0$. This shows that Q^s is singular with respect to P. Thus we have

Theorem 10.9 (Lebesgue Decomposition) *Let P and Q be two measures on the same probability space. Then $Q(A) = Q^c(A) + Q^s(A)$, where $Q^c \ll P$ and $Q^s \perp P$.*

10.3   Change of Measure for Processes

Since realizations of a Brownian motion are continuous functions, the probability space is taken to be the set of continuous functions on $[0, T]$, $\Omega = C([0, T])$. Because we have to describe the collection of sets to which the probability is assigned, the concepts of an open (closed) set is needed. These are defined in the usual way with the help of the the distance between two functions taken as $\sup_{t \leq T} |w_1(t) - w_2(t)|$. The σ-field is the one generated by open sets. Probability measures are defined on the measurable subsets of Ω. If $\omega = w_{[0,T]}$ denotes a continuous function, then there is a probability measure P on this space such that the "coordinate" process $B(t, \omega) = B(t, w_{[0,T]}) = w(t)$ is a Brownian motion, P is the Wiener measure, see Section 5.7. Note that although a Brownian motion can be defined on $[0, \infty)$, the equivalent change of measure is defined only on finite intervals $[0, T]$. A random variable on this space is a function $X : \Omega \to \mathbb{R}$, and $X(\omega) = X(w_{[0,T]})$, also known as a functional of Brownian motion. Since probabilities can be obtained as expectation of indicators, $P(A) = EI_A$, it is important to know how to calculate expectations. $E(X)$ is given as an integral with respect to the Wiener measure

$$E(X) = \int_\Omega X(w_{[0,T]}) dP. \tag{10.19}$$

In particular, if $X(w_{[0,T]}) = h(w(T))$, for some function of real argument h, then

$$E(X) = \int_\Omega h(w(T)) dP = E(h(B(T))), \tag{10.20}$$

which can be evaluated by using the $N(0, T)$ distribution of $B(T)$. Similarly, if X a function of finitely many values of w, $X(w_{[0,T]}) = h(w(t_1), \ldots, w(t_n))$, then the expectation can be calculated by using the multivariate Normal distribution of $B(t_1), \ldots, B(t_n)$. However, for a functional which depends on the whole path $w_{[0,T]}$, such as integrals of $w(t)$ or $\max_{t \leq T} w(t)$, the distribution of the functional is required (10.21). It can be calculated by using the distribution of X, F_X, as for any random variable

$$E(X) = \int_\Omega X(w_{[0,T]}) dP = \int_\mathbb{R} x dF_X(x). \tag{10.21}$$

If P and Q are equivalent, that is, they have same null sets, then there exists a random variable Λ (the Radon-Nikodym derivative $\Lambda = dQ/dP$), such that the probabilities under Q are given by $Q(A) = \int_A \Lambda dP$. Girsanov's theorem gives the form of Λ.

First we state a general result, that follows directly from Theorem 10.8, for the calculation of expectations and conditional expectations under an absolutely continuous change of measure. It is also known as the general Bayes formula.

Theorem 10.10 *Let $\Lambda(t)$ be a positive P-martingale such that $E_P(\Lambda(T)) = 1$. Define the new probability measure Q by the relation $Q(A) = \int_A \Lambda(T)dP$, (so that $dQ/dP = \Lambda(T)$). Then Q is absolutely continuous with respect to P and for a Q-integrable random variable X*

$$E_Q(X) = E_P\big(\Lambda(T)X\big). \tag{10.22}$$

$$E_Q(X|\mathcal{F}_t) = E_P\left(\frac{\Lambda(T)}{\Lambda(t)}X|\mathcal{F}_t\right), \tag{10.23}$$

and if X is $\mathcal{F}_t$ measurable, then for $s \leq t$

$$E_Q(X|\mathcal{F}_s) = E_P\left(\frac{\Lambda(t)}{\Lambda(s)}X|\mathcal{F}_s\right). \tag{10.24}$$

PROOF: It remains to show (10.24). $E_Q(X|\mathcal{F}_s) = E_P\left(\frac{\Lambda(T)}{\Lambda(s)}X|\mathcal{F}_s\right) = E_P\left(E_P(\frac{\Lambda(T)}{\Lambda(s)}X|\mathcal{F}_t)|\mathcal{F}_s\right) = E_P\left(X\frac{1}{\Lambda(s)}E_P(\Lambda(T)|\mathcal{F}_t)|\mathcal{F}_s\right) = E_P\left(\frac{\Lambda(t)}{\Lambda(s)}X|\mathcal{F}_s\right).$ □

The following result follows immediately from (10.24).

Corollary 10.11 *A process $M(t)$ is a Q-martingale if and only if $\Lambda(t)M(t)$ is a P-martingale.*

By taking $M(t) = 1$ we obtain a result is used in financial applications.

Theorem 10.12 *Let $\Lambda(t)$ be a positive P-martingale such that $E_P(\Lambda(T)) = 1$, and $dQ/dP = \Lambda(T)$. Then $1/\Lambda(t)$ is a Q-martingale.*

We show next that convergence in probability is preserved under an absolutely continuous change of measure.

Theorem 10.13 *Let $X_n \to X$ in probability P and $Q \ll P$. Then $X_n \to X$ in probability Q.*

PROOF: Denote $A_n = \{|X_n - X| > \varepsilon\}$. Then convergence in probability of X_n to X means $P(A_n) \to 0$. But $Q(A_n) = E_P(\Lambda I_{A_n})$. Since Λ is P-integrable, the result follows by dominated convergence.

□

Corollary 10.14 *The quadratic variation of a process does not change under an absolutely continuous change of the probability measure.*

PROOF: The sums $\sum_{i=0}^{n}(X(t_{i+1}) - X(t_i))^2$ approximating the quadratic variation converge in P probability to $[X, X](t)$. By the above result they converge to the same limit under an equivalent to P probability Q.

□

Theorem 10.15 (Girsanov's theorem for Brownian motion) *Let $B(t)$, $0 \leq t \leq T$, be a Brownian motion under probability measure* P. *Consider the process $W(t) = B(t) + \mu t$. Define the measure* Q *by*

$$\Lambda = \frac{d\mathrm{Q}}{d\mathrm{P}}(B_{[0,T]}) = e^{-\mu B(T) - \frac{1}{2}\mu^2 T}, \tag{10.25}$$

where $B_{[0,T]}$ denotes a path of Brownian motion on $[0, T]$. Then Q *is equivalent to* P, *and $W(t)$ is a* Q-*Brownian motion.*

$$\frac{d\mathrm{P}}{d\mathrm{Q}}(W_{[0,T]}) = \frac{1}{\Lambda} = e^{\mu W(T) - \frac{1}{2}\mu^2 T}. \tag{10.26}$$

PROOF: The proof uses Levy's characterization of Brownian motion, as a continuous martingale with quadratic variation process t. Quadratic variation is the same under P and Q by Theorem 10.13. Therefore (with a slight abuse of notation) using the fact that μt is smooth and does not contribute to the quadratic variation,

$$[W, W](t) = [B(t) + \mu t, B(t) + \mu t] = [B, B](t) = t.$$

It remains to establish that $W(t)$ is a Q-martingale. Let $\Lambda(t) = \mathrm{E}_P (\Lambda | \mathcal{F}_t)$. By the Corollary (10.11) to Theorem 10.10 it is enough to show that $\Lambda(t)W(t)$ is a P-martingale. This is done by direct calculations.

$$\mathrm{E}_P(W(t)\Lambda(t)|\mathcal{F}_s) = \mathrm{E}_P\left((B(t) + \mu t)e^{-\mu B(t) - \frac{1}{2}\mu^2 t}|\mathcal{F}_s\right) = W(s)\Lambda(s).$$

$\square$

It turns out that a drift of the form $\int_0^t H(s)ds$ with $\int_0^T H^2(s)ds < \infty$ can be removed similarly by a change of measure.

Theorem 10.16 (Girsanov's theorem for removal of drift) *Let $B(t)$ be a P-Brownian motion, and $H(t)$ is such that $X(t) = -\int_0^t H(s)dB(s)$ is defined, moreover $\mathcal{E}(X)$ is a martingale. Define an equivalent measure* Q *by*

$$\Lambda = \frac{d\mathrm{Q}}{d\mathrm{P}}(B) = e^{-\int_0^T H(s)dB(s) - \frac{1}{2}\int_0^T H^2(s)ds} = \mathcal{E}(X)(T). \tag{10.27}$$

Then the process

$$W(t) = B(t) + \int_0^t H(s)ds \text{ is a } Q\text{-Brownian motion.} \tag{10.28}$$

PROOF: The proof is similar to the previous one and we only sketch it. We show that $W(t) = B(t) + \int_0^t H(s)ds$ is a continuous Q-martingale with quadratic variation t. The result then follows by Levy's characterization. The quadratic variation of $W(t) = B(t) + \int_0^t H(s)ds$ is t, since the integral is continuous and is of finite variation. $W(t)$ is clearly continuous. To establish that $W(t)$ is a Q-martingale, by Corollary 10.11, it is enough to show that $\Lambda(t)W(t)$ is a P-martingale, with $\Lambda(t) = E_P(\Lambda|\mathcal{F}_t)$. To this end consider

$$\Lambda(t)\mathcal{E}(W)(t) = \mathcal{E}(X)_t\mathcal{E}(W)_t = \mathcal{E}(X + W + [X, W])_t = \mathcal{E}(X + B)_t,$$

where we have used the rule for the product of semimartingale exponentials (Theorem 8.13), $[X, W](t) = -\int_0^t H(s)ds$ and $W(t) = B(t) + [X, W](t)$. Since X and B are both P-martingales, so is $X + B$. Consequently $\mathcal{E}(X + B)$ is also a P-martingale. Hence $\Lambda(t)\mathcal{E}(W)_t$ is a P-martingale, and consequently $\mathcal{E}(W)_t$ is a Q-martingale. This implies that $W(t)$ is also a Q-martingale.

$\square$

Girsanov's theorem holds in n dimensions.

Theorem 10.17 *Let B be a* P *n-dimensional Brownian motion and $W = (W^1(t), \ldots, W^n(t))$, where*

$$W^i(t) = B^i(t) + \int_0^t H^i(s)ds,$$

with $H(t) = (H^1(t), H^2(t), \ldots, H^n(t))$ a regular adapted process satisfying $\int_0^T |H(s)|^2 ds < \infty$. Let

$$X(t) = -H \cdot B := -\sum_{i=1}^n \int_0^t H^i(s)dB^i(s),$$

and assume that $\mathcal{E}(X)_T$ is a martingale. Then there is an equivalent probability measure Q, such that W is a Q-Brownian motion. Q is determined by

$$\frac{dQ}{dP}(B_{[0,T]}) = \Lambda(B_{[0,T]}) = \mathcal{E}(X)(T).$$

Comment that a sufficient condition for $\mathcal{E}(X)$ to be a martingale is Theorem 8.17, $E(e^{\frac{1}{2}\int_0^T H^2(s)ds}) < \infty$, and for $\mathcal{E}(H \cdot B)$ to be a martingale is $E(e^{\frac{1}{2}\int_0^T |H(s)|^2 ds}) < \infty$.

The proof for n dimensions is similar to one dimension, using calculations from Exercise 10.5.

We give a version of Girsanov's theorem for martingales. The proof is similar to the one above, and is not given.

Theorem 10.18 *Let $M_1(t)$, $0 \le t \le T$ be a continuous P-martingale. Let $X(t)$ be a continuous P-martingale such that $\mathcal{E}(X)$ is a martingale. Define a new probability measure Q by*

$$\frac{dQ}{dP} = \Lambda = \mathcal{E}(X)(T) = e^{X(T) - \frac{1}{2}[X,X](T)}. \tag{10.29}$$

Then

$$M_2(t) = M_1(t) - [M, X](t) \tag{10.30}$$

is a continuous martingale under Q.

A sufficient condition for $\mathcal{E}(X)$ to be a martingale is Novikov's condition $E\left(e^{\frac{1}{2}[X,X]_T}\right) < \infty$, or Kazamaki's condition (8.40).

Change of Drift in Diffusions

Let $X(t)$ be a diffusion, so that with a P-Brownian motion $B(t)$, $X(t)$ satisfies the following stochastic differential equation with $\sigma(x, t) > 0$,

$$dX(t) = \mu_1(X(t), t)dt + \sigma(X(t), t)dB(t). \tag{10.31}$$

Let

$$H(t) = \frac{\mu_1(X(t), t) - \mu_2(X(t), t)}{\sigma(X(t), t)}, \tag{10.32}$$

and define Q by $dQ = \Lambda dP$ with

$$\Lambda = \frac{dQ}{dP} = \mathcal{E}\left(-\int_0^\cdot H(t)dB(t)\right)(T) = e^{-\int_0^T H(t)dB(t) - \frac{1}{2}\int_0^T H^2(t)dt}. \tag{10.33}$$

By Girsanov's theorem, provided the process $\mathcal{E}(H \cdot B)$ is a martingale, the process $W(t) = B(t) + \int_0^t H(s)ds$ is a Q-Brownian motion. But

$$dW(t) = dB(t) + H(t)dt = dB(t) + \frac{\mu_1(X(t), t) - \mu_2(X(t), t)}{\sigma(X(t), t)}dt. \tag{10.34}$$

Rearranging, we obtain the equation for $X(t)$

$$dX(t) = \mu_2(X(t), t)dt + \sigma(X(t), t)dW(t), \tag{10.35}$$

with a Q-Brownian motion $W(t)$. Thus the change of measure for a Brownian motion given above results in the change of the drift in the stochastic differential equation.

Example 10.4: (Maximum of arithmetic Brownian motion)
Let $W(t) = \mu t + B(t)$, where $B(t)$ is P-Brownian motion, and $W^*(t) = \max_{s \le t} W(s)$. When $\mu = 0$ the distribution of the maximum as well as the joint distribution are known, Theorem 3.21. We find these distributions when $\mu \ne 0$.
    Let $B(t)$ be a Q-Brownian motion, then $W(t)$ is a P-Brownian motion with $\Lambda = dP/dQ$ is given by (10.26). Let $A = \{W(T) \in I_1, W^*(T) \in I_2\}$, where I_1, I_2 are intervals on the line. Then we obtain from (10.26)

$$P((W(T), W^*(T)) \in A) = \int_A e^{\mu W(T) - \frac{1}{2}\mu^2 T}dQ = E_Q(e^{\mu W(T) - \frac{1}{2}\mu^2 T}I(A)). \tag{10.36}$$

Denote by $q_{W,W^*}(x,y)$ and $p_{W,W^*}(x,y)$ the joint density of $(W(T), W^*(T))$ under Q and P respectively. Then it follows from (10.36)

$$\int_{\{x \in I_1, y \in I_2\}} p_{W,W^*}(x,y)dxdy = \int_{\{x \in I_1, y \in I_2\}} q_{W,W^*}(x,y)e^{\mu x - \frac{1}{2}\mu^2 T}dxdy. \quad (10.37)$$

Thus

$$p_{W,W^*}(x,y) = e^{\mu x - \frac{1}{2}\mu^2 T}q_{W,W^*}(x,y). \quad (10.38)$$

The density q_{W,W^*} is given by (3.16). The joint density $p_{W,W^*}(x,y)$ can be computed (see Exercise 10.6), and the distribution of the maximum is found by $\int p_{W,W^*}(x,y)dx$.

10.4   Change of Wiener Measure

Girsanov's Theorem 10.16 states that if the Wiener measure P is changed to Q by $dQ/dP = \Lambda = \mathcal{E}(\int_0^t H(s)dB(s))$ where $B(t)$ is a P-Brownian motion, and $H(t)$ is some predictable process, then Q is equivalent to P, and $B(t) = W(t) + \int_0^t H(s)ds$ for a Q-Brownian motion W. In this section we prove the converse that the Radon-Nikodym derivative of any measure Q, equivalent to the Wiener measure P, is a stochastic exponential of $\int_0^t q(s)dB(s)$ for some predictable process q. Using the predictable representation property of Brownian martingales we prove the following result first.

Theorem 10.19 *Let* $\mathbb{F}$ *be Brownian motion filtration and* Y *be a positive random variable. If* $EY < \infty$ *then there exists a predictable process* $q(t)$ *such that* $Y = (EY)e^{\int_0^T q(t)dB(t) - \frac{1}{2}\int_0^T q^2(t)dt}$.

PROOF: Let $M(t) = E(Y|\mathcal{F}_t)$. Then $M(t), 0 \le t \le T$ is a positive uniformly integrable martingale. By Theorem 8.35 $M(t) = EY + \int_0^t H(s)dB(s)$. Define $q(t) = H(t)/M(t)$. Then we have

$$dM(t) = H(t)dB(t) = M(t)q(t)dB(t) = M(t)dX(t), \quad (10.39)$$

with $X(t) = \int_0^t q(s)dB(s)$. Thus M is a semimartingale exponential of X, $M(t) = M(0)e^{X(t) - \frac{1}{2}[X,X](t)}$ and the result follows.

□

It remains to show that q is properly defined. We know that for each t, $M(t) > 0$ with probability one. But there may be an exceptional set N_t, of probability zero, on which $M(t) = 0$. As there are uncountably many t's the union of N_t's may have a positive probability (even probability one), precluding q being finite with a positive probability. The next result shows that this is impossible.

Theorem 10.20 *Let* $M(t), 0 \le t \le T$, *be a martingale, such that for any* t, $P(M(t) > 0) = 1$. *Then* $M(t)$ *never hits zero on* $[0,T]$.

PROOF: Let $\tau = \inf\{t : M(t) = 0\}$. Using the Basic Stopping equation $EM(\tau \wedge T) = EM(T)$. So we have

$$EM(T) = EM(\tau \wedge T) = EM(T)I(\tau > T) + EM(\tau)I(\tau \leq T) = EM(T)I(\tau > T).$$

It follows that $EM(T)(1 - I(\tau > T)) = 0$. Since the random variables under the expectation are non-negative, this implies $M(T)(1 - I(\tau > T)) = 0$ a.s. Thus $I(\tau > T) = 1$ a.s. Thus there is a null set N such that $\tau > T$ outside N, or $P(M(t) > 0, \text{for all } t \leq T) = 1$. Since T was arbitrary, the argument implies that $\tau = \infty$ and zero is never hit.

□

Remark 10.2: Note that if for some stopping time τ a non-negative martingale is zero, $M(\tau) = 0$, and optional stopping holds, for all $t > \tau$ $E(M(t)|\mathcal{F}_\tau) = M(\tau) = 0$, then $M(t) = 0$ a.s. for all $t > \tau$.

Corollary 10.21 *Let* P *be the Wiener measure,* $B(t)$ *be a* P*-Brownian motion and* Q *be equivalent to* P. *Then there exists a predictable process* $q(t)$, *such that*

$$\Lambda(B_{[0,T]}) = \frac{dQ}{dP}(B_{[0,T]}) = e^{\int_0^T q(t)dB(t) - \frac{1}{2}\int_0^T q^2(t)dt}. \tag{10.40}$$

Moreover, $B(t) = W(t) + \int_0^t q(s)ds$, *where* $W(t)$ *is a* Q*-Brownian motion.*

PROOF: By the Radon-Nikodym theorem, $(dQ/dP) = \Lambda$ with $0 < \Lambda < \infty$. Since $P(\Lambda > 0) = 1$ and $E_P\Lambda = 1$, existence of the process $q(t)$ follows by Theorem 10.19. Representation for $B(t)$ as a Brownian motion with drift under Q follows from (10.27) in Girsanov's theorem.

□

(Corollary 10.21) is used in Finance, where $q(t)$ denotes the market price for risk.

10.5   Change of Measure for Point Processes

Consider a point process $N(t)$ with intensity $\lambda(t)$ (see Chapter 9 for definitions). This presumes a probability space, which can be taken as the space of right-continuous non-decreasing functions with unit jumps, and a probability measure P so that N has intensity $\lambda(t)$ under P. Girsanov's theorem asserts that there is a probability measure Q, equivalent to P, under which $N(t)$ is Poisson process with the unit rate. Thus an equivalent change of measure corresponds to a change in the intensity. If we look upon the compensator $\int_0^t \lambda(s)ds$ as the drift, then an equivalent change of measure results in a change of the drift.

Theorem 10.22 (Girsanov's theorem for Poisson processes) *Let $N(t)$ be a Poisson process with rate 1 under P, $0 \leq t \leq T$. For a constant $\lambda > 0$ define Q by*

$$\frac{dQ}{dP} = e^{(1-\lambda)T + N(T) \ln \lambda}. \tag{10.41}$$

Then under Q, N is a Poisson process with rate λ.

PROOF: Let

$$\Lambda(t) = e^{(1-\lambda)t + N(t) \ln \lambda} = e^{(1-\lambda)t} \lambda^{N(t)}. \tag{10.42}$$

It is easy to see that $\Lambda(t)$ is a P-martingale with respect to the natural filtration of the process $N(t)$. Indeed, using independence of increments of the Poisson process, we have

$$
\begin{aligned}
E_P(\Lambda(t)|\mathcal{F}_s) &= e^{(1-\lambda)t} E_P(\lambda^{N(t)}|\mathcal{F}_s) \\
&= e^{(1-\lambda)t} \lambda^{N(s)} E_P(\lambda^{N(t)-N(s)}|\mathcal{F}_s) \\
&= e^{(1-\lambda)t} \lambda^{N(s)} E(\lambda^{N(t)-N(s)}) \\
&= e^{(1-\lambda)t} \lambda^{N(s)} e^{(\lambda-1)(t-s)} = e^{(1-\lambda)s} \lambda^{N(s)} = \Lambda(s).
\end{aligned}
$$

We have used that P-distribution of $N(t) - N(s)$ is Poisson with parameter $(t-s)$, hence $E_P(\lambda^{N(t)-N(s)}) = e^{(\lambda-1)(t-s)}$. We show that under Q the increments $N(t) - N(s)$ are independent of the past and have the Poisson $\lambda(t-s)$ distribution, establishing the result. Fix $u > 0$. The conditional expectation under the new measure, Theorem 10.10, and the definition of $\Lambda(t)$ (10.42) yield

$$
\begin{aligned}
E_Q\left(e^{u(N(t)-N(s))}|\mathcal{F}_s\right) &= E_P\left(e^{u(N(t)-N(s))} \frac{\Lambda(t)}{\Lambda(s)}|\mathcal{F}_s\right) \\
&= e^{(1-\lambda)(t-s)} E_P\left(e^{(u+\log \lambda)(N(t)-N(s))}|\mathcal{F}_s\right) \\
&= e^{(1-\lambda)(t-s)} E_P\left(e^{(u+\log \lambda)(N(t)-N(s))}\right) \\
&= e^{\lambda(t-s)(e^u-1)}.
\end{aligned}
$$

$\square$

Theorem 10.23 *Let $N(t)$, $0 \leq t \leq T$, be a Poisson process with rate λ under Q. Define P by*

$$\frac{dP}{dQ} = e^{(\lambda-1)T - N(T) \ln \lambda}. \tag{10.43}$$

Then under P, $N(t)$ is a Poisson process with rate 1.

As a corollary we obtain that the measures of Poisson processes with constant rates are equivalent with Likelihood ratio given by the following theorem.

Theorem 10.24 *Let N be a Poisson process with rate $\lambda > 0$ under the probability P_λ. Then it is a Poisson process with rate μ under the equivalent measure P_μ defined by*

$$\frac{dP_\lambda}{dP_\mu} = e^{(\mu-\lambda)T - N(T)(\ln\lambda - \ln\mu)}. \tag{10.44}$$

Note that the key point in changing measure was the martingale property (under the original measure) of the Likelihood $\Lambda(t)$. This property was established directly in the proofs. Observe that the Likelihood is the stochastic exponential of the point process martingale $M(t) = N(t) - A(t)$. For example, in changing the rate from 1 to λ, $\Lambda = \mathcal{E}\big((\lambda - 1)M\big)$ with $M(t) = N(t) - t$. It turns out that a general point process $N(t)$ with intensity $\lambda(t)$ can be obtained from a Poisson process with rate 1 by a change of measure. In fact, Theorem 10.24 holds for general point processes with stochastic intensities. The form of stochastic exponential $\mathcal{E}\big((\lambda - 1) \cdot M\big)$ for non-constant λ is not hard to obtain, see Exercise 10.8,

$$\Lambda = \mathcal{E}\Big(\int (\lambda(s) - 1)dM(s)\Big)(T) = e^{\int_0^T (1 - \lambda(s))ds + \int_0^T \ln\lambda(s)dN(s)} \tag{10.45}$$

The next result establishes that a point process is a Poisson process (with a constant rate) under a suitable change of measure.

Theorem 10.25 *Let $N(t)$ be a Poisson process with rate 1 under P, and $M(t) = N(t) - t$. If for a predictable process $\lambda(s)$, $\mathcal{E}\big(\int(\lambda(s) - 1)dM(s)\big)$ is a martingale for $0 \leq t \leq T$, then under the probability measure Q defined by (10.45) $dQ/dP = \Lambda$, N is a point process with the stochastic intensity $\lambda(t)$. Conversely, if Q is absolutely continuous with respect to P then there exists a predictable process $\lambda(t)$, such that under Q, N is a point process with the stochastic intensity $\lambda(t)$.*

10.6   Likelihood Functions

When observations X are made from competing models described by the probabilities P and Q, the Likelihood is the Radon-Nikodym derivative $\Lambda = dQ/dP$.

Likelihood for Discrete Observations

Suppose that we observe a discrete random variable X, and there are two competing models for X: it can come from distribution P or Q. For the observed number x the Likelihood is given by

$$\Lambda(x) = \frac{Q(X = x)}{P(X = x)}. \tag{10.46}$$

Small values of $\Lambda(x)$ provide evidence for model P, and large values for model Q. If X is a continuous random variable with densities $f_0(x)$ under P and $f_1(x)$ under Q, then $dP = f_0(x)dx$, $dQ = f_1(x)dx$, and the Likelihood is given by

$$\Lambda(x) = \frac{f_1(x)}{f_0(x)}. \qquad (10.47)$$

If the observed data are a finite number of observations $x_1, x_2, \ldots, x_n$ then similarly with $\boldsymbol{x} = (x_1, x_2, \ldots, x_n)$

$$\Lambda(\boldsymbol{x}) = \frac{Q(\boldsymbol{X} = \boldsymbol{x})}{P(\boldsymbol{X} = \boldsymbol{x})}, \text{ or } \Lambda(\boldsymbol{x}) = \frac{f_1(\boldsymbol{x})}{f_0(\boldsymbol{x})}, \qquad (10.48)$$

depending on whether the models are discrete or continuous. Note that if one model is continuous and the other is discrete, then the corresponding measures are singular and the Likelihood does not exist.

Likelihood Ratios for Diffusions

Let X be a diffusion solving the SDE with a P-Brownian motion $B(t)$

$$dX(t) = \mu_1(X(t), t)dt + \sigma(X(t), t)dB(t), \qquad (10.49)$$

Suppose that it satisfies another equation with a Q-Brownian motion $W(t)$

$$dX(t) = \mu_2(X(t), t)dt + \sigma(X(t), t)dW(t). \qquad (10.50)$$

The Likelihood, as we have seen, is given by (10.33)

$$\Lambda(X_{[0,T]}) = \frac{dQ}{dP} = e^{\int_0^T \frac{\mu_2(X(t),t) - \mu_1(X(t),t)}{\sigma(X(t),t)} dB(t) - \frac{1}{2}\int_0^T \frac{(\mu_2(X(t),t) - \mu_1(X(t),t))^2}{\sigma^2(X(t),t)} dt}.$$
$$(10.51)$$

Since $B(t)$ is not observed directly, the Likelihood should be expressed as a function of the observed path $X_{[0,T]}$. Using equation (10.49) we obtain $dB(t) = \frac{dX(t) - \mu_1(X(t),t)dt}{\sigma(X(t),t)}$, and putting this into the Likelihood, we obtain

$$\Lambda(X)_T = \frac{dQ}{dP} = e^{\int_0^T \frac{\mu_2(X(t),t) - \mu_1(X(t),t)}{\sigma^2(X(t),t)} dX(t) - \frac{1}{2}\int_0^T \frac{\mu_2^2(X(t),t) - \mu_1^2(X(t),t)}{\sigma^2(X(t),t)} dt}. \quad (10.52)$$

Using the Likelihood a decision can be made as to what model is more appropriate for X.

Example 10.5: (Hypotheses testing)
Suppose that we observe a continuous function x_t, $0 \le t \le T$, and we want to know whether it is just White noise, the null hypothesis corresponding to probability

measure P or it is some signal contaminated by noise, the alternative hypothesis corresponding to probability measure Q.

$$H_0 : \text{ noise: } dX(t) = dB(t), \text{ and, } H_1 : \text{ signal+noise: } dX(t) = h(t)dt + dB(t)$$

Here we have $\mu_1(x) = 0$, $\mu_2(x) = h(t)$, $\sigma(x) = 1$. The Likelihood is given by

$$\Lambda(X)_T = \frac{dQ}{dP} = e^{\int_0^T h(t)dX(t) - \frac{1}{2}\int_0^T h^2(t)dt}. \tag{10.53}$$

This leads to the Likelihood ratio test of the form: conclude the presence of noise if $\Lambda \geq k$, where k is determined from setting the probability of the type one error to α.

$$P(e^{\int_0^T h(t)dB(t) - \frac{1}{2}\int_0^T h^2(t)dt} \geq k) = \alpha. \tag{10.54}$$

Example 10.6: (Estimation in Ornstein-Uhlenbeck Model)
Consider estimation of the friction parameter α in the Ornstein-Uhlenbeck model on $0 \leq t \leq T$,

$$dX(t) = -\alpha X(t)dt + \sigma dB(t).$$

Denote by P_α the measure corresponding to $X(t)$, so that P_0 corresponds to $\sigma B(t)$, $0 \leq t \leq T$. The Likelihood is given by

$$\Lambda(\alpha, X_{[0,T]}) = \frac{dP_\alpha}{dP_0} = \exp\left(\int_0^T \frac{-\alpha X(t)}{\sigma^2}dX(t) - \frac{1}{2}\int_0^T \frac{\alpha^2 X^2(t)}{\sigma^2}dt\right).$$

Maximizing log Likelihood, we find

$$\hat{\alpha} = -\frac{\int_0^T X(t)dX(t)}{\int_0^T X^2(t)dt}. \tag{10.55}$$

Remark 10.3: Let $X(t)$ and $Y(t)$ satisfy the stochastic differential equations for $0 \leq t \leq T$

$$dX(t) = \mu_X(X(t), t)dt + \sigma_X(X(t), t)dW(t), \tag{10.56}$$

and

$$dY(t) = \mu_Y(Y(t), t)dt + \sigma_Y(Y(t), t)dW(t). \tag{10.57}$$

Consider probability measures induced by these diffusions on the space of continuous functions on $[0, T]$, $C[0, T]$. It turns out that if $\sigma_X \neq \sigma_Y$ then P_X and P_Y are *singular* ("live" on different sets). It means that by observing a process continuously over an interval of time we can decide precisely from which equation it comes. This identification can be made with the help of the quadratic variation process,

$$d[X, X](t) = \sigma^2(X(t), t)dt, \text{ and } \sigma^2(X(t), t) = \frac{d[X, X](t)}{dt}, \tag{10.58}$$

and $[X, X]$ is known exactly if a path is observed continuously.

If $\sigma_X = \sigma_Y = \sigma$ then P_X and P_Y are equivalent. The Radon-Nikodym derivatives (the Likelihoods) are given by

$$\frac{dP_Y}{dP_X}(X_{[0,T]}) = e^{\int_0^T \frac{\mu_Y(X(t),t) - \mu_X(X(t),t)}{\sigma^2(X(t),t)} dX(t) - \frac{1}{2} \int_0^T \frac{\mu_Y^2(X(t),t) - \mu_X^2(X(t),t)}{\sigma^2(X(t),t)} dt},$$

(10.59)

and

$$\frac{dP_X}{dP_Y}(Y_{[0,T]}) = e^{-\int_0^T \frac{\mu_Y(Y(t),t) - \mu_X(Y(t),t)}{\sigma^2(Y(t),t)} dY(t) + \frac{1}{2} \int_0^T \frac{\mu_Y^2(Y(t),t) - \mu_X^2(Y(t),t)}{\sigma^2(Y(t),t)} dt}.$$

(10.60)

Notes. Material for this chapter is based on Karatzas and Shreve (1988), Liptser and Shirayev (1974), Lamberton and Lapeyre (1996).

10.7 Exercises

Exercise 10.1: Let P be $N(\mu_1, 1)$ and Q be $N(\mu_2, 1)$ on $\mathbb{R}$. Show that they are equivalent and that the Radon-Nikodym derivative $dQ/dP = \Lambda$ is given by $\Lambda(x) = e^{(\mu_2 - \mu_1)x + \frac{1}{2}(\mu_1^2 - \mu_2^2)}$. Give also dP/dQ.

Exercise 10.2: Show that if X has $N(\mu, 1)$ distribution under P, then there is an equivalent measure Q, such that X has $N(0, 1)$ distribution under Q. Give the Likelihood dQ/dP and also dP/dQ. Give the Q-distribution of $Y = X - \mu$.

Exercise 10.3: Y has a Lognormal $LN(\mu, \sigma^2)$ distribution. Using a change of measure calculate $EYI(Y > K)$. Hint: change measure from $N(\mu, \sigma^2)$ to $N(\mu + \sigma^2, \sigma^2)$.

Exercise 10.4: Let $X(t) = B(t) + \sin t$ for a P-Brownian motion $B(t)$. Let Q be an equivalent measure to P such $X(t)$ is a Q-Brownian motion. Give $\Lambda = dQ/dP$.

Exercise 10.5: Let $\boldsymbol{B}$ be an n-dim Brownian motion and $\boldsymbol{H}$ an adapted regular process. Let $\boldsymbol{H} \cdot \boldsymbol{B}(T) = \sum_{i=1}^n \int_0^T H^i(s) dB^i(s)$ be a martingale. Show that the martingale exponential is given by $\exp(\boldsymbol{H} \cdot \boldsymbol{B}(T) - \frac{1}{2} \int_0^T |\boldsymbol{H}(s)|^2 ds)$. Hint: show that quadratic variation of $\boldsymbol{H} \cdot \boldsymbol{B}$ is given by $\int_0^T |\boldsymbol{H}(s)|^2 ds$, where $|\boldsymbol{H}(s)|^2$ denotes the length of vector $\boldsymbol{H}(s)$.

Exercise 10.6: Let $W_\mu(t) = \mu t + B(t)$, where $B(t)$ is a P-Brownian motion. Show that

$$P(\max_{t \le T} W_\mu(t) \le y | W_\mu(T) = x) = 1 - e^{\frac{-2y(y-x)}{T}}, \quad x \le y,$$

and
$$P(\min_{t \leq T} W_\mu(t) \geq y | W_\mu(T) = x) = 1 - e^{\frac{-2y(y-x)}{T}}, \quad x \geq y.$$

Hint: use the joint distributions (10.38) and (3.16).

Exercise 10.7: Prove Theorem 10.23.

Exercise 10.8: Let $N(t)$ be Poisson process with rate 1 and $\bar{N}(t) = N(t) - t$. Show that for an adapted, continuous and bounded process $H(t)$, the process $M(t) = \int_0^t H(s) d\bar{N}(s)$ is a martingale for $0 \leq t \leq T$. Show that

$$\mathcal{E}(M)(t) = e^{-\int_0^t H(s)ds + \int_0^t \ln(1+H(s))dN(s)}.$$

Exercise 10.9: (Estimation of parameters)
Find the Likelihood corresponding to different values of μ of the process $X(t)$ given by $dX(t) = \mu X(t)dt + \sigma X(t)dB(t)$ on $[0, T]$. Give the maximum Likelihood estimator.

Exercise 10.10: Verify the martingale property of the Likelihood occurring in the change of rate in a Poisson process.

Exercise 10.11: Let $dQ = \Lambda dP$ on $\mathcal{F}_T$, $\Lambda(t) = E_P(\Lambda | \mathcal{F}_t)$ is continuous. For a P-martingale $M(t)$ find a finite variation process $A(t)$ such that $M'(t) = M(t) - A(t)$ is a Q-local martingale.

Exercise 10.12: Let B and N be respectively a Brownian motion and a Poisson process on the same space $(\Omega, \mathcal{F}, \mathbb{F}, P)$, $0 \leq t \leq T$. Define $\Lambda(t) = e^{(\ln 2)N(t) - t}$ and $dQ = \Lambda(T)dP$. Show that B is a Brownian motion under Q.

Exercise 10.13:

1. Let B and N be respectively a Brownian motion and a Poisson process on the same space $(\Omega, \mathcal{F}, \mathbb{F}, P)$, $0 \leq t \leq T$, and $X(t) = B(t) + N(t)$. Give an equivalent probability measure Q_1 such that $B(t) + t$ and $N(t) - t$ are Q_1-martingales. Deduce that $X(t)$ is a Q_1-martingale.

2. Give an equivalent probability measure Q_2 such that $B(t) + 2t$ and $N(t) - 2t$ are Q_2-martingales. Deduce that $X(t)$ is a Q_2-martingale.

3. Deduce that there are infinitely many equivalent probability measures Q such that $X(t) = B(t) + N(t)$ is a Q-martingale.

Chapter 11

Applications in Finance: Stock and FX Options

In this chapter the fundamentals of Mathematics of Option Pricing are given. The concept of arbitrage is introduced, and a martingale characterization of models that don't admit arbitrage is given, the First Fundamental Theorem of asset pricing. The theory of pricing by no-arbitrage is presented first in the Finite Market model, and then in a general Semimartingale Model, where the martingale representation property is used. Change of measure and its application as the change of numeraire are given as a corollary to Girsanov's theorem and general Bayes formula for expectations. They represent the main techniques used for pricing foreign exchange options, exotic options (asian, lookback, barrier options) and interest rates options.

11.1 Financial Derivatives and Arbitrage

A financial derivative or a contingent claim on an asset is a contract that allows purchase or sale of this asset in the future on terms that are specified in the contract. An option on stock is a basic example.

Definition 11.1 *A call option on stock is a contract that gives its holder the right to buy this stock in the future at the price K written in the contract, called the exercise price or the strike.*

A European call option allows the holder to exercise the contract (that is, to buy this stock at K) at a particular date T, called the maturity or the expiration date. An American option allows the holder to exercise the contract at any time before or at T.

287

A contract that gives its holder the right to sell stock is called a *put* option. There are two parties to a contract, the seller (writer) and the buyer (holder). The holder of the options has the right but no obligations in exercising the contract. The writer of the options has the obligation to abide by the contract, for example, they must sell the stock at price K if the call option is exercised. Denote the price of the asset (e.g. stock) at time t by $S(t)$. A contingent claim on this asset has its value at maturity specified in the contract, and it is some function of the values $S(t), 0 \leq t \leq T$. Simple contracts depend only on the value at maturity $S(T)$.

Example 11.1: (Value at maturity of European call and put options)
If you hold a call option, then you can buy one share of stock at T for K. If at time T, $S(T) < K$, you will not exercise your option, as you can buy stock cheaper than K, thus this option is worthless. If at time T, $S(T) \geq K$, then you can buy one share of stock for K and sell it immediately for $S(T)$ making profit of $S(T) - K$. Thus a European call option has the value at time T

$$C(T) = (S(T) - K)^+ = \max(0, S(T) - K). \qquad (11.1)$$

Similar considerations reveal that the value of the European put at maturity is

$$P(T) = (K - S(T))^+ = \max(0, K - S(T)). \qquad (11.2)$$

Example 11.2: (Exotic Options)
The value at maturity of Exotic Options depends prices of the asset on the whole time interval before maturity, $S(t)$, $t \leq T$.
1. Lookback Options. Lookback call pays at T, $X = (S(T) - S_{\min})^+ = S(T) - S_{\min}$ and lookback put $X = S_{\max} - S(T)$, where $S_{\min}$ and $S_{\max}$ denote the smallest and the largest values of $S(t)$ on $[0, T]$.
2. Barrier Options. The call that gets knocked out when the price falls below a certain level H (down and out call) pays at T

$$X = (S(T) - K)^+ I(\min_{0 \leq t \leq T} S(t) \geq H), \quad S(0) > H, \ K > H.$$

3. Asian Options. Payoff at T depends on the average price $\bar{S} = \frac{1}{T} \int_0^T S(u) du$ during the life of the option. Average call pays $X = (\bar{S} - K)^+$, and average put $X = (K - \bar{S})^+$. Random strike option pays $X = (S(T) - \bar{S})^+$.

Whereas the value of a financial claim at maturity can be obtained from the terms specified in the contract, it is not so clear how to obtain its value prior to maturity. This is done by the pricing theory, which is also used to manage financial risk.

If the payoff of an option depends only on the price of stock at expiration, then it is possible to graph the value of an option at expiration against the underlying price of stock. For example, the payoff function of a call option is $(x - K)^+$. Some payoff functions are given in Exercise 11.1.

Arbitrage and Fair Price

Arbitrage is defined in Finance as a strategy that allows to make a profit *out of nothing* without taking any risk. Mathematical formulation of arbitrage is given later in the context of a market model.

Example 11.3: (Futures, or Forward) Consider a contract that gives the holder 1 share of stock at time T, so its value at time T is the market price $S(T)$. Denote the price of this contract at time 0 by $C(0)$. The following argument shows that $C(0) = S(0)$, as any other value results in an arbitrage profit.

If $C(0) > S(0)$, then sell the contract and receive $C(0)$; buy the stock and pay $S(0)$. The difference $C(0) - S(0) > 0$ can be invested in a risk-free bank account with interest rate r. At time T you have the stock to deliver , and make arbitrage profit of $(C(0) - S(0))e^{rT}$. If $C(0) < S(0)$, the reverse strategy results in arbitrage profit: you buy the contract and pay $C(0)$; sell the stock and receive $S(0)$, (selling the stock when you don't hold it is called short selling and it is allowed). Invest the difference $S(0) - C(0) > 0$ in a risk-free bank account. At time T exercise the contract by buying back stock at $S(T)$. The profit is $(S(0) - C(0))e^{rT}$. Thus any price $C(0)$ different from $S(0)$ results in arbitrage profit. The only case which does not result in arbitrage profit is $C(0) = S(0)$.

The fair price in a game of chance with a profit X (negative profit is loss) is the expected profit EX. The above example also demonstrates that financial derivatives are not priced by their expectations, and in general, their arbitrage-free value is different to their fair price.

The following example uses a two-point distribution for the stock on expiration to illustrate that prices for options can not be taken as expected payoffs, and all the prices but one, lead to arbitrage opportunities.

Example 11.4: The current price of the stock is \$10. We want to price a call option on this stock with the exercise price $K = 10$ and expiration in one period. Suppose that the stock price at the end of the period can have only two values \$8 and \$12 per share, and that the riskless interest rate is 10%. Suppose the call is priced at \$1 per share. Consider the strategy: buy call option on 200 shares and sell 100 shares of stock

		$S_T = 12$	$S_T = 8$
Buy option on 200 shares	-200	400	0
Sell (short) 100 shares	1000	-1200	-800
Savings account	800	880	880
Profit	0	+80	+80

We can see that in either case $S_T = 8$ or $S_T = 12$ an arbitrage profit of \$80 is realized. It can be seen, by reversing the above strategy, that any price above \$1.36 will result in arbitrage. The price that does not lead to arbitrage is \$1.36.

Equivalence Portfolio. Pricing by No Arbitrage

The main idea in pricing by no-arbitrage arguments is to replicate the payoff of the option at maturity by a portfolio consisting of stock and bond (cash). To avoid arbitrage, the price of the option at any other time must equal the value of the replicating portfolio, which is valued by the market. Consider the one period case $T = 1$ first, and assume that in one period stock price moves up by factor u or down by d, $d < 1 < u$, and we want to price a claim C that has the values C_u and C_d on maturity. Schematically the following trees for prices of the stock and the option are drawn.

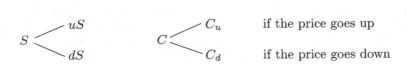

$$
\begin{array}{ll}
S \begin{cases} uS \\ dS \end{cases} & C \begin{cases} C_u & \text{if the price goes up} \\ C_d & \text{if the price goes down} \end{cases}
\end{array}
$$

Note that the values of the option C_u and C_d are known, since the values of the stock uS and dS are known by assumption. A portfolio that replicates the options consists of a shares of stock and b of bond (cash in savings account). After one period the value of this portfolio is

$$
aS_T + br = \begin{cases} auS + br & \text{if } S_T = uS \\ adS + br & \text{if } S_T = dS. \end{cases}
$$

Since this portfolio is equivalent to the option, by matching the payoff, we have a system of two linear equations for a and b.

$$
\left.\begin{array}{l}
auS + br = C_u \\
adS + br = C_d
\end{array}\right\}.
$$

$$
a = \frac{C_u - C_d}{(u - d)S}, \qquad b = \frac{uC_d - dC_u}{(u - d)r}. \tag{11.3}
$$

The price of the option must equal that of the replicating portfolio

$$
C = aS + b, \tag{11.4}
$$

with a and b from (11.3). To prove that the price is given by (11.4) consider cases when the price is above and below that value C. If an option is priced above C, then selling the option and buying the portfolio specified by a and b results in arbitrage. If the option is priced below C then buying the option and selling the portfolio results in arbitrage. It will be seen later that there are no arbitrage strategies when the option is priced at C.

Example 11.5: (continued) $K = 10$, $C_T = (S_T - K)^+$, $r = 1.1$.

$$
\begin{array}{llll}
& 12 & u = 1.2 & \qquad\qquad 2 = C_u \\
10 & & & C \\
& 8 & d = 0.8 & \qquad\qquad 0 = C_d
\end{array}
$$

$a = 0.5$, $b = -3.64$, from (11.3). Thus this option is replicated by the portfolio consisting of 0.5 shares and borrowing 3.64 dollars. Initial value of this portfolio is $C = 0.5 \cdot 10 - 3.64 = 1.36$, which gives the no-arbitrage price for the call option.

The formula for the price of the option C (11.4) can be written as

$$C = aS + b = \frac{1}{r}\left(pC_u + (1-p)C_d\right), \tag{11.5}$$

with

$$p = \frac{r - d}{u - d}. \tag{11.6}$$

It can be viewed as *the discounted expected payoff of the claim, with probability p of up and $(1 - p)$ down movements.* The probability p, calculated from the given returns of the stock by (11.6), is called the arbitrage-free or risk-neutral probability, and has nothing to do with subjective probabilities of market going up or down.

In the above example $p = \frac{1.1-0.8}{1.2-0.8} = 0.75$. So that $C = \frac{1}{1.1} 2 \cdot 0.75 = 1.36$.

Binomial Model

The one-period formula can be applied recursively to price a claim when trading is done one period after another. The tree for option's prices in a two-period model is given by

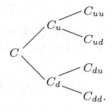

The final prices of the option are known from the assumed model for stock prices. The prices prior to expiration are obtained recursively, by using the one-period formula (11.5),

$$C_u = \frac{1}{r}(pC_{uu} + (1-p)C_{ud}), \quad C_d = \frac{1}{r}(pC_{du} + (1-p)C_{du}).$$

Using the formula (11.5) again

$$C = \frac{1}{r}(pC_u + (1-p)C_d) = \frac{1}{r^2}(p^2C_{uu} + p(1-p)C_{ud} + (1-p)pC_{du} + (1-p)^2C_{dd}).$$

In the n period model, $T = n$, if $C_{udu...du} = C_{u...ud...d}$, then continuing by induction

$$C = \frac{1}{r^n}\mathrm{E}(C_n) = \frac{1}{r^n}\sum_{i=0}^{n}\binom{n}{i}p^i(1-p)^{n-i}\underbrace{C_{u...u}}_{i}\underbrace{d....d}_{n-i} \qquad (11.7)$$

is today's price of a claim which is to be exercised n periods from now. C can be seen as the expected payoff, expressed in the now dollar value, of the final payoff C_n, when the probability of the market going up on each period is the arbitrage-free probability p. For a call option

$$\underbrace{C_{u...u}}_{i}\underbrace{d....d}_{n-i} = (u^id^{n-i}S - K)^+,$$

and C can be written by using the complimentary Binomial cumulative probability distribution function $\mathrm{Bin}(j; n, p) = \mathrm{P}(S_n > j)$

$$C = S\mathrm{Bin}(j; n, p') - r^{-n}K\mathrm{Bin}(j; n, p),$$

where $j = [\frac{\ln(K/Sd^n)}{\ln(u/d)}] + 1$ and $p' = \frac{u}{r}p$.

It is possible to obtain the option pricing formula of Black and Scholes from the Binomial formula by taking limits as the length of the trading period goes to zero and the number of trading periods n goes to infinity.

Pricing by No Arbitrage

Given a general model these are the question we ask.

1. If we have a model for evolution of prices, *how can we tell if there are arbitrage opportunities?* Not finding any is a good start, but not a proof that there are none.

2. If we know that there are no arbitrage opportunities in the market, *how do we price a claim, such as an option?*

3. *Can we price any option, or are there some that cannot be priced by arbitrage arguments?*

The answers are given by two main results, called the Fundamental Theorems of asset pricing. In what follows we outline the mathematical theory of pricing of claims in finite and general market models.

11.2 A Finite Market Model

Consider a model with one stock with price $S(t)$ at time t, and a riskless investment (bond, or cash in a savings account) with price $\beta(t)$ at time t. If the riskless rate of investment is a constant $r > 1$ then $\beta(t) = r^t\beta(0)$.

A market model is called finite if $S(t), t = 0, \ldots, T$ take finitely many values. A portfolio $(a(t), b(t))$ is the number of shares of stock and bond units held during $[t - 1, t)$. The information available after observing prices up to time t is denoted by the σ-field $\mathcal{F}_t$. The portfolio is decided on the basis of information at time $t - 1$, in other words $a(t), b(t)$ are $\mathcal{F}_{t-1}$ measurable, $t = 1, \ldots, T$, or in our terminology they are predictable processes. The change in market value of the portfolio at time t is the difference between its value after it has been established at time $t - 1$ and its value after prices at t are observed, namely

$$a(t)S(t) + b(t)\beta(t) - a(t)S(t-1) - b(t)\beta(t-1) = a(t)\Delta S(t) + b(t)\Delta\beta(t).$$

A portfolio (trading strategy) is called *self-financing* if all the changes in the portfolio are due to gains realized on investment, that is, no funds are borrowed or withdrawn from the portfolio at any time,

$$V(t) = V(0) + \sum_{i=1}^{t} \left(a(i)\Delta S(i) + b(i)\Delta\beta(i) \right), \quad t = 1, 2, \ldots, T. \tag{11.8}$$

The initial value of the portfolio $(a(t), b(t))$ is $V(0) = a(1)S(0) + b(1)\beta(0)$ and subsequent values $V(t), t = 1, \ldots, T$ are given by

$$V(t) = a(t)S(t) + b(t)\beta(t). \tag{11.9}$$

$V(t)$ represents the value of the portfolio just before time t transactions after time t price was observed. Since the market value of the portfolio $(a(t), b(t))$ at time t after $S(t)$ is announced is $a(t)S(t) + b(t)\beta(t)$, and the value of the newly setup portfolio is $a(t + 1)S(t) + b(t + 1)\beta(t)$, a self-financing strategy must satisfy

$$a(t)S(t) + b(t)\beta(t) = a(t + 1)S(t) + b(t + 1)\beta(t). \tag{11.10}$$

A strategy is called *admissible* if it is self-financing and the corresponding value process is non-negative.

A contingent claim is a non-negative random variable X on $(\Omega, \mathcal{F}_T)$. It represents an agreement which pays X at time T, for example, for a call with strike K, $X = (S(T) - K)^+$.

Definition 11.2 *A claim X is called attainable if there exists an admissible strategy replicating the claim, that is, $V(t)$ satisfies (11.8), $V(t) \geq 0$ and $V(T) = X$.*

Definition 11.3 *An arbitrage opportunity is an admissible trading strategy such that $V(0) = 0$, but $\mathrm{E}V(T) > 0$.*

Note that since $V(T) \geq 0$, $\mathrm{E}V(T) > 0$ is equivalent to $\mathrm{P}(V(T) > 0) > 0$. The following result is central to the theory.

Theorem 11.4 *Suppose there is a probability measure Q, such that the discounted stock process $Z(t) = S(t)/\beta(t)$ is a Q-martingale. Then for any admissible trading strategy the discounted value process $V(t)/\beta(t)$ is also a Q-martingale.*

Such Q is called an equivalent martingale measure (EMM) or a risk-neutral probability measure.

PROOF: Since the market is finite, the value process $V(t)$ takes only finitely many values, therefore $\mathrm{E}V(t)$ exist. The martingale property is verified as follows.

$$
\begin{aligned}
\mathrm{E}_Q\left(\frac{V(t+1)}{\beta(t+1)}|\mathcal{F}_t\right) &= \mathrm{E}_Q\big(a(t+1)Z(t+1) + b(t+1)|\mathcal{F}_t\big) \\
&= a(t+1)\mathrm{E}_Q\big(Z(t+1)|\mathcal{F}_t\big) + b(t+1) \text{ since } a(t) \text{ and } b(t) \text{ are predictable} \\
&= a(t+1)Z(t) + b(t+1) \text{ since } Z(t) \text{ is a martingale} \\
&= a(t)Z(t) + b(t) \text{ since } (a(t), b(t)) \text{ is self-financing (11.10)} \\
&= \frac{V(t)}{\beta(t)}.
\end{aligned}
$$

$\square$

This result is "nearly" the condition for no-arbitrage, since it states that if we start with zero wealth, then positive wealth cannot be created if probabilities are assigned by Q. Indeed, by the above result $V(0) = 0$ implies $\mathrm{E}_Q(V(T)) = 0$, and $Q(V(T) = 0) = 1$. However, in the definition of an arbitrage strategy the expectation is taken under the original probability measure P. To establish the result for P equivalence of probability measures is used. Recall the definition from Chapter 10.

Definition 11.5 *Two probability measures P and Q are called equivalent if they have same null sets, that is, for any set A with $\mathrm{P}(A) = 0$, $\mathrm{Q}(A) = 0$ and vice versa.*

Equivalent probability measures in a market model reflect the fact that investors agree on the space of all possible outcomes, but assign different probabilities to the outcomes. The following result gives a probabilistic condition to assure that the model does not allow for arbitrage opportunities. It states that no arbitrage is equivalent to existence of EMM.

Theorem 11.6 (First Fundamental Theorem) *A market model does not have arbitrage opportunities if and only if there exists a probability measure* Q, *equivalent to* P, *such that the discounted stock process* $Z(t) = S(t)/\beta(t)$ *is a* Q-*martingale.*

PROOF: Proof of sufficiency. Suppose there exists a probability measure Q, equivalent to P, such that the discounted stock process $Z(t) = S(t)/\beta(t)$ is a Q-martingale. Then there are no arbitrage opportunities. By Theorem 11.4 any admissible strategy with $V(0) = 0$ must have $Q(V(T) > 0) = 0$. Since Q is equivalent to P, $P(V(T) > 0) = 0$. But then $E_P(V(T)) = 0$. Thus there are no admissible strategies with $V(0) = 0$ and $E_P(V(T)) > 0$, in other words, there is no arbitrage.

A proof of necessity requires additional concepts, see, for example, Harrison and Pliska (1981).

□

Claims are priced by the replicating portfolios.

Theorem 11.7 (Pricing by No-Arbitrage) *Suppose that the market model does not admit arbitrage, and* X *is an attainable claim with maturity* T. *Then* $C(t)$, *the arbitrage-free price of* X *at time* $t \leq T$, *is given by* $V(t)$, *the value of a portfolio of any admissible strategy replicating* X. *Moreover*

$$C(t) = V(t) = E_Q \left(\frac{\beta(t)}{\beta(T)} X \big| \mathcal{F}_t \right), \tag{11.11}$$

where Q *is an equivalent martingale probability measure.*

PROOF: Since X is attainable, it is replicated by an admissible strategy with the value of the replicating portfolio $V(t)$, $0 \leq t \leq T$, and $X = V(T)$. Fix one such strategy. To avoid arbitrage, the price of X at any time $t < T$ must be given by the value of this portfolio $V(t)$, otherwise arbitrage profit is possible.

Since the model does not admit arbitrage a martingale probability measure Q exists by Theorem 11.6. The discounted value process $V(t)/\beta(t)$ is a Q-martingale by Theorem 11.4, hence by the martingale property

$$\frac{V(t)}{\beta(t)} = E_Q(\frac{V(T)}{\beta(T)} | \mathcal{F}_t). \tag{11.12}$$

But $V(T) = X$, and we have

$$\frac{V(t)}{\beta(t)} = E_Q \left(\frac{1}{\beta(T)} X \big| \mathcal{F}_t \right). \tag{11.13}$$

□

Note that $\frac{\beta(t)}{\beta(T)}X$ represents the value of the claim X in the dollars at time t, and the price of X at time $t = 0$ is given by

$$C(0) = \mathrm{E}_Q \left(\frac{1}{\beta(T)} X \right). \tag{11.14}$$

Remark 11.1: The lhs. of equation (11.13) $V(t)/\beta(t) = a(t)Z(t) + b(t)$ is determined by the portfolio $a(t), b(t)$, but its rhs. $\mathrm{E}_Q \left(\frac{1}{\beta(T)} X | \mathcal{F}_t \right)$ is determined by the measure Q and has nothing to do with a chosen portfolio. This implies that for a given martingale measure Q, and any t all self-financing portfolios replicating X have the same value, moreover this common value is also the same for different martingale probability measures. Thus equation (11.11) provides an unambiguous price for the claim X at time t.

However, if a claim X is not attainable its expectation may vary with the measure Q, see Example 11.7 below.

Now we know how to price attainable claims. If all the claims in the market are attainable, then we can price any claim.

Definition 11.8 *Market models in which any claim is attainable are called complete.*

The following result characterizes complete models in terms of the martingale measure. The proof can be found in Harrison Kreps (1979) and Harrison and Pliska (1983).

Theorem 11.9 (Completeness) *The market model is complete if and only if the martingale probability measure Q is unique.*

Example 11.6: (A complete model)
The one-step Binomial model ($t = 0, 1$) can be described by payoff vector of the stock (d, u) and of the bond (r, r). A claim X is a vector (x_1, x_2) representing the payoff of the claim when the market goes down, x_1, and up x_2. As the two vectors (r, r) and (d, u) span $\mathbf{R}^2$, any vector is a linear combination of those two, hence any claim can be replicated by a portfolio, and the model is complete. To find a martingale probability Q, we solve $\mathrm{E}_Q Z(1) = Z(0)$, or $\frac{1}{r}(pu + (1-p)d) = 1$. The unique solution for $p = \frac{r-d}{u-d}$. It is a probability if and only if $d < r < u$. Thus the existence and uniqueness of the martingale measure is verified. In this model any claim can be priced by no-arbitrage considerations.

Example 11.7: (An incomplete model)
The model where stock's payoff can take 3 possible values $(d, 1, u)$ and the bond with payoff $(1, 1, 1)$ is not complete. These vectors span a subspace of R^3, therefore not all possible returns can be replicated. For example, the claim that pays \$1 when the stock goes up and nothing in any other case has the payoff $(0, 0, 1)$ and cannot be

replicated. We now verify that there are many martingale measures. To find them we must solve $E_Q Z(1) = Z(0)$.

$$\left.\begin{array}{r} \frac{1}{r}[p_u u + p_n + p_d d] = 1 \\[2mm] p_u + p_n + p_d = 1 \end{array}\right\}$$

Any solution of this system makes $Z(t)$, $t = 0, 1$ into a martingale.

$$p_u = \frac{r-1}{u-1} + \frac{1-d}{u-1}p_d, \quad p_n = \frac{u-r}{u-1} - \frac{u-d}{u-1}p_d, \quad 0 \le p_u, p_n, p_d \le 1$$

11.3 Semimartingale Market Model

Arbitrage in Continuous Time Models

In continuous time there are different versions of the no-arbitrage concept. The main premise is the same as in discrete time, it should be impossible to make "something" out of nothing without taking risk. The difference between different versions of no-arbitrage is in the kinds of allowable (admissible) self-financing strategies that define how "something" is made.

Let $a(t)$ and $b(t)$ denote the number of shares and bond units respectively held at time t. The market value of the portfolio at time t is given by

$$V(t) = a(t)S(t) + b(t)\beta(t). \tag{11.15}$$

The change in the value of the portfolio due to change in the price of assets during dt is $a(t)dS(t) + b(t)d\beta(t)$.

Definition 11.10 *A portfolio $(a(t), b(t))$, $0 \le t \le T$, is called self-financing if the change in value comes only from the change in prices of the assets,*

$$dV(t) = a(t)dS(t) + b(t)d\beta(t), \tag{11.16}$$

$$V(t) = V(0) + \int_0^t a(u)dS(u) + \int_0^t b(u)d\beta(u). \tag{11.17}$$

It is assumed that $S(t)$ and $\beta(t)$ are semimartingales. The processes $a(t)$ and $b(t)$ must be predictable processes satisfying a certain condition for the stochastic integral to be defined, see (8.8) and (8.52).

In a general situation, when both $S(t)$ and $\beta(t)$ are stochastic, the rhs. of (11.17) can be seen as a stochastic integral with respect to the vector process $(S(t), \beta(t))$. Such integrals extend the standard definition of scalar stochastic integrals and can be defined for a larger class of integrands, due to possible interaction between the components. We don't go into details, as they are rather complicated, and refer to Shiryaev (1999), Jacod and Shiryaev (1987).

The following concepts describe no-arbitrage: No Arbitrage (NA), No Free Lunch (NFL), No Free Lunch with Bounded Risk (NFLBR), No Free Lunch with Vanishing Risk (NFLVR), No Feasible Free Lunch with Vanishing Risk (NFFLVR), see Shiryaev (1999), Kreps (1981), Delbaen and Schachermayer (1994), Kabanov (2001). We don't aim to present the Arbitrage theory in continuous time, and consider a simpler formulation, following Harrison and Pliska (1981).

We consider a model in continuous time $0 \le t \le T$ consisting of two assets, a semimartingale $S(t)$ representing the stock price process, and the savings account (or bond) $\beta(t)$, $\beta(0) = 1$. We assume $\beta(t)$ is continuous and of finite variation. The following is a central result.

Theorem 11.11 $(a(t), b(t))$ *is self-financing if and only if the discounted value process* $\frac{V(t)}{\beta(t)}$ *is a stochastic integral with respect to the discounted price process*

$$\frac{V(t)}{\beta(t)} = V(0) + \int_0^t a(u)dZ(u), \qquad (11.18)$$

where $Z(t) = S(t)/\beta(t)$.

PROOF: Using the assumption that the bond process is continuous and of finite variation, we have

$$
\begin{aligned}
d\left(\frac{V(t)}{\beta(t)}\right) &= \frac{1}{\beta(t-)}dV(t) + V(t-)d\left(\frac{1}{\beta(t)}\right) + d[V, \frac{1}{\beta}](t) \\
&= \frac{1}{\beta(t)}dV(t) + V(t-)d\left(\frac{1}{\beta(t)}\right).
\end{aligned}
$$

Using the self-financing property

$$d\left(\frac{V(t)}{\beta(t)}\right) = \frac{1}{\beta(t)}\big(a(t)dS(t) + b(t)d\beta(t)\big) + \big(a(t)S(t-) + b(t)\beta(t)\big)d\left(\frac{1}{\beta(t)}\right)$$

$$=a(t)\underbrace{\left(\frac{dS(t)}{\beta(t-)} + S(t-)d(\frac{1}{\beta(t)})\right)}_{dZ(t)=d(S(t)/\beta(t))} +b(t)\underbrace{\left(\frac{1}{\beta(t)}d\beta(t) + \beta(t)d(\frac{1}{\beta(t)})\right)}_{d(\beta(t)\cdot\frac{1}{\beta(t)})=0} = a(t)dZ(t).$$

The other direction. Assume (11.18). From $V(t) = a(t)S(t) + b(t)\beta(t)$, we find $b(t) = V(t)/\beta(t) - a(t)Z(t)$. Using (11.18),

$$b(t) = V(0) + \int_0^t a(u)dZ(u) - a(t)Z(t). \qquad (11.19)$$

Using (11.18) $V(t) = V(0)\beta(t) + \beta(t)\int_0^t a(u)dZ(u)$. Hence

$$
\begin{aligned}
dV(t) &= V(0)d(\beta(t)) + \int_0^t a(u)dZ(u)d(\beta(t)) + \beta(t)a(t)dZ(t) \\
&= V(0)d(\beta(t)) + (b(t) - V(0) + a(t)Z(t))d(\beta(t)) + \beta(t)a(t)dZ(t) \\
&= b(t)d(\beta(t)) + a(t)Z(t)d(\beta(t)) + \beta(t)a(t)dZ(t) \\
&= b(t)d(\beta(t)) + a(t)d(Z(t)\beta(t)) = b(t)d(\beta(t)) + a(t)dS(t),
\end{aligned}
$$

and self-financing property of $V(t)$ is established.

$\square$

The basis for mathematical formulation of arbitrage is the existence of the EMM, the equivalent martingale probability measure. In a general model existence of such a measure is introduced as an assumption.

EMM Assumption

There exists a martingale probability measure Q which is equivalent to the original measure P, such that the discounted price process $Z(t) = S(t)/\beta(t)$ is a Q-martingale.

We give examples of models where the EMM assumption does not and does hold.

Example 11.8: $dS(t) = .04S(t)dt$. $d\beta(t) = 0.03\beta(t)dt$. $S(t) = S(0)e^{0.04t}$, $\beta(t) = e^{0.03t}$, $S(t)e^{-0.03t} = e^{0.01t}$. Since $e^{0.01t}$ is a deterministic non-constant function, there is no probability measure Q that would make it into a martingale.

Example 11.9: $S(t) = S(0) + \int_0^t B(s)ds$, where $B(s)$ is a P-Brownian motion. By the Corollary 10.21, an equivalent change of measure results in $B(t)$ being transformed into $B(t) + q(t)$, for some process $q(t)$. Thus under the EMM Q, $S(t) = S(0) + \int_0^t B(s)ds + \int_0^t q(s)ds$. Since $S(t)$ has finite variation, it can not be a martingale, because a continuous martingale is either a constant or has infinite variation (see Theorem 7.29). Hence there is no EMM in this model.

Example 11.10: (Bachelier model)

$$
S(t) = S(0) + \mu t + \sigma B(t), \tag{11.20}
$$

for positive constants μ, σ. $\beta(t) = 1$. The EMM exists (and is unique) by the Girsanov's theorem.

Example 11.11: (Black-Scholes model)

$$
dS(t) = \mu S(t)dt + \sigma S(t)dB(t), \quad \beta(t) = e^{rt}. \tag{11.21}
$$

Solving the SDE for $S(t)$, $Z(t) = S(t)e^{-rt} = S(0)e^{(\mu-r-\frac{1}{2}\sigma^2)t+\sigma B(t)}$. This process is a martingale if and only if $\mu = r$; when $\mu = r$, it is the exponential martingale of $\sigma B(t)$, when $\mu \neq r$, it is a martingale times a non-constant deterministic function, easily seen not a martingale. Writing $dS(t) = \sigma S(t)(\frac{\mu}{\sigma}dt + dB(t))$, and using the change of drift in diffusions, there is (a unique) Q, such that $\frac{\mu}{\sigma}dt + dB(t) = \frac{r}{\sigma}dt + dW(t)$ for a Q Brownian motion $W(t)$. So $\sigma B(t) = rt + \sigma W(t) - \mu t$ Thus the equation for $Z(t)$ in terms of $W(t)$ is

$$Z(t) = S(0)e^{(\mu-r-\frac{1}{2}\sigma^2)t+\sigma B(t)} = S(0)e^{-\frac{1}{2}\sigma^2 t+\sigma W(t)},$$

verifying that Q is the EMM.

Admissible Strategies

The discounted value of a replicating self-financing portfolio $V(t)/\beta(t)$ is a stochastic integral with respect to the Q-martingale $Z(t)$, Theorem 11.11. We would like it to be a martingale, because then all its values can be determined by its final value, which is matched to the claim X. The martingale property implies $V(t)/\beta(t) = E_Q(V(T)/\beta(T)|\mathcal{F}_t) = E_Q(X/\beta(T)|\mathcal{F}_t)$.

However, a stochastic integral with respect to a martingale is only a local martingale. Thus the discounted value process of a self-financing portfolio in (11.18) is a local martingale under Q. Since it is non-negative, it is a super-martingale (Theorem 7.23). Supermartingales have non-increasing expectations. In particular, there are strategies (called suicidal) that can turn the initial investment into nothing. Adding such a strategy to any other self-financing portfolio will change the initial value without changing the final value. Thus there are self-financing strategies with the same final value but different initial values. This phenomenon is precisely the difference in the situation between the finite market model and the general model. Note that, similar to the finite market model, a self-financing strategy cannot create something out of nothing, since the expectations of such strategies are non-increasing. To eliminate the undesirable strategies from consideration, only martingale strategies are admissible. We follow Harrison and Pliska (1981).

Fix a reference EMM, equivalent martingale probability measure Q, so that the discounted stock price process $Z(t)$ is a Q-martingale; expectations are taken with respect to this measure Q.

Definition 11.12 *A predictable and self-financing strategy $(a(t), b(t))$ is called admissible if*

$$\sqrt{\int_0^t a^2(u)d[Z,Z](u)} \quad \text{is finite and locally integrable } 0 \leq t \leq T. \quad (11.22)$$

Moreover $V(t)/\beta(t)$ is a non-negative Q-martingale.

Note that condition (11.22) is needed to define the stochastic integral $\int_0^t a(u)dZ(u)$, see condition (8.8). If (11.22) holds then $\int_0^t a(u)dZ(u)$ and consequently $V(t)/\beta(t)$ are local martingales. If moreover $V(t)/\beta(t) \geq 0$ then it is a supermartingale.

Pricing of Claims

A claim is a non-negative random variable. It is attainable if it is integrable, $EX < \infty$, and there exists an admissible trading strategy such that at maturity T, $V(T) = X$. To avoid arbitrage, the value of an attainable claim at time $t < T$ must be the same as that of the replicating portfolio at t.

Theorem 11.13 *The price $C(t)$ at time t of an attainable claim X, is given by the value of an admissible replicating portfolio $V(t)$, moreover*

$$C(t) = \mathrm{E}_Q\left(\frac{\beta(t)}{\beta(T)}X|\mathcal{F}_t\right). \tag{11.23}$$

The proof follows by the martingale property of $V(t)/\beta(t)$ in exactly the same way as in the finite model case.

$$C(t) = V(t) = \beta(t)\mathrm{E}_Q\left(V(T)/\beta(T)|\mathcal{F}_t\right)$$

$$= \beta(t)\mathrm{E}_Q\left(\frac{V(T)}{\beta(T)}|\mathcal{F}_t\right) = \beta(t)\mathrm{E}_Q\left(\frac{X}{\beta(T)}|\mathcal{F}_t\right).$$

Since attainable claims can be priced, the natural question is "how can one tell whether a claim is attainable?". The following result gives an answer using the predictable representation property of the discounted stock price.

Theorem 11.14 *Let X be an integrable claim and let $M(t) = \mathrm{E}_Q\left(\frac{X}{\beta(T)}|\mathcal{F}_t\right)$, $0 \leq t \leq T$. Then X is attainable if and only if $M(t)$ can be represented in the form*

$$M(t) = M(0) + \int_0^t H(u)dZ(u)$$

for some predictable process H. Moreover $V(t)/\beta(t) = M(t)$ is the same for any admissible portfolio that replicates X.

PROOF: Suppose X is attainable, and that $(a(t), b(t))$ replicates it. By the previous result $V(t)/\beta(t) = M(t)$. It follows By (11.18) that the desired representation holds with $H(t) = a(t)$. Conversely, if $M(t) = M(0) + \int_0^t H(u)dZ(u)$, take $a(t) = H(t)$, and $b(t) = M(0) + \int_0^t H(u)dZ(u) - H(t)Z(t)$. This gives a self-financing strategy.

$\square$

Completeness of a Market Model

A market model is *complete* if any integrable claim is attainable, in other words, can be replicated by a self-financing portfolio. If a model is complete, then any claim can be priced by no-arbitrage considerations.

We know by Theorem 11.14 that for a claim to be attainable the martingale $M(t) = E_Q(\frac{X}{\beta(T)}|\mathcal{F}_t)$ should have a predictable representation with respect to the Q-martingale $Z(t) = S(t)/\beta(t)$. Recall, that the martingale $Z(t)$ has the predictable representation property if any other martingale can be represented as a stochastic integral with respect to it, see Definition 8.34. For results on predictable representations see Section 8.12. In particular, if the martingale $Z(t)$ has the predictable representation property, then all claims in the model are attainable. It turns out that the opposite is also true, moreover there is a surprising characterization, the equivalent martingale measure Q is unique.

Theorem 11.15 (Second Fundamental Theorem) *The following are equivalent:*

1. *The market model is complete.*

2. *The martingale $Z(t)$ has the predictable representation property.*

3. *The EMM Q, that makes $Z(t) = S(t)/\beta(t)$ into a martingale, is unique.*

11.4 Diffusion and the Black-Scholes Model

In this section we apply general results to the diffusion models of stock prices. In a diffusion model the stock price is assumed to satisfy

$$dS(t) = \mu(S(t))dt + \sigma(S(t))dB(t), \tag{11.24}$$

where $B(t)$ is P-Brownian motion. Bond price is assumed to be deterministic and continuous $\beta(t) = \exp(\int_0^t r(u)du)$. According to Theorem 11.13 pricing of claims is done under the martingale probability measure Q that makes $Z(t) = S(t)/\beta(t)$ into a martingale.

Theorem 11.16 *Let $H(t) = \frac{\mu(S(t))-r(t)S(t)}{\sigma(S(t))}$. Suppose that $\mathcal{E}\left(\int_0^\cdot H(t)dB(t)\right)$ is a martingale. Then the EMM exists and is unique. It is defined by*

$$\Lambda = \frac{dQ}{dP} = \mathcal{E}\left(\int_0^\cdot H(t)dB(t)\right)(T) = e^{\int_0^T H(t)dB(t)-\frac{1}{2}\int_0^T H^2(t)dt}. \tag{11.25}$$

The SDE for $S(t)$ under Q with a Brownian motion W, is

$$dS(t) = r(t)S(t)dt + \sigma(S(t))dW(t). \tag{11.26}$$

PROOF: It is easy to see by Itô's formula that $Z(t)$ satisfies

$$dZ(t) = d\Big(\frac{S(t)}{\beta(t)}\Big) = \frac{\mu(S(t)) - r(t)S(t)}{\beta(t)} dt + \frac{\sigma(S(t))}{\beta(t)} dB(t).$$

$$= \frac{\sigma(S(t))}{\beta(t)} \Big(\frac{\mu(S(t)) - r(t)S(t)}{\sigma(S(t))} dt + dB(t) \Big).$$

For $Z(t)$ to be a martingale it must have zero drift coefficient. Define Q by (11.25). Using (10.34), $\frac{\mu(S(t))-r(t)S(t)}{\sigma(S(t))} dt + dB(t) = dW(t)$ with a Q Brownian motion $W(t)$. This gives the SDE

$$dZ(t) = d\Big(\frac{S(t)}{\beta(t)}\Big) = \frac{\sigma(S(t))}{\beta(t)} dW(t).$$

Using integration by parts we obtain the SDE (11.26) under the EMM Q for $S(t)$.

□

To price claims, expectations should be calculated by using equation (11.26), and not the original one in (11.24). The effect of change of measure is the change in the drift: $\mu(x)$ is changed into rx, where r is the riskless interest rate.

Black-Scholes Model

The Black-Scholes model is the commonly accepted model for pricing of claims in the Financial Industry. The main assumptions of the model are: the riskless interest rate is a constant r, $\sigma(S(t)) = \sigma S(t)$, where the constant σ is called "volatility". The stock price processes $S(t)$ satisfies SDE

$$dS(t) = \mu S(t)dt + \sigma S(t)dB(t). \tag{11.27}$$

Using Itô's formula with $f(x) = \ln x$ (see Example 5.5) we find that the solution is given by

$$S(t) = S(0)e^{(\mu - \frac{\sigma^2}{2})t + \sigma B(t)}. \tag{11.28}$$

This model corresponds to the simplest random model for the return $R(t)$ on stock

$$dR(t) = \frac{dS(t)}{S(t)} = \mu dt + \sigma dB(t). \tag{11.29}$$

$S(t)$ is the stochastic exponential of $R(t)$, $dS(t) = S(t)dR(t)$. By Theorem 8.12 $S(t) = S(0)\mathcal{E}(R)_t = S(0)e^{R(t) - \frac{1}{2}[R,R](t)}$, giving (11.28).

The EMM Q makes $S(t)e^{-rt}$ into a martingale. By Theorem 11.16 it exists and is unique. It is obtained by letting $\frac{\mu}{\sigma}dt + dB(t) = \frac{r}{\sigma}dt + dW(t)$, for a Q-Brownian motion $W(t)$. In this case $H(t) = \frac{\mu - r}{\sigma}$, $\int_0^\cdot H(t)dB(t) = \frac{\mu - r}{\sigma}B(t)$

and $\mathcal{E}\left(\int_0^{\cdot} H(t)dB(t)\right)(T) = \mathcal{E}\left(\frac{\mu-r}{\sigma}B\right)(T)$ is the exponential martingale of Brownian motion. The SDE (11.26) under the EMM Q for $S(t)$ is

$$dS(t) = rS(t)dt + \sigma S(t)dW(t). \tag{11.30}$$

It has the solution

$$S(t) = S(0)e^{(r-\frac{\sigma^2}{2})t+\sigma W(t)}. \tag{11.31}$$

Thus under the equivalent martingale measure Q, $S(T)$ has a Lognormal distribution, $LN\left((r - \frac{\sigma^2}{2})T + \ln S(0), \sigma^2 T\right)$. The price of a claim X at time T is given by

$$C(t) = e^{-r(T-t)}E_Q(X|\mathcal{F}_t). \tag{11.32}$$

If $X = g(S(T))$, then by the Markov property of $S(t)$, $C(t) = E_Q\left(g(S(T))|\mathcal{F}_t\right) = E_Q\left(g(S(T))|S(t)\right)$. The conditional distribution under Q given $\mathcal{F}_t$ is obtained from the equation (using (11.31))

$$S(T) = S(t)e^{(r-\frac{\sigma^2}{2})(T-t)+\sigma(W(T)-W(t))},$$

and is Lognormal, $LN\left((r - \frac{\sigma^2}{2})(T - t) + \ln S(t), \sigma^2(T - t)\right)$.

Pricing a Call Option

A call option pays $X = (S(T) - K)^+$ at time T. To find its price at time $t = 0$, according to (11.32), we must calculate $E(S(T) - K)^+$, where expectation E is taken under the arbitrage-free probability Q. In the Black-Scholes model the Q-distribution of $S(T)$ is a Lognormal. Denote for brevity the parameters of this distribution by μ and σ^2, with $\mu = (r - \frac{\sigma^2}{2})T + \ln S(0)$ and new $\sigma^2 = \sigma^2 T$.

$$E(S(T)-K)^+ = E(S(T)-K)I(X > K) = ES(T)I(S(T) > K) - KQ(S(T) > K).$$

The second term is easy to calculate by using Normal probabilities as $K(1 - \Phi((\log K - \mu)/\sigma))$.

The first term can be calculated by a direct integration, or by changing measure, see Example 10.2. $S(T)$ is absorbed into the Likelihood $dQ_1/dQ = \Lambda = S(T)/ES(T)$. Now, by Lognormality of $S(T)$ write it as e^Y, with $Y \sim N(\mu, \sigma^2)$. Then $ES(T) = Ee^Y = e^{\mu+\sigma^2/2}$, and

$$\begin{aligned}
E(e^Y I(e^Y > K)) &= Ee^Y E((e^Y/Ee^Y)I(e^Y > K)) = Ee^Y E(\Lambda I(e^Y > K)) \\
&= Ee^Y E_{Q_1} I(e^Y > K) = Ee^Y Q_1(e^Y > K).
\end{aligned}$$

Since $dQ_1/dQ = e^{Y-\mu-\sigma^2/2}$, it follows by Theorem 10.4 that Q_1 is $N(\mu + \sigma^2, \sigma^2)$ distribution. Therefore

$$E(e^Y I(e^Y > K)) = e^{\mu+\sigma^2/2}(1 - \Phi((\ln K - \mu - \sigma^2)/\sigma)).$$

Finally, using $1 - \Phi(x) = \Phi(-x)$, we obtain

$$E(X - K)^+ = e^{\mu + \sigma^2/2}\Phi(\frac{\mu + \sigma^2 - \ln K}{\sigma}) - K\Phi(\frac{\mu - \ln K}{\sigma}). \qquad (11.33)$$

The price at time t of the European call option on stock with strike K and maturity T is given by (Theorem 11.13)

$$C(t) = e^{-r(T-t)}E_Q((S(T) - K)^+|\mathcal{F}_t).$$

It follows from (11.33) that $C(t)$ is given by the Black-Scholes formula

$$C(t) = S(t)\Phi(h(t)) - Ke^{-r(T-t)}\Phi\left(h(t) - \sigma\sqrt{T-t}\right), \qquad (11.34)$$

where

$$h(t) = \frac{\ln\frac{S(t)}{K} + (r + \frac{1}{2}\sigma^2)(T - t)}{\sigma\sqrt{T-t}}. \qquad (11.35)$$

Pricing of Claims by a PDE. Replicating Portfolio

Let X be a claim of the form $X = g(S(T))$. Since the stock price satisfies SDE (11.30), by the Markov property of $S(t)$ it follows from (11.32) that the price of X at time t

$$C(t) = e^{-r(T-t)}E_Q\left(g(S(T))|\mathcal{F}_t\right) = e^{-r(T-t)}E_Q\left(g(S(T))|S(t)\right). \qquad (11.36)$$

By the Feynman-Kac formula (Theorem 6.8),

$$C(x,t) = e^{-r(T-t)}E_Q\left(g(S(T))|S(t) = x\right)$$

solves the following partial differential equation (PDE)

$$\frac{1}{2}\sigma^2 x^2 \frac{\partial^2 C(x,t)}{\partial x^2} + rx\frac{\partial C(x,t)}{\partial x} + \frac{\partial C(x,t)}{\partial t} - rC = 0. \qquad (11.37)$$

The boundary condition is given by the value of the claim at maturity

$$C(x,T) = g(x), \quad x \geq 0,$$

with $g(x)$ is the value of the claim at time T when the stock price is x. When $x = 0$ equation (11.37) gives

$$C(0,t) = e^{-r(T-t)}g(0), \quad 0 \leq t \leq T,$$

(portfolio of only a bond has its value at time t as the discounted final value.) For a call option $g(x) = (x - K)^+$. This partial differential equation was solved by Black and Scholes with the Fourier transform method. The solution is the Black-Scholes formula (11.34).

Next we give an argument to find the replicating portfolio, which is also used to re-derive the PDE (11.37). It is clear that $C(t)$ in (11.36) is a function of $S(t)$ and t. Let $(a(t), b(t))$ be a replicating portfolio. Since it is self-financing its value process satisfies

$$dV(t) = a(t)dS(t) + b(t)d\beta(t). \qquad (11.38)$$

If X is an attainable claim, that is, $X = V(T)$, then the arbitrage-free value of X at time $t < T$ is the value of the portfolio

$$V(t) = a(t)S(t) + b(t)\beta(t) = C(S(t), t). \qquad (11.39)$$

Thus we have by (11.38)

$$dC(S(t), t) = dV(t) = a(t)dS(t) + b(t)d\beta(t). \qquad (11.40)$$

Assume that $C(x, t)$ is smooth enough to apply Itô's formula. Then from the SDE for the stock price (11.30) (from which $d[S, S](t) = \sigma^2 S^2(t)dt$), we have

$$dC(S(t), t) = \frac{\partial C(S(t), t)}{\partial x}dS(t) + \frac{\partial C(S(t), t)}{\partial t}dt + \frac{1}{2}\frac{\partial^2 C(S(t), t)}{\partial x^2}\sigma^2 S^2(t)dt. \qquad (11.41)$$

By equating the two expressions above we have

$$\left(a(t) - \frac{\partial C(S(t), t)}{\partial x}\right)dS(t) = \left(\frac{\partial C(S(t), t)}{\partial t} + \frac{1}{2}\frac{\partial^2 C(S(t), t)}{\partial x^2}\sigma^2 S^2(t)\right)dt - b(t)d\beta(t). \qquad (11.42)$$

The lhs has a positive quadratic variation unless $a(t) - \frac{\partial C(S(t), t)}{\partial x} = 0$, and the rhs has zero quadratic variation. For them to be equal we must have for all t

$$a(t) = \frac{\partial C}{\partial x}(S(t), t), \qquad (11.43)$$

and consequently

$$b(t)d\beta(t) = \left(\frac{\partial C(S(t), t)}{\partial t} + \frac{1}{2}\frac{\partial^2 C(S(t), t)}{\partial x^2}\sigma^2 S^2(t)\right)dt. \qquad (11.44)$$

Putting the values of $a(t)$ and $b(t)\beta(t)$ into equation (11.39), taking into account $d\beta(t) = r\beta(t)dt$, and replacing $S(t)$ by x, we obtain the PDE (11.37). The replicating portfolio is given by (11.43) and (11.44)

Using the Black-Scholes formula, as the solution of this PDE, the replicating portfolio is obtained from (11.43)

$$a(t) = S(t)\Phi(h(t)), \quad b(t) = K\Phi(h(t) - \sigma\sqrt{T-t}), \qquad (11.45)$$

where $h(t)$ is given by (11.35).

Another way to derive the Black-Scholes PDE, useful in other models, is given in Exercise 11.8.

Validity of the Assumptions

The simulated time series of a geometric Brownian motion (with $\sigma = 1$ and various μs (265 points)) look similar to the time series of observed stock prices (daily stock closing prices for period 5 Aug 91- 7 Aug 92), see Fig. 11.1. This shows that the suggested model is capable of producing realistic looking price processes, but, of course, does not prove that this is the correct model. The assumption of normality can be checked by looking at histograms and using formal statistical tests. The histogram of the BHP stock returns points to normality. However, the assumption of constant volatility does not seem to be true, and this issue is addressed next.

Implied Volatility

Denote by C^m the observed market price of an option with strike K and expiration T. The implied volatility $I_t(K, T)$ is defined as the value of the volatility parameter σ in the Black-Scholes formula (11.34) that matches the observed price, namely

$$C^{\text{BS}}(I_t(K, T)) = C^m_t(K, T), \tag{11.46}$$

where C^{BS} is given by (11.34). It has been observed that the implied volatility as a function of strike K (and of term T) is not a constant, and has a graph that looks like a smile. Models with stochastic volatility are able to reproduce the observed behaviour of implied volatilities, see Fouque *et al.* (2000).

Stochastic Volatility Models

A class of models in which the volatility parameter is not a constant, but a stochastic process itself is known as stochastic volatility models. These models were introduced to explain the smile in the implied volatility. An example of such is the Heston (1993) model, in which the stock price $S(t)$ and the volatility $v(t)$ satisfy the following SDEs under the EMM,

$$
\begin{aligned}
dS(t) &= rS(t)dt + \sqrt{v(t)}S(t)dB(t) \\
dv(t) &= \alpha(\mu - v(t))dt + \delta\sqrt{v(t)}\,dW(t), \tag{11.47}
\end{aligned}
$$

where the Brownian motions B and W are correlated. Stochastic volatility models are incomplete, see Example 8.26. Therefore a replicating portfolio involves a stock and another option. Pricing by using an EMM or replicating by a portfolio with a stock and another option, leads to a PDE for the price of an option, by using the Feynman-Kac formula with a two-dimensional diffusion process, see Fouque *et al.* (2000), p.45. Heston (1993) derived a PDE and its solution for the transform of the option price. The price itself is obtained by the inversion of the transform.

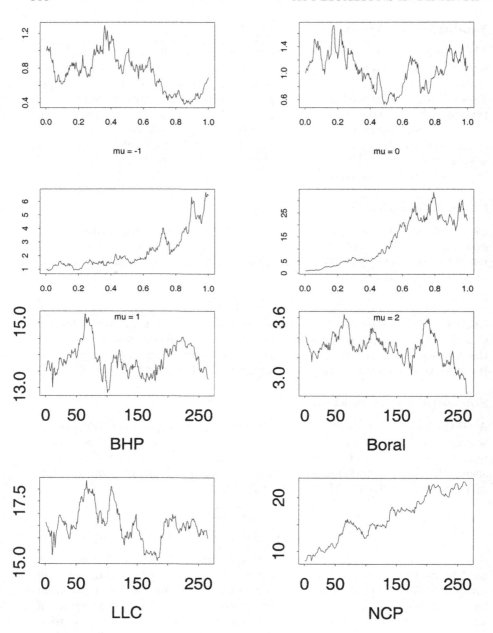

Figure 11.1: Simulated and real stock prices.

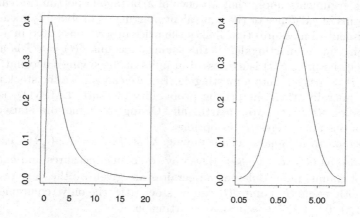

Figure 11.2: Lognormal LN(0,1) and Normal N(0,1) densities.

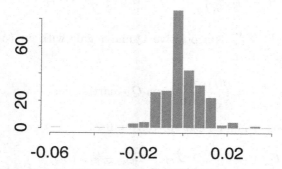

Figure 11.3: Histogram of BHP daily returns over 30 days.

11.5   Change of Numeraire

A numeraire is the asset in which values of other assets are measured. The no-arbitrage arguments imply that absence of arbitrage opportunities can be stated in terms of existence of the equivalent martingale probability measure, EMM, (risk-neutral measure) that makes the price process measured in a chosen numeraire into a martingale. If the savings account $\beta(t) = e^{rt}$ is a numeraire, then the price $S(t)$ is expressed in units of the saving's account, and Q makes $Z(t) = S(t)/e^{rt}$ into a martingale. But we can choose the stock price to be the numeraire, then the pricing probability measure, EMM Q_1, is the one that makes $\beta(t)/S(t)$ into a martingale. Change of numeraire is used for currency options and interest rates options.

The price of an attainable claim paying X at T, $C = E_Q(\frac{X}{\beta(T)})$ can be calculated also as $C = E_{Q_1}(\frac{X}{S(T)})$. The way to change measures and express the prices of claims under different numeraires are given by the next results. Note that both assets $S(t)$ and $\beta(t)$ can be stochastic, the only requirement is that $S(t)/\beta(t)$, $0 \le t \le T$ is a positive martingale.

Theorem 11.17 (Change of numeraire) *Let $S(t)/\beta(t)$, $0 \le t \le T$ be a positive Q-martingale. Define Q_1 by*

$$\frac{dQ_1}{dQ} = \Lambda(T) = \frac{S(T)/S(0)}{\beta(T)/\beta(0)}. \tag{11.48}$$

Then under Q_1, $\beta(t)/S(t)$ is a martingale. Moreover, the price of an attainable claim X at time t is related under the different numeraires by the formula

$$C(t) = E_Q(\frac{\beta(t)}{\beta(T)}X|\mathcal{F}_t) = E_{Q_1}(\frac{S(t)}{S(T)}X|\mathcal{F}_t). \tag{11.49}$$

PROOF: $\Lambda(t) = \frac{S(t)/S(0)}{\beta(t)/\beta(0)}$ is a positive Q-martingale with $E_Q(\Lambda(T)) = 1$. Therefore by Theorem 10.12

$$\frac{1}{\Lambda(t)} = \frac{\beta(t)/\beta(0)}{S(t)/S(0)} \quad \text{is a } Q_1\text{-martingale.} \tag{11.50}$$

By the general Bayes formula Theorem 10.10 (10.23)

$$E_Q(\frac{\beta(t)}{\beta(T)}X|\mathcal{F}_t) = \frac{E_{Q_1}(\frac{1}{\Lambda(T)}\frac{\beta(t)}{\beta(T)}X|\mathcal{F}_t)}{E_{Q_1}(\frac{1}{\Lambda(T)}|\mathcal{F}_t)} = \frac{E_{Q_1}(\frac{\beta(t)}{S(T)}X|\mathcal{F}_t)}{\frac{1}{\Lambda(t)}} = E_{Q_1}(\frac{S(t)}{S(T)}X|\mathcal{F}_t).$$

$\square$

A General Option Pricing Formula

As a corollary we obtain the price of a call option in a general setting.
$E_Q\big(\frac{1}{\beta(T)}(S(T) - K)^+\big) = E_Q\big(\frac{S(T)}{\beta(T)}I(S(T) > K)\big) - E_Q\big(\frac{K}{\beta(T)}I(S(T) > K)\big)$.
The first term is evaluated by changing the numeraire,
$E_Q\big(\frac{S(T)}{\beta(T)}I(S(T) > K)\big) = E_{Q_1}\big(\frac{1}{\Lambda_T}\frac{S(T)}{\beta(T)}I(S(T) > K)\big) = \frac{\beta(0)}{S(0)}Q_1(S(T) > K)$.
The second term is $KQ(S(T) > K)/\beta(T)$, when $\beta(t)$ is deterministic. Thus

$$C = \frac{\beta(0)}{S(0)}Q_1(S(T) > K) - \frac{K}{\beta(T)}Q(S(T) > K). \qquad (11.51)$$

This is a generalization of the Black-Scholes formula (Geman *et al.* (1995), Björk (1998), Klebaner (2002)). One can verify that the Black-Scholes formula is also obtained when the stock is used as the numeraire (Exercise 11.12).

If $\beta(t)$ is stochastic, then by a using T-bond as a numeraire,

$$E_Q\left(\frac{K}{\beta(T)}I(S(T) > K)\right) = P(0,T)KQ_T(S(T) > K),$$

where $P(0,T) = E_Q(1/\beta(T))$, $\Lambda(T) = \frac{1}{P(0,T)\beta(T)}$. To evaluate expectations above, we need to find the distribution under the new measure. This can be done by using the SDE under Q_1.

SDEs under a Change of Numeraire

Let $S(t)$ and $\beta(t)$ be positive processes. Let Q and Q_1 be respectively the equivalent martingale measures when $\beta(t)$ and $S(t)$ are numeraires, so that $S(t)/\beta(t)$ is a Q-martingale, and $\beta(t)/S(t)$ a Q_1-martingale. Suppose that under Q

$$\begin{aligned} dS(t) &= \mu_S(t)dt + \sigma_S(t)dB(t), \\ d\beta(t) &= \mu_\beta(t)dt + \sigma_\beta(t)dB(t), \end{aligned}$$

where B is a Q-Brownian motion, and coefficients are adapted processes. Let $X(t)$ have the SDEs under Q and Q_1, with a Q_1-Brownian motion B^1

$$\begin{aligned} dX(t) &= \mu_0(X(t))dt + \sigma(X(t))dB(t), \\ dX(t) &= \mu_1(X(t))dt + \sigma(X(t))dB^1(t), \end{aligned}$$

Since these measures are equivalent, the diffusion coefficient for X is the same, and the drift coefficients are related by

Theorem 11.18

$$\mu_1(X(t)) = \mu_0(X(t)) + \sigma(X(t))\left(\frac{\sigma_S(t)}{S(t)} - \frac{\sigma_\beta(t)}{\beta(t)}\right). \qquad (11.52)$$

PROOF: We know from Section 10.3, that Q_1 and Q are related by

$$\Lambda(T) = \frac{dQ_1}{dQ} = \mathcal{E}\left(\int_0^{\cdot} H(t)dB(t)\right)(T),$$

with $H(t) = \frac{\mu_1(X(t)) - \mu_0(X(t))}{\sigma(X(t))}$. On the other hand, by Theorem 11.17 $\frac{dQ_1}{dQ} = \Lambda(T) = \frac{S(T)/S(0)}{\beta(T)/\beta(0)}$. Moreover $\Lambda(t) = \mathrm{E}_Q(\Lambda(T)|\mathcal{F}_t) = \frac{S(t)/S(0)}{\beta(t)/\beta(0)}$ by the martingale property of (S/β). Using the exponential SDE,

$$d\Lambda(t) = \Lambda(t)H(t)dB(t) = d\left(\frac{S(t)/S(0)}{\beta(t)/\beta(0)}\right),$$

it follows that

$$\frac{\mu_1(X(t)) - \mu_0(X(t))}{\sigma(X(t))}dB(t) = \frac{\beta(t)}{S(t)}d\left(\frac{S(t)}{\beta(t)}\right).$$

Using the SDEs for S and β, and that S/β has no dt term, the result follows.
$\square$

Corollary 11.19 *The SDE for $S(t)$ under Q_1, when it is a numeraire, is*

$$dS(t) = \left(\mu_S(t) + \frac{\sigma_S^2(t)}{S(t)}\right)dt + \sigma_S(t)dB^1(t), \qquad (11.53)$$

where $B^1(t)$ is a Q_1-martingale. Moreover, in the Black-Scholes model when $\mu_S = \mu S$, and $\sigma_S = \sigma S$, the new drift coefficient is $(\mu + \sigma^2)S$.

For other results on change of numeraire see Geman *et al.* (1995).

11.6   Currency (FX) Options

Currency options involve at least two markets, one domestic and foreign markets. For simplicity we consider just one foreign market. For details on arbitrage theory in Foreign Market Derivatives see Musiela and Rutkowski (1998).

The foreign and domestic interest rates in riskless accounts are denoted by r_f and r_d, (subscripts f and d denoting foreign and domestic respectively). The EMM's exist in both markets, Q_f and Q_d. Let $U(t)$ denote the value of a foreign asset in foreign currency at time t, say JPY. Assume that $U(t)$ evolves according the Black-Scholes model, and write its equation under the EMM (risk-neutral) measure,

$$dU(t) = r_f U(t)dt + \sigma_U U(t)dB_f(t), \qquad (11.54)$$

where B_f is a Q_f-Brownian motion. A similar equation holds for assets in the domestic market under its EMM Q_d. The price process discounted by $e^{r_d t}$ in the domestic market is a Q_d-martingale, and an option on an asset in the domestic market which pays $C(T)$ (in domestic currency) at T has its time t price

$$C(t) = e^{-r_d(T-t)} E_{Q_d}(C(T)|\mathcal{F}_t).$$

A link between the two markets is provided by the exchange rate. Take $X(t)$ to be the price in the domestic currency of 1 unit of foreign currency at time t. Note that $X(t)$ itself is not an asset, but it gives the price of a foreign asset in domestic currency when multiplied by the price of the foreign asset. We assume that $X(t)$ follows the Black-Scholes SDE under the domestic EMM Q_d

$$dX(t) = \mu_X X(t)dt + \sigma_X X(t)dB_d(t),$$

and show that μ_X must be $r_d - r_f$. Take the foreign savings account $e^{r_f t}$, its value in domestic currency is $X(t)e^{r_f t}$. Thus $e^{-r_d t}X(t)e^{r_f t}$ should be a Q_d-martingale, as a discounted price process in the domestic market. But from (11.6)

$$e^{-r_d t}X(t)e^{r_f t} = S(0)e^{(r_f - r_d + \mu_X - \frac{1}{2}\sigma_X^2)t + \sigma_X B_d(t)},$$

implying $\mu_X = r_d - r_f$. Thus the equation for $X(t)$ under the domestic EMM Q_d is

$$dX(t) = (r_d - r_f)X(t)dt + \sigma_X X(t)dB_d(t), \tag{11.55}$$

with a Q_d-Brownian motion $B_d(t)$. In general, the Brownian motions B_d and B_f are correlated, i.e. for some $|\rho| < 1$, $B_f(t) = \rho B_d(t) + \sqrt{1 - \rho^2}W(t)$, with independent Brownian motions B_d and W. $d[B_d, W](t) = \rho dt$. (The model can also be set up by using two-dimensional independent Brownian motions, a (column) vector Q_d-Brownian motion $[B_d, W]$, and a matrix σ_U, the row vector $\sigma_U = [\rho, \sqrt{1 - \rho^2}]$).

The value of the foreign asset in domestic currency at time t is given by $U(t)X(t) = \tilde{U}(t)$. Therefore $e^{-r_d t}\tilde{U}(t)$, as a discounted price process, is a Q_d-martingale. One can easily see from (11.54) and (11.55) that

$$e^{-r_d t}\tilde{U}(t) = X(0)U(0)e^{-\frac{1}{2}(\sigma_U^2 + \sigma_X^2)t + \sigma_U B_f(t) + \sigma_X B_d(t)}.$$

For it to be a Q_d-martingale, it must hold $dB_f(t) + \rho\sigma_X dt = d\tilde{B}_d(t)$ is a Q_d-Brownian motion (correlated with B_d). This is accomplished by Girsanov's theorem, and the SDE for $U(t)$ under the domestic EMM Q_d becomes

$$dU(t) = (r_f - \rho\sigma_U\sigma_X)U(t)dt + \sigma_U U(t)d\tilde{B}_d(t). \tag{11.56}$$

Now it is also easy to see that the process $\tilde{U}(t)$ (the foreign asset in domestic currency) under the domestic EMM Q_d has the SDE

$$d\tilde{U}(t) = r_d\tilde{U}(t)dt + (\sigma_U d\tilde{B}_d(t) + \sigma_X dB_d(t))\tilde{U}(t). \tag{11.57}$$

Since
$$\sigma_U d\tilde{B}_d(t) + \sigma_X dB_d(t) = \bar{\sigma} d\tilde{W}_d(t),$$
for some Q_d-Brownian motion $\tilde{W}_d(t)$, with
$$\bar{\sigma}^2 = \sigma_U^2 + \sigma_X^2 + 2\rho\sigma_U\sigma_X, \qquad (11.58)$$

the volatility of $\tilde{U}(t)$ is given by the formula that combines the foreign asset volatility and the exchange rate volatility (11.58).

Options on Foreign Currency

Consider a call option on one unit of foreign currency, say on JPY with the strike price K_d in AUD. Its payoff at T in AUD is $(X(T) - K_d)^+$. Thus its price at time t is given by

$$C(t) = e^{-r_d(T-t)} E_{Q_d}((X(T) - K_d)^+|\mathcal{F}_t).$$

The conditional Q_d-distribution of $X(T)$ given $X(t)$ is Lognormal, by using (11.6), and standard calculations give the Black-Scholes currency formula (Garman-Kohlhagen (1983))

$$C(t) = X(t)e^{-r_f(T-t)}\Phi(h(t)) - K_d e^{-r_d(T-t)}\Phi(h(t) - \sigma_X\sqrt{T-t}),$$

$$h(t) = \frac{\ln\frac{X(t)}{K_d} + (r_d - r_f + \sigma_S^2/2)(T-t)}{\sigma_X\sqrt{T-t}}.$$

Other options on currency are priced similarly, by using equation (11.6).

Options on Foreign Assets Struck in Foreign Currency

Consider options on a foreign asset denominated in the domestic currency. A call option pays at time T the amount $X(T)(U(T) - K_f)^+$. Since the amount is specified in the domestic currency the pricing should be done under the domestic EMM Q_d. The following argument achieves the result. In the foreign market options are priced by the Black-Scholes formula. The price in domestic currency is obtained by multiplying by $X(t)$. A call option is priced by

$$C(t) = U(t)\left(\Phi(h(t)) - K_f e^{-r_f(T-t)}\Phi\left(h(t) - \sigma_U\sqrt{T-t}\right)\right),$$

where

$$h(t) = \frac{\ln\frac{U(t)}{K_f} + (r_f + \frac{1}{2}\sigma_U^2)(T-t)}{\sigma_U\sqrt{T-t}}.$$

Calculations of the price under Q_d are left as an exercise in change of numeraire, Exercise 11.13.

Options on Foreign Assets Struck in Domestic Currency

Consider a call option on the foreign asset $U(t)$ with strike price K_d in AUD. Its payoff at T in AUD is $(X(T)U(T) - K_d)^+ = (\tilde{U}(T) - K_d)^+$. Thus its price at time t is given by

$$C(t) = e^{-r_d(T-t)}E_{Q_d}((\tilde{U}(T) - K_d)^+|\mathcal{F}_t).$$

The conditional distribution of $\tilde{U}(T)$ given $\tilde{U}(t)$ is Lognormal, by equation (11.57), with volatility given by $\bar{\sigma}^2 = \sigma_U^2 + \sigma_X^2 + 2\rho\sigma_U\sigma_X$ from (11.58). Thus

$$C(t) = \tilde{U}(t)\Phi(h(t)) - K_d e^{-r_d(T-t)}\Phi(h(t) - \bar{\sigma}\sqrt{T-t}),$$

$$h(t) = \frac{\ln\frac{\tilde{U}(t)}{K_d} + (r_d + \frac{\bar{\sigma}^2}{2})(T-t)}{\bar{\sigma}\sqrt{T-t}}.$$

Guaranteed Exchanged Rate (Quanto) Options

A quanto option pays in domestic currency the foreign payoff converted at a predetermined rate. For example a quanto call pays at time T the amount $X(0)(U(T) - K_f)^+$, (K_f is in foreign currency). Thus its time t value is given by

$$C(t) = X(0)e^{-r_d(T-t)}E_{Q_d}((U(T) - K_f)^+|\mathcal{F}_t).$$

The conditional distribution of $U(T)$ given $U(t)$ is Lognormal, by equation (11.56), with volatility σ_U. Standard calculations give

$$C(t) = X(0)e^{-r_d(T-t)}\left(U(t)e^{\delta(T-t)}\Phi(h(t)) - K_f\Phi(h(t) - \sigma_U\sqrt{T-t})\right),$$

where $\delta = r_f - \rho\sigma_U\sigma_X$ and $h(t) = \frac{\ln\frac{U(t)}{K_f} + (\delta + \frac{\sigma_U^2}{2})(T-t)}{\sigma_U\sqrt{T-t}}$.

11.7 Asian, Lookback and Barrier Options

Asian Options

We assume that $S(t)$ satisfies the Black-Scholes SDE, which we write under the EMM Q

$$dS(t) = rS(t)dt + \sigma S(t)dB(t).$$

Asian options pay at time T, the amount $C(T)$ given by $\left(\frac{1}{T}\int_0^T S(u)du - K\right)^+$ (fixed strike K) or $\left(\frac{1}{T}\int_0^T S(t)dt - KS(T)\right)^+$ (floating strike $KS(T)$). Both kinds involve the integral average of the stock price $\bar{S} = \frac{1}{T}\int_0^T S(u)du$. The average of Lognormals is not an analytically tractable distribution, therefore direct calculations do not lead to a closed form solution for the price of the option. Pricing by using PDEs was done by Rogers and Shi (1995), and Vecer (2002). We present Vecer's approach, which rests on the idea that it is possible to replicate the integral average of the stock $\bar{S}$ by a self-financing portfolio. Let

$$
\begin{aligned}
a(t) &= \frac{1}{rT}(1 - e^{-r(T-t)}) \text{ and} \\
dV(t) &= a(t)dS(t) + r(V(t) - a(t)S(t))dt \\
&= rV(t)dt + a(t)(dS(t) - rS(t)dt), \quad (11.59)
\end{aligned}
$$

with $V(0) = a(0)S(0) = \frac{1}{rT}(1 - e^{-rT})S(0)$. It is easy to see that $V(t)$ is a self-financing portfolio (Exercise 11.14). Solving the SDE (11.59) (by looking at $d(V(t)e^{-rt})$) for $V(t)$ and using integration by parts, we obtain

$$
\begin{aligned}
V(T) &= e^{rT}V(0) + a(T)S(T) - e^{rT}a(0)S(0) - \int_0^T e^{r(T-t)}S(t)da(t) \\
&= \frac{1}{T}\int_0^T S(t)dt, \quad (11.60)
\end{aligned}
$$

because of

$$
d(e^{r(T-t)}S(t)a(t)) = e^{r(T-t)}a(t)dS(t) - re^{r(T-t)}S(t)dt + e^{r(T-t)}S(t)da(t), \text{ and}
$$

$$
a(T)S(T) - e^{rT}a(0)S(0) = \int_0^T e^{r(T-t)}a(t)(dS(t) - rS(t)dt) + \int_0^T e^{r(T-t)}S(t)da(t).
$$

The self-financing portfolio $V(t)$ consists of $a(t)$ shares and $b(t) = \frac{e^{-rT}}{T}\int_0^t S(u)du$ cash and has the value $\bar{S}$ at time T.

Consider next pricing the option with the payoff $(\bar{S} - K_1 S(T) - K_2)^+$, which encompasses both the fixed and the floating kinds (by taking one of K's as 0). To replicate such an option hold at time t $a(t)$ of the stock, start with initial wealth $V(0) = a(0)S(0) - e^{-rT}K_2$ and follow the self-financing strategy (11.59). The terminal value of this portfolio is $V(T) = \bar{S} - K_2$. The payoff of the option is

$$
(\bar{S} - K_1 S(T) - K_2)^+ = (V(T) - K_1 S(T))^+.
$$

Thus the price of the option at time t is

$$
C(t) = e^{-r(T-t)}E_Q(V(T) - K_1 S(T))^+|\mathcal{F}_t).
$$

Let $Z(t) = V(t)/S(t)$. Then proceeding,

$$
C(t) = e^{-r(T-t)}E_Q(S(T)(Z(T) - K_1)^+|\mathcal{F}_t).
$$

The multiplier $S(T)$ inside is dealt with by the change of numeraire, and the remaining expectation is that of a diffusion $Z(t)$, which satisfies the backward PDE. Using $S(t)$ as a numeraire (Theorem 11.17, (11.49)) we have

$$C(t) = S(t)\mathrm{E}_{Q_1}((Z_T - K)^+|\mathcal{F}_t). \tag{11.61}$$

By Ito's formula the SDE for $Z(t)$ under Q, when e^{rt} is a numeraire, is

$$dZ(t) = \sigma^2(Z(t) - a(t))dt + \sigma(a(t) - Z(t))dB(t).$$

By Theorem 11.18 the SDE for $Z(t)$ under the EMM Q_1, when $S(t)$ is a numeraire, is

$$dZ(t) = -\sigma(Z(t) - q(t))dB^1(t),$$

where $B^1(t)$ is a Q_1-Brownian motion, $(B^1(t) = B(t) + \sigma t)$. Now we can write a PDE for the conditional expectation in (11.61).

$$\mathrm{E}_{Q_1}((Z(T) - K)^+|\mathcal{F}_t) = \mathrm{E}_{Q_1}((Z(T) - K_1)^+|Z(t)),$$

by the Markov property. Hence

$$u(x, t) = \mathrm{E}_{Q_1}((Z(T) - K_1)^+|Z(t) = x)$$

satisfies the PDE (see Theorem 6.6)

$$\frac{\sigma^2}{2}(x - a(t))^2\frac{\partial^2 u}{\partial x^2} + \frac{\partial u}{\partial t} = 0,$$

subject to the boundary condition $u(x, T) = (x - K_1)^+$. Finally, the price of the option at time zero is given by using (11.61)

$$V(0, S(0), K_1, K_2) = S(0)u(0, Z(0)),$$

with $Z(0) = V(0)/S(0) = \frac{1}{rT}(1 - e^{-rT}) - e^{-rT}K_2/S(0)$.

Lookback Options

A lookback call pays $X = S(T) - S_*$ and a lookback put $X = S^* - S(T)$, where S_* and S^* denote the smallest and the largest values of stock on $[0, T]$. The price of a lookback put is given by

$$C = e^{-rT}\mathrm{E}_Q(S^* - S(T)) = e^{-rT}\mathrm{E}_Q(S^*) - e^{-rT}\mathrm{E}_Q(S(T)), \tag{11.62}$$

where Q is the martingale probability measure.
Since $S(t)e^{-rt}$ is a Q-martingale, $e^{-rT}\mathrm{E}_Q(S(T)) = S(0)$. To find $\mathrm{E}_Q(S^*)$, the Q-distribution of S^* is needed. The equation for $S(t)$ is given by $S(t) =$

$S(0)e^{(r-\frac{1}{2}\sigma^2)t+\sigma B(t)}$ with a Q-Brownian motion $B(t)$, see (11.31). Clearly, S^* satisfies

$$S^* = S(0)e^{\left((r-\frac{1}{2}\sigma^2)t+\sigma B(t)\right)^*}.$$

The distribution of maximum of a Brownian motion with drift is found by a change of measure, see Exercise 10.6 and Example 10.4 (10.38). By the Girsanov's theorem $(r - \frac{1}{2}\sigma^2)t + \sigma B(t) = \sigma W(t)$, for a Q_1-Brownian motion $W(t)$, with $\frac{dQ}{dQ_1}(W_{[0,T]}) = e^{cW(T)-\frac{1}{2}c^2T}$, $c = (r - \frac{1}{2}\sigma^2)/\sigma$. Clearly, $S^* = S(0)e^{W^*}$. The distribution of W^* under Q_1 is the distribution of the maximum of Brownian motion, and its distribution under Q is obtained as in Example 10.4 (see (10.38)). Therefore

$$e^{-rT}\mathrm{E}_Q(S^*) = e^{-rT}S(0)\int_{-\infty}^{\infty} e^{\sigma y}f_{W^*}(y)dy, \qquad (11.63)$$

where $f_{W^*}(y)$ is obtained from (10.38) (see also Exercise 11.15). Lookback calls are priced similarly. The price of a Lookback call is given by $S(0)$ times

$$(1+\frac{\sigma^2}{2r})\Phi\left(\frac{\ln(K/S(0)) + \frac{\sigma^2 T}{2}}{\sigma\sqrt{T}}\right) - (1-\frac{\sigma^2}{2r})e^{-rT}\Phi\left(\frac{\ln(K/S(0)) - \frac{\sigma^2 T}{2}}{\sigma\sqrt{T}}\right) - \frac{\sigma^2}{2r}.$$
$$(11.64)$$

Barrier Options

Examples of Barrier Options are down-and-out calls or down-and-in calls. There are also corresponding put options. A down-and-out call gets knocked out when the price falls below a certain level H, it has the payoff $(S(T) - K)^+I(S_* \geq H)$, $S(0) > H$, $K > H$. A down-and-in call has the payoff $(S(T) - K)^+I(S_* \leq H)$. To price these options the joint distribution of $S(T)$ and S_* is needed under Q. For example, the price of down-and-in call is given by

$$
\begin{aligned}
C &= e^{-rT}\mathrm{E}_Q((S(T) - K)^+I(S_* \leq H)) \\
&= e^{-rT}\int_{\frac{\ln(K/S(0))}{\sigma}}^{\infty}\int_{-\infty}^{\frac{\ln(H/S(0))}{\sigma}} \left(S(0)e^{\sigma x} - K\right)g(x,y)dxdy,
\end{aligned}
$$

where $g(x,y)$ is the probability density of $(W(T), W_*(T))$ under the martingale measure Q. It is found by changing measure as described above and in Example 10.4 (see also Exercise 11.16). Double barrier options have payoffs depending on $S(T)$, S_* and S^*. An example is a call which pays only if the price never goes below a certain level H_1 or above a certain level H_2 with the payoff

$$X = (S(T) - K)^+I(H_1 < S_* < S^* < H_2).$$

The joint distribution of Brownian motion and its minimum and maximum is given by Theorem 3.23. By using the change of measure described above, the joint distribution of $S(T), S_*, S^*$ can be found and double barrier options can be priced. Calculations are based on the following result.

Let $W_\mu(t) = \mu t + B(t)$. Then the probability of hitting barriers a and b on the interval $[0, t]$, given the values at the end points $W_\mu(0) = x_0$ and $W_\mu(t) = x_1$

$$P(a < \min_{0 \leq t \leq t} W_\mu(t) \leq \max_{0 \leq t \leq t} W_\mu(t) \leq b | W_\mu(0) = x, W_\mu(t) = y),$$

does not depend on μ, and is given by (e.g. Borodin and Salminen (1996))

$$P(a, b, x, y, t) = \exp\left\{\frac{(y-x)^2}{2t}\right\} \times \tag{11.65}$$

$$\sum_{n=-\infty}^{\infty} \left(\exp\left\{\frac{(y-x+2n(b-a))^2}{-2t}\right\} - \exp\left\{\frac{(y-x+2n(b-a)+2(x-a))^2}{-2t}\right\}\right)$$

if $a \leq x \leq b$, $a \leq y \leq b$, and zero otherwise. This formula becomes simpler when one of the barriers is infinite, namely if $a = -\infty$ (single high barrier) we get

$$P(-\infty, b, x, y, t) = 1 - \exp\left\{-2\frac{(b-x)(b-y)}{t}\right\} \tag{11.66}$$

if $x \leq b$, $y \leq b$, and zero otherwise. If $b = \infty$ (single low barrier) we get

$$P(a, \infty, x, y, t) = 1 - \exp\left\{-2\frac{(a-x)(a-y)}{t}\right\} \tag{11.67}$$

if $a \leq x$, $a \leq y$, and zero otherwise.

11.8 Exercises

Exercise 11.1: (Payoff Functions and Diagrams)
Graph the following payoffs.

1. A *straddle* consists of buying a call and a put with the same exercise price and expiration date.

2. A *butterfly spread* consists of buying two calls with exercises prices K_1 and K_3 and selling a call with exercise price K_2, $K_1 < K_2 < K_3$.

Exercise 11.2: (Binomial Pricing Model)

1. Give the probability space for the Binomial model.

2. Show that the stock price can be expressed as $S(t+1) = S(t)\xi_{t+1}$, where ξ_{t+1} is the return in period $t + 1$, $t = 0, 1, \ldots, n - 1$, and the variables ξ_n are independent and identically distributed with values u and d.

3. Show that the Binomial model does not admit arbitrage if and only if $d < r < u$.

4. Describe the arbitrage-free probability Q, and show that the discounted stock price S_t/r^t, $t = 0, 1 \ldots, n$ is a Q-martingale.
   5. Show that this model is complete.

5. Show that if the price of an option is given by (11.7), then arbitrage strategies do not exist.

Exercise 11.3: Verify that in the model given in Example 11.7 any attainable claim has the same price under any of the martingale measures. Give an example of an unattainable claim X and show that $\mathrm{E}_Q(X)$ is different for different martingale measures Q.

Exercise 11.4: Show that if Q is equivalent to P and $X \geq 0$, then $\mathrm{E}_P(X) > 0$ implies $\mathrm{E}_Q(X) > 0$, and vice versa.

Exercise 11.5: (Pricing in incomplete markets)

1. Show that if $M(t)$, $0 \leq t \leq T$, is a martingale under two different probability measures Q and P, then for $s < t$ $\mathrm{E}_Q(M(t)|\mathcal{F}_s) = \mathrm{E}_P(M(t)|\mathcal{F}_s)$ a.s. If in addition $M(0)$ is non-random, then $\mathrm{E}_P M(t) = \mathrm{E}_Q M(t)$.

2. Show that the price of an attainable claim X, $C(t) = \beta(t)\mathrm{E}_Q(X/\beta(T)|\mathcal{F}_t)$ is the same for all martingale measures.

Exercise 11.6: (Non-completeness in mixed models.)
In this exercise the price of an asset is modelled as a sum of a diffusion and a jump process. Take for example $X(t) = W(t) + N(t)$, with Brownian motion W and Poisson process N. Give at least two equivalent probability measures Q_1 and Q_2, such that X is a Q_i-martingale, $i = 1, 2$ (see Exercise 10.13).

Exercise 11.7: Give the Radon-Nikodym derivative Λ in the change to the EMM Q in the Black-Scholes model.

Exercise 11.8: A way to derive a PDE for the option price is based on the fact that $C(t)e^{-rt} = V(t)e^{-rt}$ is a Q-local martingale. Obtain the Black-Scholes PDE for the price of the option using the Black-Scholes model for the stock price.
Hint: expand $d(C(S(t), t)e^{-rt})$ and equate the coefficient of dt to zero.

Exercise 11.9: Derive the PDE for the price of the option in Heston's model.

Exercise 11.10: Show that the expected return on stock under the martingale probability measure Q is the same as on the riskless asset. This is the reason why the martingale measure Q is also called "risk-neutral" probability measure.

Exercise 11.11: Assume $S(t)$ evolves according the Black-Scholes model. Show that under the EMM Q_1, when $S(t)$ is a numeraire, $d(e^{rt}/S(t)) = \sigma(e^{rt}/S(t))dW_t$, where $W(t) = B(t) - \sigma t$ is a Q_1-Brownian motion. Give the Likelihood dQ_1/dQ. Give the SDE for $S(t)$ under Q_1.

Exercise 11.12: Derive the Black-Scholes formula by using the stock price as the numeraire.

Exercise 11.13: A call option on an asset in the foreign market pays at time T, $S(T)(U(T) - K)^+$ in the domestic currency, and its time t price $C(t) = e^{-r_d(T-t)}E_{Q_d}(S(T)(U(T) - K)^+|\mathcal{F}_t)$. Taking numeraire based on the Q_d-martingale $S(t)e^{-(r_d-r_f)t}$ obtain the formula for $C(t)$.

Exercise 11.14: Let $V(t) = a(t)S(t) + b(t)e^{rt}$ be a portfolio, $0 \leq t \leq T$. Show that it is self-financing if and only if $dV(t) = a(t)dS(t) + r(V(t) - a(t)S(t))dt$.

Exercise 11.15: Derive the price of a Lookback call.

Exercise 11.16: Show that the price of a down-and-in call is given by

$$e^{-rT}\left(\frac{H}{S(0)}\right)^{2r/\sigma^2-1}\left(F\Phi\left(\frac{\ln(F/K)+\frac{\sigma^2T}{2}}{\sigma\sqrt{T}}\right) - K\Phi\left(\frac{\ln(F/K)-\frac{\sigma^2T}{2}}{\sigma\sqrt{T}}\right)\right),$$

where $F = e^{rT}H^2/S(0)$.

Exercise 11.17: Assume that $S(T)/S$ is does not depend on S, where $S(T)$ is the price of stock at T and $S = S(0)$. Let T be the exercise time and K the exercise price of the call option. Show that the price of this option satisfies the following PDE

$$C = S\frac{\partial C}{\partial S} + K\frac{\partial C}{\partial K}.$$

You may assume all the necessary differentiability. Hence show that the delta of the option $\frac{\partial C}{\partial S}$ in the Black-Scholes model is given by $\Phi(h(t))$ with $h(t)$ given by (11.35).

Chapter 12

Applications in Finance: Bonds, Rates and Options

Money invested for different terms T yield a different return corresponding to the rate of interest $R(T)$. This function is called the yield curve, or the term structure of interest rates. Every day this curve changes, the time t curve denoted by $R(t, T)$. However, the rates are not traded directly, they are derived from prices of bonds traded on the bond market. This leads to construction of models for bonds and no-arbitrage pricing for bonds and their options. We present the main models used in the literature and in applications, treating in detail the Merton, Vasicek's, Heath-Jarrow-Morton (HJM) and Brace-Gatarek-Musiela (BGM) models. In our treatment we concentrate on the main mathematical techniques used in such models without going into details of their calibration.

12.1 Bonds and the Yield Curve

A \$1 bond with maturity T is a contract that guarantees the holder \$1 at T. Sometimes bonds also pay a certain amount, called a coupon, during the life of the bond, but for the theory it suffices to consider only bonds without coupons (zero-coupon bonds). Denote by $P(t, T)$ the price at time t of the bond paying \$1 at T, $P(T, T) = 1$. The yield to maturity of the bond is defined as

$$R(t, T) = -\frac{\ln P(t, T)}{T - t}, \qquad (12.1)$$

and as a function in T, is called the yield curve at time t. Assume also that a savings account paying at t instantaneous rate $r(t)$, called the spot (or short) rate, is available. \$1 invested until time t will result in

$$\beta(t) = e^{\int_0^t r(s)ds}. \tag{12.2}$$

To avoid arbitrage between bonds and savings account, a certain relation must hold between bonds and the spot rate. If there were no uncertainty, then to avoid arbitrage the following relation must hold

$$P(t,T) = e^{-\int_t^T r(s)ds}, \tag{12.3}$$

since investing either of these amounts at time t results in \$1 at time T. When the rate is random, then $\int_t^T r(s)ds$ is also random and in the future of t, whereas the price $P(t,T)$ is known at time t, and the above relation holds only "on average", equation (12.5) below.

We assume a probability model with a filtration $\mathbb{F} = \{\mathcal{F}_t\}$, $0 \leq t \leq T^*$, and adapted processes $P(t,T)$, $t \leq T \leq T^*$, and $\beta(t)$. For the extension of the no-arbitrage theory see Artzner and Delbaen (1989), Lamberton and Lapeyre (1996), Björk (1998), Musiela and Rutkowski (1998), Shiryaev (1999). In addition to the number of no-arbitrage concepts in continuous time (see Section 11.3) the continuum of bond maturities T makes the market model have infinitely many assets and produces further complication. There are different approaches, including finite portfolios, where at each time only finitely many bonds are allowed, and infinite, measure-valued portfolios. In all of the approaches the no-arbitrage condition is formulated with the help of the following assumption.

EMM Assumption

There is a probability Q (called the equivalent martingale measure), equivalent to P (the original "real-world" probability), such that simultaneously for all $T \leq T^*$, the process in t, $P(t,T)/\beta(t)$ is a martingale , $0 \leq t \leq T$.

The martingale property implies that

$$\mathrm{E}_Q\left(\frac{1}{\beta(T)} \mid \mathcal{F}_t\right) = \mathrm{E}_Q\left(\frac{P(T,T)}{\beta(T)} \mid \mathcal{F}_t\right) = \frac{P(t,T)}{\beta(t)}, \tag{12.4}$$

where $\mathcal{F}_t$ denotes the information available from the bond prices up to time t. Since $P(T,T) = 1$, we obtain the expression for the price of the bond

$$P(t,T) = \mathrm{E}_Q\left(\frac{\beta(t)}{\beta(T)} \mid \mathcal{F}_t\right) = \mathrm{E}_Q\left(e^{-\int_t^T r(s)ds} \mid \mathcal{F}_t\right). \tag{12.5}$$

It shows that the bond can be seen as a derivative on the short rate.

12.2 Models Adapted to Brownian Motion

Here we derive the SDE for bond $P(t, T)$ under Q starting only with the EMM assumption. The SDE under P is then derived as a consequence of the predictable representation property. As usual, this property points out the existence of certain processes, but does not say how to find them.

Consider a probability space with a P-Brownian motion $W(t)$ and its filtration $\mathcal{F}_t$. Assume that the spot rate process $r(t)$ *generates* $\mathcal{F}_t$, and that the bond processes $P(t, T)$, for any $T < T^*$, are adapted. The EMM assumption gives the price of the bond by the equation (12.5). The martingale

$$\frac{P(t, T)}{\beta(t)} = \mathrm{E}_Q \left(e^{-\int_0^T r(s)ds} \mid \mathcal{F}_t \right) \tag{12.6}$$

is adapted to the Brownian filtration. By Theorem 10.19 there exists an adapted process $X(t) = \int_0^t \sigma(t, T)dB(t)$, where $B(t)$ is a Q-Brownian motion, such that

$$d\left(\frac{P(t, T)}{\beta(t)}\right) = \left(\frac{P(t, T)}{\beta(t)}\right)dX(t) = \sigma(t, T)\left(\frac{P(t, T)}{\beta(t)}\right)dB(t).$$

Opening $d(P/\beta)$, we obtain the SDE for $P(t, T)$ under the EMM Q

$$\frac{dP(t, T)}{P(t, T)} = r(t)dt + \sigma(t, T)dB(t). \tag{12.7}$$

This is the pricing equation for bonds and their options.

Note that the return on savings account satisfies $d\beta(t)/\beta(t) = r(t)dt$, and the return on bond has an extra term with a Brownian motion. This makes bonds $P(t, T)$ riskier than the (also random) savings account $\beta(t)$.

We find the SDE for the bond under the original probability measure P next. Since Q is equivalent to the Wiener measure, by Corollary 10.21 there is an adapted process $q(t)$, such that

$$W(t) = B(t) + \int_0^t q(s)ds \tag{12.8}$$

is a P-Brownian motion, with $dQ/dP = e^{\int_0^T q(t)dW(t) - \frac{1}{2}\int_0^T q^2(t)dt}$. Substituting $dB(t) = dW(t) - q(t)dt$ into SDE (12.7), we obtain the SDE under P

$$\frac{dP(t, T)}{P(t, T)} = \left(r(t) - \sigma(t, T)q(t)\right)dt + \sigma(t, T)dW(t). \tag{12.9}$$

Remark 12.1: It follows from equation (12.9) that $-q(t)$ is the excess return on the bond above the riskless rate, expressed in standard units; it is known as "the market price of risk" or "risk premium". Most common assumption is that $q(t) = q$ is a constant.

12.3   Models Based on the Spot Rate

A model for term structure and pricing can be developed from a model for the spot rate. These models are specified under the real world probability measure P, and $r(t)$ is assumed to satisfy

$$dr(t) = m(r(t))dt + \sigma(r(t))dW(t), \qquad (12.10)$$

where $W(t)$ is a P-Brownian motion, m and σ are functions of a real variable. The bond price $P(t,T)$ satisfies (12.5)

$$P(t,T) = E_Q\left(e^{-\int_t^T r(s)ds}\;\bigg|\;\mathcal{F}_t\right).$$

The expectation is under Q, and the model (12.10) is specified under P. Therefore the SDE for $r(t)$ under Q is needed. Note that we could express the above expectation in terms of E_P by

$$E_P\left(e^{-\int_t^t r(s)ds + \int_t^T q(s)dW(s) - \frac{1}{2}\int_t^T q^2(s)ds}\;\bigg|\;\mathcal{F}_t\right),$$

but this expectation seems to be untractable even in simple models.

We move between P and the EMM Q by using (12.8) expressed as $dB(t) = dW(t) - q(t)dt$. Thus under Q

$$dr(t) = (m(r(t)) + \sigma(r(t))q(t))dt + \sigma(r(t))dB(t). \qquad (12.11)$$

The process $r(t)$ is also a diffusion under Q, therefore by Markov property

$$E_Q\left(e^{-\int_t^T r(s)ds}\;\bigg|\;\mathcal{F}_t\right) = E_Q\left(e^{-\int_t^T r(s)ds}\;\bigg|\;r(t)\right).$$

The last expression satisfies a PDE by the Feynman-Kac formula (Theorem 6.8). Fix T, and denote by

$$C(x,t) = E_Q\left(e^{-\int_t^T r(s)ds}\;\bigg|\;r(t) = x\right),$$

then by (6.22) it satisfies

$$\frac{1}{2}\sigma^2(x)\frac{\partial^2 C}{\partial x^2}(x,t) + (m(x) + \sigma(x)q(t))\frac{\partial C}{\partial x}(x,t) + \frac{\partial C}{\partial t}(x,t) - xC(x,t) = 0,$$
$$(12.12)$$

with the boundary condition $C(x,T) = 1$. The price of the bond is obtained from this function by

$$P(t,T) = C(r(t),t).$$

A similar PDE with suitable boundary conditions holds for options on bonds.
We list some of the well-known models for the spot rate.
The Merton model

$$dr(t) = \mu dt + \sigma dW(t). \tag{12.13}$$

The Vasicek model

$$dr(t) = b(a - r(t))dt + \sigma dW(t). \tag{12.14}$$

The Dothan model

$$dr(t) = \mu r(t)dt + \sigma dW(t). \tag{12.15}$$

The Cox-Ingersoll-Ross (CIR) model

$$dr(t) = b(a - r(t))dt + \sigma\sqrt{r(t)}dW(t). \tag{12.16}$$

The Ho-Lee model

$$dr(t) = \mu(t)dt + \sigma dW(t). \tag{12.17}$$

The Black-Derman-Toy model

$$dr(t) = \mu(t)r(t)dt + \sigma(t)dW(t). \tag{12.18}$$

The Hull-White model

$$dr(t) = b(t)(a(t) - r(t))dt + \sigma(t)dW(t). \tag{12.19}$$

The Black-Karasinski model

$$dr(t) = r(t)(a(t) - b(t)\ln r(t))dt + \sigma(t)r(t)dW(t). \tag{12.20}$$

The functions $m(r)$ and $\sigma(r)$ involve parameters that need to be estimated.
They are chosen in such way that values of bonds and options agree as close
as possible with the values observed in the market. This process is called
calibration, and we don't address it.

In what follows we derive prices of bonds and their options for some models
by using probability calculations rather than solving PDEs.

12.4 Merton's Model and Vasicek's Model

Merton's Model

The spot rate in the Merton model satisfies SDE (12.13). Its solution is

$$r(t) = r_0 + \mu t + \sigma W(t). \tag{12.21}$$

The savings account is given by

$$\beta(t) = e^{\int_0^t r(s)ds} = e^{r_0 t + \mu t^2/2} e^{\sigma \int_0^t W_s ds}.$$

Assume the constant price for risk $q(t) = q$, then by (12.11) the SDE for $r(t)$ under Q is

$$dr(t) = (\mu + \sigma q)dt + \sigma dB(t).$$

The price of the bond $P(t,T)$ is given by $P(t,T) = \mathrm{E}_Q(e^{-\int_t^T r(s)ds} \mid \mathcal{F}_t)$. Since the conditional expectation given $\mathcal{F}_t$ is needed, use the decomposition of $r(s)$ into an $\mathcal{F}_t$-measurable part, and $\mathcal{F}_t$-independent part.

$$r(s) = r(t) + (\mu + \sigma q)(s-t) + \sigma(W(s) - W(t)) = r(t) + (\mu + \sigma q)(s-t) + \sigma \hat{W}(s-t),$$

with $\hat{W}(s-t)$ independent of $\mathcal{F}_t$. Then

$$\begin{aligned} P(t,T) &= e^{-r(t)(T-t) - (\mu + \sigma q)(T-t)^2/2} \mathrm{E}_Q(e^{-\sigma \int_t^T \hat{W}(s-t)ds} \mid \mathcal{F}_t) \\ &= e^{-r(t)(T-t) - (\mu + \sigma q)(T-t)^2/2 + \sigma^2 (T-t)^3/6}, \end{aligned} \qquad (12.22)$$

where we used that the distribution of the random integral $\int_0^{T-t} \hat{W}(u)du$ is $N(0, (T-t)^3/3)$, see Example 3.6. The yield curve is given by

$$R(t,T) = -\frac{\ln P(t,T)}{T-t} = r(t) + \frac{1}{2}(\mu + \sigma q)(T-t) - \frac{1}{6}\sigma^2 (T-t)^2.$$

Since $r(t)$ has a Normal distribution, so does $R(t,T)$, so that $P(t,T)$ is Log-normal. Pricing of a call option on the bond in this model is covered in the Exercise 12.2.

Note that the yields for different maturities differ by a deterministic quantity. Therefore $R(t,T_1)$ and $R(t,T_2)$ are perfectly correlated. This is taken as a shortcoming of the model.

Vasicek's Model

The spot rate in Vasicek's model satisfies SDE (12.14). Its solution is

$$r(t) = a - e^{-bt}(a - r(0)) + \sigma \int_0^t e^{-b(t-s)} dW(s). \qquad (12.23)$$

We derive the solution below, but note that it is easy to check that (12.23) is indeed a solution, see also the Langevin SDE Example 5.6, equation (5.15). Writing the SDE (12.14) in the integral form and taking expectations (it easy to see that the Itô integral has zero mean and interchanging the integral and the expectation is justified by Fubini's theorem), we have

$$\mathrm{E}r(t) - \mathrm{E}r(0) = \int_0^t b(a - \mathrm{E}r(s))ds. \tag{12.24}$$

Put $h(t) = \mathrm{E}r(t)$. Differentiating, we obtain

$$h'(t) = b(a - h(t)).$$

This equation is solved by separating variables. Integrating from 0 to t, and performing the change of variable $u = h(s)$ we obtain

$$\mathrm{E}r(t) = a - e^{-bt}(a - r(0)). \tag{12.25}$$

Let now

$$X(t) = r(t) - \mathrm{E}r(t) = r(t) - h(t). \tag{12.26}$$

$X(t)$, clearly, satisfies

$$dX(t) = dr(t) - dh(t) = -bX(t)dt + \sigma dW(t), \tag{12.27}$$

with the initial condition $X(0) = 0$. But this is the equation for the Ornstein-Uhlenbeck process. By (5.13), $X(t) = \sigma \int_0^t e^{-b(t-s)} dW(s)$, and (12.23) follows from the equations (12.26) and (12.25).

Make two observations next. First, the long-term mean is a,

$$\lim_{t \to \infty} \mathrm{E}r(t) = a.$$

Second, the process $X(t) = r(t) - \mathrm{E}r(t)$ reverts to zero, hence $r(t)$ reverts to its mean: if $r(t)$ is above its mean, then the drift is negative, making $r(t)$ decrease; and if $r(t)$ is below its mean, then the drift is positive, making $r(t)$ increase. Mean reversion is a desirable property in modelling of rates.

To proceed to calculation of bond prices $P(t,T) = \mathrm{E}_Q(e^{-\int_t^T r(s)ds} \mid \mathcal{F}_t)$, further assumptions on the market price of risk $q(t)$ are needed. Assume $q(t) = q$ is a constant.

We move between P and the EMM Q by using (12.8), which states $dB(t) = dW(t) - qdt$. Therefore the equation for $r(t)$ under Q is given by (12.14) with a replaced by

$$a^* = a + \frac{\sigma q}{b}. \tag{12.28}$$

To calculate the Q-conditional distribution of $\int_t^T r(s)ds$ given $\mathcal{F}_t$ needed for the bond price, observe that by the Markov property of the solution, for $s > t$ the process starts at $r(t)$ and runs for $s - t$, giving

$$r(s) = a^* - e^{-b(s-t)}(a^* - r(t)) + \sigma e^{-b(s-t)} \int_0^{s-t} e^{bu} dB(u), \tag{12.29}$$

with a Q-Brownian motion $B(u)$, independent of $\mathcal{F}_t$. Thus $r(s)$ is a Gaussian process, since the function in the Itô integral is deterministic (see Theorem 4.11). The conditional distribution of $\int_t^T r(s)ds$ given $\mathcal{F}_t$ is the same as that given $r(t)$ and is a Normal distribution. Thus the calculation of the bond price involves the expectation of a Lognormal, $E_Q(e^{-\int_t^T r(s)ds} \mid r(t))$. Thus

$$P(t,T) = e^{-\mu_1 + \sigma_1^2/2}, \tag{12.30}$$

where μ_1 and σ_1^2 are the conditional mean and variance of $\int_t^T r(s)ds$ given $r(t)$. Using (12.29) and calculating directly or using (12.4) $h = a^* - h'/b$, we have

$$\begin{aligned}
\mu_1 &= E_Q\left(\int_t^T r(s)ds \,\Big|\, r(t)\right) = \int_t^T E_Q\left(r(s) \mid r(t)\right)ds \\
&= a^*(T - t) + \frac{1}{b}(a^* - r(t))(1 - e^{-b(T-t)}).
\end{aligned} \tag{12.31}$$

To calculate σ_1^2, use the representation for $r(s)$ conditional on $r(t)$, equation (12.29)

$$\sigma_1^2 = Cov\left(\int_t^T r(s)ds, \int_t^T r(u)du\right) = \int_t^T \int_t^T Cov(r(s), r(u))ds\,du.$$

Now it is not hard to calculate, by the formula for the covariance of Itô Gaussian integrals (4.26) or (4.27), that for $s > u$

$$Cov(r(s), r(u)) = \frac{\sigma^2}{2b}\left(e^{b(u-s)} - e^{-b(u+s-2t)}\right).$$

Putting this expression (and a similar one for when $s < u$) in the double integral, we obtain

$$\sigma_1^2 = \frac{\sigma^2}{b^2}(T - t) - \frac{\sigma^2}{2b^3}(3 - 4e^{-b(T-t)} - e^{-2b(T-t)}).$$

Thus denoting $R(\infty) = a^* - \frac{\sigma^2}{2b^2}$,

$$P(t,T) = e^{\frac{1}{b}(1-e^{-b(T-t)})(R(\infty)-r(t))-(T-t)R(\infty)-\frac{\sigma^2}{4b^3}(1-e^{-b(T-t)})^2}. \tag{12.32}$$

From the formula (12.32) the yield to maturity is obtained by (12.1). Vasicek (1977) obtained the formula (12.32) by solving the PDE (12.12).

Since the price of bonds is Lognormal with known mean and variance, a closed form expression for the price of an option on the bond can be obtained.

12.5 Heath-Jarrow-Morton (HJM) Model

The class of models suggested by Heath, Jarrow, and Morton (1992) is based on modelling the forward rates. These rates are implied by bonds with different maturities. By definition, forward rates $f(t, T)$, $t \leq T \leq T^*$ are defined by the relation

$$P(t, T) = e^{-\int_t^T f(t,u)du}. \tag{12.33}$$

Thus the forward rate $f(t, T)$, $t \leq T$, is the (continuously compounding) rate at time T as seen from time t,

$$f(t, T) = -\frac{\partial \ln P(t, T)}{\partial T},$$

The spot rate $r(t) = f(t, t)$. Consequently the savings account $\beta(t)$ grows according to

$$\beta(t) = e^{\int_0^t f(s,s)ds}. \tag{12.34}$$

The assumption of HJM model is that for a fixed T, the forward rate $f(t, T)$ is a diffusion in t variable, namely

$$df(t, T) = \alpha(t, T)dt + \sigma(t, T)dW(t). \tag{12.35}$$

where $W(t)$ is P-Brownian motion and processes $\alpha(t, T)$ and $\sigma(t, T)$ are adapted and continuous. $\alpha(t, T)$, $\sigma(t, T)$ and the initial conditions $f(0, T)$, are the parameters of the model.

EMM assumption

There exists an equivalent martingale probability measure (EMM), Q $\sim$ P such that for all $T \leq T^*$, $\frac{P(t,T)}{\beta(t)}$ is a Q-martingale. Assuming the existence of Q we find equations for the bonds and the rates under Q.

Bonds and Rates under Q and the No-arbitrage Condition

The EMM assumption implies that $\alpha(t, T)$ is determined by $\sigma(t, T)$ when SDE for forward rates is considered under Q.

Theorem 12.1 *Assume the forward rates satisfy SDE (12.35), the EMM assumption holds, $\frac{P(t,T)}{\beta(t)}$ is a Q-martingale, and all the conditions on the coefficients of the SDE (12.35) needed for the analysis below. Let*

$$\tau(t, T) = \int_t^T \sigma(t, u)du. \tag{12.36}$$

Then

$$\alpha(t,T) = \sigma(t,T)\tau(t,T). \tag{12.37}$$

Moreover, under the EMM Q the forward rates satisfy the SDE with a Q-Brownian motion B

$$df(t,T) = \sigma(t,T)\tau(t,T)dt + \sigma(t,T)dB(t). \tag{12.38}$$

Conversely, if the forward rates satisfy the SDE (12.38) then $\frac{P(t,T)}{\beta(t)}$ is a Q-local martingale. If the appropriate integrability conditions hold, then it is a Q-martingale.

PROOF: The idea is simple, find $d\left(\frac{P(t,T)}{\beta(t)}\right)$ and equate the coefficient of dt to zero. Let

$$X(t) = \ln P(t,T) = -\int_t^T f(t,u)du. \tag{12.39}$$

By Itô's formula

$$d\left(\frac{P(t,T)}{\beta(t)}\right) = \frac{P(t,T)}{\beta(t)}\left(dX(t) + \frac{1}{2}d[X,X](t) - r(t)dt\right). \tag{12.40}$$

It is not hard to show, see Example 12.1, that

$$dX(t) = -d\left(\int_t^T f(t,u)du\right) = -A(t,T)dt - \tau(t,T)dW(t), \tag{12.41}$$

where $A(t,T) = -r(t) + \int_t^T \alpha(t,u)du$. Thus

$$d\left(\frac{P(t,T)}{\beta(t)}\right) = \frac{P(t,T)}{\beta(t)}\left(\left(-\int_t^T \alpha(t,u)du\right)dt + \frac{1}{2}\tau^2(t,T)dt - \tau(t,T)dW(t)\right). \tag{12.42}$$

By Girsanov's theorem

$$dW(t) + \left(\frac{\int_t^T \alpha(t,u)du}{\tau(t,T)} - \frac{1}{2}\tau(t,T)\right)dt = dB(t), \tag{12.43}$$

for a Q-Brownian motion $B(t)$. This gives the SDE for the discounted bond under Q

$$d\left(\frac{P(t,T)}{\beta(t)}\right) = -\frac{P(t,T)}{\beta(t)}\tau(t,T)dB(t). \tag{12.44}$$

Thus if the model is specified under Q, $W(t) = B(t)$, then

$$\int_t^T \alpha(t,u)du = \frac{1}{2}\tau^2(t,T). \tag{12.45}$$

Differentiating in t gives the condition (12.37). The SDE (12.38) follows from (12.37).

$\square$

Corollary 12.2 *The bonds satisfy the following equations under* Q *for* $t \leq T$

$$dP(t,T) = P(t,T)\left(r(t)dt - \tau(t,T)dB(t)\right), \quad and \qquad (12.46)$$

$$P(t,T) = P(0,T)e^{-\int_0^t \tau(s,T)dB(s) - \frac{1}{2}\int_0^t \tau^2(s,T)ds + \int_0^t r(s)ds}. \qquad (12.47)$$

PROOF: Equation (12.44) shows that $\frac{P(t,T)}{\beta(t)}$ is the stochastic exponential of $-\int_0^t \tau(s,T)dB(s)$. Hence

$$\frac{P(t,T)}{\beta(t)} = \frac{P(0,T)}{\beta(0)}\mathcal{E}(-\int_0^t \tau(s,T)dB(s))$$

$$= \frac{P(0,T)}{\beta(0)}e^{-\int_0^t \tau(s,T)dB(s) - \frac{1}{2}\int_0^t \tau^2(s,T)ds}. \qquad (12.48)$$

Since $\beta(t) = e^{\int_0^t r(s)ds}$ the bond's price is given by (12.47). The SDE (12.46) follows, as the stochastic exponential SDE of $-\int_0^t \tau(s,T)dB(s) + \int_0^t r(s)ds$.

$\square$

Using (12.47) for T_1 and T_2, we obtain for $t \leq T_1 \leq T_2$ by eliminating $\int_0^t r(s)ds$

Corollary 12.3 *A relation between bonds with different maturities is given by*

$$P(t,T_2) = \frac{P(0,T_2)}{P(0,T_1)}e^{-\int_0^t (\tau(s,T_2)-\tau(s,T_1))dB(s) - \frac{1}{2}\int_0^t (\tau^2(s,T_2)-\tau^2(s,T_1))ds}P(t,T_1).$$

$$(12.49)$$

Remark 12.2:
1. Equation (12.37) is known as the no-arbitrage or EMM condition.
2. The effect of the change to the EMM Q is in the change of drift in the SDE for the forward rates, from (12.35) to (12.38).
3. The volatility of the bond is $\tau(t,T) = \int_t^T \sigma(t,s)ds$, the integrated forward volatilities by (12.46).
4. The expression (12.47) for $P(t,T)$ includes the Itô integral $\int_0^t \tau(s,T)dB(s)$, which is not observed directly. It can be obtained from $\int_t^T f(t,u)du$ using (12.38) and interchanging integrals. Integrated processes and interchanging integrals can be justified rigorously, see Heath *et al.* (1992), and in greater generality Hamza *et al.* (2002).
5. The vector case of $W(t)$ and $\sigma(t,T)$ in (12.35) gives similar formulae by using notation $\sigma(t,T)dW(t) = \sum_{i=1}^d \sigma_i(t,T)dW_i(T)$, the scalar product and replace σ^2 by the norm $|\sigma|^2$, see Exercise 5.9.

Example 12.1: Differentiation of an integral of the form $\int_t^T f(t,s)ds$. We show that (12.41) holds. Introduce $G(u,t) = \int_u^T f(t,s)ds$. We are interested in $dG(t,t)$; it is found by

$$dG(t,t) = \left(\frac{\partial}{\partial u} G(t,t) + \frac{\partial}{\partial t} G(t,t) \right) dt,$$

or

$$d \left(\int_t^T f(t,v)dv \right) = -f(t,t)dt + \int_t^T df(t,v).$$

Now, $f(t,t) = r(t)$ and, using the model for $df(t,v)$ we obtain

$$d \left(\int_t^T f(t,v)dv \right) = -r(t)dt + \left(\int_t^T \alpha(t,v)dv \right) dt + \left(\int_t^T \sigma(t,v)dv \right) dW(t).$$

Exchange of the integrals is justified in Exercise 12.3.

Example 12.2: (Ho-Lee model)
Consider the SDE for forward rates (12.38) with the simplest choice of constant volatilities $\sigma(t,T) = \sigma$=const. Then $\tau(t,T) = \int_t^T \sigma(t,u)du = \sigma(T - t)$. Thus

$$df(t,T) = \sigma^2(T - t)dt + \sigma dB(t),$$

$$f(t,T) = f(0,T) + \sigma^2 t(T - \frac{t}{2}) + \sigma B(t), \quad \text{and}$$

$$r(t) = f(t,t) = f(0,t) + \sigma^2 \frac{t^2}{2} + \sigma B(t).$$

They contain the Brownian motion, which is not observed directly. Eliminating it, $\sigma B(t) = r(t) - f(0,t) - \sigma^2 \frac{t^2}{2}$, we have

$$f(t,T) = r(t) + \sigma^2 t(T - t) + f(0,T) - f(0,t).$$

This equation shows that forward rates $f(t,T_1) - f(t,T_2) = f(0,T_1) - f(0,T_2) + \sigma^2 t(T_1 - T_2)$ differ by a deterministic term, therefore they are also perfectly correlated. $r(t)$ and $f(t,T)$ are also perfectly correlated. This seems to contradict observations in financial markets.

The bond in terms of the forward rates (12.33)

$$P(t,T) = e^{-\int_t^T f(t,u)du} = e^{-\int_t^T f(0,u)du - \sigma^2 \int_t^T t(u - \frac{t}{2})du - \sigma B(t)(T-t)}.$$

Using $-\int_t^T f(0,u)du = \ln P(0,T) - \ln P(0,t)$,

$$P(t,T) = \frac{P(0,T)}{P(0,t)} e^{-\sigma^2 tT(T-t)/2 - \sigma B(t)(T-t)}.$$

Eliminating $B(t)$, we obtain the equation for the bond in terms of the spot rate, from which the yield curve is determined

$$P(t,T) = \frac{P(0,T)}{P(0,t)} e^{-(T-t)r(t) - \sigma^2 t(T-t)^2/2 + (T-t)f(0,t)}.$$

The graphs below illustrate possible curves produced by the Ho-Lee model.

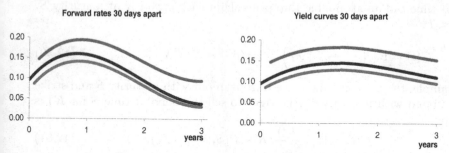

Figure 12.1: Forward curves and Yield curves.

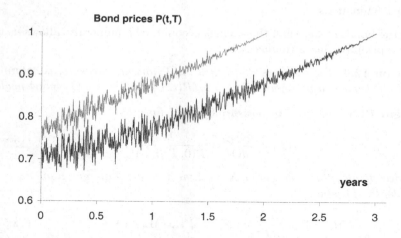

Figure 12.2: Bonds $P(t, T_i)$ for two maturities T_1 and T_2 one year apart.

12.6   Forward Measures. Bond as a Numeraire

Options on a Bond

An attainable claim is one that can be replicated by a self-financing portfolio
of bonds. The predictable representation property gives the arbitrage-free
price at time t of an attainable European claim with payoff X at maturity S,
$t \leq S \leq T$,

$$C(t) = \mathrm{E}_Q \left(\frac{\beta(t)}{\beta(S)} X \mid \mathcal{F}_t \right) = \mathrm{E}_Q (e^{-\int_t^S r(u)du} X \mid \mathcal{F}_t). \tag{12.50}$$

For example, the price at time t of a European put with maturity S and strike
K on a bond with maturity T (the right to sell the bond at time S for K) is
given by

$$C(t) = \mathrm{E}_Q \big(\frac{\beta(t)}{\beta(S)} (K - P(S,T))^+ \mid \mathcal{F}_t \big). \tag{12.51}$$

This option is in effect a cap on the interest rate over $[S, T]$ (see Section 12.7).

Forward Measures

Taking the bond, rather than the savings account, as a numeraire allows us to
simplify option pricing formulae.

Definition 12.4 *The measure* Q_T, *called the T-forward measure, is obtained
when the T-bond is a numeraire, i.e.* $\beta(t)/P(t,T)$, $t \leq T$ *is a* Q_T*-martingale.*

Theorem 12.5 *The forward measure* Q_T *is defined by*

$$\Lambda(T) = \frac{dQ_T}{dQ} = \frac{1}{P(0,T)\beta(T)}. \tag{12.52}$$

*The price of an attainable claim X at time t under different numeraire is
related by the formula,*

$$C(t) = \mathrm{E}_Q \big(\frac{\beta(t)}{\beta(T)} X \mid \mathcal{F}_t \big) = P(t,T) \mathrm{E}_{Q_T} (X \mid \mathcal{F}_t). \tag{12.53}$$

PROOF: follows directly from the change of numeraire Theorem 11.17. By the
EMM assumption $P(t,T)/\beta(t)$, $t \leq T$, is a positive Q-martingale. $Q_T := Q_1$,
with $\frac{dQ_1}{dQ} = \Lambda(T) = \frac{P(T,T)/P(0,T)}{\beta(T)/\beta(0)} = \frac{1}{P(0,T)\beta(T)}$, which is (12.52). By equation
(11.49)

$$C(t) = \mathrm{E}_Q \big(\frac{\beta(t)}{\beta(T)} X \mid \mathcal{F}_t \big) = \mathrm{E}_{Q_T} \big(\frac{P(t,T)}{P(T,T)} X \mid \mathcal{F}_t \big) = P(t,T) \mathrm{E}_{Q_T} (X \mid \mathcal{F}_t).$$

$\square$

Remark 12.3: $E_{Q_T}(X|\mathcal{F}_t) = F(t)$ is called the forward price at t for the date T of an attainable claim X. These prices are Q_T-martingales, see Doob-Levy martingale, Theorem 7.9.

Forward Measure in HJM

Theorem 12.6 *The process*

$$dB^T(t) = dB(t) + \tau(t, T)dt \tag{12.54}$$

is a Q_T-Brownian motion. Moreover, the the forward rates $f(t, T)$ and the bond $P(t, T)$ satisfy the following SDEs under Q_T

$$df(t, T) = \sigma(t, T)dB^T(t). \tag{12.55}$$

$$dP(t, T) = P(t, T)\left((r(t) + \tau^2(t, T))dt - \tau(t, T)dB^T(t)\right). \tag{12.56}$$

PROOF: The T-forward martingale measure Q_T is obtained by using $\Lambda(T) = \frac{1}{P(0, T)\beta(T)}$. Then

$$\Lambda_t = E(\Lambda|\mathcal{F}_t) = \frac{P(t, T)}{P(0, T)\beta(t)} = \mathcal{E}(-\int_0^t \tau(s, T)dB(s)).$$

Therefore by the Girsanov's theorem $dB^T(t) = dB(t) + \tau(t, T)dt$ is a Q_T-Brownian motion. Under Q the forward rates satisfy the SDE (12.38)

$$df(t, T) = \sigma(t, T)\tau(t, T)dt + \sigma(t, T)dB(t).$$

The SDE under Q_T is obtained by replacing $B(t)$ with $B^T(t)$,

$$df(t, T) = \sigma(t, T)\left(\tau(t, T)dt + dB(t)\right) = \sigma(t, T)dB^T(t).$$

The SDE under Q_T for the bond prices is obtained similarly from the SDE (12.46) under Q.

□

The above result shows that under the forward measure $f(t, T)$ is a Q_T-local martingale. It also shows that $\tau(t, T)$ is the volatility of $P(t, T)$ under the forward measure Q_T.

Distributions of the Bond in HJM with Deterministic Volatilities

The following result, needed for option pricing, gives the conditional distribution of $P(T, T + \delta)$. As a corollary distributions of $P(t, T)$ are obtained.

Theorem 12.7 *Suppose that the forward rates satisfy the HJM model with a deterministic $\sigma(t,T)$. Then for any $0 \le t < T$ and any $\delta > 0$, the Q-conditional distribution of $P(T, T+\delta)$ given $\mathcal{F}_t$ is Lognormal with mean*

$$E_Q(\ln P(T, T+\delta)|\mathcal{F}_t) = \ln \frac{P(t, T+\delta)}{P(t, T)} - \frac{1}{2} \int_t^T (\tau^2(s, T+\delta) - \tau^2(s, T))ds,$$

and variance

$$\gamma^2(t) = Var(\ln P(T, T+\delta)|\mathcal{F}_t) = \int_t^T (\tau(s, T+\delta) - \tau(s, T))^2 ds. \quad (12.57)$$

PROOF:　By letting $t = T_1 = T$ and $T_2 = T + \delta$ in equation (12.49) , the expression for $P(T, T+\delta)$ is obtained,

$$P(T, T+\delta) = \frac{P(0, T+\delta)}{P(0, T)} e^{-\int_0^T (\tau(s, T+\delta) - \tau(s, T))dB(s) - \frac{1}{2}\int_0^T (\tau^2(s, T+\delta) - \tau^2(s, T))ds}.$$

$$(12.58)$$

To find its conditional distribution given $\mathcal{F}_t$, separate the $\mathcal{F}_t$-measurable term,

$$
\begin{aligned}
P(T, T+\delta) &= \frac{P(0, T+\delta)}{P(0, T)} e^{-\int_0^t (\tau(s, T+\delta) - \tau(s, T))dB(s) - \frac{1}{2}\int_0^t (\tau^2(s, T+\delta) - \tau^2(s, T))ds} \\
&\quad \times e^{-\int_t^T (\tau(s, T+\delta) - \tau(s, T))dB(s) - \frac{1}{2}\int_t^T (\tau^2(s, T+\delta) - \tau^2(s, T))ds} \\
&= \frac{P(t, T+\delta)}{P(t, T)} e^{-\int_t^T (\tau(s, T+\delta) - \tau(s, T))dB(s) - \frac{1}{2}\int_t^T (\tau^2(s, T+\delta) - \tau^2(s, T))ds}. \quad (12.59)
\end{aligned}
$$

If $\tau(t, T)$ is non-random, then the exponential term is independent of $\mathcal{F}_t$. Hence the desired conditional distribution given $\mathcal{F}_t$ is Lognormal with the mean and variance as stated.

□

Corollary 12.8 *The conditional distribution of $P(T, T+\delta)$ given $\mathcal{F}_t$ under the forward measure $Q_{T+\delta}$ is Lognormal with mean*

$$E_{Q_{T+\delta}}(\ln P(T, T+\delta)|\mathcal{F}_t) = \ln \frac{P(t, T+\delta)}{P(t, T)} + \frac{1}{2} \int_t^T (\tau(s, T+\delta) - \tau(s, T))^2 ds,$$

and variance $\gamma^2(t)$ (12.57).

PROOF:　Use equation (12.59) for the conditional representation, together with (12.54), a $Q_{T+\delta}$-Brownian motion $dB^{T+\delta}(t) = dB(t) + \tau(t, T+\delta)dt$, to have

$$P(T, T+\delta) = \frac{P(t, T+\delta)}{P(t, T)} e^{-\int_t^T (\tau(s, T+\delta) - \tau(s, T))dB^{T+\delta}(s) + \frac{1}{2}\int_t^T (\tau(s, T+\delta) - \tau(s, T))^2 ds}.$$

$$(12.60)$$

□

By taking $t = T = 0$, $T + \delta = T$ in Theorem 12.7 and its corollary we obtain

Corollary 12.9 *In the case of constant volatilities the* Q*-distribution and* Q_T*-distribution of* $P(t, T)$ *are Lognormal.*

The Lognormality of $P(t, T)$ can also be obtained directly from equations (12.47) and (12.38).

12.7 Options, Caps and Floors

We give common options on interest rates and show how they relate to options on bonds.

Cap and Caplets

A *cap* is a contract that gives its holder the right to pay the rate of interest smaller of the two, the floating rate, and rate k, specified in the contract. A party holding the cap will never pay rate exceeding k, the rate of payment is capped at k. Since the payments are done at a sequence of payments dates $T_1, T_2, \ldots, T_n$, called a *tenor*, with $T_{i+1} = T_i + \delta$ (e.g. $\delta = \frac{1}{4}$ of a year), the rate is capped over intervals of time of length δ. Thus a cap is a collection of caplets.

Figure 12.3: Payment dates and simple rates.

Consider a caplet over $[T, T + \delta]$. Without the caplet, the holder of a loan must pay at time $T + \delta$ an interest payment of $f\delta$, where f is the floating, simple rate over the interval $[T, T + \delta]$. If $f > k$, then a caplet allows the holder to pay $k\delta$. Thus the caplet is worth $f\delta - k\delta$ at time $T + \delta$. If $f < k$, then the caplet is worthless. Therefore, the caplet's worth to the holder is $(f - k)^+\delta$. In other words, a caplet pays to its holder the amount $(f - k)^+\delta$ at time $T + \delta$. Therefore a caplet is a call option on the rate f, and its price at time t, as any other option, is given by the expected discounted payoff at maturity under the EMM Q,

$$\text{Caplet}(t) = \text{E}_Q \left(\frac{\beta(t)}{\beta(T + \delta)} (f - k)^+ \delta \mid \mathcal{F}_t \right). \tag{12.61}$$

To evaluate this expectation we need to know the distribution of f under Q. One way to find the distribution of the simple floating rate is to relate it to the bond over the same time interval. By definition, $\frac{1}{P(T, T+\delta)} = 1 + f\delta$. This relation is justified as the amounts obtained at time $T + \delta$ when \$1 invested at time T in the bond and in the investment account with a simple rate f. Thus

$$f = \frac{1}{\delta} \left(\frac{1}{P(T, T+\delta)} - 1 \right). \tag{12.62}$$

This is a basic relation between the rates that appear in option contracts and bonds. It gives the distribution of f in terms of that of $P(T, T + \delta)$.

Caplet as a Put Option on the Bond

We show next that a caplet is in effect a put option on the bond. From the basic relation (EMM) $P(T, T + \delta) = \mathrm{E}(\frac{\beta(T)}{\beta(T+\delta)} \mid \mathcal{F}_T)$. Proceeding from (12.61) by the law of double expectation, with $\mathrm{E} = \mathrm{E}_Q$

$$
\begin{aligned}
\mathrm{Caplet}(t) &= \mathrm{E}(\mathrm{E}(\frac{\beta(t)\beta(T)}{\beta(T)\beta(T+\delta)}(\frac{1}{P(T,T+\delta)} - 1 - k\delta)^+ \mid \mathcal{F}_T) \mid \mathcal{F}_t) \\
&= \mathrm{E}(\frac{\beta(t)}{\beta(T)}(\frac{1}{P(T,T+\delta)} - 1 - k\delta)^+ \mathrm{E}(\frac{\beta(T)}{\beta(T+\delta)} \mid \mathcal{F}_T) \mid \mathcal{F}_t) \\
&= (1 + k\delta)\mathrm{E}(\frac{\beta(t)}{\beta(T)}(\frac{1}{(1+k\delta)} - P(T,T+\delta))^+ \mid \mathcal{F}_t). \tag{12.63}
\end{aligned}
$$

Thus a caplet is a put option on $P(T, T+\delta)$ with strike $\frac{1}{(1+k\delta)}$, and exercise time T. In practical modelling, as in the HJM model with deterministic volatilities, the distribution of $P(T, T + \delta)$ is Lognormal, giving rise to the Black-Scholes type formula for a caplet, Black's (1976) formula.

The price of caplet is easier to evaluate by using a forward measure (Theorem 12.5). Take $P(t, T+\delta)$ as a numeraire, which corresponds to $T+\delta$-forward measure $Q_{T+\delta}$,

$$
\begin{aligned}
\mathrm{Caplet}(t) &= P(t, T+\delta)\mathrm{E}_{Q_{T+\delta}}((f - k)^+ \delta \mid \mathcal{F}_t) \tag{12.64} \\
&= P(t, T+\delta)\mathrm{E}_{Q_{T+\delta}}\left((\frac{1}{P(T,T+\delta)} - 1 - k\delta)^+ \mid \mathcal{F}_t \right).
\end{aligned}
$$

Caplet Pricing in HJM model

The price of a caplet, which caps the rate at k over the time interval $[T, T+\delta]$, at time t is given by equation (12.63) under the EMM Q, and by (12.64) under the forward EMM $Q_{T+\delta}$. These can be evaluated in closed form when the volatilities are non-random.

For evaluating the expectation in the caplet formula (12.64), note that if X is Lognormal $e^{N(\mu, \sigma^2)}$, then $1/X = e^{N(-\mu, \sigma^2)}$ is also Lognormal. This allows us to price the caplet by doing standard calculations for $E(X - K)^+$ giving Black's caplet formula.

The price of a Cap is then found by using forward measures

$$\text{Cap}(t) = \sum_{i=1}^{n} \text{Caplet}_i(t) = \sum_{i=1}^{n} P(t, T_i) E_{Q_{T_i}} ((f_{i-1} - k)^+ \delta \mid \mathcal{F}_t).$$ (12.65)

The Cap pricing formula is given in Exercise 12.6.

A floor is a contract that gives its holder the right to receive the larger of the two rates, the rate specified in the contract k, and the floating simple rate f. Floors are priced similarly to Caps with floorlets being $(k - f_{i-1})^+$.

12.8 Brace-Gatarek-Musiela (BGM) Model

In financial markets Black-Scholes like formulae are used for everything: bonds, rates, etc. To make the practice consistent with the theory, Brace, Gatarek and Musiela introduced a class of models, which can be seen as a subclass of HJM models, where instead of the forward rates $f(t, T)$, the LIBOR rates are modelled (Brace *et al.*(1997), Musiela and Rutkowski (1998)). In BGM models the rates are Lognormal under forward measures, the fact that implies option pricing formulae consistent with market use.

LIBOR

The time t forward δ-LIBOR (London Inter-Bank Offer Rate) is the simple rate of interest on $[T, T + \delta]$

$$L(t, T) = \frac{1}{\delta} \left(\frac{P(t, T)}{P(t, T + \delta)} - 1 \right),$$ (12.66)

note that $L(T, T) = f$ is the rate that gets capped, see (12.62).

Theorem 12.10 *The SDE for $L(t, T)$ under the forward $Q_{T+\delta}$ measure, when $P(t, T + \delta)$ is the numeraire, is*

$$dL(t, T) = L(t, T) \left(\frac{1 + L(t, T)\delta}{L(t, T)\delta} \right) (\tau(t, T + \delta) - \tau(t, T)) dB^{T+\delta}(t).$$ (12.67)

PROOF: By Corollary 12.3 to equation (12.49),

$$\frac{P(t, T)}{P(t, T + \delta)} = \frac{P(0, T)}{P(0, T + \delta)} e^{-\int_0^t (\tau(s,T) - \tau(s,T+\delta)) dB(s) - \frac{1}{2} \int_0^t (\tau^2(s,T) - \tau^2(s,T+\delta)) ds}.$$

(12.68)

$dB^{T+\delta}(t) = dB(t) + \tau(t, T + \delta)dt$, is a $Q_{T+\delta}$-Brownian motion (12.54), giving

$$\frac{P(t,T)}{P(t,T+\delta)} = \frac{P(0,T)}{P(0,T+\delta)}e^{-\int_0^t (\tau(s,T)-\tau(s,T+\delta))dB^{T+\delta}(s) - \frac{1}{2}\int_0^t (\tau(s,T)-\tau(s,T+\delta))^2 ds}.$$

(12.69)

$$= \frac{P(0,T)}{P(0,T+\delta)}\mathcal{E}\left(\int_0^t (\tau(s,T+\delta) - \tau(s,T))dB^{T+\delta}(s)\right).$$

Using the stochastic exponential equation, it follows that

$$d\left(\frac{P(t,T)}{P(t,T+\delta)}\right) = \left(\frac{P(t,T)}{P(t,T+\delta)}\right)(\tau(t,T+\delta) - \tau(t,T))dB^{T+\delta}(t). \quad (12.70)$$

Finally, using the definition of $L(t, T)$, the SDE (12.67) is obtained.

□

Choose now the HJM volatility $\sigma(t, s)$, such that $\gamma(t, T)$ is deterministic

$$\gamma(t,T) = \left(\frac{1 + L(t,T)\delta}{L(t,T)\delta}\right)(\tau(t,T+\delta) - \tau(t,T)) = \frac{1 + L(t,T)\delta}{L(t,T)\delta}\int_T^{T+\delta} \sigma(t,s)ds.$$

(12.71)

Corollary 12.11 *Let $\gamma(t, T)$ be deterministic such that $\int_0^T \gamma^2(s, T)ds < \infty$. Then $L(t, T)$ for $t \leq T$, solves the SDE*

$$dL(t,T) = L(t,T)\gamma(t,T)dB^{T+\delta}(t), \quad (12.72)$$

with the initial condition $L(0, T) = \frac{1}{\delta}(\frac{P(0,T)}{P(0,T+\delta)} - 1)$. Thus $L(t, T)$ is Lognormal under the forward measure $Q_{T+\delta}$, moreover it is a martingale.

$$L(T,T) = L(t,T)e^{\int_t^T \gamma(s,T)dB^{T+\delta}(s) - \frac{1}{2}\int_t^T \gamma^2(s,T)ds}, \quad (12.73)$$

and the conditional distribution of $L(T, T)$ given $\mathcal{F}_t$ is Lognormal with mean $\ln L(t, T) - \frac{1}{2}\int_t^T \gamma^2(s, T)ds$ and variance $\int_t^T \gamma^2(s, T)ds$.

We now prove that the choice of forward volatilities specified above is possible.

Theorem 12.12 *Let $\gamma(t, T)$, $t \leq T$ be given and such that the Itô integral $\int_0^T \gamma(s, T)dB(s)$ is defined. Then there exist forward rates volatilities $\sigma(t, T)$, such that the integrated volatility $\int_T^{T+\delta} \sigma(s, u)du$ is determined uniquely, and (12.72) holds.*

PROOF: (12.72) is equivalent to

$$L(t,T) = L(0,T)\mathcal{E}\left(\int_0^t \gamma(s,T)dB^{T+\delta}(s)\right), \quad t \leq T. \quad (12.74)$$

By the definition of $L(t,T)$, and (12.69) we have

$$
\begin{aligned}
L(t,T) &= \frac{1}{\delta}\left(\frac{P(t,T)}{P(t,T+\delta)} - 1\right) \\
&= \frac{1}{\delta}\left(\frac{P(0,T)}{P(0,T+\delta)}\mathcal{E}\left(\int_0^t (\tau(s,T+\delta) - \tau(s,T))dB^{T+\delta}(s)\right) - 1\right).
\end{aligned}
$$

Equating this to (12.74) we obtain the equation for stochastic exponentials

$$
\mathcal{E}\left(\int_0^t (\tau(s,T+\delta) - \tau(s,T))dB^{T+\delta}(s)\right) = (1-c)\mathcal{E}\left(\int_0^t \gamma(s,T)dB^{T+\delta}(s)\right) + c,
$$
(12.75)

with $c = \frac{P(0,T+\delta)}{P(0,T)}$. Now, using the stochastic logarithm (Theorem 5.3)

$$
\int_0^t \left(\int_T^{T+\delta} \sigma(s,u)du\right)dB^{T+\delta}(s) = \mathcal{L}\left((1-c)\mathcal{E}\left(\int_0^t \gamma(s,T)dB^{T+\delta}(s)\right) + c\right),
$$
(12.76)

from which the integrated volatility $\int_T^{T+\delta}\sigma(s,u)du$ and a suitable process $\sigma(t,T)$ can be found (see Exercise 12.7).

$\square$

Caplet in BGM

Caplet is a Call on LIBOR, it pays $\delta(L(T,T) - k)^+$ at $T + \delta$. By (12.53) its price at time t

$$
C(t) = P(t,T+\delta)\mathbb{E}_{Q_{T+\delta}}(\delta(L(T,T) - k)^+|\mathcal{F}_t).
$$
(12.77)

Using that $L(T,T)$ is Lognormal under the forward measure $Q_{T+\delta}$ (12.73), the caplet is priced by Black's caplet formula (agrees with the market) by evaluating under the forward measure.

$$
C(t) = P(t,T+\delta)\left(L(t,T)N(h_1) - kN(h_2)\right),
$$
(12.78)

$$
h_{1,2} = \frac{\ln\frac{L(t,T)}{k} \pm \frac{1}{2}\int_t^T \gamma^2(s,T)ds}{\sqrt{\int_t^T \gamma^2(s,T)ds}}.
$$

SDEs for Forward LIBOR under Different Measures

Consider now a sequence of dates $T_0, T_1, \ldots, T_n$, and denote by Q_{T_k} the forward measure corresponding to the numeraire $P(t,T_k)$. Corollary 12.11 states that $L(t,T_{k-1})$ for $t \leq T_{k-1}$ satisfies the SDE

$$
dL(t,T_{k-1}) = \gamma(t,T_{k-1})L(t,T_{k-1})dB^{T_k}(t),
$$
(12.79)

where $B^{T_k}(t)$ is a Q_{T_k}-Brownian motion.

An SDE for all of the rates under a single measure is sometimes required (for a swaption), it is given by

Theorem 12.13 *For a given i, the SDE for the forward LIBOR $L(t, T_{k-1})$ on $[T_{k-1}, T_k]$ under the forward measure Q_{T_i}, is given by (12.79) for $i = k$, by (12.80) for $i > k$, and (12.81) $i < k$.*

$$
dL(t, T_{k-1}) = -L(t, T_{k-1}) \sum_{j=k}^{i-1} \frac{\delta\gamma(t, T_{k-1})\gamma(t, T_j)L(t, T_j)}{1 + \delta L(t, T_j)} dt
$$
$$
+ \gamma(t, T_{k-1})L(t, T_{k-1})dB^{T_i}(t). \tag{12.80}
$$

$$
dL(t, T_{k-1}) = L(t, T_{k-1}) \sum_{j=i}^{k-1} \frac{\delta\gamma(t, T_{k-1})\gamma(t, T_j)L(t, T_j)}{1 + \delta L(t, T_j)} dt
$$
$$
+ \gamma(t, T_{k-1})L(t, T_{k-1})dB^{T_i}(t). \tag{12.81}
$$

PROOF: We establish the relationship between different forward measures as well as corresponding Brownian motions. By (12.54) the Brownian motion under the forward measure Q_{T_k} satisfies

$$
dB^{T_k}(t) = dB(t) + \tau(t, T_k)dt. \tag{12.82}
$$

Hence, using this with $k - 1$, we obtain that B^{T_k} and $B^{T_{k-1}}$ are related by

$$
dB^{T_k}(t) = dB^{T_{k-1}}(t) + (\tau(t, T_k) - \tau(t, T_{k-1}))dt.
$$

By (12.71), from the choice of γ,

$$
\tau(t, T_k) - \tau(t, T_{k-1}) = \gamma(t, T_{k-1}) \frac{L(t, T_{k-1})\delta}{1 + L(t, T_{k-1})\delta},
$$

giving the relationship for the SDEs for LIBOR

$$
dB^{T_k}(t) = dB^{T_{k-1}}(t) + \gamma(t, T_{k-1}) \frac{L(t, T_{k-1})\delta}{1 + L(t, T_{k-1})\delta} dt.
$$

Now fix $i > k$. Using the above relation iteratively from i to k, we obtain

$$
dB^{T_i}(t) = dB^{T_k}(t) + \sum_{j=k}^{i-1} \gamma(t, T_j) \frac{L(t, T_j)\delta}{1 + L(t, T_j)\delta} dt. \tag{12.83}
$$

Replace B^{T_k} in the SDE (12.79) for $L(t, T_{k-1})$ under Q_{T_k}, $dL(t, T_{k-1}) = \gamma(t, T_{k-1})L(t, T_{k-1})dB^{T_k}(t)$, by B^{T_i} with the drift from (12.83) to obtain (12.80). The case $i < k$ is proved similarly.

$\square$

Another proof can be done by using the result on the SDE under a new numeraire, Theorem 11.18, with $\beta(t) = P(t, T_k)$ and $S(t) = P(t, T_i)$.

Choice of Bond Volatilities

We comment briefly on the choice of bond volatilities. For a sequence of dates equation (12.71) implies from the choice of γ,

$$\tau(t, T_k) - \tau(t, T_{k-1}) = \gamma(t, T_{k-1}) \frac{L(t, T_{k-1})\delta}{1 + L(t, T_{k-1})\delta}. \qquad (12.84)$$

Therefore the bond volatilities $\tau(t, T_k)$ can be obtained recursively,

$$\tau(t, T_k) = \tau(t, T_{k-1}) + \gamma(t, T_{k-1}) \frac{L(t, T_{k-1})\delta}{1 + L(t, T_{k-1})\delta} = \sum_{j=0}^{k-1} \gamma(t, T_j) \frac{L(t, T_j)\delta}{1 + L(t, T_j)\delta}. \qquad (12.85)$$

In practice $\gamma(t, T)$ is taken to be a function of T only (Rebonato, Brace), for example $\gamma(t, T) = (a + be^{cT})$.

12.9 Swaps and Swaptions

A basic instrument in the interest rate market is the payer swap, in which the floating rate is swapped in arrears (at T_i) against a fixed rate k at n intervals of length $\delta = T_i - T_{i-1}$, $i = 1, 2 \ldots, n$. The other party in the swap enters a receiver swap, in which a fixed rate is swapped against the floating. By a swap we shall mean only a payer swap. A swap value at time t, by the basic relation (12.50) is given by

$$\text{Swap}(t, T_0, k) = \mathrm{E}_Q \sum_{i=1}^{n} \delta \left(\frac{\beta(t)}{\beta(T_i)} (L(T_{i-1}, T_{i-1}) - k) \,\middle|\, \mathcal{F}_t \right).$$

This can be written by using forward measures (Theorem 12.5)

$$\text{Swap}(t, T_0, k) = \delta \sum_{i=1}^{n} P(t, T_i) \mathrm{E}_{Q_{T_i}} \left((L(T_{i-1}, T_{i-1}) - k) \,\middle|\, \mathcal{F}_t \right)$$

$$= \delta \sum_{i=1}^{n} P(t, T_i)(L(t, T_{i-1}) - k), \qquad (12.86)$$

as under the forward measure Q_{T_i}, $L(t, T_{i-1})$ is a martingale (Corollary 12.11).

A swap agreement is worthless at initiation. The forward swap rate is that fixed rate of interest which makes a swap worthless. Namely, the swap rate at time t for the date T_0 solves, by definition, $\text{Swap}(t, T_0, k) = 0$. Hence

$$k(t, T_0) = \frac{\sum_{i=1}^{n} P(t, T_i) L(t, T_{i-1})}{\sum_{i=1}^{n} P(t, T_i)}. \qquad (12.87)$$

Thus the t-value of the swap in terms of this rate is

$$\text{Swap}(t, T_0, k) = \delta \left(\sum_{i=1}^n P(t, T_i) \right) (k(t, T_0) - k). \tag{12.88}$$

Other expressions for the Swap are given in Exercises 12.10 and 12.10.

A payer swaption maturing (with exercise) at $T = T_0$ gives its holder the right to enter a payer swap at time T_0. A swaption, Swaption(t, T_0), delivers at time T_0 a swap, Swap(T_0, T_0), when the value of that swap is positive. This shows the value of the swaption at maturity is $(\text{Swap}(T_0, T_0))^+$. Thus its value at time $t \le T_0$ is

$$
\begin{aligned}
\text{Swaption}(t, T_0) &= \text{E}_Q \left(\frac{\beta(t)}{\beta(T_0)} (\text{Swap}(T_0, T_0))^+ \Big| \mathcal{F}_t \right) & (12.89) \\
&= \delta \text{E}_Q \left(\frac{\beta(t)}{\beta(T_0)} \left(\sum_{i=1}^n P(T_0, T_i) \right) (k(T_0, T_0) - k)^+ \Big| \mathcal{F}_t \right).
\end{aligned}
$$

Consider taking $\sum_{i=1}^n P(t, T_i)$ as the numeraire instead of $\beta(t)$. The process $\sum_{i=1}^n P(t, T_i)/\beta(t)$ is a Q-martingale, as a sum of martingales. By Theorem 11.17, the new measure $\hat{Q}_{T_0}$ defined by $\Lambda(T) = \dfrac{\sum_{i=1}^n P(T_0, T_i)/\sum_{i=1}^n P(0, T_i)}{\beta(T_0)}$, as its Radon-Nikodym derivative with respect to Q gives (a call on the swap rate)

$$\text{Swaption}(t, T_0) = \delta \left(\sum_{i=1}^n P(t, T_i) \right) \text{E}_{\hat{Q}_{T_0}} \left((k(T_0, T_0) - k)^+ \Big| \mathcal{F}_t \right). \tag{12.90}$$

The distribution of the swap rate under the swap-rate measure $\hat{Q}_{T_0}$ is approximately Lognormal. Differentiating, simplifying and approximating $k(t, T_0)$ in (12.87) leads to the SDE for the swap rate

$$dk(t, T_0) = \hat{\sigma}(t, T_0) k(t, T_0) d\hat{B}_{T_0},$$

where $\hat{B}_{T_0}$ is a $\hat{Q}_{T_0}$-Brownian motion, and

$$\hat{\sigma}(t, T_0) = \sum_{i=1}^n w_i \gamma(t, T_{i-1}), \quad w_i = \frac{\frac{P(0, T_i)}{P(0, T_0)} L(0, T_{i-1})}{\sum_{i=1}^n \frac{P(0, T_i)}{P(0, T_0)} L(0, T_{i-1})}.$$

The expression for the swaption (12.90) integrates to the Black Swaption Formula as used in the market.

Another way to evaluate a swaption is by simulation. For details on analytic approximations and simulations see Brace $et\ al.$ (1997), Brace (1998), Musiela and Rutkowski (1998).

Remark 12.4: We have presented a brief mathematical treatment of models for rates and bonds based on diffusion processes. There is a large amount of literature on models based on processes with jumps. For jumps in the spot rate see Borovkov *et al.* (2003) and references therein. HJM with jumps were studied by Shirakawa (1991), and more generally by Björk, Kabanov and Runggaldier (BKR) (1996) and Björk, Di Masi, Kabanov and Runggaldier (BDMKR) (1997). Kennedy (1994) considered a Gaussian Markov field model. HJM and BDMKR models can be criticized for being an infinite-dimensional diffusion driven by a finite number of independent noise processes. Cont (1998) suggests modelling the forward curves by an infinite-dimensional diffusion driven by an infinite-dimensional Brownian motion. This approach is included in Random Fields models, such as Brownian and Poisson sheet models, also known as models with space-time noise. The most general model that allows for existence of EMM is given in Hamza *et al.* (2002); it includes Gaussian and Poisson random field models.

12.10   Exercises

Exercise 12.1: Show that a European call option on the T-bond is given by $C(t) = P(t,T)Q_T(P(s,T) > K \mid \mathcal{F}_t) - KP(t,s)Q_s(P(s,T) > K \mid \mathcal{F}_t)$, where s is the exercise time of the option and Q_s, Q_T are s and T-forward measures.

Exercise 12.2: Show that a European call option on the bond in the Merton model is given by

$$P(t,T)\Phi\left(\frac{\ln\frac{P(t,T)}{KP(t,s)} + \frac{\sigma^2(T-s)^2(s-t)}{2}}{\sigma(T-s)\sqrt{s-t}}\right) - KP(t,s)\Phi\left(\frac{\ln\frac{P(t,T)}{KP(t,s)} - \frac{\sigma^2(T-s)^2(s-t)}{2}}{\sigma(T-s)\sqrt{s-t}}\right).$$

Exercise 12.3: (Stochastic Fubini Theorem)
Let $H(t,s)$ be continuous $0 \le t, s \le T$, and for any fixed s, $H(t,s)$ as a process in t, $0 \le t \le T$ is adapted to the Brownian motion filtration $\mathcal{F}_t$. Assume $\int_0^T H^2(t,s)dt < \infty$, so that for each s the Itô integral $X(s) = \int_0^T H(t,s)dW(t)$ is defined. Since $H(t,s)$ is continuous, $Y(t) = \int_0^T H(t,s)ds$ is defined and it is continuous and adapted. Assume

$$\int_0^T \mathrm{E}\left(\int_0^T H^2(t,s)dt\right)ds < \infty.$$

1. Show that $\int_0^T \mathrm{E}|X(s)|ds \le \int_0^T \left(\mathrm{E}\int_0^T H^2(t,s)dt\right)^{1/2}ds < \infty$, consequently $\int_0^T X(s)ds$ exists.

2. If $0 = t_0 < t_1 < \ldots < t_n = T$ is a partition of $[0, T]$, show that

$$\int_0^T \left(\sum_{i=0}^{n-1} H(t_i, s) \big(W(t_{i+1}) - W(t_i) \big) \right) ds = \sum_{i=0}^{n-1} \left(\int_0^T H(t_i, s) ds \right) \big(W(t_{i+1}) - W(t_i) \big).$$

3. By taking the limits as the partition shrinks, show that

$$\int_0^T X(s) ds = \int_0^T Y(t) dW(t),$$

in other words the order of integration can be interchanged

$$\int_0^T \left(\int_0^T H(t, s) dW(t) \right) ds = \int_0^T \left(\int_0^T H(t, s) ds \right) dW(t).$$

Exercise 12.4: (One factor HJM) Show that the correlation between the forward rates $f(t, T_1)$ and $f(t, T_2)$ in the HJM model with deterministic volatilities $\sigma(t, T)$ is given by

$$\rho(T_1, T_2) = Corr(f(t, T_1), f(t, T_2)) = \frac{\int_0^t \sigma(s, T_1) \sigma(s, T_2) ds}{\sqrt{\int_0^t \sigma^2(s, T_1) ds \int_0^t \sigma^2(s, T_2) ds}}.$$

Give the class of volatilities for which the correlation is one.

Exercise 12.5: Find the forward rates $f(t, T)$ in Vasicek's model. Give the price of a cap.

Exercise 12.6: (Caps Pricing Market Formula)
Show that in the HJM model with a deterministic $\sigma(t, T)$ the price of a cap with trading dates $T_i = T + i\delta$, $i = 1, \ldots, n$, and strike rate k is given by

$$\sum_{i=1}^{n} \left(P(t, T_{i-1}) \Phi(-h_{i-1}(t)) - (1 + K\delta) P(t, T_i) \Phi(-h_{i-1}(t) - \gamma_{i-1}(t)) \right),$$

where $\gamma_{i-1}^2(t) = Var(\ln P(T_{i-1}, T_i)) = \int_t^{T_{i-1}} |\tau(s, T_i) - \tau(s, T_{i-1})|^2 ds$, with $\tau(t, T) = \int_t^T \sigma(t, s) ds$ and $h_{i-1}(t) = \frac{1}{\gamma_{i-1}(t)} \left(\ln \frac{(1+k\delta) P(t, T_i)}{P(t, T_{i-1})} - \frac{1}{2} \gamma_{i-1}^2(t) \right)$.

Exercise 12.7: Let $0 < c < 1$, and $\gamma(t)$ a bounded deterministic function be given. Show that there is a process $\beta(s)$, such that

$$c + (1 - c) \mathcal{E}\left(\int_0^t \gamma(s) dB(s) \right) = \mathcal{E}\left(\int_0^t \beta(s) dB(s) \right).$$

Hence deduce the existence of forward rates volatilities $\sigma(t, T)$ in HJM from specification of the forward LIBOR volatilities $\gamma(t, T)$ in BGM.

Exercise 12.8: (Two factor and higher HJM Models)
Two-factor HJM is given by the SDE

$$df(t,T) = \alpha(t,T)dt + \sigma_1(t,T)dW_1(t) + \sigma_2(t,T)dW_2(t),$$

where W_1 and W_2 are independent Brownian motions.
1. Give the stochastic differential equation for the log of the bond prices, and show that

$$\frac{P(t,T)}{\beta(t)} = P(0,T)e^{-\int_0^t A_1(u,T)du - \int_0^t \tau_1(t,T)dW_1(u) - \int_0^t \tau_2(t,T)dW_2(u)},$$

with $\tau_i(t,T) = \int_t^T \sigma_i(t,s)ds,$, $i = 1, 2$, and $A_1(t,T) = \int_t^T \alpha(t,s)ds$.
2. Using the same proof as in the one-factor model, show that the no-arbitrage condition is given by

$$\alpha(t,T) = \sigma_1(t,T) \int_t^T \sigma_1(t,s)ds + \sigma_2(t,T) \int_t^T \sigma_2(t,s)ds.$$

Exercise 12.9: Show that a swap can be written as

$$\text{Swap}(t,T_0,k) = P(t,T_0) - P(t,T_n) - k\delta \sum_{i=1}^n P(t,T_i).$$

Exercise 12.10: Denote by $b(t) = \delta \sum_{i=1}^n P(t,T_i)$. Show that for $0 < t \leq T_0$
$\text{Swap}(t,T_0,k) = P(t,T_0) - P(t,T_n) - kb(t)$, and that the swap rate

$$k(t) = \frac{P(t,T_0) - P(t,T_n)}{b(t)}.$$

Exercise 12.11: (Jamshidian (1996) Swaptions Pricing Formula)
Assume that the swap rate $k(t) > 0$, and that $v^2(t) = \int_t^T \frac{1}{k^2(s)}d[k,k](s)$ is deterministic. Show that

$$\text{Swaption}(t) = b(t)\left(\alpha_+(t)k(t) - k\alpha_-(t)\right),$$

where

$$\alpha_\pm(t) = \Phi\left(\frac{\ln k(t)/k}{v(t)} \pm \frac{v(t)}{2}\right).$$

Chapter 13

Applications in Biology

In this chapter applications of stochastic calculus to population models are given. Diffusion models, Markov Jump process models, age-dependent non-Markov models and stochastic models for competition of species are presented. Diffusion Models are used in various areas of Biology as models for population growth and genetic evolution. Birth and Death processes are random walks in continuous time and are met in various applications. A novel approach to the age-dependent branching (Bellman-Harris) process is given by treating it as a simple measure-valued process. The stochastic model for interacting populations that generalizes the Lotka-Volterra prey-predator model is treated by using a semimartingale representation. It is possible to formulate these models as stochastic differential or integral equations. We demonstrate how results on stochastic differential equations and martingales presented in earlier chapters are applied for their analysis.

13.1 Feller's Branching Diffusion

A simple branching process is a model in which individuals reproduce independently of each other and of the history of the process. The continuous approximation to branching process is the branching diffusion. It is given by the stochastic differential equation for the population size $X(t)$, $0 < X(t) < \infty$,

$$dX(t) = \alpha X(t)dt + \sigma\sqrt{X(t)}dB(t). \tag{13.1}$$

In this model the infinitesimal drift and variance are proportional to the population size. The corresponding forward (Kolmogorov or Fokker-Plank) equation for the probability density of $X(t)$ is

$$\frac{\partial p(t,x)}{\partial t} = -\alpha\frac{\partial xp(t,x)}{\partial x} + \frac{\sigma^2}{2}\frac{\partial xp(t,x)}{\partial x^2}. \tag{13.2}$$

Analysis of the process was done by solving the partial differential equation (13.2) (Feller (1951)). Here we demonstrate the stochastic calculus approach by obtaining the information directly from the stochastic equation (13.1). First we prove that a solution of (13.1) is always non-negative, moreover once it hits 0 it stays at 0 forever.

Theorem 13.1 *Let $X(t)$ solve (13.1) and $X(0) > 0$. Then $P(X(t) \geq 0$ for all $t \geq 0) = 1$, moreover if $\tau = \inf\{t : X(t) = 0\}$, then $X(t) = 0$ for all $t \geq \tau$.*

PROOF: Consider first (13.1) with $X(0) = 0$. Clearly, $X(t) \equiv 0$ is a solution. Conditions of the Yamada-Watanabe result on the existence and uniqueness are satisfied, see Theorem 5.5. Therefore solution is unique, and $X(t) \equiv 0$ is the only solution. Consider now (13.1) with $X(0) > 0$. The first hitting time of zero τ is a stopping time with $X(\tau) = 0$. By the strong Markov property $X(t)$ for $t > \tau$ is the solution to (13.1) with zero initial condition, hence it is zero for all $t > \tau$. Thus if $X(0) > 0$, the process is positive for all $t < \tau$, and zero for all $t \geq \tau$.

$\square$

The next result describes the exponential growth of the population.

Theorem 13.2 *Let $X(t)$ solve (13.1) and $X(0) > 0$. Then $EX(t) = X(0)e^{\alpha t}$. $X(t)e^{-\alpha t}$ is a non-negative martingale which converges almost surely to a non-degenerate limit as $t \to \infty$.*

PROOF: First we show that $X(t)$ is integrable. Since Itô integrals are local martingales, $\int_0^t \sqrt{X(s)}dB(s)$ is a local martingale. Let T_n be a localizing sequence, so that $\int_0^{t \wedge T_n} \sqrt{X(s)}dB(s)$ is a martingale in t for any fixed n. Then using (13.1) we can write

$$X(t \wedge T_n) = X(0) + \alpha \int_0^{t \wedge T_n} X(s)ds + \sigma \int_0^{t \wedge T_n} \sqrt{X(s)}dB(s). \qquad (13.3)$$

Taking expectation, we obtain

$$EX(t \wedge T_n) = X(0) + \alpha E \int_0^{t \wedge T_n} X(s)ds. \qquad (13.4)$$

Since X is non-negative, $\int_0^{t \wedge T_n} X(s)ds$ is increasing to $\int_0^t X(s)ds$. Therefore by monotone convergence,

$$E \int_0^{t \wedge T_n} X(s)ds \to E \int_0^t X(s)ds,$$

as $n \to \infty$. Therefore $\lim_{n \to \infty} \mathrm{E}X(t \wedge T_n) = X(0) + \mathrm{E}\int_0^t X(s)ds$. Since $X(t \wedge T_n) \to X(t)$ we obtain by Fatou's lemma (noting that if a limit exists, then $\lim \inf = \lim$)

$$
\begin{aligned}
\mathrm{E}X(t) &= \mathrm{E}\lim_{n \to \infty} X(t \wedge T_n) = \mathrm{E}\liminf_{n \to \infty} X(t \wedge T_n) \\
&\leq \liminf_{n \to \infty} \mathrm{E}X(t \wedge T_n) = \lim_{n \to \infty} \mathrm{E}X(t \wedge T_n) = X(0) + \mathrm{E}\int_0^t X(s)ds.
\end{aligned}
$$

Using Gronwall's inequality (Theorem 1.20) we obtain that the expectation is finite,

$$\mathrm{E}X(t) \leq (1 + X(0))e^{\alpha t}. \tag{13.5}$$

Now we can show that the local martingale $\int_0^t \sqrt{X(s)}dB(s)$ is a true martingale. Consider its quadratic variation

$$\mathrm{E}\Big[\int_0^\cdot \sqrt{X(s)}dB(s), \int_0^\cdot \sqrt{X(s)}dB(s)\Big](t) = \mathrm{E}\int_0^t X(s)ds < Ce^{\alpha t}.$$

Thus by Theorem 7.35 $\int_0^t \sqrt{X(s)}dB(s)$ is a martingale. Now we can take expectations in (13.1). Differentiating with respect to t and solving the resulting equation, we find $\mathrm{E}X(t) = X(0)e^{\alpha t}$.

To prove the second statement, use integration by parts for $X(t)e^{-\alpha t}$ to obtain

$$U(t) = X(t)e^{-\alpha t} = X(0) + \sigma\int_0^t e^{-\alpha s}\sqrt{X(s)}dB(s),$$

and $U(t)$ is a local martingale.

$$\mathrm{E}[U,U](t) = \mathrm{E}\int_0^t e^{-2\alpha s}X(s)ds = \int_0^t e^{-\alpha s}ds < \infty,$$

and $\mathrm{E}[U,U](\infty) < \infty$. Therefore $U(t)$ is a martingale, moreover, it is square integrable on $[0,\infty)$. Since it is uniformly integrable, it converges to a non-degenerate limit.

$\square$

The next result uses diffusion theory to find the probability of extinction.

Theorem 13.3 *Let $X(t)$ solve (13.1) and $X(0) = x > 0$. Then the probability of ultimate extinction is* $e^{-\frac{2\alpha}{\sigma^2}x}$ *if $\alpha > 0$, and 1 if $\alpha \leq 0$*

PROOF: Let $T_0 = \tau = \inf\{t : X(t) = 0\}$ be the first time of hitting zero, and T_b the first time of hitting $b > 0$. By the formula (6.52) on exit probabilities

$$\mathrm{P}_x(T_0 < T_b) = \frac{S(b) - S(x)}{S(b) - S(0)}, \tag{13.6}$$

where $S(x)$ is the scale function, see (6.50). In this example

$$S(x) = \int_{x_1}^{x} e^{-\int_{x_0}^{u} \frac{2\alpha}{\sigma^2} dy} du, \tag{13.7}$$

where x_0 and x_1 are some positive constants. Simplifying (13.7) we obtain from (13.6)

$$P_x(T_0 < T_b) = \frac{e^{-cx} - e^{-cb}}{1 - e^{-cb}}, \tag{13.8}$$

with $c = 2\alpha/\sigma^2$. The probability of ultimate extinction is obtained by taking limit as $b \to \infty$ in (13.8), that is,

$$P_x(T_0 < T_\infty) = \lim_{b \to \infty} P_x(T_0 < T_b) = \begin{cases} e^{-\frac{2\alpha}{\sigma^2} x} & \text{if } \alpha > 0, \\ 1 & \text{if } \alpha \le 0. \end{cases} \tag{13.9}$$

where $T_\infty = \lim_{b\to\infty} T_b$ is the explosion time, which is infinite if the explosion does not occur.

$\square$

Branching diffusion is related to a stopped Brownian motion with drift by a change of time. See Example 7.17.

13.2   Wright-Fisher Diffusion

In population dynamics frequencies of genes or alleles are studied. It is assumed for simplicity that the population size N is fixed and individuals are of two types: A and a. If individuals of type A mutate to type a with the rate γ_1/N and individuals of type a mutate to type A with the rate γ_2/N, then it is possible to approximate the frequency of type A individuals $X(t)$ by the Wright-Fisher diffusion, given by the stochastic equation

$$dX(t) = \big(-\gamma_1 X(t) + \gamma_2(1 - X(t)) \big) dt + \sqrt{X(t)(1 - X(t))} dB(t), \tag{13.10}$$

with $0 < X(0) < 1$. For complete description of the process its behaviour at the boundaries 0 and 1 should also be specified. When $X(t) = 0$ or 1, then all the individuals at time t are of the same type. Consider first the case of no mutation: $\gamma_1 = \gamma_2 = 0$. Then the equation for $X(t)$ is

$$dX(t) = \sqrt{X(t)(1 - X(t))} dB(t), \tag{13.11}$$

with $0 < X(0) = x < 1$. The scale function is given by $S(x) = x$, consequently for $0 \le a < x < b \le 1$

$$P_x(T_b < T_a) = \frac{x - a}{b - a}. \tag{13.12}$$

The process is stopped when it reaches 0 or 1, because at that time and at all future times all the individuals are of the same type. This phenomenon is called fixation. The probability that fixation at A occurs having started with proportion x of type A individuals is x, and with the complementary probability fixation at a occurs.

The expected time to fixation is found by Theorem 6.16 as the solution to $Lv = -1$ with boundary conditions $v(0) = v(1) = 0$, and $Lv = \frac{x(1-x)}{2}v''$. Solving this equation (using $\int \ln x = x \ln x - x$), we obtain that the expected time to fixation having started with proportion x of type A individuals is given by

$$v(x) = \mathrm{E}_x \tau = -2\big((1-x)\ln(1-x) + x\ln x\big). \qquad (13.13)$$

In the model with one-way mutation when, for example $\gamma_2 = 0$, $\gamma_1 = \gamma$, $X(t)$ satisfies

$$dX(t) = -\gamma X(t)dt + \sqrt{X(t)(1 - X(t))}dB(t), \qquad (13.14)$$

with $0 < X(0) < 1$. The process is stopped once it reaches the boundaries 0 or 1. In this case the scale function is given by $S(x) = \big(1 - (1-x)^{1-2\gamma}\big)/(1-2\gamma)$ if $\gamma \neq 1/2$ and $S(x) = -\log(1-x)$ when $\gamma = 1/2$. Note that by continuity of paths $T_b \uparrow T_1$ as $b \uparrow 1$, and it is easy to see that if $\gamma \geq 1/2$ then

$$\mathrm{P}_x(T_1 < T_0) = \lim_{b \uparrow 1} \mathrm{P}_x(T_b < T_0) = 0. \qquad (13.15)$$

It is possible to see by Theorem 6.16 that the expected time to fixation is finite. Thus fixation at type a occurs with probability 1. If $\gamma < 1/2$ then expected time to fixation is finite, but there is a positive probability that fixation at type A also occurs.

In the model with two-way mutation both $\gamma_1, \gamma_2 > 0$. Analysis of this model is done by using the diffusion processes techniques described in Chapter 6, but it is too involved to be given here in detail. The important feature of this model is that fixation does not occur and $X(t)$ admits a stationary distribution. Stationary distributions can be found by formula (6.69). We find

$$\pi(x) = \frac{C}{\sigma^2(x)} \exp\left(\int_{x_0}^x \frac{2\mu(y)}{\sigma^2(y)}dy\right) = C(1-x)^{2\gamma_1 - 1}x^{2\gamma_2 - 1}, \qquad (13.16)$$

which is the density of a Beta distribution. $C = \Gamma(2\gamma_1)\Gamma(2\gamma_2)/\Gamma(2\gamma_1 + 2\gamma_2)$.

Diffusion models find frequent use in population genetics, see for example, Ewens (1979). For more information on Wright-Fisher diffusion see, for example, Karlin and Taylor (1981), Chapter 15.

Remark 13.1: The Theory of weak convergence is used to show that a diffusion approximation to alleles frequencies is valid. This theory is not presented here but can be found in many texts, see for example Liptser and Shiryayev (1989), Ethier and Kurtz (1986).

13.3   Birth-Death Processes

A Birth-Death process is a model of a developing in time population of particles. Each particle lives for a random length of time at the end of which it splits into two particles or dies. If there are x particles in the population, then the lifespan of a particle has an exponential distribution with parameter $a(x)$, the split into two occurs with probability $p(x)$ and with the complimentary probability $1 - p(x)$ a particle dies producing no offspring.

Denote by $X(t)$ the population size at time t. The change in the population occurs only at the death of a particle, and from state x the process jumps to $x + 1$ if a split occurs or to $x - 1$ if a particle dies without splitting. Thus the jump variables $\xi(x)$ take values 1 and -1 with probabilities $p(x)$ and $1 - p(x)$ respectively. Using the fact that the minimum of exponentially distributed random variables is exponentially distributed, we can see that the process stays at x for an exponentially distributed length of time with parameter $\lambda(x) = xa(x)$.

Usually, the Birth-Death process is described in terms of birth and death rates; in a population of size x, a particle is born at rate $b(x)$ and dies at the rate $d(x)$. These refer to the infinitesimal probabilities of population increasing and decreasing by one, namely for an integer x

$$\begin{aligned} \mathrm{P}(X(t+\delta) = x + 1 | X(t) = x) &= b(x)\delta + o(\delta), \\ \mathrm{P}(X(t+\delta) = x - 1 | X(t) = x) &= d(x)\delta + o(\delta), \end{aligned}$$

and with the complimentary probability no births or deaths happen in $(t, t+\delta)$ and the process remains unchanged

$$\mathrm{P}(X(t+\delta) = x | X(t) = x) = 1 - (b(x) + d(x))\delta + o(\delta).$$

Here $o(\delta)$ denotes a function that $\lim_{\delta \to 0} o(\delta)/\delta = 0$. Once population reaches zero it stays there forever.

$$\mathrm{P}(X(t+\delta) = 0 | X(t) = 0) = 1.$$

It can be seen that these assumptions lead to the model of a Jump Markov process with the holding time parameter at x

$$\lambda(x) = b(x) + d(x), \tag{13.17}$$

and the jump from x

$$\xi(x) = \left\{ \begin{array}{ll} +1 & \text{with probability } \frac{b(x)}{b(x)+d(x)} \\ -1 & \text{with probability } \frac{d(x)}{b(x)+d(x)} \end{array} \right. \tag{13.18}$$

The first two moments of the jumps are easily computed to be

$$m(x) = \frac{b(x) - d(x)}{b(x) + d(x)} \text{ and } v(x) = 1. \tag{13.19}$$

By Theorem 9.15 the compensator of $X(t)$ is given in terms of

$$\lambda(x)m(x) = b(x) - d(x),$$

which is the survival rate in the population. Thus the Birth-Death process $X(t)$ can be written as

$$X(t) = X(0) + \int_0^t \big(b(X(s)) - d(X(s))\big)ds + M(t), \tag{13.20}$$

where $M(t)$ is a local martingale with the sharp bracket

$$\langle M, M \rangle (t) = \int_0^t \big(b(X(s)) + d(X(s))\big)ds. \tag{13.21}$$

Since $E|\xi(x)| = 1$, it follows by Theorem 9.16 that if the linear growth condition

$$b(x) + d(x) \leq C(1 + x), \tag{13.22}$$

holds then the local martingale M in representation (13.20) is a square integrable martingale.

Sometimes it is convenient to describe the model in terms of the *individual* birth and death rates. In a population of size x, each particle is born at the rate $\beta(x)$ and dies at the rate $\gamma(x)$. The individual rates relate to the population rates by

$$b(x) = x\beta(x), \text{ and } d(x) = x\gamma(x). \tag{13.23}$$

Introduce the individual survival rate $\alpha(x) = \beta(x) - \gamma(x)$. Then the stochastic equation (13.20) becomes

$$X(t) = X(0) + \int_0^t \alpha(X(s))X(s)ds + M(t). \tag{13.24}$$

Birth-Death Processes with Linear Rates

Suppose that the *individual* birth and death rates β and γ are constants. This corresponds to linear rates $b(x) = \beta x$, $d(x) = \gamma x$ (see (13.23)). The linear growth condition 13.22 is satisfied, and we have from equation (13.20)

$$X(t) = X(0) + \alpha \int_0^t X(s)ds + M(t), \tag{13.25}$$

where $\alpha = \beta - \gamma$ is the survival rate and $M(t)$ is a martingale. Moreover, by (13.21)

$$\langle M, M \rangle (t) = (\beta + \gamma) \int_0^t X(s) ds. \tag{13.26}$$

Taking expectations in (13.25)

$$EX(t) = X(0) + \alpha \int_0^t EX(s) ds, \tag{13.27}$$

and solving this equation, we find

$$EX(t) = X_0 e^{\alpha t}. \tag{13.28}$$

Using integration by parts (note that $e^{\alpha t}$ has zero covariation with $X(t)$), we have

$$d\left(X(t) e^{-\alpha t}\right) = e^{-\alpha t} dX(t) - \alpha e^{-\alpha t} X(t-) dt. \tag{13.29}$$

Since $X(t-) dt$ can be replaced by $X(t) dt$, by using the equation (13.25) in its differential form we obtain

$$d\left(X(t) e^{-\alpha t}\right) = e^{-\alpha t} dX(t) - \alpha e^{-\alpha t} X(t) dt = e^{-\alpha t} dM(t). \tag{13.30}$$

Thus

$$X(t) e^{-\alpha t} = X(0) + \int_0^t e^{-\alpha s} dM(s). \tag{13.31}$$

Using the rule for the sharp bracket of the integral and equation (13.26), we find

$$\left\langle \int_0^{\cdot} e^{-\alpha s} dM(s), \int_0^{\cdot} e^{-\alpha s} dM(s) \right\rangle (t) = \int_0^t e^{-2\alpha s} d\langle M, M \rangle (s)$$

$$= (\beta + \gamma) \int_0^t e^{-2\alpha s} X(s) ds.$$

Since $EX(t) = X(0) e^{\alpha t}$, by taking expectations it follows that

$$E \left\langle \int_0^{\cdot} e^{-\alpha s} dM(s), \int_0^{\cdot} e^{-\alpha s} dM(s) \right\rangle (\infty) < \infty.$$

Consequently $X(t) e^{-\alpha t}$ is a square integrable martingale on $[0, \infty)$.

$$Var(X(t) e^{-\alpha t}) = E \left\langle \int_0^t e^{-\alpha s} dM(s) \right\rangle (t) = X(0) \frac{\beta + \gamma}{\alpha} \left(1 - e^{-\alpha t}\right). \tag{13.32}$$

Since $X(t) e^{-\alpha t}$ is square integrable on $[0, \infty)$, it converges almost surely as $t \to \infty$ to a non-degenerate limit.

Processes with Stabilizing Reproduction

Consider the case when the rates stabilize as population size increases, namely $\beta(x) \to \beta$ and $\gamma(x) \to \gamma$ as $x \to \infty$. Then, clearly, $\alpha(x) = \beta(x) - \gamma(x) \to \alpha = \beta - \gamma$. Depending on the value of α radically distinct modes of behaviour occur. If $\alpha < 0$ then the process dies out with probability one. If $\alpha > 0$ then the process tends to infinity with a positive probability. If the rate of convergence of $\alpha(x)$ to α is fast enough, then exponential growth persists as in the classical case. Let $\varepsilon(x) = \alpha(x) - \alpha$. Then under some technical conditions, the following condition is necessary and sufficient for exponential growth,

$$\int_1^\infty \frac{\varepsilon(x)}{x} dx < \infty. \tag{13.33}$$

If (13.33) holds then $X(t)e^{-\alpha t}$ converges almost surely and in the mean square to a non-degenerate limit.

The case $\alpha = 0$ provides examples of slowly growing populations, such as those with a linear rate of growth. Let $\alpha(x) > 0$ and $\alpha(x) \downarrow 0$. Among such processes there are some that become extinct with probability one, but there are others that have a positive probability of growing to infinity. Consider the case when $\alpha(x) = c/x$ when $x > 0$. It is possible to show that if $c > 1/2$ then $q = P(X(t) \to \infty) > 0$. Stochastic equation (13.24) becomes

$$X(t) = X(0) + c \int_0^t I(X(s) > 0) ds + M(t). \tag{13.34}$$

By taking expectations in (13.34) one can see that $\lim_{t\to\infty} EX(t)/t = cq$, and the mean of such processes grows linearly. By using Theorem 9.19 and Exercise 9.5 it is possible to show that for any k, $E(X^k(t))/t^k$ converges to the k-th moment of a gamma distribution times q. This implies that on the set $\{X(t) \to \infty\}$, $X(t)/t$ converges in distribution to a gamma distribution. In other words, such processes grow linearly, and not exponentially. By changing the rate of convergence of $\alpha(x)$ other polynomial growth rates of the population can be obtained.

Similar results are available for population-dependent Markov Branching processes that generalize Birth-Death processes by allowing a random number of offspring at the end of a particle's life. For details of the stochastic equation approach to Markov Jump processes see Klebaner (1994), where such processes were studied as randomly perturbed linear differential equations; Hamza and Klebaner (1995).

13.4   Branching Processes

In this Section we look at a population model known as the age-dependent Bellman-Harris Branching process, see Harris (1963), Athreya and Ney (1972). This model generalizes the Birth and Death process model in two respects: firstly the lifespan of individuals need not have the exponential distribution, and secondly, more than one particle can be born. In the age-dependent model a particle lives for a random length of time with distribution G, and at the end of its life leaves a random number of children Y, with mean $m = EY$. All particles reproduce and die independently of each other. Denote by $Z(t)$ the number of particles alive at time t. It can be seen that unless the lifetime distribution is exponential, the process $Z(t)$ is not Markovian, and its analysis is usually done by using Renewal Theory (Harris (1963)). Here we apply stochastic calculus to obtain a limiting result for the population size counted in a special way (by reproductive value).

Consider a collection of individuals with ages $(a^1, \ldots, a^z) = A$. It is convenient to look at the vector of ages $A = (a^1, \ldots, a^z)$ as a counting measure A, defined by $A(B) = \sum_{i=1}^{z} 1_B(a^i)$, for any Borel set B in $\mathbb{R}^+$. For a function f on $\mathbb{R}$ the following notations are used

$$(f, A) = \int f(x) A(dx) = \sum_{i=1}^{z} f(a^i).$$

The population size process in this notation is $Z(t) = (1, A_t)$. In this approach the process of ages A_t is a measure-valued process, although simpler than studied e.g. in Dawson (1993). To convert a measure-valued process into a scalar valued, test functions are used. Test functions used on the space of measures are of the form $F((f, \mu))$, where F and f are functions on $\mathbb{R}$. Let $h(a) = \frac{G'(a)}{1-G(a)}$ be rate of dying at age a. E with and without the subscript A denotes the expectation when the processes starts with individuals aged $(a^1, \ldots, a^z) = A$. The following result is obtained by direct calculations.

Theorem 13.4 *For a bounded differentiable function F on $\mathbb{R}^+$ and a continuously differentiable function f on $\mathbb{R}^+$, the following limit exists*

$$\lim_{t \to 0} \frac{1}{t} E_A \Big\{ F((f, A_t)) - F((f, A)) \Big\} = \mathcal{G}F((f, A)), \qquad (13.35)$$

$$\mathcal{G}F((f, A)) = F'((f, A))(f', A) \qquad (13.36)$$

$$+ \sum_{i=1}^{z} h(a^j)\{E_A \left(F(Y_j f(0) + (f, A) - f(a^j)) \right) - F((f, A))\},$$

The operator $\mathcal{G}$ in (13.36) defines a generator of a measure-valued branching process in which the movement of the particles is deterministic, namely shift. The following result gives the integral equation (SDE) for the age process under the test functions.

Theorem 13.5 *For a bounded C^1 function F on $\mathbb{R}$ and a C^1 function f on $\mathbb{R}^+$*

$$F((f, A_t)) = F((f, A_0)) + \int_0^t \mathcal{G}F((f, A_s))ds + M_t^{F,f}, \qquad (13.37)$$

where $M_t^{F,f}$ is a local martingale with the sharp bracket given by

$$\langle M^{F,f}, M^{F,f}\rangle_t = \int_0^t \mathcal{G}F^2((f, A_s))ds - 2\int_0^t F((f, A_s))\mathcal{G}F((f, A_s))ds. \quad (13.38)$$

Consequently,

$$\mathrm{E}_A(M_t^{F,f})^2 = \mathrm{E}_A\Big(\int_0^t \mathcal{G}F^2((f, A_s))ds - 2\int_0^t F((f, A_s))\mathcal{G}F((f, A_s))ds\Big),$$

provided $\mathrm{E}_A(M_t^{F,f})^2$ exists.

PROOF: The first statement is obtained by Dynkin's formula. Expression (13.38) is obtained by letting $U_t = F((f, A_t))$ and an application of the following result. □

Theorem 13.6 *Let U_t be a locally square-integrable semi-martingale such that $U_t = U_0 + A_t + M_t$, where A_t is a predictable process of locally finite variation and M_t is a locally square-integrable local martingale, $A_0 = M_0 = 0$. Let $U_t^2 = U_0^2 + B_t + N_t$, where B_t is a predictable process and N_t is a local martingale. Then*

$$\langle M, M\rangle_t = B_t - 2\int_0^t U_{s-}dA_s - \sum_{s \le t}(A_s - A_{s-})^2. \qquad (13.39)$$

Of course, if A is continuous (as in our applications) the last term in the above formula (13.39) vanishes.

PROOF: By the definition of the quadratic variation process (or integration by parts)

$$U_t^2 = U_0^2 + 2\int_0^t U_{s-}dU_s + [U, U]_t = U_0^2 + 2\int_0^t U_{s-}dA_s + 2\int_0^t U_{s-}dM_s + [U, U]_t.$$

Using the representation for U_t^2 given in the conditions of the lemma, we obtain that $[U, U]_t - B_t + 2 \int_0^t U_{s-} dA_s$ is a local martingale. Since $[A, A]_t = \sum_{s \le t} (A_s - A_{s-})^2$, the result follows.

□

Let $v^2 = \mathrm{E}(Y^2)$ be the second moment of the offspring distribution. Applying Dynkin's formula to the function $F(u) = u$ (and writing M^f for $M^{1,f}$), we obtain the following

Theorem 13.7 *For a C^1 function f on $\mathbb{R}^+$*

$$(f, A_t) = (f, A_0) + \int_0^t (Lf, A_s)ds + M_t^f, \tag{13.40}$$

where the linear operator L is defined by

$$Lf = f' - hf + mhf(0), \tag{13.41}$$

and M_t^f is a local square integrable martingale with the sharp bracket given by

$$\langle M^f, M^f \rangle_t = \int_0^t \left(f^2(0)v^2 h + hf^2 - 2f(0)mhf, A_s \right) ds. \tag{13.42}$$

PROOF: The first statement is Dynkin's formula for $F(u) = u$. This function is unbounded and the standard formula cannot be applied directly. However it can be applied by taking smooth bounded functions that agree with u on bounded intervals, $F_n(u) = u$ for $u \le n$, and the sequence of stopping times $T_n = \inf\{(f, A_t) > n\}$ as a localizing sequence. The form of the operator L follows from (13.36). Similarly (13.42) follows from (13.38) by taking $F(u) = u^2$.

□

By taking f to be a constant, $f(u) = 1$, Theorem 13.7 yields the following corollary for the population size at time t, $Z(t) = (1, A_t)$.

Corollary 13.8 *The compensator of $Z(t)$ is given by $(m - 1) \int_0^t (h, A_s)ds$.*

It is useful to be able to take expectations in Dynkin's formula. The next result gives a sufficient condition.

Theorem 13.9 *Let $f \ge 0$ be a C^1 function on $\mathbb{R}^+$ that satisfies*

$$|(Lf, A)| \le C(1 + (f, A)), \tag{13.43}$$

for some constant C and any A, and assume that (f, A_0) is integrable. Then (f, A_t) and M_t^f in (13.37) are also integrable with $\mathrm{E}M_t^f = 0$.

PROOF: Let T_n be a localizing sequence, then from (13.37)

$$(f, A_{t \wedge T_n}) = (f, A_0) + \int_0^{t \wedge T_n} (Lf, A_s)ds + M_{t \wedge T_n}^f, \tag{13.44}$$

where $M_{t \wedge T_n}^f$ is a martingale. Taking expectations we have

$$\mathrm{E}(f, A_{t \wedge T_n}) = \mathrm{E}(f, A_0) + \mathrm{E} \int_0^{t \wedge T_n} (Lf, A_s)ds. \tag{13.45}$$

By Condition (13.43),

$$\begin{aligned}
|\mathrm{E} \int_0^{t \wedge T_n} (Lf, A_s)ds| &\leq \mathrm{E} \int_0^{t \wedge T_n} |(Lf, A_s)|ds \\
&\leq Ct + C\mathrm{E} \int_0^{t \wedge T_n} (f, A_s)ds.
\end{aligned}$$

Thus we have from (13.45)

$$\begin{aligned}
\mathrm{E}(f, A_{t \wedge T_n}) &\leq \mathrm{E}(f, A_0) + Ct + C\mathrm{E} \int_0^{t \wedge T_n} (f, A_s)ds \\
&\leq \mathrm{E}(f, A_0) + Ct + C\mathrm{E} \int_0^t I(s \leq T_n)(f, A_s)ds \\
&\leq \mathrm{E}(f, A_0) + Ct + C \int_0^t \mathrm{E}(f, A_{s \wedge T_n})ds.
\end{aligned}$$

It now follows by Gronwall's inequality (Theorem 1.20) that

$$\mathrm{E}(f, A_{t \wedge T_n}) \leq \mathrm{E}(f, A_0) + Ct + te^{Ct} < \infty. \tag{13.46}$$

Taking $n \to \infty$, we conclude that $\mathrm{E}(f, A_t) < \infty$ by Fatou's lemma. Thus (f, A_t) is integrable, as it is non-negative. Now by Condition (13.43)

$$\mathrm{E}| \int_0^t (Lf, A_s)ds| \leq \int_0^t \mathrm{E}|(Lf, A_s)|ds \leq \int_0^t C(1 + \mathrm{E}(f, A_s))ds < \infty. \tag{13.47}$$

It follows from (13.47) that $\int_0^t (Lf, A_s)ds$ and its variation process $\int_0^t |(Lf, A_s)|ds$ are both integrable, and from (13.37) that

$$M_t^f = (f, A_t) - (f, A_0) - \int_0^t (Lf, A_s)ds \tag{13.48}$$

is integrable with zero mean.

$\square$

For simplicity we assume that $G(u) < 1$ for all $u \in \mathbb{R}^+$. Equation (13.40) can be analyzed through the eigenvalue problem for the operator L.

Theorem 13.10 *Let L be the operator in (13.41). Then the equation*

$$Lq = rq \tag{13.49}$$

has a solution q_r for any r. The corresponding eigenfunction (normed so that $q(0) = 1$) is given by

$$q_r(u) = \frac{me^{ru}}{1 - G(u)} \left(1 - \int_0^u e^{-rs} dG(s)\right). \tag{13.50}$$

Proof: Since eigenfunctions are determined up to a multiplicative constant, we can take $q(0) = 1$. Equation (13.49) is a first order linear differential equation, and solving it we obtain the solution (13.50).

Theorem 13.11 *Let q_r be a postive eigenfunction of L corresponding to the eigenvalue r. Then $Q_r(t) = e^{-rt}(q_r, A_t)$ is a positive martingale.*

PROOF:　Using (13.40) and the fact that q_r is an eigenfunction for L, we have

$$(q_r, A_t) = (q_r, A_0) + r \int_0^t (q_r, A_s) ds + M_t^{q_r}, \tag{13.51}$$

where $M_t^{q_r}$ is a local martingale. The functions q_r clearly satisfy condition (13.43). Therefore (q_r, A_t) is integrable, and it follows from (13.51) by taking expectations that

$$E(q_r, A_t) = e^{rt} E(q_r, A_0). \tag{13.52}$$

Using integration by parts for $e^{-rt}(q_r, A_t)$, we obtain from (13.51) that

$$dQ_r(t) = d(e^{-rt}(q_r, A_t)) = e^{-rt} dM_t^{q_r},$$

and

$$Q_r(t) = (q_r, A_0) + \int_0^t e^{-rs} dM_s^{q_r} \tag{13.53}$$

is a local martingale as an integral with respect to the local martingale M^{q_r}. Since a positive local martingale is a supermartingale, and $Q_r(t) \geq 0$, $Q_r(t)$ is a super-martingale. But from (13.52) it follows that $Q_r(t)$ has a constant mean. Thus the supermartingale $Q_r(t)$ is a martingale.

$\square$

The Malthusian parameter α is defined as the value of r which satisfies

$$m \int_0^\infty e^{-ru} dG(u) = 1. \tag{13.54}$$

We assume that the Malthusian parameter α exists and is *positive*, in this case the process is called supercritical.

Theorem 13.12 *There is only one bounded positive eigenfunction V, the reproductive value function, corresponding to the eigenvalue α which is the Malthusian parameter,*

$$V(u) = \frac{me^{\alpha u}}{1 - G(u)} \int_u^\infty e^{-\alpha s} dG(s).$$ (13.55)

PROOF: It follows that for $r > \alpha$, $m \int_0^\infty e^{-ru} dG(u) < 1$ and the eigenfunction q_r in (13.50) is positive and grows exponentially fast or faster. For $r < \alpha$, the eigenfunction q_r takes negative values. When $r = \alpha$, $q^\alpha = V$ in (13.55). To see that it is bounded by m, replace $e^{-\alpha s}$ by its largest value $e^{-\alpha u}$.

□

This is the main result

Theorem 13.13 $W_t = e^{-\alpha t}(V, A_t)$ *is a positive square integrable martingale, and therefore converges almost surely and in L^2 to the non-degenerate limit $W \geq 0$, $EW = (V, A_0) > 0$ and $P(W > 0) > 0$.*

PROOF: That W_t is a martingale follows by Theorem 13.11. It is positive therefore it converges almost surely to a limit $W \geq 0$ by the martingale convergence Theorem 7.11. To show that the limit is not identically zero, we show that convergence is also in L^2, i.e. the second moments (and the first moments) converge, implying that $EW = \lim_{t \to \infty} EW_t = EW_0 = (V, A_0)$.

It follows from (13.42) that

$$\langle M^V, M^V \rangle_t = \int_0^t \left((\sigma^2 + (m - V)^2)h, A_s \right) ds.$$ (13.56)

By (13.53)

$$W_t = (V, A_0) + \int_0^t e^{-\alpha s} dM_s^V,$$ (13.57)

and we obtain that

$$\langle W, W \rangle_t = \int_0^t e^{-2\alpha s} d\langle M^V, M^V \rangle_s = \int_0^t e^{-2\alpha s} \left((\sigma^2 + (m - V)^2)h, A_s \right) ds.$$ (13.58)

Now it is easy to check that there is a constant C and r, $\alpha \leq r < 2\alpha$, such that

$$\left((\sigma^2 + (m - V)^2)h, A_s \right) \leq C(q_r, A_s),$$ (13.59)

and using Theorem 13.11,

$$E \int_0^\infty e^{-2\alpha s} \left((\sigma^2 + (m - V)^2)h, A_s \right) ds < C \int_0^\infty e^{(r - 2\alpha)s} < \infty.$$ (13.60)

This implies from (13.58) that $\mathrm{E}\langle W, W\rangle_\infty < \infty$. Therefore W_t is a square integrable martingale (see Theorem 8.27) and the result follows.

$\square$

The martingale $\{W_t\}$ is given in Harris (1963). The convergence of the population numbers $Z(t)e^{-\alpha t}$ is obtained from that of W_t by using a Tauberian theorem. For details and extensions of the model to the population-dependent case see Jagers and Klebaner (2000).

13.5   Stochastic Lotka-Volterra Model

Deterministic Lotka-Volterra system

The Lotka-Volterra system of ordinary differential equations, (Lotka (1925), Volterra (1926))

$$\dot{x}_t = \mathbf{a}x_t - \mathbf{b}x_t y_t$$
$$\dot{y}_t = \mathbf{c}x_t y_t - \mathbf{d}y_t, \tag{13.61}$$

with positive x_0, y_0 and positive parameters $\mathbf{a}$, $\mathbf{b}$, $\mathbf{c}$, $\mathbf{d}$ describes a behaviour of a prey-predator system in terms of the prey and predator "intensities" x_t and y_t. Here, $\mathbf{a}$ is the rate of increase of prey in the absence of predator, $\mathbf{d}$ is a rate of decrease of predator in the absence of prey while the rate of decrease in prey is proportional to the number of predators $\mathbf{b}y_t$, and similarly the rate of increase in predator is proportional to the number of prey $\mathbf{c}x_t$. The system (13.61) is one of the simplest non-linear systems.

Since the population numbers are discrete, a description of the predator-prey model in terms of continuous intensities x_t, y_t is based implicitly on a natural assumption that the numbers of both populations are large, and the intensities are obtained by a normalization of population numbers by a large parameter K. Thus (13.61) is an approximation, an asymptotic description of the interaction between the predator and the prey. Although this model may capture some essential elements in that interaction, it is not suitable to answer questions of extinction of populations, as the extinction never occurs in the deterministic model, see Figure 13.5 for the pair x_t, y_t in the phase plane.

We introduce here a probabilistic model which has as its limit the deterministic Lotka-Volterra model, evolves in continuous time according to the same local interactions and allows us to evaluate asymptotically the time for extinction of prey species.

There is a vast amount of literature on the Lotka-Volterra model and a history of research on stochastic perturbations of this system exact, approximate and numerical.

The system (13.61) possesses the first integral which is a closed orbit in the first quadrant of phase plane x, y. It is given by

$$r(x, y) = \mathbf{c}x - \mathbf{d}\log x + \mathbf{b}y - \mathbf{a}\log y + r_0, \qquad (13.62)$$

where r_0 is an arbitrary constant. It depends only on the initial point (x_0, y_0), see Figure below.

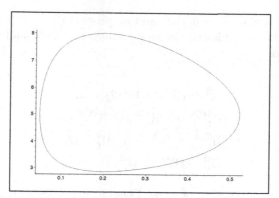

Figure 13.1: First integral $r(x, y)$, $x_0 = 0.3, y_0 = 3, \mathbf{a} = 5, \mathbf{b} = 1, \mathbf{c} = 5, \mathbf{d} = 1$.

Stochastic Lotka-Volterra system

Let X_t and Y_t be numbers of prey and predators at time t. We start with simple equations for prey-predator populations

$$X_t = X_0 + \pi'_t - \pi''_t$$
$$Y_t = Y_0 + \widehat{\pi}'_t - \widehat{\pi}''_t \qquad (13.63)$$

where π'_t is the number of prey born up to time t, π''_t is the number of prey killed up to time t, $\widehat{\pi}'_t$ is the number of predators born up to time t, $\widehat{\pi}''_t$ is the number of predators that have died up to time t.

We assume that $\pi'_t, \pi''_t, \widehat{\pi}'_t, \widehat{\pi}''_t$ are Poisson processes with the following state-dependent random rates $\mathbf{a}X_t$, $\frac{\mathbf{b}}{K}X_tY_t$, $\frac{\mathbf{c}}{K}X_tY_t$, $\mathbf{d}Y_t$ respectively and disjoint jumps (the latter assumption reflects the fact that in a short time interval $(t, t + \delta t)$ only one prey might be born and only one might be killed, only one predator might be born and only one might die, with the above-mentioned intensities. Moreover all these events are disjoint in time).

Assume $X_0 = Kx_0$ and $Y_0 = Kx_0$ for some fixed positive x_0, y_0 and a large integer parameter K. Introduce the prey and predator populations normalized by K

$$x_t^K = \frac{X_t}{K}, \quad \text{and} \quad y_t^K = \frac{Y_t}{K}.$$

It terms of x_t^K and y_t^K the introduced intensities for Poisson processes can be written as $\mathbf{a}Kx_t^K$, $\mathbf{b}Kx_t^K y_t^K$, $\mathbf{c}Kx_t^K y_t^K$, $\mathbf{d}Ky_t^K$.

Existence

In this section, we show that the random process (X_t, Y_t) is well defined by equations (13.63). To this end, let us introduce four independent sequences of Poisson processes

$$\Pi_t^{\mathbf{a}} = (\Pi_t^{\mathbf{a}}(1), \Pi_t^{\mathbf{a}}(2), \ldots),$$
$$\Pi_t^{\mathbf{b}/K} = (\Pi_t^{\mathbf{b}/K}(1), \Pi_t^{\mathbf{b}/K}(2), \ldots),$$
$$\Pi_t^{\mathbf{c}/K} = (\Pi_t^{\mathbf{c}/K}(1), \Pi_t^{\mathbf{c}/K}(2), \ldots),$$
$$\Pi_t^{\mathbf{d}} = (\Pi_t^{\mathbf{d}}(1), \Pi_t^{\mathbf{d}}(2), \ldots).$$

Each of them is a sequence of i.i.d. Poisson processes with rates $\mathbf{a}, \frac{\mathbf{b}}{K}, \frac{\mathbf{c}}{K}, \mathbf{d}$ respectively. Define the processes (X_t, Y_t) by the system of Itô equations

$$X_t = X_0 + \int_0^t \sum_{n \geq 1} I(X_{s-} \geq n) d\Pi_s^{\mathbf{a}}(n) - \int_0^t \sum_{n \geq 1} I(X_{s-} Y_{s-} \geq n) d\Pi_s^{\mathbf{b}/K}(n),$$

$$Y_t = Y_0 + \int_0^t \sum_{n \geq 1} I(X_{s-} Y_{s-} \geq n) d\Pi_s^{\mathbf{c}/K}(n) - \int_0^t \sum_{n \geq 1} I(Y_{s-} \geq n) d\Pi_s^{\mathbf{d}}(n),$$

$$\tag{13.64}$$

governed by these Poisson processes, which obviously has a unique solution until the time of explosion, on the time interval $[0, T_\infty)$, where

$$T_\infty = \inf\{t > 0 : X_t \vee Y_t = \infty\}.$$

The Poisson processes with state-dependent rates in (13.63) are obtained as follows

$$\pi_t' = \int_0^t \sum_{n \geq 1} I(X_{s-} \geq n) d\Pi_s^{\mathbf{a}}(n), \quad \pi_t'' = \int_0^t \sum_{n \geq 1} I(X_{s-} Y_{s-} \geq n) d\Pi_s^{\mathbf{b}/K}(n),$$

$$\widehat{\pi}_t' = \int_0^t \sum_{n \geq 1} I(X_{s-} Y_{s-} \geq n) d\Pi_s^{\mathbf{c}/K}(n), \quad \widehat{\pi}_t'' = \int_0^t \sum_{n \geq 1} I(Y_{s-} \geq n) d\Pi_s^{\mathbf{d}}(n).$$

It is easy to see that $\pi'_t, \pi''_t, \widehat{\pi}'_t, \widehat{\pi}''_t$ have the required properties. Their jumps are of size one. Since all Poisson processes are independent, their jumps are disjoint, so that the jumps of $\pi'_t, \pi''_t, \widehat{\pi}'_t, \widehat{\pi}''_t$ are disjoint as well. To describe their intensities introduce a stochastic basis $(\Omega, \mathcal{F}, \mathbb{F} = (\mathcal{F}_t)_{t \geq 0}, \mathrm{P})$, the filtration $\mathbb{F}$ is generated by all Poisson processes and satisfies the general conditions. Then, obviously, the random process

$$A'_t = \int_0^t \sum_{n \geq 1} I(X_s \geq n) \mathbf{a} ds$$

is adapted and continuous, hence predictable. Using $\int_0^t f(s-)ds = \int_0^t f(s)ds$,

$$\pi'_t - A'_t = \int_0^t \sum_{n \geq 1} I(X_{s-} \geq n) d\Pi^{\mathbf{a}}_s(n) - \int_0^t \sum_{n \geq 1} I(X_{s-} \geq n) \mathbf{a} ds$$

$$= \int_0^t \left(\sum_{n \geq 1} I(X_{s-} \geq n) \right) d\left(\Pi^{\mathbf{a}}_s(n) - \mathbf{a} ds\right).$$

Thus $\pi'_t - A'_t$ is an integral with respect to a martingale, hence is a local martingale. Therefore A'_t is the compensator of π'_t. Since X_s is an integer-valued random variable, $\sum_{n \geq 1} I(X_s \geq n) = X_s$, giving that $A'_t = \int_0^t \mathbf{a} X_s ds$ and the intensity of π'_t is $\mathbf{a} X_t$, as claimed. Analogously, other compensators are

$$A''_t = \int_0^t \frac{\mathbf{b}}{K} X_s Y_s ds, \quad \widehat{A}'_t = \int_0^t \frac{\mathbf{c}}{K} X_s Y_s ds, \quad \widehat{A}''_t = \int_0^t \mathbf{d} Y_s ds,$$

and thus all other intensities have the required form.

We now show that the process $(X_t, Y_t)_{t \geq 0}$ does not explode.

Theorem 13.14

$$\mathrm{P}(T_\infty = \infty) = 1.$$

PROOF: Set $T^X_n = \inf\{t > 0 : X_t \geq n\}$, $n \geq 1$ and denote by $T^X_\infty = \lim_{n \to \infty} T^X_n$. Using (13.64) we obtain

$$\mathrm{E} X_{t \wedge T^X_n} \leq X_0 + \int_0^t \mathbf{a} \mathrm{E} X_{s \wedge T^X_n} ds.$$

By Gronwall's inequality, $\mathrm{E} X_{T^X_n \wedge T} \leq X_0 e^{\mathbf{a} T}$ for every $T > 0$. Hence by the Fatou's lemma $\mathrm{E} X_{T^X_\infty \wedge T} \leq X_0 e^{\mathbf{a} T}$. Consequently, for all $T > 0$

$$\mathrm{P}(T^X_\infty \leq T) = 0.$$

Set $T_\ell^Y = \inf\{t : Y_t \geq \ell\}$, $\ell \geq 1$ and denote by $T_\infty^Y = \lim_{\ell \to \infty} T_\ell^Y$. Using (13.64) we obtain

$$\mathrm{E}Y_{t \wedge T_n^X \wedge T_\ell^Y} \leq Y_0 + \int_0^t \frac{\mathbf{c}}{K} \mathrm{E}\big(X_{s \wedge T_n^X \wedge T_\ell^Y} Y_{s \wedge T_n^X \wedge T_\ell^Y}\big) ds$$

$$\leq Y_0 + \int_0^t \frac{\mathbf{c}}{K} n \mathrm{E}Y_{s \wedge T_n^X \wedge T_\ell^Y} ds.$$

Hence, by Gronwall's inequality, for every $T > 0$, $\mathrm{E}Y_{T \wedge T_n^x \wedge T_m^y} \leq Y_0 e^{\mathbf{c}nT}$ and by the Fatou's lemma,

$$\mathrm{E}Y_{T \wedge T_n^X \wedge T_\infty^Y} \leq Y_0 e^{\frac{\mathbf{c}}{K} nT}, \ n \geq 1.$$

Consequently, $\mathrm{P}(T_\infty^Y \leq T_n^X \wedge T) = 0, \forall\, T > 0, n \geq 1$ and, since $T_n^X \nearrow \infty$, as $n \to \infty$, we obtain
$$\mathrm{P}(T_\infty^Y \leq T) = 0.$$
Since $T_\infty = T_\infty^X \wedge T_\infty^Y$, $\mathrm{P}(T_\infty \leq T) \leq \mathrm{P}(T_\infty^X \leq T) + \mathrm{P}(T_\infty^Y \leq T)$, and we have $\mathrm{P}(T_\infty \leq T) = 0$ for any $T > 0$.

$\square$

Corollary 13.15 *For $T_n = \inf\{t : X_t \vee Y_t \geq n\}$, and for all $T > 0$*

$$\lim_{n \to \infty} \mathrm{P}(T_n \leq T) = 0.$$

The above description of the model allows us to claim that (X_t, Y_t) is a continuous-time pure jump Markov process with jumps of two possible sizes in both coordinates: "1" and "-1" and infinitesimal transition probabilities (as $\delta t \to 0$)

$$\mathrm{P}\Big(X_{t+\delta t} = X_t + 1 \big| X_t, Y_t\Big) = \mathbf{a}X_t \delta t + o(\delta t)$$

$$\mathrm{P}\Big(X_{t+\delta t} = X_t - 1 \big| X_t, Y_t\Big) = \frac{\mathbf{b}}{K} X_t Y_t \delta t + o(\delta t)$$

$$\mathrm{P}\Big(Y_{t+\delta t} = Y_t + 1 \big| X_t, Y_t\Big) = \frac{\mathbf{c}}{K} X_t Y_t \delta t + o(\delta t)$$

$$\mathrm{P}\Big(Y_{t+\delta t} = Y_t - 1 \big| X_t, Y_t\Big) = \mathbf{d}Y_t \delta t + o(\delta t).$$

Semimartingale decomposition for (x_t^K, y_t^K)

Let $A_t', A_t'', \widehat{A}_t', \widehat{A}_t''$ be the compensators of $\pi_t', \pi_t'', \widehat{\pi}_t', \widehat{\pi}_t''$ defined above. Introduce martingales

$$M_t' = \pi_t' - A_t', \ M_t'' = \pi_t'' - A_t'', \ \widehat{M}_t' = \widehat{\pi}_t' - \widehat{A}_t', \ \widehat{M}_t'' = \widehat{\pi}_t'' - \widehat{A}_t'',$$

and also normalized martingales

$$m_t^K = \frac{M_t' - M_t''}{K} \quad \text{and} \quad \widehat{m}_t^K = \frac{\widehat{M}_t' - \widehat{M}_t''}{K}. \tag{13.65}$$

Then, from (13.64) it follows that the process (x_t^K, y_t^K) admits the semimartingale decomposition

$$x_t^K = x_0 + \int_0^t [\mathbf{a} x_s^K - \mathbf{b} x_s^K y_s^K] ds + m_t^K,$$

$$y_t^K = y_0 + \int_0^t [\mathbf{c} x_s^K y_s^K - \mathbf{d} y_s^K] ds + \widehat{m}_t^K, \tag{13.66}$$

which is a stochastic analogue (in integral form) of the equations (13.61).

In the sequel we need quadratic variations of the martingales in (13.66). By Theorem 9.3 all martingales are locally square integrable and possess the predictable quadratic variations

$$\langle M', M' \rangle_t = A_t', \quad \langle M'', M'' \rangle_t = A_t'' \quad \text{and} \quad \langle \widehat{M}', \widehat{M}' \rangle_t = \widehat{A}_t', \quad \langle \widehat{M}'', \widehat{M}'' \rangle_t = \widehat{A}_t'', \tag{13.67}$$

and zero covariations $\langle M', M'' \rangle_t \equiv 0, \dots, \langle \widehat{M}', \widehat{M}'' \rangle_t \equiv 0$, since the jumps of $\pi_t', \pi_t'', \widehat{\pi}_t', \widehat{\pi}_t''$ are disjoint. Hence we obtain the sharp brackets of the martingales in equations (13.66),

$$\langle m^K, \widehat{m}^K \rangle_t \equiv 0,$$

$$\langle m^K, m^K \rangle_t = \frac{1}{K} \int_0^t (\mathbf{a} x_s^K + \mathbf{b} x_s^K y_s^K) ds,$$

$$\langle \widehat{m}^K, \widehat{m}^K \rangle_t = \frac{1}{K} \int_0^t (\mathbf{c} x_s^K y_s^K + \mathbf{d} y_s^K) ds. \tag{13.68}$$

Note that the stochastic equations above do not satisfy the linear growth condition in x_t^K, y_t^K. Nevertheless, the solution exists for all t.

Deterministic (Fluid) approximation

We now show that the Lotka-Volterra equations (13.61) describe a limit (also known as fluid approximation) for the family (x_t^K, y_t^K) as parameter $K \to \infty$. Results on the fluid approximation for discontinuous Markov processes can be found in Kurtz (1981).

Theorem 13.16 *For any $T > 0$ and $\eta > 0$*

$$\lim_{K \to \infty} \mathrm{P}\Big(\sup_{t \leq T} (|x_t^K - x_t| + |y_t^K - y_t|) > \eta \Big) = 0.$$

PROOF: Set
$$T_n^K = \inf\{t : x_t^K \vee y_t^K \geq n\}. \tag{13.69}$$

By Corollary 13.15, since $T_n^K = T_{nK}$,

$$\lim_{n \to \infty} \limsup_{K \to \infty} P\big(T_n^K \leq T\big) = 0. \tag{13.70}$$

Hence, it suffices to show that for every $n \geq 1$,

$$\lim_{K \to \infty} P\Big(\sup_{t \leq T_n^K \wedge T} \big(|x_t^K - x_t| + |y_t^K - y_t|\big) > \eta \Big) = 0. \tag{13.71}$$

Since $\sup\limits_{t \leq T_n^K \wedge T} (x_t^K \vee y_t^K) \leq n + 1$, there is a constant L_n, depending on n and T, such that for $t \leq T_n^K \wedge T$

$$\big|(\mathbf{a}x_t^K - \mathbf{b}x_t^K y_t^K) - (\mathbf{a}x_t - \mathbf{b}x_t y_t)\big| \leq L_n\big(|x_t^K - x_t| + |y_t^K - y_t|\big)$$
$$\big|(\mathbf{c}x_t^K y_t^K - \mathbf{d}y_t^K) - (\mathbf{c}x_t y_t - \mathbf{d}y_t)\big| \leq L_n\big(|x_t^K - x_t| + |y_t^K - y_t|\big).$$

These inequalities and (13.61), (13.66) imply

$$|x_{T_n^K \wedge T}^K - x_{T_n^K \wedge T}| + |y_{T_n^K \wedge T}^K - y_{T_n^K \wedge T}|$$
$$\leq 2L_n \int_0^t \big(|x_{s \wedge T_n^K}^K - x_{s \wedge T_n^K}| + |y_{s \wedge T_n^K}^K - y_{s \wedge T_n^K}|\big) ds$$
$$+ \sup_{t \leq T_n^K \wedge T} |m_t^K| + \sup_{t \leq T_n^K \wedge T} |\widehat{m}_t^K|.$$

Now, by Gronwall's inequality we find

$$\sup_{t \leq T_n^K \wedge T} \big(|x_t^K - x_t| + |y_t^K - y_t|\big) \leq e^{2L_n T} \Big(\sup_{t \leq T_n^K \wedge T} |m_t^K| + \sup_{t \leq T_n^K \wedge T} |\widehat{m}_t^K| \Big).$$

Therefore (13.71) holds if both $\sup\limits_{t \leq T_n^K \wedge T} |m_t^K|$ and $\sup\limits_{t \leq T_n^K \wedge T} |\widehat{m}_t^K|$ converge in probability to zero as $K \to \infty$. By (13.68) and definition (13.69) of T_n^K,

$$\mathrm{E}\langle m^K \rangle_{T_n^K \wedge T} \leq \frac{1}{K}(\mathbf{a}(n + 1) + \mathbf{b}(n + 1)^2)T.$$

Thus by Doob's inequality for martingales (8.47)

$$\mathrm{E}\left(\big(\sup_{t \leq T_n^K \wedge T} |m_t^K| \big)^2 \right) \leq 4\mathrm{E}\left(\langle m^K, m^K \rangle_{T_n^K \wedge T} \right) \to 0.$$

as $K \to \infty$. This implies $\sup_{t \le T_n^K \wedge T} |m_t^K| \to 0$ as $K \to \infty$. The second term $\sup_{t \le T_n^K \wedge T} |\widehat{m}_t^K|$ is treated similarly, and the proof is complete.

□

Using the stochastic Lotka-Volterra model one can evaluate asymptotically, when K is large, the time to extinction of prey species, as well as the likely trajectory to extinction. One such trajectory is in the Figure below. This analysis is done by using the Large Deviations Theory, see Klebaner and Liptser (2001).

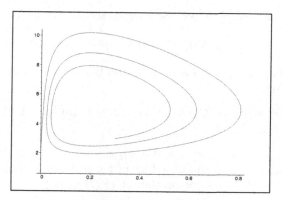

Figure 13.2: A likely path to extinction.

13.6 Exercises

Exercise 13.1: Find the expected time to extinction ET_0 in the Branching diffusion with $\alpha < 0$. Hint: use Theorem 6.16 and formula (6.98) to find $E(T_0 \wedge T_b)$. $ET_0 = \lim_{b \to \infty} E(T_0 \wedge T_b)$ by monotone convergence.

Exercise 13.2: Let $X(t)$ be a branching diffusion satisfying SDE (13.1).

1. Let $c = 2\alpha/\sigma^2$. Show that e^{-cX_t} is a martingale.

2. Let T be the time to extinction, $T = \inf\{t : X_t = 0\}$, where $T = \infty$ if $X_t > 0$ for all $t \ge 0$, and let $q_t(x) = P_x(T \le t)$ be the probability that extinction occurs by time t when the initial population size is x. Prove that $q_t(x) \le e^{-cx}$.

Exercise 13.3: A model for population growth is given by the following SDE $dX(t) = 2X(t)dt + \sqrt{X(t)}dB(t)$, and $X(0) = x > 0$. Find the probability that the population doubles its initial size x before it becomes extinct. Show that when the initial population size x is small then this probability is approximately $1/2$, but when x is large this probability is nearly one.

Exercise 13.4: Check that the distribution in (13.16) in the Wright-Fisher diffusion is stationary.

Exercise 13.5: Let a deterministic growth model be given by the differential equation
$$dx(t) = g(x(t))dt, \quad x_0 > 0, \quad t \geq 0,$$
for a positive function $g(x)$ and consider its stochastic analogue
$$dX(t) = g(X(t))dt + \sigma(X(t))dB(t), \quad X(0) > 0.$$

One way to analyze the stochastic equation is by comparison with the deterministic solution.

1. Find $G(x)$ such that $G(x(t)) = G(x(0)) + t$ and consider $Y(t) = G(X(t))$. Find $dY(t)$.

2. Let $g(x) = x^r$, $0 \leq r < 1$, and $\sigma(x)/x^r \to 0$ as $x \to \infty$. Give conditions on $g(x)$ and $\sigma^2(x)$ so that the Law of Large Numbers holds for $Y(t)$, that is, $Y(t)/t \to 1$, as $t \to \infty$ on the set $\{Y(t) \to \infty\}$.

A systematic analysis of this model is given in Keller *et al.* (1987).

Exercise 13.6: Let $L_1, L_2, \ldots, L_x$ be independent exponentially distributed random variables with parameter $a(x)$, (they represent the lifespan of particles in the population of size x). Show that $\min(L_1, L_2, \ldots, L_x)$ has an exponential distribution with parameter $xa(x)$. (In a branching model the change in the population size occurs when a particle dies, that is, at $\min(L_1, L_2, \ldots, L_x)$).

Exercise 13.7: (Birth-Death processes stochastic representation)
Let $N_k^\lambda(t)$ and $N_k^\mu(t)$, $k \geq 1$, be two independent sequences of independent Poisson processes with rates λ and μ. Let $X(0) > 0$ and for $t > 0$ $X(t)$ satisfies

$$X(t) = X(0) + \int_0^t \sum_{k \geq 1} I(X(s-) \geq k)dN_k^\lambda(s) - \int_0^t \sum_{k \geq 1} I(X(s-) \geq k)dN_k^\mu(s).$$

1. Show that $X(t)$ is a Birth-Death process and identify the rates.

2. Give the semimartingale decomposition for $X(t)$.

Chapter 14

Applications in Engineering and Physics

In this chapter methods of Stochastic Calculus are applied to the Filtering problem in Engineering and Random Oscillators in Physics. The Filtering problem consists of finding the best estimator of a signal when observations are contaminated by noise. For a number of classical equations of motions in Physics we find stationary densities when the motion is subjected to random excitations.

14.1 Filtering

The filtering problem is the problem of estimation of a signal contaminated by noise. Let $Y(t)$ be the observation process, and $\mathcal{F}_t^Y$ denote the information available by observing the process up to time t, that is $\mathcal{F}_t^Y = \sigma(Y(s), s \leq t)$. The observation $Y(t)$ at time t is the result of a deterministic transformation of the signal process $X(s)$, $s \leq t$, typically a linear transformation, to which a random noise is added. The filtering problem is to find the "best" estimate $\pi_X(t)$ of the signal $X(t)$ on the basis of all the observations $Y(s)$, $s \leq t$, or $\mathcal{F}_t^Y$. The "best" is understood in the sense of the smallest estimation error $\mathrm{E}\left((X(t) - Z(t))^2 | \mathcal{F}_t^Y\right)$, when $Z(t)$ varies over all $\mathcal{F}_t^Y$-measurable processes. Denote for an adapted process $h(t)$

$$\pi_t(h) = \mathrm{E}(h(t)|\mathcal{F}_t^Y). \tag{14.1}$$

Then by Theorem 2.26 (see also Exercise 14.2) the filtering problem is in finding $\pi_t(X)$.

Two main results of stochastic calculus are used to solve the filtering problem, Levy's characterization of Brownian motion and the predictable representation property of Brownian filtration.

General Non-linear Filtering Model

Let $(\Omega, \mathcal{F}, \mathrm{P})$ be a complete probability space, and let $(\mathcal{F}_t)$, $0 \leq t \leq T$, be a non-decreasing family of right continuous σ-algebras of $\mathcal{F}$, satisfying the usual conditions, and supporting adapted processes $X(t)$ and $Y(t)$.

We aim at finding $\pi_t(h)$ for an adapted process $h(t)$. In particular, we can find $\mathrm{E}(g(X(t))|\mathcal{F}_t^Y)$, for a function of a real variable g. When g ranges over a set of test functions, the conditional distribution of $X(t)$ given $\mathcal{F}_t^Y$ is obtained. Moreover by taking $g(x) = x$, we obtain the best estimator $\pi_X(t)$. We assume that the process $h(t)$, which we want to filter and the observation process $Y(t)$ satisfy the SDEs

$$
\begin{aligned}
dh(t) &= H(t)dt + dM(t), \\
dY(t) &= A(t)dt + B(Y(t))dW(t),
\end{aligned} \qquad (14.2)
$$

where:

- The process $M(t)$ is a $\mathcal{F}_t$-martingale,

- The process $W(t)$ is a $\mathcal{F}_t$-Brownian motion,

- The processes $H(t)$ and $A(t)$ are random, satisfying with probability one $\int_0^T |H(t)|dt < \infty$, $\int_0^t |A(t)|dt < \infty$,

- $\sup_{t \leq T} \mathrm{E}(h^2(t)) < \infty$, $\int_0^T \mathrm{E}H^2(t)dt < \infty$, $\int_0^T \mathrm{E}A^2(t)dt < \infty$,

- The diffusion coefficient of the observation process $Y(t)$ is a function $B(y)$ of $Y(t)$ only, and not X. $B^2(y) \geq C > 0$, and B^2 satisfies Lipschitz and the linear growth conditions.

Theorem 14.1 *For each* t, $0 \leq t \leq T$,

$$
d\pi_t(h) = \pi_t(H)dt + \left(\pi_t(D) + \frac{\pi_t(hA) - \pi_t(h)\pi_t(A)}{B(Y(t))}\right)d\overline{W}(t), \qquad (14.3)
$$

where $d\overline{W}(t) = \frac{dY(t) - \pi_t(A)}{B(Y(t))}$ *is an* $\mathcal{F}_t^Y$*-Brownian motion and* $D(t) = \frac{d\langle M, W \rangle(t)}{dt}$.

$\overline{W}(t)$ is called the innovation process.
We outline the main ideas of the proof. By (14.2)

$$
h(t) = h(0) + \int_0^t H(s)ds + M(t) \text{ and} \qquad (14.4)
$$

$$
\mathrm{E}(h(t) \mid \mathcal{F}_t^Y) = \mathrm{E}(h(0)|\mathcal{F}_t^Y) + \mathrm{E}\left(\int_0^t H(s)ds \mid \mathcal{F}_t^Y\right)ds + \mathrm{E}(M(t)| \mathcal{F}_t^Y).
$$

The difficulty in calculation of the conditional expectations given $\mathcal{F}_t^Y$ is that the processes, as well as the σ-fields, both depend on t.

Theorem 14.2 *Let* $A(t)$ *be an* $\mathcal{F}_t$-*adapted process such that* $\int_0^T E(|A(t)|)dt < \infty$, *and* $V(t) = \int_0^t A(s)ds$. *Then*

$$\pi_t(V) - \int_0^t \pi_s(A)ds = \mathrm{E}\left(\int_0^t A(s)ds \,\Big|\, \mathcal{F}_t^Y \right) - \int_0^t \mathrm{E}(A(s) \mid \mathcal{F}_s^Y)ds$$

is a $\mathcal{F}_t^Y$-*martingale.*

PROOF: Let $\tau \leq T$ be an $\mathcal{F}_t^Y$-stopping time. By Theorem 7.17 it is enough to show that $E(\pi_\tau(V)) = E(\int_0^\tau \pi_s(A)ds)$.

$E(\pi_\tau(V)) = E(V(\tau))$ by the law of double expectation

$$= E(\int_0^\tau A(s)ds) = \int_0^T E(I(s \leq \tau)A(s))ds$$

$$= \int_0^T E(I(s \leq \tau)\pi_s(A))ds \text{ since } I(s \leq \tau) \text{ is } \mathcal{F}_s^Y\text{-measurable}$$

$$= E(\int_0^\tau \pi_s(A)ds).$$

$\square$

Note that by the conditional version of Fubini's theorem,

$$\mathrm{E}\left(\int_0^t A(s)ds \,\Big|\, \mathcal{G} \right) = \int_0^t \mathrm{E}(A(s) \,|\mathcal{G})ds. \tag{14.5}$$

Verification of the next result is straightforward and is left as an Exercise 14.4.

Theorem 14.3 *Let* $M(t)$ *be a* $\mathcal{F}_t$-*martingale. Then* $\pi_t(M)$ *is an* $\mathcal{F}_t^Y$-*martingale.*

Corollary 14.4

$$\pi_t(h) = \pi_t(0) + \int_0^t \pi_s(H)ds + M_1(t) + M_2(t) + M_3(t), \tag{14.6}$$

where $M_i(t)$ *are martingales null at zero;* $M_1(t) = \mathrm{E}(h(X(0))|\mathcal{F}_t^Y) - h(0)$, $M_2(t) = \mathrm{E}\left(\int_0^t H(s)ds \mid \mathcal{F}_t^Y \right) ds - \int_0^t \pi_s(H)ds$, $M_3(t) = \mathrm{E}(M(t)| \mathcal{F}_t^Y)$.

PROOF: Using (14.4), the first term $\mathrm{E}(h(X(0))|\mathcal{F}_t^Y)$ is a (Doob-Levy) martingale. The second term is a martingale by Theorem 14.2, the third by the above Theorem 14.3.

$\square$

We want to use a representation of $\mathcal{F}_t^Y$ martingales. This is done by using the innovation process.

Theorem 14.5 *The innovation process* $\overline{W}(t) = \int_0^t \frac{dY(s)-\pi_s(A)ds}{B(Y(s))}$ *is an* $\mathcal{F}_t^Y$-*Brownian motion. Moreover,*

$$dY(t) = \pi_A(t)dt + B(Y(t))d\overline{W}(t). \tag{14.7}$$

PROOF: By using (14.2)

$$\overline{W}(t) = \int_0^t \frac{A(s) - E(A(s)|\mathcal{F}_s^Y)}{B(Y(s))}ds + W(t). \tag{14.8}$$

For $t > t'$, write

$$
\begin{aligned}
E\left(\overline{W}(t) - \overline{W}(t')|\mathcal{F}_{t'}^Y\right) &= E\left(W(t) - W(t')|\mathcal{F}_{t'}^Y\right) \\
&+ \int_{t'}^t E\left(\frac{A(s) - E(A(s)|\mathcal{F}_s^Y)}{B(Y(s))} \,\Big|\, \mathcal{F}_{t'}^Y\right)ds.
\end{aligned}
$$

The rhs is zero, the first term by Theorem 14.3, and the second by the conditional version of Fubini's theorem. Thus $\overline{W}(t)$ is an $\mathcal{F}_t^Y$-martingale. It is clearly continuous. It follows from (14.8) that $[\overline{W}, \overline{W}](t) = [W, W](t) = t$. By Levy's characterization Theorem the claim follows.

$\square$

Now, if

$$\mathcal{F}^{\overline{W}} = \mathcal{F}_t^Y, \tag{14.9}$$

then conditional expectations given $\mathcal{F}_t^Y$ are the same given $\mathcal{F}_t^{\overline{W}}$, and Theorem 8.35 on representation of martingales with respect to a Brownian filtration can be used for the martingales $M_i(t)$ in (14.6). Comment that for (14.9) to hold it is sufficient that the SDE for $Y(t)$ (14.7) has unique weak solution. Theorem 8.35 and its corollary (8.67) give that there are predictable processes $g_i(s)$, such that

$$M_i(t) = \int_0^T g_i(s)d\overline{W}(s), \ \text{with} \ g_i(t) = \frac{d\langle M_i, \overline{W}\rangle(t)}{dt}.$$

It is possible to show that

$$g_1(t) + g_2(t) + g_3(t) = \pi_t(D) + \frac{\pi_t(hA) - \pi_t(h)\pi_t(A)}{B(Y(t))},$$

and the result follows from (14.6). For details see Liptser and Shiryaev (2001) p. 319-325.

Filtering of Diffusions

Let $(X(t), Y(t))$, $0 \le t \le T$, be a diffusion process with with respect to independent Brownian motions $W_i(t)$, $i = 1, 2$; $\mathcal{F}_t = \sigma(W_1(s), W_2(s), s \le t)$, and

$$
\begin{aligned}
dX(t) &= a(X(t))dt + b(X(t))dW_1(t) \\
dY(t) &= A(X(t))dt + B(Y(t))dW_2(t).
\end{aligned} \tag{14.10}
$$

Assume that the coefficients satisfy the Lipschitz condition, for any of the functions a, A, b, B, e.g. $|a(x') - a(x'')| \le K|x' - x''|$; and $B^2(y) \ge C > 0$.

Let $h = h(X(t))$. The function h is assumed to be twice continuously differentiable. We apply Theorem 14.1 to $h(X(t))$. By Itô's formula, Theorem 6.1, we have

$$
h(X(t)) = h(X(0)) + \int_0^t Lh(X(s))ds + \int_0^t h'(X(s))b(X(s))dW_1(s),
$$

where

$$
Lh(x) = h'(x)a(x) + \frac{1}{2}h''(x)b^2(x).
$$

By Theorem 14.1

$$
\pi_t(h) = \pi_0(h) + \int_0^t \pi_s(Lh)ds + \int_0^t \frac{\pi_s(Ah) - \pi_s(A)\pi_s(h)}{B(Y(s))}d\overline{W}(s), \tag{14.11}
$$

where

$$
\overline{W}(t) = \int_0^t \frac{dY(s) - \pi_s(A)ds}{B(Y(s))}.
$$

The linear case can be solved and is given in the next section.

Kalman-Bucy Filter

Assume that the signal and the observation processes satisfy linear SDEs with time time-dependent non-random coefficients

$$
dX(t) = a(t)X(t)dt + b(t)dW_1(t), \tag{14.12}
$$

$$
dY(t) = A(t)X(t)dt + B(t)dW_2(t), \tag{14.13}
$$

with two independent Brownian motions (W_1, W_2), and initial conditions $X(0)$, $Y(0)$. Due to linearity this case admits a closed form solution for the processes X and Y and also for the optimal estimator of $X(t)$ given $\mathcal{F}_t^Y$. In the linear case it is easy to solve the above SDEs and verify that the processes $X(t)$, $Y(t)$ are jointly Gaussian. It is convenient in this case to use notation $\widehat{X}(t) = \pi_t(X) = \mathrm{E}(X(t)|\mathcal{F}_t^Y)$.

Theorem 14.6 *Suppose that the signal $X(t)$ and the observation $Y(t)$ are given by (14.12) and (14.13). Then the best estimator $\widehat{X}(t) = E(X(t)|\mathcal{F}_t^Y)$ satisfies the following SDE*

$$d\widehat{X}(t) = \left(a(t) - v(t)\frac{A^2(t)}{B^2(t)}\right)\widehat{X}(t)dt + v(t)\frac{A(t)}{B^2(t)}dY(t), \qquad (14.14)$$

where $v(t) = E(X(t) - \widehat{X}(t))^2$ is the squared estimation error. It satisfies the Riccati ordinary differential equation

$$\frac{dv(t)}{dt} = 2a(t)v(t) + b^2(t) - \frac{A^2(t)v^2(t)}{B^2(t)}, \qquad (14.15)$$

with initial conditions $X(0)$ and $v(0) = Var(X(0)) - \frac{Cov^2(X(0),Y(0))}{var(Y(0))}$.

PROOF: Apply Theorem 14.1 with $h(x) = x$ (see also equation (14.11)) to have

$$d\widehat{X}(t) = a(t)\widehat{X}(s)dt + \frac{A(t)}{B^2(t)}\left(\pi_t(h^2) - \left(\pi_t(h)\right)^2\right)\left(dY(t) - A(t)\widehat{X}(t)dt\right). \quad (14.16)$$

Note that
$$\pi_t(h^2) - \left(\pi_t(h)\right)^2 = E\left((X(t) - \widehat{X}(t))^2|\mathcal{F}_t^Y\right).$$

Because the processes X and Y are jointly Gaussian, $\widehat{X}(t)$ is the orthogonal projection, and $X(t) - \widehat{X}(t)$ is orthogonal (uncorrelated) to $Y(s)$, $s \leq t$. But if Gaussian random variables are uncorrelated, they are independent (see Theorem 2.19). Thus $X(t) - \widehat{X}(t)$ is independent of $\mathcal{F}_t^Y$ and we have that $v(t)$ is deterministic,

$$v(t) = E\left((X(t) - \widehat{X}(t))^2|\mathcal{F}_t^Y\right) = E(X(t) - \widehat{X}(t))^2.$$

The initial value $v(0)$ is obtained by the Theorem 2.25 on Normal correlation, as stated in the Theorem. Let $\delta(t) = X(t) - \widehat{X}(t)$. Then $v(t) = E(\delta^2(t))$. The SDE for $\delta^2(t)$ is obtained from SDEs (14.12) and (14.16) as follows

$$d\delta(t) = a(t)\delta(t)dt + b(t)dW_1(t) - \frac{A^2(t)v(t)}{B^2(t)}\delta(t)dt - \frac{A(t)v(t)}{B(t)}dW_2(t).$$

Applying now Itô's formula to $\delta^2(t)$, we find

$$\begin{aligned}
d\delta^2(t) &= \left(2\left(a(t) - \frac{A^2(t)v(t)}{B^2(t)}\right)\delta^2(t) + \left(b^2(t) + \frac{A^2(t)v^2(t)}{B^2(t)}\right)\right)dt \\
&\quad + 2\delta(t)\left(b(t)dW_1(t) - \frac{A(t)v(t)}{B(t)}dW_2(t)\right). \qquad (14.17)
\end{aligned}$$

Writing (14.17) in integral form and taking expectation

$$
\begin{aligned}
v(t) &= v(0) + \int_0^t \left(2\left(a(s) - \frac{A^2(s)v(s)}{B^2(s)}\right)v(s) + \left(b^2(s) + \frac{A^2(s)v^2(s)}{B^2(s)}\right)\right)ds \\
&= v(0) + \int_0^t \left(2a(s) + b^2(s) - \frac{A^2(s)v^2(s)}{B^2(s)}\right)ds,
\end{aligned}
\tag{14.18}
$$

which establishes (14.15).

$\square$

The multi-dimensional case is similar, with the only difference being that the equation (14.15) is the matrix Riccati equation for the estimation error covariance matrix.

The Kalman-Bucy filter allows on-line implementation of the above equations, which are used recursively to compute $\hat{X}(t + \Delta t)$ from the previous values of $\hat{X}(t)$ and $v(t)$.

Example 14.1: (Model with constant coefficients) Consider the case of constant coefficients,

$$
dX(t) = aX(t)dt + dW_1(t),
$$
$$
dY(t) = cX(t)dt + dW_2(t).
$$

In this case

$$
d\hat{X}(t) = \left(a - v(t)c^2\right)\hat{X}(t)dt + cv(t)dY(t),
\tag{14.19}
$$

and $v(t)$ satisfies the Riccati equation

$$
\frac{dv(t)}{dt} = 2av(t) + 1 - c^2v^2(t).
\tag{14.20}
$$

This equation has an explicit solution:

$$
v(t) = \frac{\gamma\alpha e^{\lambda t} + \beta}{\gamma e^{\lambda t} + 1},
\tag{14.21}
$$

where α and β are the roots of $1 + 2ax - c^2x^2$, assumed to be $\alpha > 0$, $\beta < 0$, $\lambda = c^2(\alpha - \beta)$, and $\gamma = (\sigma^2 - \beta)/(\alpha - \sigma^2)$, with $\sigma^2 = Var(X(0))$. Using $v(t)$ above the optimal estimator $\hat{X}(t)$ is found from (14.19).

For more general results see, for example, Liptser and Shiryayev (2001), Rogers and Williams (1990), Oksendal (1995). There is a vast amount of literature on filtering, see, for example, Kallianpur G. (1980), Krishnan V. (1984) and references therein.

14.2 Random Oscillators

Second order differential equations

$$\ddot{x} + h(x, \dot{x}) = 0.$$

are used to describe variety of physical phenomena, and oscillations are one of them.

Example 14.2: (Harmonic oscillator)
The autonomous vibrating system is governed by $\ddot{x} + x = 0$, $x(0) = 0$, $\dot{x}(0) = 1$, where $x(t)$ denotes the displacement from the static equilibrium position. It has the solution $x(t) = \sin(t)$. The trajectories of this system in the phase space $x_1 = x, x_2 = \dot{x}$ are closed circles.

Example 14.3: (Pendulum)
The undampened pendulum is governed by $\ddot{x} + a\sin x = 0$, where $x(t)$ denotes the angular displacement from the equilibrium. Its solution cannot be obtained in terms of elementary functions, but can be given in the phase plane, $(\dot{x})^2 = 2a\cos x + C$.

Example 14.4: (Van der Pol oscillator)
In some systems the large oscillations are dampened, whereas small ones are boosted (negative damping). Such motion is governed by the Van der Pol equation

$$\ddot{x} - a(1 - x^2)\dot{x} + x = 0, \tag{14.22}$$

where $a > 0$. Its solution cannot be obtained in terms of elementary functions, even in the phase plane.

Example 14.5: (Rayleigh oscillator)

$$\ddot{x} - a(1 - \dot{x}^2)\dot{x} + x = 0, \tag{14.23}$$

where $a > 0$.

We consider random excitations of such systems by White noise (in applied language) of the form $\sum_{i=1}^{2} f_i(x, \dot{x})\dot{W}_i(t)$, where $\dot{W}_i(t)$, $i = 1, 2$, are White noises with delta-type correlation functions

$$E\left(\dot{W}_i(t)\dot{W}_j(t + \tau)\right) = 2\pi K_{ij}\delta(\tau)dt.$$

Thus the randomly perturbed equation has the form

$$\ddot{X} + h(x, \dot{X}) = \sum_i f_i(X, \dot{X})\dot{W}_i(t).$$

The white noise is formally the derivative of Brownian motion. Since Brownian motion is nowhere differentiable, the above system has only a formal meaning.

A rigorous meaning to such equations is given by a system of first order Itô stochastic equations (a single vector valued equation). The representation as a system of two first order equations follows the same idea as in the deterministic case by letting $x_1 = x$ and $x_2 = \dot{x}$. Using Theorem 5.20 on conversion of Stratanovich SDEs into Itô SDEs, we end up with the following Itô system of stochastic differential equations

$$dX_1 = X_2 dt,$$

$$dX_2 = \left(-h(X_1, X_2) + \pi \sum_{j,k} K_{jk} f_j(X_1, X_2) \frac{\partial f_k}{\partial x_2}(X_1, X_2) \right) dt +$$

$$\sum_{i=0}^{2} f_i(X_1, X_2) dW_i(t).$$

The above system is a two-dimensional diffusion, and it has the same generator as the system below driven by a single Brownian motion $B(t)$,

$$dX_1 = X_2 dt,$$
$$dX_2 = A(X_1, X_2) dt + G(X_1, X_2) dB(t), \tag{14.24}$$

where

$$A(x_1, x_2) = -h(x_1, x_2) + \pi \sum_{j,k} K_{jk} f_j(x_1, x_2) \frac{\partial f_k(x_1, x_2)}{\partial x_2},$$

$$G(x_1, x_2) = (2\pi)^{1/2} \left(\sum_{j,k} K_{jk} f_j(x_1, x_2) f_k(x_1, x_2) \right)^{1/2}.$$

The system (14.24) is a rigorous mathematical model of random oscillators, and this form is the starting point of our analysis. Solutions to the resulting Fokker-Planck equations give densities of invariant measures or stationary distributions for such random systems. The corresponding Fokker-Planck equation has the form

$$x_2 \frac{\partial}{\partial x_1} p_s + \frac{\partial}{\partial x_2}(A p_s) - \frac{1}{2}\frac{\partial^2}{\partial x_2^2}(G^2 p_s) = 0,$$

where $p_s(x_1, x_2)$ is the density of the invariant measure. When a solution to the Fokker-Planck equation has a finite integral, then it is a stationary distribution of the process. However, they may not exist, especially in systems with trajectories approaching infinity. In such cases invariant measures which are not probability distributions may exist and provide information about the underlying dynamics.

To illustrate, consider a stochastic differential equations of order two with no noise in $\dot{x}$

$$\ddot{X} + h(X, \dot{X}) = \sigma \dot{B}.$$

The corresponding Itô system is given by

$$
\begin{aligned}
dX_1 &= X_2 dt, \\
dX_2 &= -h(X_1, X_2) dt + \sigma dB(t).
\end{aligned}
$$

In the notation of Section 6.10, the dimension $n = 2$ and there is only one Brownian motion, $d = 1$, so that $\mathbf{X}(t) = (X_1(t), X_2(t))^T$, $b(\mathbf{X}(t)) = (b_1(\mathbf{X}(t)), b_2(\mathbf{X}(t)))^T = (X_2(t), -h(X_1(t), X_2(t)))^T$ $\sigma(\mathbf{X}(t)) = (0, \sigma)^T$. The diffusion matrix

$$a = \sigma\sigma^T = \begin{bmatrix} 0 & 0 \\ 0 & \sigma^2 \end{bmatrix},$$

hence the generator is given by

$$
\begin{aligned}
L &= \sum_{i=1}^{2} b_i \frac{\partial}{\partial x_i} + \frac{1}{2} \sum_{i=1}^{2} \sum_{j=1}^{2} a_{ij} \frac{\partial^2}{\partial x_i \partial x_j} \\
&= x_2 \frac{\partial}{\partial x_1} - h(x_1, x_2) \frac{\partial}{\partial x_2} + \frac{1}{2} \sigma^2 \frac{\partial^2}{\partial x_2^2}.
\end{aligned}
\tag{14.25}
$$

An important example is provided by linear equations

$$\ddot{X} + a\dot{X} + bX = \sigma\dot{B}, \tag{14.26}$$

with constant coefficients. Solutions to these equations can be written in a general formula.

$$\mathbf{X}(t) = (\exp(Ft)) \left(\mathbf{X}(0) + \int_0^t (\exp(-Fs))(0, \sigma)^T dB(s) \right),$$

where $F = \begin{bmatrix} 0 & 1 \\ -b & -a \end{bmatrix}$ and $\exp(Ft)$ stands for the matrix exponential (see, for example, Gard (1988)).

Solutions to the Fokker-Plank equation corresponding to some linear systems (14.26) are given in Bezen and Klebaner (1996).

Non-linear Systems

For some systems the Fokker-Plank equation can be solved by the method of detailed balance. We give examples of such systems, for details see Bezen and Klebaner (1996). In what follows K_0, K_1 denote scaling constants.

Duffing Equation

Consider additive stochastic perturbations of the Duffing equation:

$$\ddot{X} + a\dot{X} + X + bX^3 = \dot{W}.$$

$$p_s(x_1, x_2) = \exp\left(-\frac{a}{2\pi K_0}(x_2^2 + \frac{1}{2}bx_1^4 + x_1^2)\right).$$

Typical phase portraits and densities of invariant measures are shown in Figure 14.1.

Random Oscillator

Consider the following oscillator with additive noise

$$\ddot{X} - a(1 - X^2 - \dot{X}^2)\dot{X} + X = \dot{W}.$$

The density of the invariant measure is given by

$$p_s = \exp\left(-\frac{a(x_1^2 + x_2^2)^2}{4\pi K_0}\right).$$

It follows that when $a = 0$ the surface representing the invariant density is a plane. When $a > 0$ a fourth order surface with respect to x_1, x_2 has maximum in a curve representing the limit cycle of the deterministic equation.

Consider now the same equation with parametric noise of the form

$$\ddot{X} - a(1 - X^2 - \dot{X}^2)\dot{X} + X = \dot{W}_0 + (X^2 + \dot{X}^2)\dot{W}_1.$$

The density of the invariant measure is given by

$$
\begin{aligned}
p_s ={} & (K_0 + 2K_1x_1^2x_2^2 + K_1x_1^4 + K_1x_2^4)^{-\frac{\sqrt{K_0}(a+2\pi K_1)}{4}} \\
& \exp\left(\frac{2a\arctan\frac{\sqrt{K_1}(x_1^2+x_2^2)}{\sqrt{K_0}}}{4\pi\sqrt{K_0K_1}}\right).
\end{aligned}
$$

Typical phase portraits and densities of the invariant measures are shown in Figure 14.2.

A System with a Cylindric Phase Plane

Consider random perturbations to a system with a cylindric phase $-\pi \leq x < \pi$, $-\infty < \dot{x} < \infty$,

$$\ddot{X} + a\dot{X} + b + \sin(X) = \dot{W}.$$

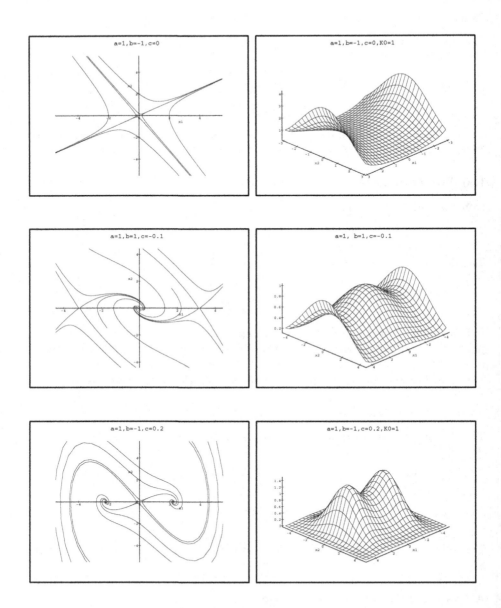

Figure 14.1: The Duffing equation.

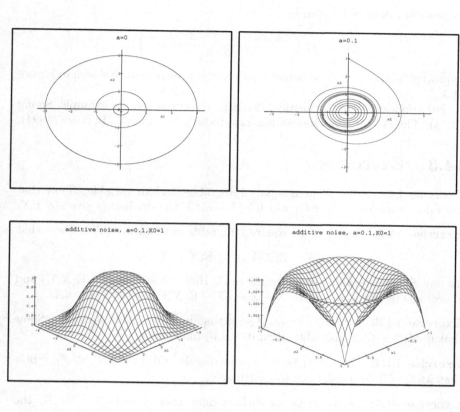

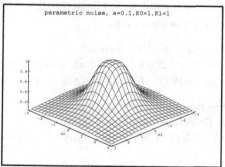

Figure 14.2: Random oscillator.

Its invariant density is given by

$$p_s = \exp\left(-\frac{a(x_2^2 + 2bx_1 - 2\cos x_1)}{2\pi K_0}\right).$$

Typical phase portraits and densities of invariant measures are shown in Figure 14.3.

For applications of stochastic differential equations see, for example, Soong (1973). The Fokker-Planck equation was studied, for example, in Soize (1994).

14.3 Exercises

Exercise 14.1: Let X be a square integrable random variable. Show that the value of the constant c for which $E(X - c)^2$ is the smallest is given by EX.

Exercise 14.2: Let X, Y be square integrable random variables. Show that

$$E(X - E(X|Y))^2 \le E(X - Z)^2$$

for any $\mathcal{F}^Y$-measurable random variable Z. Hint: show that $X - E(X|Y)$ and Z are uncorrelated, and write $X - Z = (X - E(X|Y)) + (E(X|Y) - Z)$.

Exercise 14.3: Let $M(t)$ be an $\mathcal{F}_t$-martingale and σ-fields $\mathcal{G}_t \subset \mathcal{F}_t$. Show that if $M(t)$ is $\mathcal{G}_t$-measurable, then it is a $\mathcal{G}_t$-martingale.

Exercise 14.4: Let $M(t)$ be an $\mathcal{F}_t$-martingale and σ-fields $\mathcal{G}_t \subset \mathcal{F}_t$. Show that $\hat{M}(t) = E(M(t)|\,\mathcal{G}_t)$ is a $\mathcal{G}_t$-martingale.

Exercise 14.5: Let $W(t)$ be an $\mathcal{F}_t$-Brownian motion, and $\mathcal{G}_t \subset \mathcal{F}_t$. By the Exercise 14.4 $\widehat{W}(t) = E(W(t)|\,\mathcal{G}_t)$ is a $\mathcal{G}_t$-martingale. Give an example of $\mathcal{G}_t$, such that $\hat{W}(t)$ is not a Brownian motion.

Exercise 14.6: (Observation of a constant)
Let the signal be a constant $X(t) = c$, for all t, and the observation process satisfy $dY(t) = X(t)dt + dW(t)$. Give the Kalman-Bucy filter and find $\widehat{X}(t)$.

Exercise 14.7: (Observation of Brownian motion)
Let the signal be a Brownian motion $X(t) = W_1(t)$, and the observation process satisfy $dY(t) = X(t)dt + dW_2(t)$. Give the Kalman-Bucy filter and find $\widehat{X}(t)$.

Exercise 14.8: (Filtering of indirectly observed stock prices)
Let the signal follows the Black-Scholes model $X(t) = X(0)\exp(\sigma W_1(t) + \mu t)$, and observation process satisfy $dY(t) = X(t)dt + dW_2(t)$. Give the Kalman-Bucy filter and find $\widehat{X}(t)$.

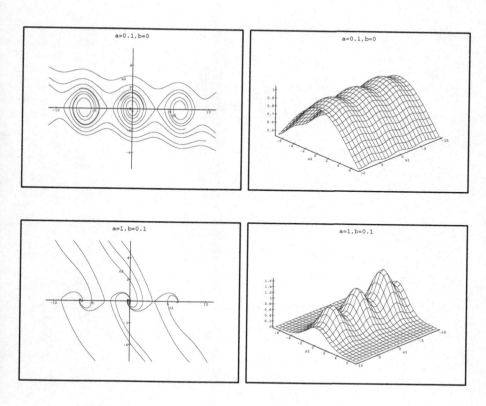

Figure 14.3: The system with a cylindric phase plane.

Solutions to Selected Exercises

Exercises to Chapter 3

Exercise 3.1: $X = \mu + AZ$ for the vector μ and the vector of independent standard Normal random variables Z. For $t = (t_1, \ldots, t_n)$

$$E(e^{itX}) = E(e^{it(\mu + AZ)}) = e^{it\mu}E(e^{it(AZ)}) = e^{it\mu}E(e^{i(tA)Z}) = e^{it\mu}\varphi(tA),$$

where φ is the characteristic function of the vector Z. By independence
$\varphi(u) = E(e^{iuZ}) = \prod_{j=1}^{n} E(e^{iu_j Z_j}) = \prod_{j=1}^{n} e^{-u_j^2/2} = e^{-\frac{1}{2}\sum_{j=1}^{n} u_j^2}$. Hence
$\varphi(tA) = e^{-\frac{1}{2}\sum_{j=1}^{n}(tA)_j^2} = e^{-\frac{1}{2}(tA)(tA)^T} = e^{-\frac{1}{2}tAA^Tt^T} = e^{-\frac{1}{2}t\Sigma t^T}$. Finally
$E(e^{itX}) = e^{it\mu}e^{-\frac{1}{2}t\Sigma t^T}$.

Exercise 3.2: $EX = \int_0^\infty x\,dF(x) = \int_0^\infty \int_0^x dt\,dF(x) = \int_0^\infty \int_t^\infty dF(x)dt = \int_0^\infty (1 - F(t))dt$.

Exercise 3.3: If $f(t)$ is non-increasing then $\int_0^\infty f(t)dt = \sum_{n=0}^\infty \int_n^{n+1} f(t)dt \leq \sum_{n=0}^\infty f(n)$. Now apply the previous result.

Exercise 3.9: For $x \geq 0$, by using the distribution of $M(t) = \max_{s \leq t} B(s)$
$P(|B(t)| > x) = P(B(t) > x) + P(B(t) < -x) = 2P(B(t) > x) = P(M(t) > x)$.

Exercise 3.10: By Theorem 3.18, $E(T_x^r) = \int_0^\infty t^r f(t)dt = \frac{|x|}{\sqrt{2\pi}} \int_0^\infty t^{r - \frac{3}{2}} e^{-\frac{x^2}{2t}} dt$
$= \frac{|x|}{\sqrt{2\pi}} \int_0^\infty s^{-r-\frac{1}{2}} e^{-\frac{x^2}{2}s} ds$. The integral converges at infinity for any r. At zero it converges only for $r + \frac{1}{2} < 1$.

Exercise 3.11: $f_M(y) = \int_{-\infty}^\infty f_{B,M}(x, y)dx = \int_{-\infty}^y \sqrt{\frac{2}{\pi}} \frac{(2y-x)}{t^{3/2}} e^{\frac{-(2y-x)^2}{2t}} dx$
$= \sqrt{\frac{2}{\pi}} \frac{1}{t^{1/2}} \int_{-\infty}^y de^{\frac{-(2y-x)^2}{2t}} = \sqrt{\frac{2}{\pi t}} e^{\frac{-y^2}{2t}} = 2f_B(y)$, by (3.16).

Exercise 3.12: $\min_{s \le t} B(s) = -\max_{s \le t} -B(s)$. Let $W(t) = -B(t)$, then it is also a Brownian motion and we have $P(B(t) \ge x, \min_{s \le t} B(s) \le y) = P(W(t) \le -x, \max_{s \le t} W(s) \ge -y) = 1 - \Phi(\frac{-2y+x}{\sqrt{t}})$. With $\phi = \Phi'$, for $y \le 0, x \ge y$, $f_{B,m}(x,y) = -\frac{\partial^2}{\partial x \partial y} P(B(t) \ge x, m(t) \le y) = \frac{2}{t} \phi'(\frac{-2y+x}{\sqrt{t}}) = \frac{2}{t} \frac{1}{\sqrt{2\pi}} e^{-\frac{(-2y+x)^2}{2t}} \left(\frac{-2y+x}{\sqrt{t}}\right) = \sqrt{\frac{2}{\pi}} \frac{(x-2y)}{t^{3/2}} e^{\frac{-(2y-x)^2}{2t}}$.

Exercise 3.13: Let $a > 0$. Let $D = \{(x,y) : y - x > a\}$. Then we have
$P(M(t) - B(t) > a) = \int \int_D f_{B,M}(x,y)dxdy = \int_0^\infty \int_{-\infty}^{y-a} f_{B,M}(x,y)dxdy$
$= \int_0^\infty \int_{-\infty}^{y-a} \sqrt{\frac{2}{\pi}} \frac{(2y-x)}{t^{3/2}} e^{\frac{-(2y-x)^2}{2t}} dxdy = \sqrt{\frac{2}{t\pi}} \int_0^\infty (-\int_{-\infty}^{y-a} (\frac{\partial}{\partial x} e^{\frac{-(2y-x)^2}{2t}})dx)dy =$
$\sqrt{\frac{2}{t\pi}} \int_0^\infty e^{\frac{-(y+a)^2}{2t}} dy = 2 \int_0^\infty \frac{1}{\sqrt{2\pi t}} e^{\frac{-(y+a)^2}{2t}} dy = 2 \int_a^\infty \frac{1}{\sqrt{2\pi t}} e^{\frac{-u^2}{2t}} du$
$= 2P(B(t) > a)$.

Exercise 3.14: $T_2 = \inf\{t > 0 : B(t) = 0\}$. Since any interval $(0, \epsilon)$ contains a zero of Brownian motion, $T_2 = 0$, the second zero is also zero.

Exercise 3.15: By the above argument, any zero of Brownian motion is a limit from the right of other zeroes. By the definition of T, it is a zero of X, but is not a limit of other zeroes. Thus X is not a Brownian motion.

Exercise 3.16: $\limsup_{t \to \infty} \frac{tB(1/t)}{\sqrt{2t \ln \ln t}} = 1$. $\frac{tB(1/t)}{\sqrt{2t \ln \ln t}} = \frac{B(1/t)}{\sqrt{2(1/t) \ln(-\ln(1/t))}}$. Thus with $\tau = 1/t$, $\limsup_{\tau \to 0} \frac{B(\tau)}{\sqrt{2\tau \ln(-\ln \tau)}} = 1$. Similar for $\liminf$.

Exercise 3.17: $(B(e^{2\alpha t_1}), B(e^{2\alpha t_2}), \ldots, B(e^{2\alpha t_n}))$ is a Gaussian vector. The finite-dimensional distributions of $X(t) = e^{-\alpha t} B(e^{2\alpha t})$ are obtained by multiplying this vector by a non-random diagonal matrix A, with the diagonal elements $(e^{-\alpha t_1}, e^{-\alpha t_2}, \ldots, e^{-\alpha t_n})$. Therefore finite-dimensional distributions of X are multivariate Normal. The mean is zero. Let $s < t$
$Cov(X(s), X(t)) = e^{-\alpha s} e^{-\alpha t} E(B(e^{2\alpha s})B(e^{2\alpha t})) = e^{-\alpha(s+t)} e^{2\alpha s} = e^{-\alpha(t-s)}$.
Note that this Gaussian process has correlated increments.

Exercise 3.18: The process $X(t) = e^{-\alpha t} B(e^{2\alpha t})$ has the given mean and covariance functions and is continuous. Since a Gaussian process is determined by these two functions this is the required version.

Exercise 3.19: $E(e^{uS_{n+1}}|S_n) = E(e^{uS_n} e^{u\xi_{n+1}}|S_n) = E(e^{u\xi_{n+1}})E(e^{uS_n}|S_n) = e^{u^2/2} e^{uS_n}$. Multiply both sides by $e^{-(n+1)u^2/2}$ for the martingale property.

Exercise 3.21: If $X_1 = 3$, then X_2 must be 1, implying that $X_3 = 2$, and cannot be 3, so that $P(X_3 = 3|X_2 = 1 \text{ or } 2, X_1 = 3) = 0$. Using standard calculations of conditional probabilities we can see that $P(X_3 = 3|X_2 = 1 \text{ or } 2) = 1/2$.

Exercise 3.22: 1. If $p = 1$ then $X(t) = aX(t-1) + Z(t).P(X(t) \in A|\mathcal{F}_{t-1}) = P(aX(t-1) + Z(t) \in A|\mathcal{F}_{t-1}) = P(aX(t-1) + Z(t) \in A|X(t-1))$. If $p \geq 2$ then the distribution of $X(t)$ depends on both $X(t-1)$ and $X(t-2)$.

3. The one-step transition probability function is found by
$$P(X(t) \leq y|X(t-1) = x) = P(aX(t-1) + Z(t) \leq y|X(t-1) = x)$$
$$= P(ax + Z(t) \leq y|X(t-1) = x) = P(Z(t) \leq y - ax) = \Phi(\tfrac{y-ax}{\sigma}).$$ The one-step transition probability density function is $\frac{\partial}{\partial y}\Phi(\frac{y-ax}{\sigma}) = \frac{1}{\sigma}f(\frac{y-ax}{\sigma})$.

Exercises to Chapter 4

Exercise 4.1: The necessary condition is $\int_0^t (t-s)^{-2\alpha}ds < \infty$. This holds if and only if $\alpha < 1/2$.

Exercise 4.2: By (4.3), $X(t) = \sum_{i=0}^n \xi_i I_{(t_i, t_{i+1}]}(t)$ for $t > 0$. Hence $\int_0^t X(s)dB(s) = \int_0^T X(s)I_{(0,t]}(s)dB(s) = \sum_{i=0}^n \xi_i(B(t_{i+1} \wedge t) - B(t_i \wedge t))$, where $t_i \wedge t = \min(t_i, t)$. The Itô integral is continuous as a sum of continuous functions.

Exercise 4.3: The first statement follows by using moment generating functions $e^{\mu_n t + \sigma_n^2 t^2/2}$ converges implies that $\mu_n \to \mu$ and $\sigma_n^2 \to \sigma^2 \geq 0$. If $\sigma^2 = 0$, then the limit is a constant μ, otherwise, the limit is $N(\mu, \sigma^2)$.

An Itô integral of a nonrandom function is a limit of approximating Itô integrals of simple nonrandom functions $X_n(t)$. Since $X_n(t)$ takes finitely many nonrandom values, $\int_0^T X_n(t)dB(t)$ has a Normal distribution with mean zero and variance $\int_0^T X_n^2(t)dt$. By the first statement $\int_0^T X(t)dB(t)$ has Normal distribution with mean zero and variance $\int_0^T X^2(t)dt$. An alternative derivation of this fact is done by using the martingale exponential of the martingale $u\int_0^t X(s)dB(s)$.

Exercise 4.5: $M(T)$ has a Normal distribution, $EM^2(T) < \infty$. By Jensen's inequality for conditional expectation (p. 45) with $g(x) = x^2$, $E(M^2(T)|\mathcal{F}_t) \geq (E(M(T)|\mathcal{F}_t))^2 = M^2(t)$. Therefore $E(M^2(t)) \leq E(M^2(T))$ and $M(t)$ is a square-integrable martingale. The covariance between $M(s)$ and $M(t) - M(s)$ for $s < t$ is zero, because by the martingale property $E(M(t) - M(s)|\mathcal{F}_s) = 0$ and $E(M(s)(M(t) - M(s)) = EE(M(s)(M(t) - M(s))|\mathcal{F}_s)$
$$= E(M(s)E(M(t) - M(s)|\mathcal{F}_s)) = 0.$$ Now, if jointly Gaussian variables are uncorrelated, they are independent. Thus the increment $M(t) - M(s)$ is independent of $M(s)$.

Let $M(t) = \int_0^t X(s)dB(s)$. Since $X(s)$ is nonrandom, $E\int_0^T X^2(s)ds = \int_0^T X^2(s)ds < \infty$ by assumption. Thus the Itô integral $M(t)$ is a martingale. It is also Gaussian by the Exercise 4.3. Thus it is a square-integrable martingale with independent increments.

Exercise 4.6: Take $f(x) = x^2$, then Itô's formula gives
$dX^2(t) = 2X(t)dX(t) + d[X, X](t)$, $[X, X](t) = X^2(t) - X^2(0) - 2\int_0^t X(s)dX(s)$.

Exercise 4.7: Let $f(x) = \sqrt{x}$, then Itô's formula gives
$df(X(t)) = f'(X(t))dX(t) + \frac{1}{2}f''(X(t))d[X, X](t)$
$= \frac{1}{2\sqrt{X(t)}}dX(t) - \frac{1}{8X(t)\sqrt{X(t)}}\sigma^2(X(t))dt$. Rearranging we obtain
$dY(t) = \frac{1}{2}\left[bY(t) + \frac{c-1}{Y(t)}\right]dt + dB(t)$.

Exercise 4.10: Let $f(x, t) = txe^{-x}$. Then if $x = B(t)$, we have $X(t)/Y(t) = f(B(t), t)$. Using Itô's formula (with partial derivatives denoted by subscripts)
$df(B(t), t) = f_x(B(t), t)dB(t) + \frac{1}{2}f_{xx}(B(t), t)d[B, B](t) + f_t(B(t), t)dt$
$t(1 - B(t))e^{-B(t)}dB(t) + \frac{1}{2}t(B(t) - 2)e^{-B(t)}dt + B(t)e^{-B(t)}dt$.

Exercise 4.13: $E|M(t)| = E|B^3(t) - 3tB(t)| \leq E|B^3(t)| + 3t|B(t)| < \infty$, since a Normal distribution has all moments. Use the expansion $(a + b)^3 = a^3 + 3a^2b + 3ab^2 + b^3$ with decomposition $B(t+s) = B(t) + (B(t+s) - B(t))$. Take $a = B(t)$, $b = B(t+s) - B(t)$, and use the the fact that $E(B(t+s) - B(t))^3 = 0$, $E(B(t + s) - B(t))^2 = s$, to get the martingale property. Using Itô's formula
$dM(t) = d(B^3(t) - 3tB(t)) = 3B^2(t)dB(t) + \frac{1}{2}6B^2(t)dt - 3B(t)dt - 3tdB(t)$
$= (3B^2(t) - 3t)dB(t)$. Since $\int_0^T E(3B^2(t) - 3t)^2dt < \infty$, $M(t)$ is a martingale on $[0, T]$ for any T.

Exercise 4.15: $df(B_1(t), \ldots, B_n(t)) = \sum \frac{\partial f}{\partial x_i}(B)dB_i(t) + \frac{1}{2}\sum_i \frac{\partial^2 f}{\partial x_i^2}(B)dt$
$= \nabla f \cdot dB + \frac{1}{2}\nabla \cdot \nabla f dt$, where "$\cdot$" is the scalar product of vectors and $B = (B_1(t), \ldots, B_n(t))$ with $dB = (dB_1(t), \ldots, dB_n(t))$. The operator $\nabla \cdot \nabla = \Delta = \sum_i \frac{\partial^2}{\partial x_i^2}$ is the Laplacian, so that the Itô's formula becomes
$df(B(t)) = \frac{1}{2}\Delta f(B(t))dt + \nabla f \cdot dB$.

Exercise 4.16: Denote $\phi(x) = \Phi'(x)$ then $\frac{\partial}{\partial x}\Phi(\frac{x}{\sqrt{T-t}}) = \phi(\frac{x}{\sqrt{T-t}})\frac{1}{\sqrt{T-t}}$,
$\frac{\partial^2}{\partial x^2}\Phi(\frac{x}{\sqrt{T-t}}) = \phi'(\frac{x}{\sqrt{T-t}})\frac{1}{T-t} = -\frac{x}{(T-t)^{3/2}}\phi(\frac{x}{\sqrt{T-t}})$, using $\phi'(x) = -x\phi(x)$ and
$\frac{\partial}{\partial t}\Phi(\frac{x}{\sqrt{T-t}}) = -\frac{1}{2}\phi(\frac{x}{\sqrt{T-t}})\frac{x}{(T-t)^{3/2}}$. Thus by Itô's formula for all $0 \leq t < T$,
$d\Phi(\frac{B(t)}{\sqrt{T-t}}) = \phi(\frac{B(t)}{\sqrt{T-t}})\frac{1}{\sqrt{T-t}}dB(t)$ and $\Phi(\frac{B(t)}{\sqrt{T-t}}) = \frac{1}{2} + \int_0^t \phi(\frac{B(s)}{\sqrt{T-s}})\frac{1}{\sqrt{T-s}}dB(s)$.
Since $\phi(\frac{B(s)}{\sqrt{T-s}})\frac{1}{\sqrt{T-s}} \leq \frac{1}{\sqrt{T-s}}$, the Itô integral above is a martingale. Thus for all $t < T$ and $s < t$, $X(t) = \Phi(\frac{B(t)}{\sqrt{T-t}})$ satisfies $E(X(t)|\mathcal{F}_s) = X(s)$.

Next, as $t \to T$, $X(t) \to Y = I(B(T) > 0) + \frac{1}{2}I(B(T) = 0)$. The martingale property holds also for $t = T$ by dominated convergence.

Exercise 4.18: $dX(t) = tdB(t) + B(t)dt$, $d[X, X](t) = (dX(t))^2 = t^2dt$, $[X, X](t) = \int_0^t s^2 ds = t^3/3$.

Exercise 4.19: $X(t) = tB(t) - \int_0^t sdB(s) = \int_0^t sdB(s) + \int_0^t B(s)ds - \int_0^t sdB(s)$ $= \int_0^t B(s)ds$. Thus $X(t)$ is differentiable and of finite variation. Hence $[X, X](t) = 0$. Itô integrals of the form $\int_0^t h(s)dB(s)$ have a positive quadratic variation. In this case the Itô integral is of the form $\int_0^t h(t, s)dB(s)$.

Exercises to Chapter 5

Exercise 5.1: By Example 4.5, $\int_0^t b(s)dB(s)$ is a Gaussian process. $X(t) = \int_0^t a(s)ds + \int_0^t b(s)dB(s)$ is also Gaussian as $\int_0^t a(s)ds$ is non-random.

Exercise 5.3: $dX(t) = X(t)(B(t)dt + B(t)dB(t))$. $X(t) = \mathcal{E}(R)(t)$, where $R(t) = \int_0^t B(s)ds + \int_0^t B(s)dB(s)$. Thus $X(t) = e^{\int_0^t (B(s) - \frac{1}{2}B^2(s))ds + \int_0^t B(s)dB(s)}$.

Exercise 5.4: Let $dM(t) = B(t)dB(t)$. Then $dX(t) = X(t)dt + dM(t)$, which is a Langevin type SDE. Solving similarly to Example 5.6 we obtain $X(t) = e^{-t}(1 + \int_0^t e^s dM(s)) = e^{-t}(1 + \int_0^t e^s B(s)dB(s))$.
The SDE is not of the diffusion type as $\sigma(t) = B(t)$. By introducing $Y(t) = B(t)$, it is a diffusion in two dimensions.

Exercise 5.5: Let $U(t) = \mathcal{E}(B)(t)$. Then $dU(t) = U(t)dB(t)$. $dU^2(t) = 2U(t)dU(t) + d[U, U](t) = 2U^2(t)dB(t) + U^2(t)dt = U^2(t)(2B(t) + dt)$. So that $U^2(t) = \mathcal{E}(2B(t) + t)$.

Exercise 5.6: $dX(t) = X(t)(X(t)dt + dB(t))$. If $dY(t) = X(t)dt + dB(t)$ then $X(t) = \mathcal{E}(Y)(t)$.

Exercise 5.10: By definition, $P(y, t, x, s) = P(B(t) + t \leq y | B(s) + s = x)$ $= P(B(t) - B(s) + t - s \leq y - x | B(s) + s = x) = P(B(t) - B(s) + t - s \leq y - x)$ by independence of increments. $B(t) - B(s)$ has $N(0, t - s)$ distribution, and $P(y, t, x, s) = \Phi\left(\frac{y - x - t + s}{\sqrt{t - s}}\right)$.

Exercise 5.12: $X(t) = \int_0^t \sqrt{X(s) + 1}dB(s)$. Assuming it is a martingale $EX(t) = 0$, $EX^2(t) = E\left(\int_0^t \sqrt{X(s) + 1}dB(s)\right)^2 = E\left(\int_0^t (X(s) + 1)ds\right) = t$. $de^{uX(t)} = ue^{uX(t)}dX(t) + \frac{1}{2}u^2 e^{uX(t)}d[X, X](t)$. $d[X, X](t) = (X(t) + 1)dt$, and $de^{uX(t)} = ue^{uX(t)}\sqrt{X(t) + 1}dB(t) + \frac{1}{2}u^2 e^{uX(t)}(X(t) + 1)dt$. Taking expectation $m(t) = 1 + \frac{1}{2}u^2 E\left(\int_0^t e^{uX(s)}(X(s) + 1)ds\right) = 1 + \frac{1}{2}u^2 \int_0^t E\left(e^{uX(s)}X(s)\right)ds + \frac{1}{2}u^2 \int_0^t E\left(e^{uX(s)}\right)ds$. Thus $\frac{\partial m}{\partial t} = \frac{u^2}{2}E\left(e^{uX(t)}X(t)\right) + \frac{u^2}{2}Ee^{uX(t)}$ and the desired PDE follows.

Exercises to Chapter 6

Exercise 6.1: Denote $L = \frac{1}{2}\frac{\partial^2}{\partial x}$ and $f(x,t) = e^{ux - u^2 t/2}$. Then $Lf = \frac{1}{2}u^2 f$ and $\frac{\partial f}{\partial t} = -\frac{1}{2}u^2 f$. Thus $Lf + \frac{\partial f}{\partial t} = 0$. By Ito's formula and Corollary 6.4 $f(B(t), t) = e^{uB(t) - u^2 t/2}$ is a martingale. Since f solves $Lf + \frac{\partial f}{\partial t} = 0$ for any fixed u, take partial derivative $\frac{\partial}{\partial u}$ and interchange the order of differentiation to have that for any fixed u, $\frac{\partial}{\partial u}Lf + \frac{\partial}{\partial u}\frac{\partial f}{\partial t} = L(\frac{\partial f}{\partial u}) + \frac{\partial}{\partial t}\frac{\partial f}{\partial u} = 0$, so that $\frac{\partial f}{\partial u}$ solves the backward equation for any fixed u and in particular for $u = 0$. Calculating the derivatives we get the result.

Exercise 6.2: The generator is given by (6.30) $L = \frac{\sigma^2}{2}\frac{d^2}{dx^2} - \alpha\frac{d}{dx}$. The backward equation is $\frac{\sigma^2}{2}\frac{\partial^2 f}{\partial x^2} - \alpha\frac{\partial f}{\partial x} = \frac{\partial f}{\partial t}$. The fundamental solution is given by the probability density of the solution to the SDE $X(t)$.
$p(t, x, y) = \frac{\partial}{\partial y}P(t, x, y)$, where $P(t, x, y) = P(X(t) \leq y | X(0) = x)$. By (5.13)
$P(t, x, y) = P(xe^{-\alpha t} + e^{-\alpha t}\int_0^t e^{\alpha s}dB(s) \leq y) = P(\int_0^t e^{\alpha s}dB(s) \leq ye^{\alpha t} - x)$
$= \Phi\left(\frac{\sqrt{\alpha}(ye^{\alpha t} - x)}{\sqrt{e^{\alpha t} - 1}}\right)$ since $\int_0^t e^{\alpha s}dB(s)$ has $N(0, \frac{1}{\alpha}(e^{\alpha t} - 1))$ distribution. Thus
$p(t, x, y) = \frac{\partial}{\partial y}P(t, x, y) = \phi\left(\frac{\sqrt{\alpha}(ye^{\alpha t} - x)}{\sqrt{e^{\alpha t} - 1}}\right)\frac{\sqrt{\alpha}e^{\alpha t}}{\sqrt{e^{\alpha t} - 1}}$ with ϕ denoting the density of $N(0, 1)$.

Exercise 6.4: $L = \frac{1}{2}\frac{\partial^2}{\partial x^2} + cx\frac{\partial}{\partial x}$. Take $f(x) = x^2$, then $Lf(x) = 1 + 2cx^2$ and by Theorem 6.11 $X^2(t) - \int_0^t (Lf + \frac{\partial f}{\partial t})(X(s))ds = X^2(t) - t - 2c\int_0^t X^2(s)ds$ is a martingale.

Exercise 6.6: Using Corollary 6.4 or Itô's formula, we have $df(B(t) + t) = f'(B(t) + t)dB(t) + f'(B(t) + t)dt + \frac{1}{2}f''(B(t) + t)dt$. A necessary condition for $f(B(t) + t)$ to be a martingale is that dt term is zero. This gives $f'(x) + \frac{1}{2}f''(x) = 0$. For example, take $f(x) = e^{-2x}$ and check directly that $e^{-2(B(t)+t)}$ is a martingale. ($e^{-2(B(t)+t)} = \mathcal{E}(2B)(t)$).

Exercise 6.7: $df(X(t), t) = \frac{\partial f}{\partial x}dX(t) + \frac{\partial^2 f}{\partial x^2}d[X, X](t) + \frac{\partial f}{\partial t}dt$. The term with $dB(t)$ is given by $\sigma(X(t), t)\frac{\partial f}{\partial x}(X(t), t)dB(t)$. Thus the PDE for f is $\sigma(x, t)\frac{\partial f}{\partial x}(x, t) = 1$.

Exercise 6.8: By letting $v' = y$ we obtain a first order differential equation $y' + \frac{2\mu}{\sigma^2}y + \frac{2}{\sigma^2} = 0$. Use the integrating factor $V(x) = \exp\left(\int_a^x \frac{2\mu(s)}{\sigma^2(s)}ds\right)$, to obtain $(yV)' = -\frac{2}{\sigma^2}V$ and $y = -\frac{1}{V}\int \frac{2}{\sigma^2}V$. The result follows.

Exercise 6.9: The scale function (6.50) $S'(x) = e^{-\int^x \frac{2\mu}{\sigma^2}ds} = e^{-\frac{2\mu}{\sigma^2}x}$ gives $S(x) = Ce^{-\frac{2\mu}{\sigma^2}x}$. $P_x(T_b < T_a) = \frac{S(x) - S(a)}{S(b) - S(a)} = (e^{-\frac{2\mu}{\sigma^2}x} - e^{-\frac{2\mu}{\sigma^2}a})/(e^{-\frac{2\mu}{\sigma^2}b} - e^{-\frac{2\mu}{\sigma^2}a})$.

Exercise 6.10: The generator of Brownian motion is $\frac{1}{2}\frac{\partial^2 f}{\partial x^2}$. By Theorem 6.6 $f(x,t) = E(B^2(T)|B(t) = x) = E((x + B(T) - B(t))^2|B(t) = x)$. By independence of increments it is the same as $E(x + B(T) - B(t))^2 = x^2 + T - t$.

Exercise 6.12: Check condition 2, Theorem 6.23 to decide convergence of the integral $\int_1^\infty \exp\left(-\int_1^x \frac{2s^2}{\sigma^2 s^{2\alpha}}ds\right)\left(\int_1^x \frac{1}{\sigma^2 y^{2\alpha}}\exp(\int_1^y \frac{2s^2}{\sigma^2 s^{2\alpha}}ds)dy\right)dx$. $\int_1^x \frac{2s^2}{\sigma^2 s^{2\alpha}}ds = \frac{2}{\sigma^2}\frac{x^{3-2\alpha}-1}{3-2\alpha}$. When $3 - 2\alpha > 0$ then $\exp(-x^{3-2\alpha}/\sigma^2)\int_1^x \frac{\exp(y^{3-2\alpha}/\sigma^2)}{\sigma^2 y^{2\alpha}}dy \sim \frac{C}{x^2}$ as $x \to \infty$ and $\int_1^\infty \exp(-x^{3-2\alpha}/\sigma^2)\int_1^x \frac{\exp(y^{3-2\alpha}/\sigma^2)}{\sigma^2 y^{2\alpha}}dydx$ converges. Consequently, the process explodes. When $3 - 2\alpha < 0$ then the integral diverges and there is no explosion. The case $\alpha = 3/2$ needs further analysis.

Exercise 6.13: Check conditions of Theorem 6.28. $\mu(x) = 0$, $\sigma(x) = 1$. $I_1 = \int_{-\infty}^{x_0} 1du = \infty$, $I_2 = \int_{x_0}^\infty 1du = \infty$. Hence $B(t)$ is recurrent. For the process $B(t)+t$, $\mu(x) = 1$ and $\sigma(x) = 1$. $I_2 = \int_{x_0}^\infty \exp\left(-2(u-x_0)\right)du = \frac{1}{2} < \infty$. Thus $B(t) + t$ is transient. The Ornstein-Uhlenbeck process is left to the reader.

Exercise 6.15: For $n = 2$, $S(x) = \ln x$. So that for $0 < y < x$, $P_x(T_y < T_b) = \frac{\ln x - \ln b}{\ln y - \ln b}$. Since $X(t)$ does not explode, $T_b \uparrow \infty$ as $b \uparrow \infty$. Therefore $P_x(T_y < \infty) = \lim_{b\to\infty} P_x(T_y < T_b) = \lim_{b\to\infty} \frac{\ln x - \ln b}{\ln y - \ln b} = 1$. For $n \geq 3$, $S(x) = \frac{1-x^{1-n/2}}{1-n/2}$. $P_x(T_y < T_b) = \frac{b^{1-n/2}-x^{1-n/2}}{b^{1-n/2}-y^{1-n/2}} \to (\frac{y}{x})^{n/2-1} < 1$. The explosion test shows that $X(t)$ does not explode, and $\lim_{b\to\infty} T_b = \infty$.

Exercise 6.17: Check (6.67). Since $1 = \int_\alpha^\beta p(t,x,y)dy = \int_\alpha^\beta p(t,y,x)dy$, for any C, $C = \int_\alpha^\beta Cp(t,y,x)dy$. Since C is a distribution on (α,β), it must be the uniform density.

Exercise 6.18: To classify 0 as a boundary, calculate L_1, L_2 and L_3, Remark 6.5, see also Theorem 6.29. $\mu(x) = b(a-x)$, $\sigma^2(x) = \sigma^2 x$. For constants C_1, C_2 $C_1 \int_0^c u^{-2ba/\sigma^2} du \leq L_1 \leq C_2 \int_0^c u^{-2ba/\sigma^2} du$, which shows that L_1 converges if $2ba/\sigma^2 < 1$. Thus 0 is a natural boundary iff $2ba/\sigma^2 \geq 1$. $L_2 < \infty$. $L_3 < \infty$ if $ba > 0$. So if $0 < 2ba/\sigma^2 < 1$, 0 is a regular boundary point, and if $ba < 0$, 0 is an absorbing boundary.

Exercises to Chapter 7

Exercise 7.1: $\mathcal{G}_t \subseteq \mathcal{F}_t$. Let $s < t$. By the smoothing property of conditional expectation 3, p. 55 (double expectation), $E(M(t)|\mathcal{G}_s) = E(E(M(t)|\mathcal{F}_s)|\mathcal{G}_s) = E(M(s)|\mathcal{G}_s) = M(s)$.

Exercise 7.3: By convexity, property 6 p. 55 of conditional expectation, $E(g(X(t))|\mathcal{F}_s) \geq g(E(X(t)|\mathcal{F}_s))$. By the submartingale property $E(X(t)|\mathcal{F}_s) \geq X(s)$. Since g is non-decreasing the result follows.

Exercise 7.4: Since square-integrable martingales are uniformly integrable (Corollary 7.8), there is Y such that $M(t) = \mathrm{E}(Y|\mathcal{F}_t)$. $Y = \lim_{t\to\infty} M(t)$. By Fatou's lemma (p. 41) $\mathrm{E}(Y^2) \le \liminf_{t\to\infty} \mathrm{E}(M^2(t)) < \infty$.

Exercise 7.5: Let τ be the first time $B(t)$ hits a or b when $B(0) = x$, $a < x < b$. τ is finite by Theorem 3.13. By stopping martingale $B(t)$ we find that $\mathrm{P}(B(\tau) = b) = (x-a)/(b-a)$, $\mathrm{P}(B(\tau) = a) = 1 - \mathrm{P}(B(\tau) = b)$, Example 7.6. Stopping $M(t)$, $\mathrm{E}M(\tau \wedge t) = M(0) = x^2$, or $\mathrm{E}B^2(\tau \wedge t) - \mathrm{E}(\tau \wedge t) = x^2$. Take $t \to \infty$. $\tau \wedge t \to \tau$, $B^2(\tau \wedge t) \to B^2(\tau)$, and by dominated convergence $\mathrm{E}B^2(\tau) - \mathrm{E}(\tau) = x^2$. But $\mathrm{E}B^2(\tau) = a^2\mathrm{P}(B(\tau) = a) + b^2\mathrm{P}(B(\tau) = b) = a^2\frac{b-x}{b-a} + b^2\frac{x-a}{b-a}$. Thus $\mathrm{E}(\tau) = \mathrm{E}B^2(\tau) - x^2$, and the result follows.

Exercise 7.6: Let $\tau = \inf\{t : B(t) - t/2 = a \text{ or } b\}$. As in the previous exercise, we obtain $\mathrm{E}M(\tau) = M(0) = e^x$ or $e^a\mathrm{P}(M(\tau) = e^a) + e^b(1 - \mathrm{P}(M(\tau) = e^b)) = e^x$. This gives $\mathrm{P}(B(\tau) - \tau/2 = a) = \mathrm{P}(M(\tau) = e^a) = \frac{e^b - e^x}{e^b - e^a}$.

Exercise 7.8: When the game is fair $p = q = 1/2$, use formula (7.17) to see that $u \to 1$. If $p \ne q$, use (7.20) to see that $u \to 1$ when $p < q$, and $u \to (q/p)^x$ when $p > q$.

Exercise 7.12: $\int_0^T \operatorname{sign}^2(B(s))ds = T < \infty$. Thus $X(t)$ is a martingale. $[X, X](t) = \int_0^t \operatorname{sign}^2(B(s))ds = t$. By Levy's theorem X is a Brownian motion.

Exercise 7.13: $[M, M](t) = \int_0^t e^{2s}ds = \frac{1}{2}(e^{2t} - 1)$. Its inverse function is $g(t) = \frac{1}{2}\ln(2t + 1)$. $M(g(t))$ is a Brownian motion by the DDS Theorem 7.37.

Exercise 7.16: $X(t) = X(0) + A(t) + M(t) = X(0) + \int_0^t \mu(s)ds + \int_0^t \sigma(s)dB_s$ is a local martingale. Therefore $A(t) = X(t) - M(t)$ is a local martingale as a difference of two local martingales. $A(t)$ is continuous. By using Corollary 7.30, $A(t)$ has infinite variation unless it is a constant. Since $A(t)$ is of finite variation it must be zero. The result follows.

Exercise 7.17: By Itô's formula
$f(B(t), t) - \int_0^t \left(\frac{\partial f}{\partial t}(B(s), s) + \frac{\partial^2 f}{\partial x^2}(B(s), s)\right)ds = \int_0^t \frac{\partial f}{\partial x}(B(s), s)dB(s)$. The rhs. is a continuous local martingale, and the lhs. is of finite variation. This can only happen when the local martingale is a constant. Thus $\frac{\partial f}{\partial x} = 0$, and f is a constant in x, hence a function of t alone.

Exercise 7.18: We know $Y(t) = \frac{1}{2}B^2(t) - \frac{1}{2}t$, and $B^2(t) = 2Y(t) + t$. Thus $B(t) = \operatorname{sign}(B(t))\sqrt{2Y(t) + t}$. Therefore $dY(t) = \operatorname{sign}(B(t))\sqrt{2Y(t) + t}dB(t)$ or $dY(t) = \sqrt{2Y(t) + t}dW(t)$, where $dW(t) = \operatorname{sign}(B(t))dB(t)$, W is a Brownian motion. Weak Uniqueness follows by Theorem 5.11. Alternatively, one can prove it directly by calculating of moments of $Y(t)$, $\mu_n(t) = \mathrm{E}Y^n(t)$ (by using Ito's formula $\mu_n(t) = \frac{n(n-1)}{2}\int_0^t (s\mu_{n-2}(s) + 2\mu_{n-1}(s))ds$ and checking Carleman's condition $\sum(\mu_n)^{-1/n} = \infty$, see Feller (1971).

Exercises to Chapter 8

Exercise 8.1: $I_{(\tau_1,\tau_2]}(t) = I_{[0,\tau_2]}(t) - I_{[0,\tau_1]}(t)$ is a sum of two left-continuous adapted functions. Let τ be a stopping time. Then $X(t) = I_{[0,\tau]}(t)$ is adapted. $\{X(t) = 0\} = \{t > \tau\} = \bigcup_{n=1}^{\infty}\{\tau \leq t - 1/n\} \in \mathcal{F}_t$, since $\{\tau \leq t - 1/n\} \in \mathcal{F}_{t-1/n} \subset \mathcal{F}_t$. See p. 54.

Exercise 8.2: The left-continuous modification of the process $H(t - \delta + \epsilon)$ is adapted and as $\epsilon \to 0$ tends to $H(t - \delta)$.

Exercise 8.4: Let $X(t) = \int_0^t H(s)dM(s)$, then it is a local martingale as a stochastic integral with respect to a local martingale. $[X,X](t) = \int_0^t H^2(s)d[M,M](s)$. Thus $E[X,X](T) < \infty$. By Theorem 7.35. this condition implies X is a square integrable martingale.

Exercise 8.5: $E(\int_0^1 N(t-)dM(t)) = 0$, so that $Var(\int_0^1 N(t-)dM(t)) = E(\int_0^1 N(t-)dM(t))^2 = E\int_0^1 N^2(t-)d[M,M](t)$. $[M,M](t) = [N,N](t) = N(t)$, so that $Var(\int_0^1 N(t-)dM(t)) = E\int_0^1 N^2(t-)dN(t) = E(\sum_{\tau_i\leq 1} N^2(\tau_{i-1}))$, where τ_i denotes the time of the i-th jump of N. But $N(\tau_i) = i$, so that $Var(\int_0^1 N(t-)dM(t)) = E(\sum_{\tau_i\leq 1}(i-1)^2 = E(\sum_{i=1}^{N(1)}(i-1)^2) = E(\sum_{k=0}^{N(1)-1} k^2)$ $= E(2N^3(1) - 3N^2(1) + N(1))/6$, since $\tau_i \leq 1$ is equivalent to $i \leq N(1)$, and $\sum_{k=0}^n k^2 = (2(n+1)^3 - 3(n+1)^2 + n + 1)/6$. Alternatively $\int_0^1 N^2(t-)dN(t)$ can be obtained by using formula (1.20) p. 12. Compute moments using the mgf $m(s) = Ee^{sN(1)} = e^{-1}e^{e^s}$. $m'(0) = EN(1) = 1$, $m''(0) = EN^2(1) = 2$, $m^{(3)}(0) = EN^3(1) = 5$, giving $Var(\int_0^1 N(t-)dM(t)) = 5/6$.

Exercise 8.6: 1. $S \wedge T$ and $S \vee T$ are stopping times, because $\{S \wedge T > t\} = \{S > t\} \cap \{T > t\} \in \mathcal{F}_t$, $\{S \vee T \leq t\} = \{S \leq t\} \cap \{T \leq t\} \in \mathcal{F}_t$.
2. The events $\{S = T\}$, $\{S \leq T\}$ and $\{S < T\}$ are in $\mathcal{F}_S$.
3. $\mathcal{F}_S \cap \{S \leq T\} \subset \mathcal{F}_T \cap \{S \leq T\}$. See p. 53 Theorem 2.38.

Exercise 8.9: $M(t)$ is a continuous martingale, $[M,M](t) = \int_0^t H^2(s)ds = t$. By Levy's theorem it is a Brownian motion. If $M(t)$ is a Brownian motion, then $[M,M](t) = \int_0^t H^2(s)ds = t$ for all t. Taking derivatives $H^2(t) = 1$ Lebesgue a.s.

Exercise 8.10: The proof is not easy if done from basic properties. But it is easy using Levy's theorem. $M(t)$ can be written as $B(t) = \int_0^t dB(s)$, $B(t\wedge T) = \int_0^t I_{[0,T]}(s)dB(s)$, $M(t) = \int_0^t \left(2I_{[0,T]}(s) - 1\right)dB(s)$. Exercise 8.1 shows that $I_{[0,T]}(s)$ is predictable. Clearly, $M(t)$ is a continuous local martingale as an Itô integral. It is also a martingale, as $E\int_0^t(2I_{[0,T]}(s) - 1)^2ds \leq 9t < \infty$. $[M,M](t) = \int_0^t(2I_{[0,T]}(s) - 1)^2ds$.

If $T \geq t$, then for all $s \leq t$, $I_{[0,T]}(s) = 1$ implying $[M, M](t) = t$. If $T < t$, then $[M, M](t) = \int_0^T (2I_{[0,T]}(s) - 1)^2 ds + \int_T^t (-1)^2 ds = T + (t - T) = t$. Thus $[M, M](t) = t$, for any T, and M is a Brownian motion by Levy's theorem.

Exercise 8.11: $d(B(t)M(t)) = B(t)dM(t) + M(t)dB(t) + d[B, M](t)$. Since M is of finite variation, by the property 7 of quadratic variation (8.19) $[B, M](t) = \sum_{s \leq t} \Delta B(s) \Delta M(s) = 0$, since by continuity of Brownian motion $\Delta B(s) = 0$. Thus $B(t)M(t) = \int_0^t B(s)dM(s) + \int_0^t M(s)dB(s)$. Both stochastic integrals satisfy the condition to be a martingale, i.e. $\mathrm{E} \int_0^T B^2(s)d[M, M](s) = \mathrm{E} \int_0^T B^2(s)dN(s) = \mathrm{E} \sum_{\tau_i \leq T} B^2(\tau_i) = \mathrm{E} \sum_{i=1}^{N(T)} \tau_i < T\mathrm{E}N(T) = T^2 < \infty$.
$\mathrm{E} \int_0^T M^2(s)d[B, B](s) = \mathrm{E} \int_0^T M^2(s)ds = \int_0^T Var(N(s))ds = \int_0^T s \, ds < \infty$.
The other two processes are martingales by similar arguments, verifying that a purely discontinuous martingale is orthogonal to any continuous martingale, i.e. their product is a martingale, p. 233.

Exercise 8.12: $dX(t) = \mu X(t)dt + aX(t-)\lambda dt + aX(t-)(dN(t) - \lambda dt) + \sigma X(t)dB(t)$. $aX(t-)(dN(t) - \lambda dt) + \sigma X(t)dB(t) = dM(t)$ is a martingale, as a sum of stochastic integrals with respect to martingales. $\int X(t-)dt = \int X(t)dt$. Thus $X(t)$ is a martingale when $\mu = -a\lambda$.

Exercise 8.13: $B^5(1) = \int_0^1 \left(5B^4(t) + 30(1-t)B^2(t) + 15(1-t)^2\right) dB(t)$. Use Ito's formula for $B^5(t)$ and that the following functions $x^5 - 10tx^3 + 15t^2 x$ and $x^3 - 3xt$ produce martingales.

Exercise 8.15: $\int_0^1 (\mathrm{sign}(B(s)) - H(s))dB(s) = 0$. The proposition follows from $\mathrm{E} \int_0^1 (\mathrm{sign}(B(s)) - H(s))^2 ds = 0$.

Exercise 8.17: We show the continuous case, the case with jumps is similar. $\mathcal{E}(X)(t) = e^{X(t) - \frac{1}{2}[X,X](t)}$. By (8.82) $X(t) = \ln \left| \frac{U(t)}{U(0)} \right| + \frac{1}{2} \int_0^t \frac{d[U,U](s)}{U^2(s)}$. By Itô's formula for $\ln U(t)$ we obtain $[X, X](t) = \int_0^t \frac{d[U,U](s)}{U^2(s)}$. Thus $\mathcal{E}(X)(t) = U(t)$.

Exercises to Chapter 9

Exercise 9.1: Change the order of integration.

Exercise 9.3: $\mathrm{P}(T_1 = x) = p(1-p)^{x-1}$, $x = 1, 2, \ldots$. Use the lack of memory property of Geometric distribution, $\mathrm{P}(T_1 \leq t + c | T_1 \geq t) = \mathrm{P}(T_1 \leq c)$ to see that the compensator of $N(t)$ is $p[t]$, where $[t]$ stands for the integer part of t.

Exercise 9.4: Repeat the proof of Theorem 9.6 and condition on $\mathcal{F}_{T_n-}$.

Exercise 9.5: Apply Theorem 9.18 to $f(x) = x^k$.

Exercises to Chapter 10

Exercise 10.2: By Theorem 10.3 the equivalent measure Q is given by $dQ/dP = e^{-\mu X + \mu^2/2}$ and $dP/dQ = e^{\mu X - \mu^2/2}$.

Exercise 10.3: Since $Y > 0$ and $E(Y/EY) = 1$, let $\Lambda = Y/EY$ and define $dQ/dP = \Lambda = e^{X-\mu-\sigma^2/2}$. By Theorem 10.4, the Q distribution of Y is $N(\mu + \sigma^2, \sigma^2)$. Thus $EYI(Y > K) = EYE\Lambda I(Y > K) = EYE_Q I(Y > K) = e^{\mu + \sigma^2/2} Q(Y > K)$. Finally, $EYI(Y > K) = e^{\mu + \sigma^2/2} \Phi((\mu + \sigma^2/2 - K)/\sigma)$, using $1 - \Phi(x) = \Phi(-x)$.

Exercise 10.4: $X(t) = B(t) + \int_0^t \cos s\, ds$. Thus by Girsanov's Theorem 10.16
$dQ/dP = e^{-\int_0^T \cos s\, dB(s) - \frac{1}{2}\int_0^T \cos^2 s\, ds}$

Exercise 10.5: With $X = \boldsymbol{H} \cdot \boldsymbol{B}$, $\mathcal{E}(X)(T) = e^{X(T) - X(0) - \frac{1}{2}[X,X](T)}$.
$[\boldsymbol{H} \cdot \boldsymbol{B}, \boldsymbol{H} \cdot \boldsymbol{B}](T) = \sum_{i,j=1}^{d} [H^i \cdot B^i, H^j \cdot B^j] = \sum_{i,j=1}^{d} \int_0^T H^i(t) H^j(t) d[B^i, B^j](t)$
$\sum_{i=1}^{d} \int_0^T (H^i(t))^2 dt = \int_0^T |\boldsymbol{H}(s)|^2 ds$, since $[B^i, B^j](t) = 0$ for $i \neq j$. Thus
$\mathcal{E}(\boldsymbol{H} \cdot \boldsymbol{B})(T) = e^{\sum_{i=1}^{n} \int_0^T H^i(s) dB^i(s) - \frac{1}{2} \sum_{i=1}^{n} \int_0^T (H^i(s))^2 ds}$
$e^{\int_0^T \boldsymbol{H}(s) d\boldsymbol{B}(s) - \frac{1}{2} \int_0^T |\boldsymbol{H}(s)|^2 ds}$.

Exercise 10.8: Since $H(t)$ is bounded $E \int_0^T H^2(t) d[\bar{N}, \bar{N}](t) = E \int_0^T H^2(t) dt$ is finite and $M(t)$ is a martingale. By (9.5)
$\mathcal{E}(M)(t) = e^{\int_0^t H(s) d\bar{N}(s)} \prod_{s \leq t} (1 + \Delta(H \cdot \bar{N})(s)) e^{-\Delta(H \cdot \bar{N})(s)}$. But the jumps of the integral occur at the points of jumps of $\bar{N}$ and $\Delta \bar{N} = \Delta N$, so that $\Delta(H \cdot N)(s) = H(s)\Delta N(s)$. Next, $\prod_{s \leq t} e^{-\Delta(H \cdot \bar{N})(s)} = e^{-\sum_{s \leq t} H(s)\Delta N(s)} = e^{-\int_0^t H(s) dN(s)}$. Proceeding, $\prod_{s \leq t} (1 + \Delta(H \cdot \bar{N})(s)) = e^{\sum_{s \leq t} \ln(1 + H(s))\Delta N(s)} = e^{\int_0^t \ln(1 + H(s)) dN(s)}$, and the result follows.

Exercise 10.9: Using (10.52) or (10.51) with $\mu_i(x,t) = \mu_i x$, $i = 1, 2$, $\sigma(x,t) = \sigma x$, P corresponds to μ_1, $B(t)$ is a P-Brownian motion.
$\Lambda(X)_T = \frac{dQ}{dP} = \mathcal{E}(\frac{\mu_2 - \mu_1}{\sigma} B)(T) = e^{\frac{\mu_2 - \mu_1}{\sigma} B(T) - \frac{(\mu_2 - \mu_1)^2}{2\sigma^2} T}$. Replace $B(T)$ by its expression as a function of X. In this case $B(T) = (\ln(\frac{X(T)}{X(0)}) - (\mu_1 - \frac{1}{2}\sigma^2)T)/\sigma$.

Exercise 10.11: By Corollary 10.11 $M'(t)$ is a Q-martingale if and only if $M'(t)\Lambda(t)$ is a P-martingale. $d(M'(t)\Lambda(t)) = M'(t)d\Lambda(t) + \Lambda(t)dM'(t) + d[M', \Lambda](t) = M'(t)d\Lambda(t) + \Lambda(t)dM(t) - \Lambda(t)dA(t) + d[M', \Lambda](t)$. The first two terms are P(local) martingales, as stochastic integrals with respect to martingales ($\Lambda(t)$ is a P-martingale). Thus $-\Lambda(t)dA(t) + d[M', \Lambda](t) = 0$. But $[A, \Lambda](t) = 0$, since A is continuous and finite variation. Thus $dA(t) = d[M, \Lambda](t)/\Lambda(t)$.

Exercise 10.13:

1. Using Girsanov's Theorem 10.15, there exists an equivalent measure Q_1 (defined by $\Lambda = dQ/dP = e^{-T/2-B(T)}$) such that $W(t) = B(t) + t$ is a Q_1-martingale. We show that under Q_1, $N(t)$ remains to be a Poisson process with rate 1. This follows by independence of $N(t)$ and $B(t)$ under P,
$$E_Q e^{uN(t)} = E_P(\Lambda e^{uN(t)}) = E_P(e^{-T/2-B(T)}e^{uN(t)}) = E_P(e^{-T/2-B(T)})E_P(e^{uN(t)})$$
$= E_P(e^{uN(t)})$, since $E_P\Lambda = 1$. This shows that one-dimensional distributions of $N(t)$ are Poisson (1). Similarly, the increments of $N(t)$ under Q_1 are Poisson and independent of the past. The statement follows.

2. Let $\Lambda = \Lambda(T) = \frac{dQ_2}{dP} = \Lambda_1\Lambda_2 = e^{-T-2B(T)}e^{-T+N(T)\ln 2}$
$= e^{-2T-2B(T)+N(T)\ln 2}$. Then $B(t)+2t$ is a Q_2-Brownian motion, and $N(t)$ is a Q-Poisson process with rate 2. To see this $E_Q(X) = E_P(\Lambda X) = E_P(\Lambda_1\Lambda_2 X)$, for any X and $E_P\Lambda_1 = E_P\Lambda_2 = 1$. So under Q_2, $N(t) - 2t$ is a martingale, and $B(t) + 2t$ is a martingale, thus $X(t) = B(t) + N(t)$ is a Q_2-martingale.

3. The probability measures defined as follows for $a > 0$ are equivalent to P, $dQ_a = e^{-a^2T/2-aB(T)}e^{(1-a)T+N(T)\ln a} = e^{(1-a-a^2/2)T-aB(T)+N(T)\ln a}dP$, and make the processes $B(t) + at$ into a Q_a-Brownian motion, and $N(t)$ into a Q_a-Poisson process with rate a.

Exercises to Chapter 11

Exercise 11.3: If $X = aS_1 + b$ for some a and b then $E_Q(X/r) = aE_Q(S_1/r) + b = aS_0 + b$ the same for any Q. Note that since the vectors S and β are not collinear, the representation of X by a portfolio (a, b) is unique. Take the claim that pays \$1 when the stock goes up and nothing in any other case, then this claim is unattainable. $E_Q(X) = 1.5(0.2 + p_d)$ depends on the choice of p_d.

Exercise 11.5: 1. $E_Q(M(t)|\mathcal{F}_s) = E_P(M(t)|\mathcal{F}_s) = M(s)$ a.s. Take $s = 0$, $E_P M(t) = E_Q M(t) = M(0)$.

2. If a claim is attainable, $X = V(T)$, where $V(t)/\beta(t)$ is a Q-martingale. Its price at time t is $C(t) = V(t)$, which does not depend on the choice of Q. By the martingale property $V(t) = \beta(t)E_Q(V(T)/\beta(T)|\mathcal{F}_t)$. This shows that $C(t) = \beta(t)E_Q(X/\beta(T)|\mathcal{F}_t)$ is the same for all Q's.

Exercise 11.10: Consider continuous time. $d(\frac{S(t)}{\beta(t)}) = \frac{dS(t)}{\beta(t)} + S(t)d(\frac{1}{\beta(t)})$.
$dR(t) = \frac{dS(t)}{S(t)} = \frac{\beta(t)}{S(t)}d(\frac{S(t)}{\beta(t)}) - \beta(t)d(\frac{1}{\beta(t)})$. $R(t) = \int_0^t \frac{\beta(u)}{S(u)}d(\frac{S(u)}{\beta(u)}) + \int_0^t \frac{d\beta(s)}{\beta(s)}$. The first integral is a martingale, as a stochastic integral with respect to a Q-martingale. Thus $E_Q R(t) = \int_0^t \frac{d\beta(s)}{\beta(s)}$, which is the risk-free return. In discrete time the statement easily follows from the martingale property of $S(t)/\beta(t)$.

Exercise 11.11: Let $X(t) = S(t)e^{-rt}$. Then it is easy to see that $X(t)$ satisfies the SDE $dX(t) = \sigma X(t)dB(t)$, for a Q-Brownian motion $B(t)$. By

Itô's formula $d(\frac{1}{X(t)}) = -\sigma\frac{1}{X(t)}(dB(t) - \sigma dt) = -\sigma\frac{1}{X(t)}dW(t)$, with $dW(t) = dB(t) - \sigma dt$. $W(t)$ is a Q-Brownian motion with drift. Using the change of measure $dQ_1/dQ = e^{\sigma B(T) - \sigma^2 T/2}$, by Girsanov's theorem $W(t)$ is a Q_1-Brownian motion. Finally, $-W(t)$ is also a Q_1-Brownian motion, and we can remove the minus sign in the SDE.

Exercise 11.12: By the Theorem 11.7, $C(t) = E_{Q_1}(\frac{S(t)}{S(T)}(S_T - K)^+|\mathcal{F}_t) = S_t E_{Q_1}((1 - \frac{K}{S(T)})^+|\mathcal{F}_t)$. The conditional distribution of $1/S(T)$ given $\mathcal{F}_t$ under Q_1 is obtained by using the SDE for $1/S(t)$. With $Y(t) = 1/X(t) = e^{rt}/S(t)$, we have from above $dY(t) = \sigma Y(t)dW(t)$ and $1/S(t) = Y(t)e^{-rt}$. Using the product rule $d(\frac{1}{S(t)}) = \frac{1}{S(t)}(-rdt + \sigma dW(t))$. Thus $1/S(t)$ is a stochastic exponential, and $1/S(T) = (1/S(t))e^{(T-t)(-r-\sigma^2/2)+\sigma(W(T)-W(t))}$. Thus given $\mathcal{F}_t$, the Q_1 conditional distribution of $1/S(T)$ is Lognormal with mean and variance $-\ln S(t) - (T-t)(r + \sigma^2/2)$ and $\sigma^2(T-t)$. Using calculations for the $E(1-X)^+$ similar to p. 313, we recover the Black-Scholes formula.

Exercise 11.14: From the first equation $b(t) = (V(t) - a(t)S(t))e^{-rt}$. Putting it in the self-financing condition $dV(t) = a(t)dS(t) + b(t)d(e^{rt})$, we get the SDE for $V(t)$.

For the other direction, let $b(t)$ be as above. Then $V(t) = a(t)S(t) + b(t)e^{rt}$, moreover the SDE for $V(t)$ gives the self-financing condition.

Exercise 11.17: Let $F(y) = e^{-rT}E((S_T/S - y)^+)$, then $C = SF(K/S)$. Now, $\partial C/\partial S = F(K/S) - KF'(K/S)/S$ and $\partial C/\partial K = F'(K/S)$. Thus $S\partial C/\partial S + K\partial C/\partial K = SF(K/S) = C$. The expression for $\frac{\partial C}{\partial S}$ in the Black-Scholes model follows from the Black-Scholes formula.

Exercises to Chapter 12

Exercise 12.1: Use Theorem 12.5, $C(t) = E_Q(\frac{\beta(t)}{\beta(s)}(P(s,T) - K)^+|\mathcal{F}_t)$ $= E_Q(\frac{\beta(t)}{\beta(s)}P(s,T)I(P(s,T) > K)|\mathcal{F}_t) - KE_Q(\frac{\beta(t)}{\beta(s)}I(P(s,T) > K)|\mathcal{F}_t)$. For the first term take Λ_2 based on the Q-martingale $P(s,T)/\beta(s)$, $s \leq T$, which corresponds to numeraire $P(t,T)$, or T-forward measure. Since the expectation must be unity, $\Lambda_2 = \frac{P(s,T)}{\beta(s)P(0,T)}$. (Formula (12.53)). Then the first term is $E_Q(\frac{P(s,T)\beta(t)}{\beta(s)}X|\mathcal{F}_t) = P(t,T)E_{Q_2}(X|\mathcal{F}_t)$. For the second term consider $\Lambda_1 = 1/(P(0,s)\beta(s))$ based on the martingale $P(t,s)/\beta(t)$, $t \leq s$, which corresponds to numeraire $P(t,s)$, or s-forward measure. Then for any X, $E_Q(\frac{\beta_t}{\beta_s}X|\mathcal{F}_t) = P(t,s)E_{Q_1}(X|\mathcal{F}_t)$. This gives for the second term $KP(t,s)Q_1(P(s,T) > K|\mathcal{F}_t)$, and the result follows.

Exercise 12.3: 1. Follows by $E|X| \leq \sqrt{EX^2}$.

2. By additivity of the integral $\int_0^T \left(\sum_{i=0}^{n-1} H(t_i, s)\big(W(t_{i+1}) - W(t_i)\big) \right) ds = \sum_{i=0}^{n-1} \left(\int_0^T H(t_i, s)ds \right) \big(W(t_{i+1}) - W(t_i)\big)$.

3. Let $X_n(s) = \sum_{i=0}^{n-1} H(t_i, s)\big(W(t_{i+1}) - W(t_i)\big)$. Then $X_n(s)$ is an approximation to the Itô integral $X(s) = \int_0^T H(t, s)dW(t)$. $EX^2(s) = \int_0^T EH^2(t, s)ds < \infty$, and under the stated conditions $\int_0^T E(X_n(s) - X(s))^2 ds$ converges to zero. This implies converges in L^2 and in probability of $\int_0^T X_n(s)ds$ to $\int_0^T X(s)ds$. The rhs $\sum_{i=0}^{n-1} \left(\int_0^T H(t_i, s)ds \right) \big(W(t_{i+1}) - W(t_i)\big)$ is an approximation of Itô integral of $Y(t)$, and converges to it in probability.

Exercise 12.5: Use the formula for the bond (12.32). $f(t, T) = -\frac{\partial \ln P(t,T)}{\partial T}$. A cap is a sum of caplets, which are priced by (12.64).

Exercise 12.7: Let $Y(t) = \int_0^t \gamma(s)dB(s)$ and $X(t) = \int_0^t \beta(s)dB(s)$ for some $\beta(s)$ to be determined. Then $\mathcal{E}(X)(t) = c + (1 - c)\mathcal{E}(Y)(t)$. Thus

$$dX(t) = \frac{d\mathcal{E}(X)(t)}{\mathcal{E}(X)(t)} = \frac{(1-c)d\mathcal{E}(Y)(t)}{c + (1-c)\mathcal{E}(Y)(t)} = \frac{(1-c)\mathcal{E}(Y)(t)\gamma(t)}{c + (1-c)\mathcal{E}(Y)(t)}dB(t), \text{ and}$$

$$\beta(t) = \frac{(1-c)\mathcal{E}(Y)(t)\gamma(t)}{c + (1-c)\mathcal{E}(Y)(t)} = \frac{(1-c)\gamma(t)e^{\int_0^t \gamma(s)dB(s) - \frac{1}{2}\int_0^t \gamma^2(s)ds}}{c + (1-c)e^{\int_0^t \gamma(s)dB(s) - \frac{1}{2}\int_0^t \gamma^2(s)ds}}.$$

Existence of forward rates volatilities follows from equation (12.76).

Exercise 12.9: At T_i the following exchange is made: the amount received is $f_{i-1}(T_i - T_{i-1})$ and paid out $k(T_i - T_{i-1})$. The resulting amount at time T_i is $1/P(T_{i-1}, T_i) - 1 - k\delta$, using $1/P(T_{i-1}, T_i) = 1 + f_{i-1}(T_i - T_{i-1})$. Thus the value at time t of the swap is

$$\text{Swap}(t, T_0, k) = \sum_{i=1}^n E_Q \left(\frac{\beta(t)}{\beta(T_i)} \left(\frac{1}{P(T_{i-1}, T_i)} - 1 - k\delta \right) \mid \mathcal{F}_t \right).$$

Using the martingale property of the discounted bonds (12.4), we obtain that $E_Q \left(\frac{\beta(t)}{\beta(T_i)} | \mathcal{F}_{T_{i-1}} \right) = P(T_{i-1}, T_i)$. The result is obtained by conditioning on $\mathcal{F}_{T_{i-1}}$ in the above sum.

Exercise 12.10: Follows from the previous exercise.

Exercise 12.11: Consider the portfolio of bonds that at any time $t < T$ is long $\alpha_+(t)$ of the T_0-bond, short $\alpha_+(t)$ of the T_n-bond, and for each i, $1 \leq i \leq n$

short $\alpha_-(t)\delta k$ of the T_i-bond. The value of this portfolio at time $t < T$ is
$C(t) = \alpha_+(t)(P(t,T_0) - P(t,T_n)) - k\alpha_-(t)b(t)$, with $b(t) = \delta \sum_{i=1}^{n} P(t,T_i)$.
By using the expression for the swap rate $C(t) = b(t)(\alpha_+(t)k(t) - K\alpha_-(t))$. It
can be seen that this portfolio has the correct final value and is self-financing,
i.e. $dC(t) = \alpha_+(t)d(P(t,T_0) - P(t,T_n)) - k\alpha_-(t)db(t)$.

Exercises to Chapter 13

Exercise 13.1: Use Theorem 6.16 and formula (6.98) to find $E(T_0 \wedge T_b)$.
$ET_0 = \lim_{b\to\infty} E(T_0 \wedge T_b)$ by monotone convergence.

Exercise 13.2: 1. $d(e^{-cX(t)}) = -ce^{-cX(t)}dX(t) + \frac{c^2}{2}e^{-cX(t)}d[X,X](t) =$
$-ce^{-cX(t)}\sigma\sqrt{X}(t)dB(t)$. By Theorem 13.1 $X(t) \geq 0$ for all t. The function
$e^{-cx}\sqrt{x}$ is bounded for $x \geq 0$. Thus $e^{-cX(t)}$ is martingale as an Itô integral of
a bounded process, Theorem 4.7.
2. Let $\tau = T \wedge t$, then τ is a bounded stopping time. Applying Optional
stopping, we have $e^{-cx} = Ee^{-cX(0)} = Ee^{-cX(\tau)} = E(e^{-cX(\tau)}I(\tau = T)) +$
$E(e^{-cX(\tau)}I(\tau = t)) \geq E(e^{-cX(\tau)}I(\tau = T)) = P(T \leq t)$.

Exercise 13.5: 1. Let $G(x) = \int_a^x du/g(u)$. Then $G(x(t)) = G(x(0))+t$. Now
$dY(t) = dt + \frac{\sigma(X(t))}{g(X(t))}dB(t) - \frac{\sigma^2(X(t))g'(X(t))}{g^2(X(t))}dt$.
2. $G(x) = x^{1-r}/(c(1-r))$ and $Y(t) = t - \int_0^t \frac{\sigma^2(X(s))r}{c(X(s))^{r+1}}ds + \int_0^t \frac{\sigma(X(s))}{(cX(s))^r}dB(s)$.
By the growth condition, $E(\int_0^t \frac{\sigma(X(s))}{(cX(s))^r}dB(s))^2 = \int_0^t E(\frac{\sigma(X(s))}{(cX(s))^r})^2 ds \leq Ct$. It
follows $E(\frac{1}{t}\int_0^t \frac{\sigma(X(s))}{(cX(s))^r}dB(s))^2 \to 0$, and implies $\frac{1}{t}\int_0^t \frac{\sigma(X(s))}{(cX(s))^r}dB(s) \to 0$ in
probability. It follows by the L'Hospital rule that $\frac{1}{t}\int_0^t \frac{\sigma^2(X(s))r}{c(X(s))^{r+1}}ds \to 0$ on the
set $\{X(t) \to \infty\}$. The LLN for $Y(t)$ now follows.

Exercise 13.6: $P(\min(L_1, L_2, \ldots, L_x) > t) = P(L_1 > t, L_2 > t, \ldots, L_x > t)$
$= (P(L_1 > t))^x = (e^{-a(x)t})^x = e^{-xa(x)t}$.

Exercise 13.7: 1. $\pi_1(t) = \int_0^t \sum_{k\geq 1} I(X(s-) \geq k)dN_k^\lambda(s)$ has jumps of size
one. Verify that $A_t^1 = \int_0^t \lambda X_s ds$ is its compensator. Similarly for the process
$\pi_2(t) = \int_0^t \sum_{k\geq 1} I(X(s-) \geq k)dN_k^\mu(s)$.
2. $X(t) = X(0) + A_1(t) - A_2(t) + \pi_1(t) - A_1(t) - \pi_2(t) + A_2(t)$.

Exercises to Chapter 14

Exercise 14.3: $E(M(t)|\mathcal{G}_s) = E(E(M(t)|\mathcal{F}_s)|\mathcal{G}_s) = E(M(s)|\mathcal{G}_s) = M(s)$.

Exercise 14.5: Let $\mathcal{G}_t$ be the trivial σ-field, then $\widehat{W}(t) = E(W(t)|\mathcal{G}_t) = 0$.

Exercise 14.6: Using formally Theorem 14.6 with $a(t) = b(t) = 0$ and $A(t) = B(t) = 1$, $d\widehat{X}(t) = -v(t)\widehat{X}(t)dt + v(t)dY(t)$, and $dv(t) = -v^2(t)dt$. Hence $d\widehat{X}(t) = -(1/t)\widehat{X}(t)dt + (1/t)dY(t)$. We obtain $v(t) = 1/t$ and $\widehat{X}(t) = Y(t)/t$. These can be obtained directly using $Y(t) = ct + W(t)$.

Exercise 14.8: $dX(t) = (\mu + \sigma^2/2)X(t)dt + \sigma X(t)dB(t)$. Apply Theorem 14.6 with $a(t) = \mu + \sigma^2/2$, $b(t) = \sigma$, $A(t) = B(t) = 1$, see Example 14.1.

References

1. Artzner P. and Delbaen F. "Term structure of interest rates: The martingale approach", *Adv. Appl. Math.* 10(1989), 95-129.

2. Bezen and Klebaner (1996) "Stationary solutions and stability of second order random differential equations," *Physica A* 233/234, 809-823.

3. Bhattacharya R.N. "Criteria for recurrence and existence of invariant measures for multi-dimensional diffusions", *Ann. Probab.* 6(1978), 541-553.

4. Björk T., Kabanov Y., and Runggaldier W. (1996) "Bond market structure in the presence of marked point processes", *Math. Finance* 7, 211-239.

5. Björk T., Di Masi G., Kabanov Y. and Runggaldier W. (1997) "Toward a general theory of bond markets", *Finance and Stochastics* 2, 141-174.

6. Björk T. *Arbitrage Pricing Theory in Continuous Time*, Oxford University Press, 1998.

7. Borodin A.N. and Salminen P. *Handbook of Brownian Motion*. Birkhäuser, 1996.

8. Borovkov K., Klebaner F.C., Virag E. (2003) "Random Step Functions Model for Interest Rates", *Finance and Stochastics* 7 , 123-143.

9. Brace A. (1998) "Swaprate Volatilities in BGM," FMMA NOTES.

10. Brace A., Gatarek D., Musiela M. (1997) "The Market Model of Interest Rate Dynamics" *Math. Finance* 7, No 2, 127-156.

11. Breiman L. *Probability*. Addison-Wesley, 1968.

12. Chung K.L. *Markov Chains with Stationary Transition Probabilities*. Springer-Verlag, 1960.

13. Chung K.L. *Lectures from Markov Processes to Brownian Motion.* Springer-Verlag, 1982.

14. Chung K.L. and Williams R.J. *Introduction to Stochastic Integration.* Birkhäuser, 1983.

15. Cont R. (1998) "Modeling term structure dynamics: an infinite dimensional approach", *Lyapunov Institue Workshop on Mathematical Finance*, Rocquencourt.

16. Dawson D.A. *Measure-valued Markov Processes*, Lecture Notes in Math., Springer, 1993.

17. Delbaen F. and Schachermayer W. (1994) "A general version of the fundamental theorem of asset pricing", *Math. Annalen* 300, 463-520.

18. Dellacherie C. and Meyer P.A. *Probabilities and Potential B.* North-Holland, 1982.

19. Dieudonne J. *Foundations of Modern Analysis.* Academic Press, 1960.

20. Dudley R.M. *Real Analysis and Probability.* Wadsworth, 1989.

21. Durrett R. *Brownian Motion and Martingales in Analysis.* Wadsworth, 1984.

22. Durrett R. *Probability: Theory and Examples.* Wadsworth, 1991.

23. Dynkin E.B. *Markov Processes.* Springer-Verlag, 1965.

24. Ethier S.N. and Kurtz T.G. *Markov Processes.* Wiley, 1986.

25. Ewens W.J. *Mathematical Population Genetics.* Springer-Verlag, 1979.

26. Feller W. (1951) "Diffusion processes in Genetics," *Second Berkely Symposium on Math. Stat. and Probab.*, 227-246.

27. Feller W. *An Introduction to Probability Theory and Its Applications. Volume 2.* Wiley, 1971.

28. Freedman D. *Brownian Motion and Diffusion.* Springer-Verlag, 1971, 1983.

29. Friedman A. *Stochastic Differential Equations and Applications. Volume 1.* Academic Press, 1975.

30. Friedman A. *Stochastic Differential Equations and Applications. Volume 2.* Academic Press, 1976.

31. Fouque J-P., Papanicolaou G., Sircar K.R. *Derivatives in Financial Markets with Stochastic Volatility.* Cambridge University Press, 2000.

32. Gard T.C. *Introduction to Stochastic Differential Equations.* Marcel Dekker, 1988.

33. Geman H., El Karoui N., and Rochet J.C. (1995) "Changes of numeraire, changes of probability measure and option pricing", *J. Appl. Probab.* 32, 443-458.

34. Gihman I.I. and Skorohod A.V. *Stochastic Differential Equations.* Springer-Verlag, 1972.

35. Grimmet G. and Stirzaker D. *Probability and Random Processes.* Oxford University Press, 1987.

36. Hamza K. and Klebaner F.C. (1995) "Representation of Markov chains as stochastic differential equations", *Lecture Notes in Statistics* 99, Springer-Verlag, 144-151.

37. Hamza K. and Klebaner F.C. (1995) "Conditions for integrability of Markov Chains," *J. Appl. Probab.*, 32, 541-547.

38. Hamza K., Jacka S.D. and Klebaner F.C. (2005) "The EMM conditions in a general model for interest rates", *Applied Probability* to appear.

39. Harris T.E. *The Theory of Branching Processes.* Springer 1963.

40. Harrison J.M. and Kreps D. (1979) "Martingales and arbitrage in multiperiod security markets," *J. Econ. Theory* 20, 381-408.

41. Harrison J.M. and Pliska S.R. (1981) "Martingales and stochastic integrals in the theory of continuous trading", *Stoch. Proc. Appl.* 11, 215-260.

42. Harrison J.M. and Pliska S.R. (1983) "A stochastic model of continuous trading: Complete markets", *Stoch. Proc. Appl.* 15, 313-316.

43. Hasminskii, R.Z. *Stochastic Stability of Differential Equations.* Sijthoff and Noordhoff, 1980.

44. Heath D., Jarrow R., and Morton A. (1992) "Bond pricing and the Term Structure of interest rates: a new methodology for contingent claims valuation", *Econometrica* 60, 77-105.

45. Heston S. (1993). "A Closed-Form Solution for Options with Stochastic Volatility with Applications to Bond and Currency Options", *Review-of-Financial-Studies* 13 (3), 585-625.

46. Hobson E.W. *The Theory of Functions of a Real Variable.* Cambridge University Press, 1921.

47. Ito K. and McKean H.P. *Diffusion Processes and Their Sample Paths.* Springer-Verlag, 1965.

48. Jacod J. and Shiryaev A.N. *Limit Theorems for Stochastic Processes.* Springer-Verlag, 1987.

49. Jagers P. and Klebaner F.C. (2000) "Population-size-dependent and age-dependent branching processes", *Stoch. Proc. Appl.,* 87, 235-254.

50. Jamshidian F. (1996) "Sorting out swaptions", *Risk* 9(3).

51. Kabanov Yu. M. (2001) "Arbitrage Theory", in *Option Pricing, Interest Rates and Risk Management,* E. Jouini, J. Cvitanic, M. Musiela, Eds., 3-42.

52. Kallenberg O. (1996) "On existence of universal functional solutions to classical SDEs," *Ann. Probab.* 24, 196-205.

53. Kallianpur G. *Stochastic Filtering Theory.* Springer-Verlag 1980.

54. Kallsen J. and Shiryaev A.N. (2002) "The cumulant process and Esscher'r change of measure," *Finance and Stochastics,* 6, 397-428.

55. Karatzas I. and Shreve S.E. *Brownian Motion and Stochastic Calculus.* Springer-Verlag, 1988.

56. Karlin S. and Taylor H.M. *A First Course in Stochastic Processes.* Academic Press, 1975.

57. Karlin S. and Taylor H.M. *A Second Course in Stochastic Processes.* Academic Press, 1981.

58. Karr A.F. *Point Processes and Their Statistical Inference.* Marcel Dekker, 1986.

59. Keller G., Kersting G., and Rösler U. (1984) "On the asymptotic behaviour of solutions of stochastic differential equations", *Z. Wahrscheinlichkeitsth.* 68, 163-189.

60. Kennedy P.D. (1994) "The Term Structure of Interest Rates as a Gaussian Markov Field", *Math. Finance* 4, 247-258.

61. Kersting G. and Klebaner F.C. (1995) "Sharp conditions for nonexplosions and explosions in Markov Jump processes", *Ann. Probab.* 23, 268-272.

62. Kersting G. and Klebaner F.C. (1996) "Explosions in Markov Jump processes and submartingale convergence", *Lecture Notes in Statistics* 114, 127-136, Springer-Verlag.

63. Klebaner F.C. (1994) "Asymptotic behaviour of Markov population processes with asymptotically linear rate of change", *J. Appl. Probab.* 31, 614-625.

64. Klebaner F.C. (2002) "Option price when the stock is a semimartingale", *Elect. Comm. in Probab.* 7, 79-83. Correction, 8.

65. Klebaner F.C. and Liptser R. (2001) "Asymptotic analysis and extinction in randomly perturbed Lotka-Volterra model", *Ann. Appl. Probab.* 11, 4, 1263-1291.

66. Kreps D.M. (1981) "Arbitrage and equilibrium in economies with infinitely many commodities", *J. Math. Economics* 8, 15-35.

67. Krishnan V. *Nonlinear Filtering and Smoothing: An Introduction to Martingales, Stochastic integrals, and Estimation.* Wiley 1984.

68. Kurtz, T.G. *Approximation of Population Processes.* Society for Industrial and Applied Mathematics, 1981.

69. Lamberton D. and Lapeyre B. *Introduction to Stochastic Calculus Applied to Finance.* Chapman and Hall, 1996.

70. Liptser R.S. and Shiryayev A.N. *Theory of Martingales.* Kluwer, 1989.

71. Liptser R.S. and Shiryayev A.N. *Statistics of Random Processes I and II.* Springer-Verlag, 1974. Second Edition 2001.

72. Loeve, M. *Probability Theory I and II.* Springer-Verlag, 1978.

73. Métivier M. *Semimartingales.* de Gruyter, 1982.

74. Molchanov S.A. (1975) "Diffusion Processes and Riemannian Geometry," *Russian Math. Surveys* 30, 1-63.

75. Muisela M., and Rutkowski M. *Martingale Methods in Financial Modelling.* Springer, 1998.

76. Oksendal B. *Stochastic Differential Equations.* Springer-Verlag, 1995.

77. Pinsky R. *Positive Harmonic Functions and Diffusion.* Cambridge University Press, 1995.

78. Protter P. *Stochastic Integration and Differential equations.* Springer-Verlag, 1992.

79. Rebonato R. *Modern Pricing of Interest-Rate Derivatives.* Princeton University Press, 2002.

80. Revuz D. and Yor M. *Continuous martingales and Brownian Motion.* Springer-Verlag, 1991. Third edition 2nd printing 2001.

81. Rogers L.C.G. and Williams D. *Diffusions, Markov Processes, and Martingales. Volume 2. Ito Calculus.* Wiley, 1990.

82. Rogers L.C.G. and Williams D. *Diffusions, Markov Processes, and Martingales. Volume 1. Foundations.* Wiley, 1994.

83. Rogers L.C.G. and Shi Z. (1995) "The value of an Asian option," *J. Appl. Probab.* 32, 1077-1088.

84. Saks S. *Theory of the Integral.* Dover, 1964.

85. Shiryaev A.N. *Essentials of Stochastic Finance.* World Scientific, Singapore, 1999.

86. Soize C. *The Fokker-Plank Equation For Stochastic Dynamical Systems and Its Explicit Steady State Solutions.* World Scientific, 1994.

87. Soong T.T. *Random Differential equations in Science and Engineering.* Academic Press, 1973.

88. Stroock D., and Varadhan S.R.S. *Multidimensional Diffusion Processes.* Springer-Verlag, 1979.

89. Vasicek O. (1977) "An equilibrium characterization of the Term Structure," *J. Fin. Econ.* 5, 177-188.

90. Vecer J. (2002) "Unified Asian Pricing", *Risk* 15, 6, 113-116.

91. Yamada T. and Watanabe S. (1971) "On the uniqueness of solutions of stochastic differential equations", *J. Math. Kyoto Univ.* 11, 155-167.

Index

413

AUTOMATED DATABASE APPLICATIONS TESTING

Specification Representation for
Automated Reasoning

SERIES IN MACHINE PERCEPTION AND ARTIFICIAL INTELLIGENCE*

Editors: **H. Bunke** (Univ. Bern, Switzerland)
P. S. P. Wang (Northeastern Univ., USA)

*For the complete list of titles in this series, please write to the Publisher.

Series in Machine Perception and Artificial Intelligence – Vol. 76

AUTOMATED DATABASE APPLICATIONS TESTING

Specification Representation for Automated Reasoning

Rana Farid Mikhail, Donald Berndt & Abraham Kandel

University of South Florida, USA

World Scientific

NEW JERSEY · LONDON · SINGAPORE · BEIJING · SHANGHAI · HONG KONG · TAIPEI · CHENNAI

Published by

World Scientific Publishing Co. Pte. Ltd.

5 Toh Tuck Link, Singapore 596224

USA office: 27 Warren Street, Suite 401-402, Hackensack, NJ 07601

UK office: 57 Shelton Street, Covent Garden, London WC2H 9HE

British Library Cataloguing-in-Publication Data
A catalogue record for this book is available from the British Library.

Series in Machine Perception and Artificial Intelligence — Vol. 76
AUTOMATED DATABASE APPLICATIONS TESTING
Specification Representation for Automated Reasoning

ISBN-13 978-981-283-728-8
ISBN-10 981-283-728-0

Printed in Singapore.

Dedication Page

To my beloved father, Dr. Farid Halim Mikhail - All I am, I owe to you. Your words of wisdom will remain in my heart and mind as long as I live.

To my husband and best friend, Emad - You're the best part of my day - every day!

To my mother Nadia and brother Ramy, every day I appreciate you more and learn how blessed I am, to have you as my family.

Rana Mikhail

To my mother Theresa Ciaravino Berndt, who lived a life of faith and family.

Don Berndt

Dedicated to the memory of my dear brother Professor Shmuel Kandel.

Abraham Kandel

Preface

In this book we introduce SpecDB, a database created to represent and host software specifications in a machine-readable format. The specifications represented in SpecDB are for the purpose of unit testing database operations. A structured representation aids in the processes of both automated software testing and software code generation, based on the actual software specifications. We describe the design of SpecDB, the underlying database that can hold the specifications required for unit testing database operations.

Specifications can be fed directly into SpecDB, or, if available, the formal specifications can be translated to the SpecDB representation. An algorithm that translates formal specifications to the SpecDB representation is described. The Z formal specification language has been chosen as an example for the translation algorithm. The outcome of the translation algorithm is a set of machine-readable formal specifications.

To demonstrate the use of SpecDB, two automated tools are presented. The first automatically generates database constraints from represented business rules in SpecDB. This constraint generator gives the advantage of enforcing some business rules at the database level for better data quality. The second automated application of SpecDB is a reverse engineering tool that logs the actual execution of the program from the code. By Automatically comparing the output of this tool to the specifications in SpecDB, errors of commission are highlighted that might otherwise not be identified. Some errors of commission including coding unspecified behavior together with correct coding of the specifications cannot be discovered through black box testing techniques, since these techniques cannot observe what other modifications or outputs have happened in the background. For example, black box, functional testing techniques cannot identify an error if the software being tested produced the correct specified output but more over, sent classified data to insecure locations. Accordingly, the decision of whether a software application passed a test depends on whether it coded all the specifications and only the specifications for that unit. Automated tools, using the reverse engineering application introduced in this book, can thus automatically make the decision whether the software passed a test or not based on the provided specifications.

This book is intended for to the members of the software testing team and developers of software testing tools. System analysts and designers could also benefit from the ideas presented in this book. Although the title of the book caters for database applications, yet, those interested in software applications that do not involve database systems can also benfit greatly from the concepts in this book, since applications based on database systems are a superset of those that do not access databases.

Acknowledgments

The authors would like to acknowledge the National Institute for Systems Test and Productivity (NISTP) at the University of South Florida for funding parts of this work and other related projects, under sponsorship from the USA Space and Naval Warfare Systems Command (N00039-01-1-2248). NISTP supported a multi-year effort focused on investigating the intersection of computational intelligence and software quality assurance. We have been fortunate to be part of a changing group of interdisciplinary researchers that were free to pursue a variety of somewhat unusual projects. We are thankful for the opportunity.

Contents

Chapter 1

Introduction

1.1 The Need for Testing

One of the common but very true jokes about the software industry tells of a software entrepreneur proudly stating that if the automobile industry had advanced in the past decade with the same rate as that of the computer industry, we would all be driving $25 cars that run for a 1,000 miles per gallon of gasoline. To which an automobile CEO replies, that if cars were like today's software, they would crash twice a day for no reason, and when you call for service, they would tell you to shut down the engine, as well as all windows, get out of the car, lock it, and start over! Indeed, in the past decade, computer hardware has advanced significantly offering much higher processing capability and speed. Consequently, the complexity and size of software has increased exponentially too. Software is vital to almost every industry and we are becoming ever more dependent on our computers to perform every day tasks. As the complexity and size of software grow, so does the need to insure the reliability and precision of software outputs. Software errors can be fatal in numerous industries; including airlines, medical and the military. Billions of dollars are spent on software and software testing and maintenance annually by the DoD. Software errors result in hundreds of millions in losses annually from the DoD alone [1] (software errors are defined in Section 1.3.1).

Accordingly, a lot of attention is now directed towards assuring software reliability, accuracy, security and usefulness. It is crucial to dedicate a lot of resources, including time, effort and a budget during the software lifecycle to verify that the end product is useful, usable and accomplishes the tasks it was built to perform.

1

1.2   Why Does Software have Errors?

Software errors result from many reasons, from misunderstanding of the requirements, to poor design; also from tired or careless programmers to rushed deadlines [2]. Often testing is delayed until later in the software development life cycle, and little time is given for testing, debugging, and re-testing. Coverage gaps are also a problem; this happens when not all parts of the software are tested.

Some software failures happen due to a specific scenario, a sequence of events that might have been tested individually but not in that particular sequence, that causes the software failures. The operating environment, various connected hardware peripherals, other concurrently running software and different operating systems could also be a source of software failures [4].

It was proven that complete testing is impossible [3]. Meaning that it is impossible to test all values of each possible input, valid and invalid, with their respective combinations for a software application with considerable size and complexity; the number of test cases will be astronomical and the time needed for testing and verification will be almost endless. For example, a very simple program might take only three inputs, an integer Age, between 0 and 99, a character, Initial, between A and Z, and another integer representing a Month between the integers 1 and 12.

```
Age: integer [0..99]
Initial: char [A..Z]
Month: integer [1..12]
```

For this simple program, there are 100 possible valid values for Age, 26 valid values for Initial, and 12 valid values for Month. Accordingly there are 100*26*12 = 31,200 possible combinations of valid values. If complete testing was required, there would be 31,200 test cases that should be carried out to test the operation of this program on valid values alone. Even if the 31,200 test cases were run, this would not be considered complete testing. The number of test cases required for complete testing will be much greater, when the invalid values are added. Invalid values could be in one, two or all three inputs. The number will increase exponentially if we want to create test cases for all these cases; where only one of the three variables contains an invalid value, a combination of two of the variables are invalid, or all three. For example, the Age variable is an integer, one can enter any integer from the negative integer boundary to the positive integer boundary. Also, one can try to enter characters, real numbers or any other value for the Age integer value and test the program behavior and how it handles exceptions. The possibilities and combinations of all those cases for this very simple three input program are thus almost endless. Also it is impossible to completely test all values of real numbers, even in a small range, and for strings representing names for example, etc. Some transient errors are revealed in complex distributed or embedded control systems after prolonged usage. This may be a result of a memory overflow, or minute system clock discrepancies that accumulate to a noticeable difference after prolonged usage. Such errors are very difficult to recognize [77], and require high volume system testing or long sequence testing techniques, and extended random regression testing [79]. Hence there is a need for better software testing techniques, which focus on testing parts of the input domain, yet, attain good coverage to verify software quality.

1.3 Software Testing Definitions

1.3.1 *Software Errors, Faults and Failures*

A software error is a mismatch between the program and its specifications, provided that the specifications are accurate and complete. A software application is considered erroneous if it fails to be useful. A software fault is defined as the execution of an error [5]. Faults of commission are those resulting from incorrectly coding specifications, or coding unspecified requirements. Faults of omission result from failing to implement specified requirements. A software failure happens when a fault of commission is executed while running the software application [5]. From Figure 1.1, the goal of testing is to make sure that $S \cap P = S = P$. If the above equation is reached in testing a software application, the resulting product is very well tested and functional, provided that the specifications are correct, accurate, complete and do not contain any contradictions.

1.3.2 *Software Testing*

Software testing is becoming an important area of research aimed at producing more reliable systems and minimizing programming errors [78]. Since it is impossible to completely test how a software application performs on all values of all of its inputs, as described in Section 1.2, a good strategy to test

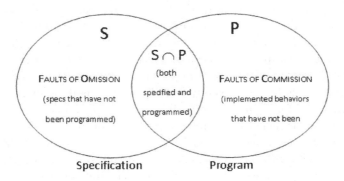

Fig. 1.1   The Relationship Between a Program Specification and its Implementation [5]

software has to be followed in order to choose test cases that test the software without coverage gaps. However, there is much more to good testing than running a program several times to check if it functions correctly [6]. In order to systematically test software, we need to model the environment in which the software is intended to run, select test scenarios that attain good system coverage, execute and evaluate the test cases, correct the errors, and measure the testing progress [4]. Some faults may result from correcting other errors, and hence regression testing is often used, which entails re-running all test cases after an error has been corrected to verify that the fix did not inject any errors in a part of the software that was previously running correctly. Software testing has been defined in many ways. Some authors define it as the process of running a software application to verify its correct execution of its valid and complete specifications, and verify that it runs correctly in the environment it was created for [5, 86]. Other authors argue that test scenarios that do not reveal errors or failures are also of great importance, since the testing team is assuring by such scenarios that the system actually implemented the required specifications correctly [6]. Software can also fail by executing too slowly or using too much memory; reasonable execution speed and memory usage are obvious requirements that might not be in the specification documents. A number of good references for software testing are [5, 6, 553]. Accordingly, it is crucial to create test cases that reflect a good coverage of all areas of the software and determine its usability.

1.4 When Should Testing Start in the Software Lifecycle?

One of the most important and useful practices in software engineering is the incorporation of software testing from the very beginning of the software lifecycle. The software life cycle is described in more detail in [7, 8, 9]. Waiting until the software is implemented and then starting to test the software was proven in the last decade by many authors and software team managers to be a very costly and difficult task to manage and carryout. Testing should start as soon as the analysts finish writing the first draft of the specification document describing what the required system should be built to accomplish. Testers can then begin identifying test cases, which are called system level test cases. System level test cases are very easily interpreted and basically describe in a broad spectrum what the main functionalities of the software should be. By reading the system level test cases, both the development team and the end users can reach an early agreement and common understanding of the functionality of the software under construction.

After agreeing on the specifications, and the system level test cases, the designers can then start defining the details of the system, and answer many questions from as broad as what programming language should be used to the minute details of the functionality of each sub-routine. Testers have a very important role to play during the design phase. During that phase testers should generate the integration level and unit level test cases. Integration level test cases are targeted towards testing the correctness of the interaction between the different sub-routines (methods, functions, procedures, etc.) and the correctness of the interaction between the different modules or subsystems of the software under construction [5].

Once implementation starts, the programmers should have both the design and the test cases that will be exercised against their implementation. Having the complete set of test cases, that cover all possible outputs of the functionality a programmer is implementing, will help the programmer to understand the design and implement it correctly. There is a strong possibility that the programmer will make sure the implemented code passes the given tests before submitting the code for integration with the rest of the system. There are many advantages for the above approach of starting testing early in the software lifecycle. In this situation, when the programmer is given both the design and test cases, testing starts very early in the software lifecycle. A test case will be executed multiple times to check that the system passes it correctly at the unit level as well as the system level after integration with the rest of the system modules. There will be no ambiguity in what exactly the programmer should implement, even if the programmer misinterprets the design and specified functionality, reading the test cases, and their respective expected output, will clear much of the ambiguity, and will assure a higher level of software quality, compared to that when only a design is given to the programmer. Once the implementation phase is over, the majority of the testing phase will have been implemented already. The cost of integration and system level testing will be minimal compared to that of delaying testing until the last phase in the software lifecycle. Also the overall quality of the implemented software will be much higher than might result given the same resources of time and money, but leaving testing till after implementation. It is worth noting that the waterfall model of software development where all the specifications of the software are known from the beginning is not what is generally followed in real world software development environments. Test cases should be created for all the specifications as they are gathered for the different builds or versions of the software under construction. Testing should be carried out for each of those versions to make sure that the specifications thus far collected, are all implemented correctly. Specifications should be tested to make sure that they do not contradict previously specified requirements. Google, Microsoft, and Sun Microsystems are investing $7.5 million to fund research on alternative software development models to the waterfall model with higher reliability and faster deployment, at UC Berkeley [101].

In conclusion, it is essential that the testing process starts from the beginning of the software lifecycle and be concurrent with all the phases as a better software engineering practice.

1.5   Types of Testing Techniques

There are two approaches to identifying test cases: functional and structural. Functional testing, also often referred to as black box testing, is the process of testing the executable program without any reference to the source code of the software. It is called functional since we treat the software as a function, we provide inputs, and observe its output. Functional test cases can be reused even when the code changes, is translated into another language, or its structure differs. Another advantage of functional testing is that the test cases can be created during the early phases of development, even before implementation starts, since it is based on the specifications. There are a number of functional testing techniques including boundary value analysis, equivalence class testing, and decision table-based testing. The reader can refer to [5] for a good coverage of those three techniques. Other black box testing techniques are domain driven, stress driven, specification driven, risk driven, random / statistical, function, regression, scenario, use case, user and exploratory testing. The reader can refer to [7] for an explanation of these techniques. In structural testing on the other hand, also referred to as white or clear box testing, test cases are created to execute specific branches and paths in the source code. Structural testing techniques include decision-to-decision and define-use testing. The advantage of structural testing is in its ability to confirm that all paths in the source code have been executed in test cases, and thus those testing techniques can be used as a coverage matrix to guide other functional testing techniques.

1.6 UML

To better understand some types of today's automated testing tools that are targeted towards testing in the requirements definition and design phases, one should understand the Unified Modeling Language (UML). The reader can refer to [11, 12, 13, 14] for UML syntax and diagrams. The majority of software development organizations today, use UML in describing a system design. System designers represent the design of a system using UML, a set of diagrams that both technical and non-technical system end users can understand and use as a basis for discussions and reassurance that the software under development is what the end-users actually have in mind. Those UML diagrams are then used as a basis for more detailed system design, implementation and testing. In their book, Business Modeling with UML [11], Eriksson and Penker describe the nine different UML diagrams and the use of each for business modeling. They state that 'UML standardizes notation for describing a process, but it doesn't standardize a process for producing those descriptions' [11]. This leads to the first objective, namely, to create a standard representation for storing UML or software specifications represented by a formal specification languages like Z, since UML diagrams are not machine interpretable.

1.7   Formal Specification Languages

Formal specifications, also known as formal methods, are a mathematical-like notation used to describe the specifications, i.e. what a system should execute, without necessarily going into the details of how the requirements should be implemented. Formal specifications where developed primarily to result in accurate, unambiguous software specifications to replace informal natural language specifications for better engineered software. The reader can refer to [15-24, 65] for a detailed description of formal methods, use, semantics and syntax.

1.8   Current Testing Technologies and Tools

1.8.1   *Types of Automated Testing Tools*

Depending on where in the software lifecycle a development team would like to introduce an automated testing tool, the types of tools can be divided into tools for the analysis and specifications gathering phase, others for the requirements definition phase, design phase, implementation phase, and finally quality assurance and debugging phase. There are also metrics and testing organization tools to aid in the process of managing the testing effort. Some tools are targeted for specific purposes like test data generators [57], simulators and prototyping tools.

There are dozens of automated software testing tools in the market today. The tools range in their functionality between functional-black box testing, to source code analysis tools. A good list can be found at the Aptest: Software Testing Resources and Tools website [25]. A lot of the available tools record and play back test scenarios the tester inputs while doing exploratory or scenario testing. Regression testing is the most commonly automated testing technique.

Some of the tools are targeted towards detecting run-time errors and memory leaks. Among those tools are Purify from Rational Software, C Verifier from PolySpace, SWAT from Coverity, Insure++ from Parasoft and GlowCode from Electric Software. Other tools are test case organization tools, to help in the organization of the testing process. There are a number of automated test data generators and a lot of research in the automation of test data creation [69-72, 80-85, 87-95]. Test data generators generate data to simulate real life data in operational systems for the purpose of testing. Simulators of test environments are also available as well as automated stress and load testing tools. The following sections include a more detailed look at some of the features and underlying technology in several of the most widely used tools.

1.8.2 *Design and Visual Modeling Tools*

Design and visual modeling tools are mainly used in the analysis and design phases. These tools help the development team as well as the end users to reach a common understanding of the functionality of the system under construction. Some of the available tools also generate the system backbone from the design generated by the visual modeling tools, including the database tables, classes, modules, etc. The programmers can then write the code in the designated locations. Rational Rose and Oracle Designer are two of the common design and visual modeling tools available in the market.

1.8.2.1 *Rational Rose*

Rational Rose provides a channel to communicate a system's architecture between various parties of the development team, from analysts and designers to developers and other team members. Rational Rose allows the designers to visually map system architecture to unified models and diagrams using models similar to UML. The diagrams go in their detailed level only to define the interaction between the different parts of the software. You can write code in the subroutine definition area, and thus when you choose to generate the code from Rational Rose, the code that you wrote will also exist where it should in your application. However, Rational Rose does not go into the details of the design or interpret the functionality of the software as it does with the higher level design. Accordingly, the testing tools that are now available and produced by Rational, do not depend on the specified functionality, but on the backbone, and the actual implementation and user scenarios.

1.8.2.2 *Oracle Designer*

Oracle is one of the leading companies in database and information systems solutions. One of the products offered by Oracle is Oracle Designer. As stated by Oracle, Designer is "a complete toolset to model, generate, and capture the requirements and design of enterprise applications" [26]. Oracle Designer has many modeling tools to help in the design and analysis capture of a database application. Among the tools included in Designer are the process Modeler, Dataflow Diagrammer, function Hierarchy Diagrammer, Entity relationship Diagrammer, Design Editor, Dependency Manager and Matrix Diagrammer. None of the above tools captures the detailed design and the functionality of the application, in a way that could be interpreted by an automated testing tool. The diagrams and matrixes are used to communicate system requirements and design between the different development team members, but not for an automated testing tool, that can intelligently understand the specifications and design of the software under construction and accordingly generate good test cases to test it. In the design editor, one could design the forms and the components of the application's windows, etc. One can also determine what type of triggers run, and what triggers them, but the code and the meaning of the implementation in those triggers can either be written in plain English as comments for the programmers or can be written in actual PL-SQL code. However, what is written in the triggers or stored procedures in designer, whether it is the actual code or plain English description, is stored as is to allow for it to exist after the user chooses to generate the application from the stored design in Oracle Designer.

Designer automatically generates a script for the specified database tables, constraints, modules for the application, and also forms, triggers, etc. Designer thus stores the structure of the application, whether it is the database schema architecture or the higher level process interaction and application dataflow and user interface. Accordingly, Oracle Designer lacks the feature that is proposed in this book to create a well defined representation of the detailed design of the application under construction.

1.8.3 *Automated Testing Tools*

It was essential to survey the current automated testing tools in today's market, in order to identify where improvements can be made. The conclusions reached after surveying the market guided much of this research. The products of the most popular vendors of automated testing tools are discussed in this section, including Mercury Interactive, Rational Software, Segue and Compuware.

In the United Kingdom, Mercury Interactive holds at least 50% of the market share in software testing tools. They have a number of automated test tools, including TestDirector [45], WinRunner [46], QuickTest [47] and LoadRunner [48].

Rational Software offers the most complete lifecycle toolset, including testing for the windows platform. They are the market leaders in object-oriented development tools. The leading testing tools available from Rational Software include Rational Robot, Functional Tester, Performance Tester and Rational Purify. Rational Robot is a test management tool that gives test cases for objects such as lists, menus, and bitmaps. It also provides test cases specific to objects in the development environment. As described by Rational [41] it is a "General purpose test automation tool for client/server applications". Functional Tester, on the other hand is an "automated functional and regression testing of Java, Web and VS.NET WinForm-based applications", as described by Rational. It aids the process of test script writing and managing.[42]. Performance Tester verifies application response time and scalability, it aids in emulating a real time web-based environment where lots of users are concurrently using the software. Rational Purify aids in the detection if memory leak and memory corruption for Linux and UNIX [44].

Compuware's major product is QARun. Similar to other automated testing tools, QARun depends on scenario testing, by recording the user's actions and respective system responses into test scripts to test some of the application's functions [52]. Compuware also has a load testing tool.

Segue is another organization that provides some comparable software testing tools to Mercury Interactive and Rational Software. Among the products that it offers are SilkCentral Test Manager, SilkTest 5, and SilkPerformer. Comparable to Rational Robot, SilkCentral is a test management and organization tool [49]. SilkTest is comparable to QuickTest, and Rational Functional Tester. It is a regression testing tool that records and plays back the test cases [50]. SilkPerformer is an automated load, stress and performance testing tool [51], comparable to Rational's Performance Tester and Mercury's LoadRunner.

All four major automated testing tools vendors have products similar in functionality to the products of their competitors. The tools available are mainly categorized into four major types; namely, test management, capture and replay, functional and regression testing and performance or stress testing tools. Table 1.1 summarizes the functionality of the different testing tools. Test management tools like Mercury's TestDirector, Rational TestManager, SilkCentral by Segue and QACenter by Compuware all organize the testing process. Test management tools as described by Mercury, thus provide a consistent, repeatable process for gathering requirements, planning and scheduling tests, analyzing results, and managing defects and

Organization	Scenario Management	Capture and Replay	Functional & Regression Testing	Performance Testing
Mercury Interactive	TestDirector	WinRunner	QuickTest	LoadRunner
Rational Software	TestManager	Rational Robot	Functional Tester	Performance Tester, Purify
Segue	SilkCentral	SilkTest	SilkTest5	SilkPerformer
Compuware	QACenter	QARun	TestPartner	QACenter

issues [45]. Testers can create test cases and log them in the test management tools to organize the testing process.

Capture and replay tools usually automatically generate test scripts from test scenarios captured during a tester's navigation through the software. Examples of the commonly used capture and playback tools are Mercury's WinRunner, Rational Robot, SilkTest by Segue, and QARun by Compuware.

Functional and regression testing tools are also common tools in software test automation. Regression testing tools run test scripts after issues have been resolved, to determine whether the fixes caused other problems in other parts of the software that were previously running correctly. Examples of regression and functional testing tools are Mercury's QuickTest, Rational Functional Tester, Segue's SilkTest5, and Compuware's TestPartner.

Performance and stress testing tools emulate an environment with thousands of users concurrently using the system. Examples of performance and stress testing tools are Mercury LoadRunner, Rational Performance Tester, Segue SilkPerformer and Compuware QACenter. Rational Purify checks for memory leaks.

1.8.4 *Disadvantage of Record and Playback Test Automation Tools*

In Section 1.8.3, the functionality of some of the most common software testing tools were outlined. However, none of those tools automatically and intelligently create test cases based on an understanding of what the software under construction is intended to execute. Some of those available tools record and playback test scenarios produced by the navigation of testers through the product while testing the product. In this case, coverage relies on how well the tester or user goes through the different and possibly numerous system scenarios. The system responses have to be judged by the tester who determines whether the system passed the test or not, based on his understanding and knowledge of the specifications. The tester thus, also acts as the oracle who decides whether the system passed that specific test, generated by the scenario in question.

1.9 Related Literature

A number of researchers in the field of software testing emphasize the need and the advantages of having a machine readable mapping of software specifications [31, 34, 38]. Other researchers describe techniques to compare software to its formal specifications as a good means of software testing [35,39]. Other techniques, such as those found in [37, 40] can be deployed to work efficiently and automatically, given a software's specification in the form of a machine-readable format, as that described in this book.

In order to automatically test database operations, Chays et. al. in [27, 28] derived the software specifications from the database schema and program code, and a tester inputting relevant heuristics. The authors noted that it is not possible to fully automate the testing and verification process unless the formal specification is given in full, which is the focus of this book. A similar approach to testing automation was illustrated by [36], in which the authors depended on programmed embedded SQL statements in the code, assuming it to be the correct source upon which to generate test cases. The weakness of this approach and the latter, lie in the fact that the test cases are generated based on the resulting software not from the specifications, as would be the result of adopting the techniques in this book.

In [33], Hung expanded on the work of Chays et al. in [28], designing an input generator for database applications. Hung emphasized the need to compute the expected output automatically. Similar to Chays et al. in [28], Hung generated test cases based on the testers values for the system inputs in their respective equivalence classes. The test cases where not generated automatically based on the specifications. The research in [58, 59, 60, 61] did not address the problem of automatically checking the results of the tests to verify whether the results are correct compared to the specifications. A number of researchers investigated the automation of testing web database applications, including [62]. Also regression testing of database applications was researched in [68]. Define-use path testing was used in [63] to test database driven applications. In [64] Cabal and Tuya tested the SE-LECT operation in database applications. Automated test case generation techniques where introduced in [32] to test the GUI of applications.

1.9.1 *AGENDA: A Test Generator for Database Applications*

Chays, Deng and Frankl carried out a research titled "AGENDA: A Test Generator for Relational Database Applications" [27, 28, 54, 55, 56]. The tests generated in their research were targeted towards database operations or application queries (insert, delete, update and select statements). There were no test cases targeted to test the application code itself, in which these operations are embedded. The tests generated in Chays' research depend on database input and output states, and do not compare the input or output to those specified by the design. That research does not tackle other types of operations where the output database state is not changed, but rather files, printed reports, magnetic card programming, etc. are the forms of output. Real world applications that use numerous types of inputs and outputs, besides database state changes will benefit only partially from AGENDA. AGENDA depends on the tester to specify whether the test passed or not. Accordingly, the tester is the oracle for AGENDA. AGENDA generates tests based on application code not the specifications and design of the software, and thus would not recognize errors of omission (parts that are designed but never programmed). When the tests are derived from the actual code, the tool will also not recognize errors of commission where the programmers wrote parts of code that were never included in the design document. Finally, AGENDA uses two types of functional testing techniques, namely, boundary value analysis and equivalence class testing. Not all faults in the software are revealed by these two types of testing alone. Boundary value analysis and equivalence class testing are good techniques to use in combination with other techniques to best test the software.

1.9.2 *Executable Software Specifications*

One of the major goals of this book is to represent software specification in a way that they can be reasoned with by automated software testing tools, to generate intelligent test cases. Intelligent test cases are a minimal set of useful test cases that collectively cover all the software and are capable of identifying errors. Accordingly, it was necessary to review the literature for automated software specifications. In [96], Corning describes how Microsoft internally test their systems by using a technique that automates the execution of human entered software test case specifications. Accordingly, what is automated is the generation of test cases based on the specifications of each of the test cases. In other words, the test cases are not created using a computationally intelligent technique to generate interesting and important test cases, because the software specifications are not executable (though the test cases specifications are executable). In this situation, the testers are the ones that create test cases, which can be automatically re-run with different parameters. The parameters of each test case are the test case specifications, or test specifications in short. Those parameters are executable. Accordingly, automated tools can read the specification of a test case, including the input variable names, possible values, pre-conditions, as well as other factors, and create multiple test cases that match those specifications. Those multiple test cases can then be run automatically.

Formal specifications are not readily executable [97]. Fuchs believes that formal specifications should be executable if possible [98] and argues against Hayes and Jones in [97] who wanted to make the distinction between human-readable specifications and machine-executable prototyping. Fuchs also indicates that the executable specifications in his research [98] are at the same high level of abstraction, the aim of the possibly executable specifications is for rapid prototyping and to facilitate the validation process. In the early nineteen nineties, automated testing and automated test case generation was not the purpose of those techniques of trying to execute the specifications, the arguments were mainly targeted towards rapid prototyping, not automated testing.

In [99, 100], the authors describe a mechanism to transform formal specifications to executable code. The result is not however, a means of representing the specifications to an automated tool to reason about and generate intelligently test cases from it.

1.10   Objectives of this Book

After surveying today's market for available software testing tools, none were found to generate the tests independently from specifications. The available tools are either code-coverage tools or capture-playback tools . They all rely on the implemented software either to create tests based on white box testing and the implemented code in question, or on the testers to provide good scenarios or test cases and their corresponding expected output. Accordingly, one can fairly make the assumption that currently the term "automatic testing" is misused to refer to automatic test execution and often misapplied to techniques that only partially automate portions of the testing process. After surveying the available tools, it was found that none of the tools in fact automatically generates the test cases for the software; meaning that, none of the available tools creates test cases based on its ability to interpret what the software under construction is implemented to achieve. The only tools close to automatically generating test cases, are those for stress testing [29], which create test cases to see how the software will react when a huge number of users log-in at the same time for example, or for database systems, creating a huge database with very big tables, etc. These tools, do not test the actual functionality of the software and whether it performs its tasks correctly or not, they just generate and execute stress tests. Basically, none of the available tools can understand the specifications and the software design, and accordingly create its own set of test cases that intelligently test and achieve good test coverage of the software.

The main reason why such an intelligent test case generator does not exist is that there does not exist a means of representing the specifications, requirements, and design in such a way that it could be interpreted, understood and used by an intelligent tool to generate test cases. Creating this machine readable and machine-interpretable representation of a system specifications and design is one of the main goals of this book. The new concepts introduced in this book are discussed in this section.

The goal of the concepts introduced in this book is thus to create a unified, machine readable representation of a software application's unit's design and specifications; whereby an automated testing tool could interpret and understand the software's units' design and specifications and automatically generate intelligent test cases to test the specified units of the software and determine whether the actual output matches the expected specified output; and to use the machine readable specifications in automated testing tools to show its use. One of those tools is to automatically generate constraints and enforce business rules for database applications at the database level. Another one of the automated testing tools is to use the specification representation to show errors of commission using a reverse engineering approach.

An automatic test case generator using the unified specification representation (SpecDB) will guarantee good coverage of the software, since it can generate test cases for all the specified functionalities of the software, utilizing the specifications not the implementation. Such an intelligent test case generating tool can easily find errors of omission, where the programmers did not implement parts of the software. It can also spot errors of commission caused by programming unspecified functionality in the software. Both of the latter types of errors would never be spotted if the tests were not generated based on the program specifications and design. Finally, such an intelligent test-case generator can spot faults of commission caused by errors in programming a specified functionality incorrectly.

The goal of the concepts introduced in this book should not be misinterpreted as a requirements verifier. Requirements verifiers test the requirements document for clarity, consistency and completeness [30], however, they do not check the actual implemented software based on these requirements. A first step before starting to represent the specifications and the design in a way interpretable by a testing tool is to use a requirements verifier to check the accuracy of the specifications and the requirements [66]. Hence, the goal of the book is to represent (formal) software specifications in a machine readable format to aid in the testing and test result verification processes and to create useful testing tools that make use of the machine readable software specifications to intelligently guide the testing process and assure better software quality.

Accordingly, the first objective is to create a standard process for producing and storing UML or formal specifications and other software specification and design constructs and detailed functionality specification. In order to create an intelligent automated test case generation tool for software testing purposes, there has to be a way for the tool to understand what the software is built for, its functionalities, inputs, outputs, processing scenarios, etc. Accordingly, the objective of this book is to provide that standard way of representing specifications, in a way that could be interpreted by an automated test case generator. This representation of specifications is detailed in Chapter 2, SpecDB: A Database Representation of Software Specifications.

This book includes extensions to the work by Chays, Deng and Frankl on AGENDA[27,28]. One of the uses of SpecDB, is to create an automated tool to generate intelligent test cases from SpecDB based on its understanding of the functionality of the software under test, and test the application units containing both imperative-like code (programming constructs like if, while, etc. and abstract data types, etc.) and also including the database operations. This proposed solution can compare actual program outputs to the specified design, stored in SpecDB. Therefore, the tool can automatically make the decision whether or not the test passed, based on its understanding of the available querable design, represented in SpecDB. Based on the specification, automated testing tools using SpecDB can spot errors of omission and commission immediately. In their research [27], Chays and his research team, wrote: "Unless there is a formal specification of the intended behavior of the application, including a specification of how the database state will be changed, it is not possible to fully automate checking of the application's results." Again, representing the formal specification they referred to, in a machine readable format, is the first objective of of this book, to make it possible to automate the checking of the software's results.

The second objective is to use the representation of the design and specifications to create an intelligent business rules generator for database applications. The proposed representation of the analysis and design of software will be based on a database structure that can be easily queried. Business rules and constraints will automatically be generated to enforce those constraints at the database level, not the application level.

The final part of the book is dedicated to studying the feasibility and effectiveness of structural test case generation, based on the level of detail given by the software specification and design. From the actual implementation, a reverse engineering approach is used to show all the software output, and post conditions after a test is executed. The output and post conditions are compared to the specified behavior if represented in SpecDB and accordingly, errors are automatically spotted and code segments that caused the errors are isolated. The automated structural testing techniques used in the reverse engineering automated testing approach are dataflow testing and path testing.

In Figure 1.2 the main concepts introduced in this book are represented. The design of SpecDB is detailed in Chapter 1.10. The software specifications are stored in SpecDB manually or automatically using the formal specification translation algorithm in Chapter 2.5.1. The usefulness of the SpecDB representation for automatic constraint generation and testing based on reverse engineering is demonstrated using two applications described in Chapters 3.5 and 4.3, automatic constraints generator and the reverse engineering testing tool, respectively. Finally in Chapter 5.4, enhancements to available tools in the literature like AGENDA, using SpecDB are proposed.

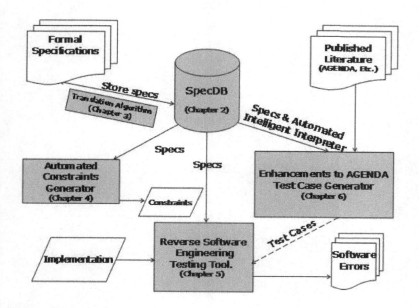

Fig. 1.2   Entities and Chapter Layout

Chapter 2

SpecDB: A Database Design for Software Specifications

2.1 Introduction

From the market survey of existing testing tools in Chapter 1, it was concluded that the current automated tools in the market lack an understanding of the functionality of the software under construction and testing. The literature review also showed how much of the research on automated tools can benefit from machine-readable specifications, which can reason with the specifications and automatically generate intelligent test cases and automatically determine whether the software passed the test based on its understanding of the specifications and design. In this chapter, one of the main models of this book is presented, namely SpecDB, a database that stores part of the software specifications to aid in the process of automated testing.

The design of SpecDB introduced in this book focuses only on the specifications required for unit testing database operations, and is not, in its current state a full representation of the complete specifications of a software system under construction. However, the techniques used in this chapter can be extended to include more of the software's specification for other types of testing; for example, security, performance, systems or integration testing, etc. In the following section, the motivation for the research in, and the development of the specification representation database, SpecDB is emphasized.

2.2 The Advantages of a Database Representation of Specifications

Since neither UML diagrams nor written formal specifications can be used as an input to automated testing tools, another channel is required to relay the specifications to automated tools. UML is not executable, nor machine readable. On the other hand, formal specifications are a complete representation, but are difficult to write and not used frequently. A database representation is chosen because it has the benefit of being machine readable and also not as hard to use as formal languages. Once the specifications of the software under construction are expressed in a machine-readable way, the specifications may be used and reasoned about by automated tools to test or generate the code for software applications automatically.

In order to create a software tool, that automatically tests parts of another software under construction, or automatically generates parts of the code for another specified software; the former has to have a good understanding of what the latter is intended to do. Accordingly, the automated software code generating / testing tool has to be given, in a machine-readable way, the specifications of the other software being built. The best way a software's specification and design can be given in a machine-readable format is through an adaptation of the already established formal methods. Hence the idea of representing specifications in a machine readable format evolved. In this chapter a standard methodology is introduced which represents software specifications in a machine readable format for the purpose of automated code generation and automated software testing. A database representation of the specifications was chosen as opposed to other forms or representation for the advantage of the inherently scalable nature of database designs. The design of SpecDB is scalable and it is easy to modify or append to the design in order to accommodate more forms of specifications. It is then very simple to update the queries created for previous versions of SpecDB, to accommodate the new changes. Runtime errors for database queries are not fatal, since they do not handle memory allocations like linked list representations. Another advantage is the use of the already stable and reliable database management systems available in the market, and the ability to create constrains and runtime triggers on database tables. Such constraints can aid in discovering contradictory, incomplete or ambiguous specifications.

Another use of SpecDB is to automatically represent programmed code and as an output, show the specifications of the programmed code. With this reverse engineering approach, SpecDB can also be used with already existing software. The advantage of this approach is to automatically test programmed software, giving the actual behavior for each operation. The analysts, designers and testers can then review all the software execution results, hence showing faults of commission that might not otherwise be spotted by normal testing techniques that deem a test successful if the specified behavior occurs. However, in some cases, the software errors lie in the fact that the programmed software does more than is specified. For example, an e-commerce application might perform its task of ordering an item, and also (as an unwanted side effect) send credit card information or passwords to another non-secure location! The latter security leak cannot be spotted by conventional black box functional testing techniques. Using SpecDB in this reverse engineering approach will spot those faults of commission as will be shown in Chapter 4.3.

Code translation from one programming language to another can be done using the reverse engineering tool and SpecDB. Code from one language can be fed into SpecDB, the specifications can then be used to automatically generate code in another language.

2.3 The Specification Database Design: SpecDB

SpecDB is a database. The entities of the SpecDB database store the (formal) specifications of a software, necessary for unit testing. SpecDB is designed to efficiently and accurately represent those parts of the specifications shown in this chapter.

In Section 2.4, the reader will see how SpecDB covers and has a means to represent different forms of specifications. SpecDB is also designed to be scalable, generic, upgradeable and modifiable. Since the model is a database, it is easy to add more entities and/or more entity attributes to represent a specification that cannot otherwise be represented in the given SpecDB design in this chapter. As an example of SpecDB's scalability, in Chapter 4 two tables are added and an existing table in SpecDB is modified to accomplish the task of an automated constraint generator using SpecDB.

The specification representation of SpecDB is designed to be the basis for the process of automated software generation and testing. Automated software code generation and/or testing tools can use SpecDB to understand what a software is designed to do, and hence can accurately test such a software by creating intelligent test cases with good coverage, and compare actual software behavior to its corresponding specified software behavior. With the use of SpecDB, the processes of automated software code generation and automated software testing can become more intelligent, efficient and reliable. In the design of SpecDB, constructs for both formal specifications and programming were taken into consideration. These rules and constructs can be found in [8, 9, 11-24, 65].

Figure 2.1 shows an overall diagram of SpecDB's design. Throughout Section 2.4 are figures showing parts of the relational database schema of the design of SpecDB. The relations between the entities are shown. Also the types of the fields in each entity are shown, numeric types are identified by the symbol "789", text or character types are identified by "A", and date or time types are identified by a calendar card. Table keys are identified by a "#" symbol, and whether each field is optional or mandatory is identified by "O" or "*" respectively. There are some fields that are constrained to a list of valid values, for example, Is_Constant is a Boolean value that can be only T for true or F for false. On the left of the field name, if the value of the field is constrained to a list of valid values, a relation symbol is viewed. Figures 2.2 and 2.3 are focused in parts of Figure 2.1, for better readability. Throughout Section 2.4 each entitiy is described in detail with focused images.

2.4 Entities for Specific Software Requirements

In this section, each of the entities of the SpecDB design is defined and explained in detail, including examples of values for each attribute and how to store the different components of a software specification. First, general entities are defined, among those are the operators, List_of_Tables, Table Description, Types, Types List of valid values (LOV) and restrictions. The SQL scripts for the creation of the entities of SpecDB are provided within each section, their corresponding constraints are listed in Appendix A . The reader can refer to those constraints to see the allowable values of the attributes, this may aid in the understanding of the meaning of the attributes in each entity in SpecDB and what each represents.

2.4.1 *Operator and Valid Operands*

In this section, the entities of SpecDB that hold operators and valid operands are explained. The Operator table holds the operators of op-

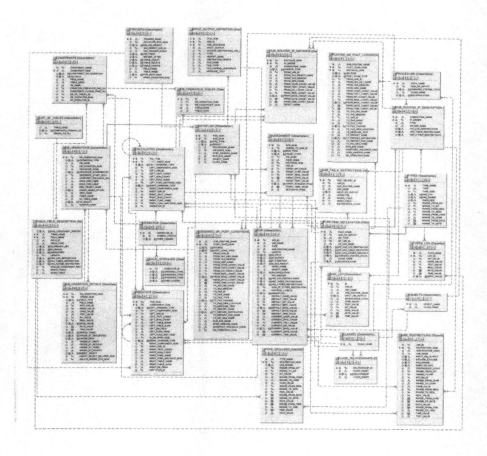

Fig. 2.1   An Overall Diagram of the SpecDB Design

erations used in any software (formal) specification or code. Operators
differ from one language to another, and hence SpecDB identifies them
with unique IDs to form a generic outcome that is not language dependent.

Examples of operator symbols are: /, MOD, AND, NOT, etc. If an operator is overloaded, records are added to the Valid_Operands table. The
subroutine that defines the operator overloading operation should be stored
in the Function_Declaration table (described in Section 2.4.5), with a subroutine name, the same as the operator symbol or name.

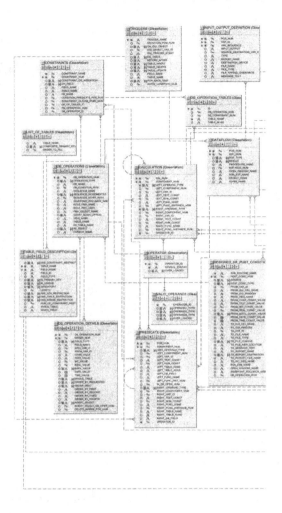

Fig. 2.2   A Focused Diagram of the General SpecDB Design (Left Section)

```
Prompt Creating Table 'Operator'
Create Table Operator
 (Operator_Id Number(2) Not Null
 ,Symbol_Syntax Varchar2(5) Not Null
 ,Over_Loaded Varchar2(1) Default 'F')
```

The Over_Loaded attribute in both the Operator and Valid_Operands
tables is a Boolean variable that can be True or False. If more than three

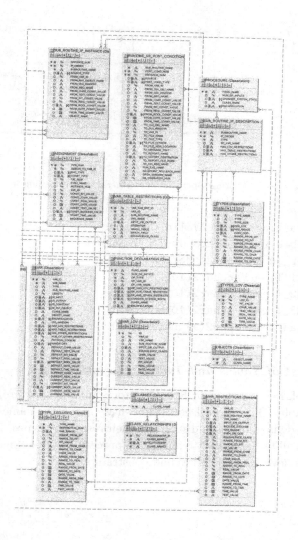

Fig. 2.3 A Focused Diagram of the General SpecDB Design (Right Section)

operands are involved in the operator overloading process, another attribute
to represent the Operand4_Type can be added to the Valid_Operands table.
The types of the operands have to reference records in the Types table
(described in Section 2.4.3).

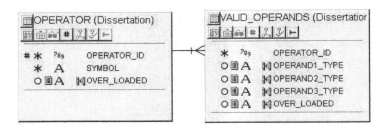

Fig. 2.4   Representing Operators and Operator Overloading in SpecDB

```
Prompt Creating Table 'Valid_Operands'
Create Table Valid_Operands
 (Operator_Id Number(2) Not Null
 ,Operand1_Type Varchar2(10)
 ,Operand2_Type Varchar2(10)
 ,Operand3_Type Varchar2(10)
 ,Over_Loaded Varchar2(1) Default 'F')
```

2.4.2   *List_of_Tables and Table Descriptions*

The List_of_Tables entity in SpecDB stores all the software/user-defined sets and tables. The Table_Description entity stores the names of the fields of the user-defined tables and sets, and their respective types. In Chapter 4, this table will be extended, to aid in the process of the automated creation of database level business rules and constraints.

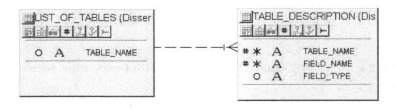

Fig. 2.5   Representing User Tables and Sets and their Description in SpecDB

```
Prompt Creating Table 'List_Of_Tables'
Create Table List_Of_Tables
 (Table_Name Varchar2(240)
 ,Composite_Primary_Key Varchar2(1) Default 'T'
 ,Order_To_Fill Number)
```

2.4.3 *Types, Types_LOV and Type Restrictions*

The Types entity stores the non-standard user-defined types. This entity stores types that are not sets, arrays, linked lists, or any type designed for data storage and retrieval. SpecDB represents the latter storage types as user-defined tables in the database, and stores their descriptions in the List_of_Tables and Table_Descriptions entities (described in Section 2.4.2). The Types entity stores user-defined types like month names, days of the week, etc. Each of those user-defined types has a basic type, e.g. String, Integer, etc. Also types can have size restrictions, e.g. a String of 20 characters. SpecDB allows for user-defined restrictions on types too. User defined restrictions can be one of the following:

```
Prompt Creating Table 'Types'
Create Table Types
 (Type_Name Varchar2(20) Not Null
 ,Type Varchar2(10)
 ,Type_Size Integer
 ,Has_Lov Varchar2(1) Default 'T'
 ,Has_Range Varchar2(1) Default 'T'
 ,Even_Odd Varchar2(1)
 ,Range_From_Int Integer
 ,Range_To_Int Integer
 ,Range_From_Real Real
 ,Range_To_Real Real
 ,Range_From_Char Char
 ,Range_To_Char Char
 ,Range_From_Date Date
 ,Range_To_Date Date)
```

2.4.3.1 *List of Valid Values (LOV)*

In the first kind of type-value restrictions, the values of a variable of a specific type are constrained to only values in a defined list of valid values.

A type can be restricted to a list of valid values, i.e. any variables of the specified type can only take on one of the values in the LOV. For example, the type Month_Name can be either January, February, etc. In this case the Has_LOV attribute value of the Types entity for the Month_Name record is set to T for true and the valid months names are placed in the Types_LOV entity, each in a separate record.

```
Prompt Creating Primary Key On 'Types_Lov'
Alter Table Types_Lov
 Add (Constraint Tlov_Pk Primary Key
 (Lov_Id))
```

2.4.3.2 *Other Restrictions*

Types can also be restricted to a set of valid values that are better specified by ranges. In this case, the ranges are stored in the Types table in their respective fields to define the allowable range. For example, for a Month_Number type, valid values are between the range of 1 o 12. Hence the Range_From_Int in the Types Table in the record defining the months is set to 1 and Range_To_Int is set to 12. A range can also be specified, with inclusions, e.g. all even integers between 0 and 100 including 0. Also exclusions can be defined from certain defined ranges by using the Type_Excluded_Ranges Table.

```
Prompt Creating Table 'Type_Restrictions'
Create Table Type_Restrictions
 (Type_Name Varchar2(20) Not Null
 ,Restriction# Integer Not Null
 ,Include_Exclude Varchar(1)
 ,Has_Range Char
 ,Even_Or_Odd Varchar2(1)
 ,Range_From_Int Integer
 ,Range_To_Int Integer
 ,Int_Value Integer
 ,Range_From_Char Char
 ,Range_To_Char Char
 ,Char_Value Char
 ,Range_From_Real Real
 ,Range_To_Real Real
 ,Real_Value Real
```

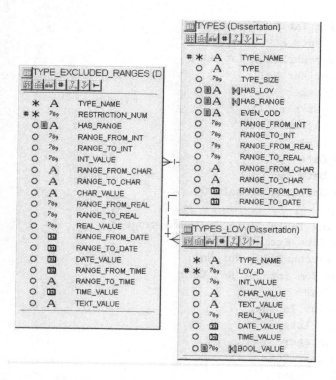

Fig. 2.6   Representing Types, Type Restrictions and Types Valid Values in SpecDB

```
,Range_From_Date Date
,Range_To_Date Date
,Date_Value Date
,Range_From_Time Date
,Range_To_Time Varchar2(240)
,Time_Value Date
,Text_Value Varchar2(240))
```

The Types_LOV table contains a record for each value in the list of
valid values for each type with LOV restrictions. The table has been de-
normalized to include values of all basic types, based on the "type" attribute
value in the Types table, the value of the LOV will be stored in the relevant
attribute. For example if the type of the Month_Number Type is Integer,
then there will be twelve records in the Types_LOV table, each record will
have a unique value between 1 and 12 stored in the Int_Value attribute.

```
Prompt Creating Table 'Types_Lov'
Create Table Types_Lov
 (Type_Name Varchar2(20) Not Null
 ,Lov_Id Number Not Null
 ,Int_Value Integer
 ,Char_Value Varchar2(1)
 ,Text_Value Varchar2(240)
 ,Real_Value Real
 ,Date_Value Date
 ,Time_Value Date
 ,Bool_Value Smallint Default 0)

Prompt Creating Table 'Type_Excluded_Ranges'
Create Table Type_Excluded_Ranges
 (Type_Name Varchar2(20) Not Null
 ,Restriction_Num Integer Not Null
 ,Has_Range Char
 ,Range_From_Int Integer
 ,Range_To_Int Integer
 ,Int_Value Integer
 ,Range_From_Char Char
 ,Range_To_Char Char
 ,Char_Value Char
 ,Range_From_Real Real
 ,Range_To_Real Real
 ,Real_Value Real
 ,Range_From_Date Date
 ,Range_To_Date Date
 ,Date_Value Date
 ,Range_From_Time Date
 ,Range_To_Time Varchar2(240)
 ,Time_Value Date
 ,Text_Value Varchar2(240))
```

2.4.4 *Variables and Restrictions on Variable Values*

In SpecDB, the Var entity stores the specifications of a used variable. Including, its type, that can be also a user-defined type already defined in the Types entity. The SpecDB design, takes into account the fact that in the same software, many different variables can have the same name, but defined in different contexts or subroutines. This entity stores qualities each variable should possess. This aids automated testing tools in creating test cases that make sure that in all the software's execution cycle and in all its states, each variable maintains its designed properties. If a variable is part of a class or an object, this can be specified, together with its encapsulation property; public, private or protected.

```
Prompt Creating Table 'Var'
Create Table Var
 (Var_Id Integer Not Null
 ,Var_Name Varchar2(15) Not Null
 ,Type Varchar2(10)
 ,Sub_Routine_Name Varchar2(30)
 ,Length Integer
 ,Is_Input Varchar2(1) Default 'T'
 ,Is_Output Varchar2(1) Default 'T'
 ,Is_Constant Varchar2(1) Default 'T'
 ,Is_Gui_Object Varchar2(1) Default 'T'
 ,Class_Name Varchar2(240)
 ,Object_Name Varchar2(240)
 ,Encapsulation Varchar2(10)
 ,Gui_Window Varchar2(240)
 ,Has_Lov_Restrictions Varchar2(1) Default 'T'
 ,Has_Table_Restrictions Varchar2(1) Default 'T'
 ,Has_Other_Restrictions Varchar2(1) Default 'T'
 ,Num_Of_Other_Restrictions Integer
 ,Physical_Logical Varchar2(1)
 ,Mandatory Varchar2(1) Default 'T'
 ,Default_Char_Value Varchar2(1)
 ,Default_Text_Value Varchar2(100)
 ,Default_Int_Value Integer
 ,Default_Real_Value Real
 ,Default_Bool_Value Varchar2(1) Default 'T'
 ,Default_Date_Value Date
 ,Default_Time_Value Date
 ,Current_Char_Value Varchar2(1)
 ,Current_Real_Value Real
 ,Current_Text_Value Varchar2(100)
```

```
,Current_Int_Value Integer
,Current_Bool_Value Varchar2(1) Default 'T'
,Current_Date_Value Date
,Current_Time_Value Date)
```

Variables can have pre-defined types, but in turn also have their own restrictions. SpecDB has the flexibility to define restrictions on both types and variables. A variable can have a LOV restriction or other restrictions as the ones previously discussed for the types restrictions.

A variable can also have a Table Restriction, i.e. the variable can only take on values that are stored in a specific field in a user-defined table. The Var_LOV table stores preset values that are preset before a software is executed, e.g. names of the months of the year. On the other hand, a variable can be restricted to take on values that are determined as the software executes, these values are stored in user-defined tables; e.g. names of people in an address book, where the names are continuously added as the address book is used. Those restrictions are stored in the Var_Table_Restrictions entity. SpecDB gives the flexibility for a variable to be restricted to take on only values from another table, or, by exclusion, to take any value that does not currently exist in the specified table. This is done by specifying the operator in the Var_Table_Restrictions entity.

```
Prompt Creating Table 'Var_Table_Restrictions'
Create Table Var_Table_Restrictions
 (Var_Tab_Rst_Id Integer Not Null
 ,Var_Id Integer Not Null
 ,Sub_Routine_Name Varchar2(240)
 ,Var_Name Varchar2(30)
 ,Input_Or_Output Varchar2(1)
 ,Operator Varchar2(240) Not Null
 ,Which_Table Varchar2(240) Not Null
 ,Which_Field Varchar2(240)
 ,Equivalence_Class Varchar2(240))


Prompt Creating Table 'Var_Restrictions'
Create Table Var_Restrictions
 (Var_Id Integer
 ,Restriction_Num Integer Not Null
 ,Sub_Routine_Name Varchar2(50)
```

Fig. 2.7 Representing Variables, Variable Restrictions and Variable Valid Values in SpecDB

```
,Var_Name Varchar2(30)
,Input_Or_Output Char
,Include_Exclude Varchar(1)
,Has_Range Char
,Even_Or_Odd Varchar2(1)
,Equivalence_Class Varchar2(240)
,Range_From_Int Integer
,Range_To_Int Integer
,Int_Value Integer
,Range_From_Char Char
,Range_To_Char Char
,Char_Value Char
,Range_From_Real Real
,Range_To_Real Real
,Real_Value Real
,Range_From_Date Date
,Range_To_Date Date
```

```
,Date_Value Date
,Range_From_Time Date
,Range_To_Time Varchar2(240)
,Time_Value Date
,Text_Value Varchar2(240))

Prompt Creating Table 'Var_LOV'
Create Table Var_Lov
 (Id Number Not Null
 ,Var_Id Integer
 ,Var_Name Varchar2(30)
 ,Sub_Routine_Name Varchar2(240)
 ,Input_Or_Output Char
 ,Equivalence_Class Varchar2(240)
 ,Char_Value Varchar2(1)
 ,Text_Value Varchar2(100)
 ,Int_Value Integer
 ,Real_Value Real
 ,Date_Value Date
 ,Time_Value Date)
```

2.4.5 *Procedure, Function, and Subroutine Descriptions*

Subroutines, or units are a major part of software specifications as well as the actual program code, and the focus of this book. In the SpecDB database, procedures differ from functions in the fact that functions return a value, whereas procedures are subroutines that do not return a value, yet can change the system state. If a procedure should change the state of the system, then the value of the Changes_System_State field, is set to true. Automated testing tools can use this specification to make sure that the programmed subroutines are implemented as designed. The inputs for both procedures and functions are placed in the Subroutine_IP_Description entity. Inputs are represented with their respective orders using the IP_Order field.

```
Prompt Creating Table 'Procedure'
Create Table Procedure
 (Proc_Name Varchar2(30) Not Null
 ,Num_Of_Inputs Integer
 ,Changes_System_State Varchar2(1) Default 'T'
```

```
,Class_Name Varchar2(240)
,Encapsulation Varchar2(10))
```

For a function
that takes two inputs, such as Calculate_Interest_Only_Monthly_Mortgage
(Loan_Amount, Rate), the record in the Function_Declaration table will be: (Calculate_Interest_Only_Monthly_Mortgage, 2, Real, 68,
Monthly_Mortgage, F, F, F, F, ...) where F stands for False. There
should be two records in the Sub_Routine_IP_Description table: (Calcu-
late_Interest_Only_Monthly_Mortgage, 1, Real, Loan_Amount, F, F, T) and
(Calculate_Interest_Only_Monthly_Mortgage, 2, Real, Rate, F, F, T). The
two inputs can have a restriction to be greater than zero. In this case two
records have to be placed in the Var_Restrictions table to describe those
restrictions.

```
Prompt Creating Table 'Sub_Routine_IP_Description'
Create Table Sub_Routine_Ip_Description
 (Subroutine_Name Varchar2(30) Not Null
 ,Ip_Order Integer Not Null
 ,Type Varchar2(240) Not Null
 ,Sr_Var_Name Varchar2(240)
 ,Has_Lov_Restrictions Char
 ,Has_Table_Restrictions Char
 ,Has_Other_Restrictions Char)


Prompt Creating Table 'Function_Declaration'
Create Table Function_Declaration
 (Func_Name Varchar2(240) Not Null
 ,Num_Of_Inputs Integer
 ,Op_Type Varchar2(240) Not Null
 ,Op_Var_Id Integer
 ,Op_Var_Name Varchar2(30)
 ,Op_Has_Lov_Restrictions Varchar2(1) Default 'T'
 ,Op_Has_Table_Restictions Varchar2(1) Default 'T'
 ,Op_Has_Other_Restrictions Varchar2(1) Default 'T'
 ,Changes_System_State Varchar2(1) Default 'T'
 ,Class_Name Varchar2(240)
 ,Encapsulation Varchar2(10) )
```

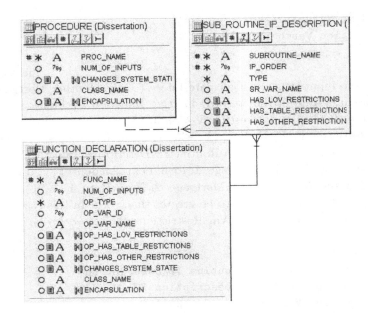

Fig. 2.8   Representing Units (Subroutines) in SpecDB

The output value returned by a function (OP) can also have restrictions. SpecDB provides a place to store these pre-determined restrictions on the valid allowable values a function can return.

2.4.6   *Subroutine Parameters and Post-Conditions*

Each time a subroutine is invoked during the life of the software runtime, it is given different parameters. The Subroutine_IP_Instance is where the values, or the sources for the values of the parameters are specified for each one of those instances. For example, if on the seventh call to Calculate_Interest_Only_Monthly_Mortgage in the program specifications or program, the input is taken from the GUI text fields (:HouseValue, :NegotiatedRate), in the Interest_Only window. Accordingly, the records in the Subroutine_IP_Instance table will be (7,1, Calculate_Interest_Only_Monthly_Mortgage, GUI, , :HouseValue, Interest_Only, ...).

```
Prompt Creating Table 'Sub_Routine_Ip_Instance'
Create Table Sub_Routine_Ip_Instance
```

```
(Instance_Num Integer Not Null
,Ip_Order Integer Not Null
,Subroutine_Name Varchar2(30) Not Null
,Source_Type Varchar2(5)
,From_Var_Id Integer
,From_Gui_Object_Name Varchar2(50)
,From_Gui_Window Varchar2(50)
,From_Seq_Name Varchar2(30)
,From_Char_Const_Value Varchar2(1)
,From_Text_Const_Value Varchar2(100)
,From_Int_Const_Value Integer
,From_Real_Const_Value Real
,From_Bool_Const_Value Varchar2(1) Default 'T'
,From_Date_Const_Value Date
,From_Time_Const_Value Date
,Object_Name Varchar2(240) )
```

The Designed_SR_Post_Conditions, holds the designed post conditions for subroutines, functions and procedures. Those will be compared to the actual post conditions during the testing phase to spot out software errors if there exists discrepancies between the designed subroutine post conditions in this table and the actual runtime subroutine post conditions, declared in the Runtime_SR_Post_Conditions table.

The Type of the post condition could be a modification to or execution of one of the following: GUI, DB, File, Report, Sequence, Message, Database Privilege or Role, Global Variable, Commit command, Savepoint command, Rollback, Run another program, Open a window or terminate the program execution. The source and the destination of the post condition are specified in this table.

If the subroutine should result in a database operation being executed, the unique ID for that operation is placed in the DB_Operation_Num field, and the actual specifications for that operation is placed in the DB_operations and DB_Operation_Details tables.

```
Prompt Creating Table 'Runtime_Sr_Post_Conditions'
Create Table Runtime_Sr_Post_Conditions
 (Sub_Routine_Name Varchar2(50) Not Null
 ,Post_Cond_Num Integer Not Null
 ,Instance_Num Integer Not Null
```

```
,Source Varchar2(5)
,Post_Cond_Type Varchar2(10)
,From_Var_Id Integer
,From_Gui_Obj_Name Varchar2(30)
,From_Gui_Window Varchar2(30)
,From_Seq_Name Varchar2(20)
,From_Char_Const_Value Varchar2(1)
,From_Text_Const_Value Varchar2(100)
,From_Int_Const_Value Integer
,From_Real_Const_Value Real
,From_Bool_Const_Value Varchar2(1) Default 'T'
,From_Date_Const_Value Date
,From_Time_Const_Value Date
,To_Gui_Obj_Name Varchar2(30)
,To_Gui_Window Varchar2(30)
,To_Var_Id Integer
,To_File_Name Varchar2(100)
,To_File_Type Varchar2(10)
,To_File_Change Varchar2(10)
,To_File_New_Location Varchar2(100)
,To_Message_Text Varchar2(240)
,To_Report_Name Varchar2(100)
,To_Report_Destination Varchar2(10)
,To_Report_File_Name Varchar2(100)
,To_Inc_Seq_Name Varchar2(30)
,Run_Exe_Name Varchar2(100)
,Savepoint_Rollback_Name Varchar2(20)
,Open_Window_Name Varchar2(40)
,Db_Operation_Num Integer )
```

The Runtime_SR_Post_Conditions table is a duplicate of the De-signed_SR_Post_ Conditions, with the difference that actual post conditions that occur due to the execution of a function or procedure are stored in the latter. Each time a subroutine is instantiated during runtime, an instance number is generated, and the input parameters used to invoke the subroutine, is stored in the Subroutine_IP_Description table.

The post conditions of that instance are stored in the Runtime_SR_Post_Conditions table, using that same instance number. The fields in this table are exactly the same as those of the De-

Fig. 2.9 Representing Subroutines Instances, and Post Conditions in SpecDB

signed_SR_Post_Conditions table, with the difference that the field Instance number is added, and is part of the key to this table. Also the data in the records indicate the actual runtime post_condition behavior after a subroutine is run, as opposed to the designed or the expected, detailed in the Designed_SR_Post_Conditions table. This table will be populated after the software is implemented and executed. An automated testing tool can make use of this table.

```
Prompt Creating Table 'Designed_Sr_Post_Conditions'
Create Table Designed_Sr_Post_Conditions
 (Sub_Routine_Name Varchar2(50) Not Null
 ,Post_Cond_Num Integer Not Null
 ,Source Varchar2(5)
 ,Post_Cond_Type Varchar2(10)
 ,From_Var_Id Integer
 ,From_Gui_Obj_Name Varchar2(30)
 ,From_Gui_Window Varchar2(30)
```

```
,From_Seq_Name Varchar2(20)
,From_Char_Const_Value Varchar2(1)
,From_Text_Const_Value Varchar2(100)
,From_Int_Const_Value Integer
,From_Real_Const_Value Real
,From_Bool_Const_Value Varchar2(1) Default 'T'
,From_Date_Const_Value Date
,From_Time_Const_Value Date
,To_Gui_Obj_Name Varchar2(30)
,To_Gui_Window Varchar2(30)
,To_Var_Id Integer
,To_File_Name Varchar2(100)
,To_File_Type Varchar2(10)
,To_File_Change Varchar2(10)
,To_File_New_Location Varchar2(100)
,To_Message_Text Varchar2(240)
,To_Report_Name Varchar2(100)
,To_Report_Destination Varchar2(10)
,To_Report_File_Name Varchar2(100)
,To_Inc_Seq_Name Varchar2(30)
,Run_Exe_Name Varchar2(100)
,Open_Window_Name Varchar2(40)
,Savepoint_Rollback_Name Varchar2(30)
,Db_Operation_Num Integer )
```

2.4.7 *Dataflow, Triggers and Input/Output Definitions*

In order to specify the order of operations of a unit in an application, the operations have to be given serial numbers. Accordingly, as a preparation stage for representing the specifications in SpecDB, unique dataflow numbers are given to operations in the (formal) specifications. Accordingly, SpecDB assumes that the program is numbered. Every line of the (formal) specification of a software as well as every line of code generated is given a unique serial number. SpecDB assumes that every line of code is a simple program statement, unnecessary bundling of code or specification on the same line is forbidden in this representation, to simplify the representation and enhance readability from SpecDB by automated tools. Accordingly, a preparation phase to simplify the specifications or code is necessary. The dataflow table stores what each line of the (formal) specification presents.

SpecDB can also be used to represent actual lines of programmed code and automatically compare the actual behavior to the specified behavior using this reverse engineering approach where the Dataflow table would contain the actual programmed lines of code.

```
Prompt Creating Table 'Dataflow'
Create Table Dataflow
 (Pos_Num Integer Not Null
 ,Suc_Num Integer Not Null
 ,Pos_Type Varchar2(9) Default 'Process'
 ,Result Varchar2(1) Default 'T'
 ,Procedure_Name Varchar2(240)
 ,Instance_Num Integer
 ,Open_Window_Name Varchar2(240)
 ,Run_Exe_Name Varchar2(240)
 ,Object_Name Varchar2(240)
 ,Class_Name Varchar2(240) )
```

Each position is identified by a number, a successor number, and a type. The position type can be one of the following: initial, begin, input, assignment, decision, while loop, for loop, run exe, output, subroutine call, database, open window, construct object, destruct object, call object method, end, end loop, end for, or terminal. According to the type of the position, the successor number can change. A position number can have many successor numbers and vice versa. For example, a decision point can have two results (true or false), and hence two successors, depending on the result.

Database management systems have structures similar to subroutines called triggers. The difference between triggers and other subroutines is that procedures or functions are called explicitly, sometimes with arguments, whereas triggers are executed implicitly when a pre-specified event occurs, also there are no arguments passed to triggers. For our design specification purpose, the trigger table stores the events that cause triggers to execute. Those cannot be part of the normal dataflow table, since they do not follow the same dataflow process as the rest of the software specification or code. However, this table will refer to the dataflow table once a trigger is defined, to adopt the same technique used to describe a normal subroutine body, to describe the trigger body. Also what is named smart trigger is described in this table. Smart triggers execute once a pre-specified event occurs to a graphical user interface (GUI) object. For example, when the mouse pointer enters a specific area in a window or when a button right click occurs, etc. If the LHS var is assigned to a GUI object value, the var_id is the place to identify that GUI object, since all GUI used vars are listed in the VAR table.

```
Prompt Creating Table 'Input_Output_Definition'
Create Table Input_Output_Definition
 (Pos_Num Integer Not Null
 ,Var_Id Number(3) Not Null
 ,Var_Sequence Integer Not Null
 ,Input_Output Varchar2(1)
 ,Source_Destination_Var_Id Integer
 ,Type Varchar2(240)
 ,Report_Name Varchar2(240)
 ,Destination_Device Varchar2(240)
 ,File_Name Varchar2(240)
 ,File_Type Varchar2(240)
 ,File_Append_Overwrite Varchar2(1)
 ,Message_Text Varchar2(240))
```

The Input_Output_Definition table describes how inputs and outputs to and from the software are represented. Inputs and outputs can be represented in many forms, file, database, through the GUI or to output devices like a printer (where a report is printed), or from other input devices. If the input or output is a database operation, then in the dataflow table, the line of code will be of type database operation, not input or output, and hence this type of input or output will not be represented in this Input_Output_Definition table.

```
Prompt Creating Table 'Triggers'
Create Table Triggers
 (Trigger_Name Varchar2(30) Not Null
 ,Dataflow_Pos_Num Integer
 ,On_Gui_Object Varchar2(1) Default 'T'
 ,Gui_Object_Var_Id Integer
 ,Gui_Trigger_Event Varchar2(240)
 ,On_Table Varchar2(1) Default 'T'
 ,Before_After Varchar2(1)
 ,Table_Insert Varchar2(1) Default 'T'
```

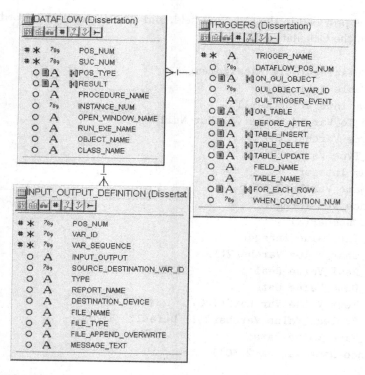

Fig. 2.10   Representing Dataflow, Triggers and System Inputs and Outputs in SpecDB

```
,Table_Delete Varchar2(1) Default 'T'
,Table_Update Varchar2(1) Default 'T'
,Field_Name Varchar2(20)
,Table_Name Varchar2(30)
,For_Each_Row Varchar2(1) Default 'T'
,When_Condition_Num Integer)
```

2.4.8   *Assignment and Calculation*

Assignment operations and calculations are discussed in this section. A variable can be assigned to a constant or another variable, e.g. a:=65; b:=c. A variable can also be assigned to a calculation or a predicate statement resulting in a true or false value, e.g. a:=5+b; e:= (a AND b) OR (c XOR d). If a variable is assigned to a calculation, then there might be a number of records in the Calculation table describing this calculation. A calculation identifier is provided in the Cal_Num field, and then at least one record is inserted in the Calculations table.

```
Prompt Creating Table 'Assignment'
Create Table Assignment
 (Pos_Num Integer Not Null
 ,Assign_To_Var_Id Number(3) Not Null
 ,Rhs_Type Varchar2(5)
 ,Const_Type Varchar2(10)
 ,Cal_Num Integer
 ,Func_Name Varchar2(100)
 ,Instance_Num Integer
 ,Var_Id Number(3)
 ,Const_Int_Value Integer
 ,Const_Char_Value Varchar2(1)
 ,Const_Real_Value Real
 ,Const_Date_Value Date
 ,Const_Text_Value Varchar2(100)
 ,Const_Boolean_Value Varchar2(1) Default 'T'
 ,Const_Time_Value Date
 ,Sequence_Name Varchar2(30))
```

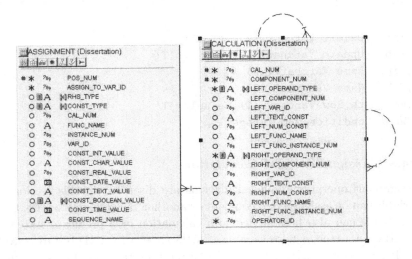

Fig. 2.11   Representing Assignment Operations and Calculations in SpecDB

In the Calculation table, complex calculations are stored in the form of simple components. Each component involves only two operands and an operator. An operand can be simple like a variable or a constant or can be complex, and consists in itself of components. An example of a Simple Calculation is a+b. The LHS is a variable, and thus "a" will be stored in the Left_Var_ID field. "b" will be stored in the Right_Var_ID field, and the operator ID for "+" will be stored in the Operator_ID field. Also a+109; a¡b; a - mod(b); are all simple calculations. Examples of a complex calculations are: (a / b) + c; (U OR V) AND (Not W), are all complex calculations.

```
Prompt Creating Table 'Calculation'
Create Table Calculation
 (Cal_Num Integer Not Null
 ,Component_Num Integer Not Null
 ,Left_Operand_Type Varchar2(5) Default 'Comp' Not Null
 ,Left_Component_Num Integer
 ,Left_Var_Id Integer
 ,Left_Text_Const Varchar2(240)
 ,Left_Num_Const Float(12)
 ,Left_Func_Name Varchar2(50)
 ,Left_Func_Instance_Num Integer
 ,Right_Operand_Type Varchar2(5) Default 'Comp' Not Null
 ,Right_Component_Num Integer .
 ,Right_Var_Id Integer
 ,Right_Text_Const Varchar2(240)
 ,Right_Num_Const Float(12)
 ,Right_Func_Name Varchar2(50)
 ,Right_Func_Instance_Num Integer
 ,Operator_Id Integer Not Null)
```

In the first example, $(a/b) + c, (a/b)$ is one component, and c is another simple component. Accordingly, there will be two records for this calculation in the Calculation table. One record with Component_Num = 1, where Left_Operand_Type="Var", Left_var_ID is that of "a", and Right_Operand_Type = "Var", and Right_Var_ID is that of "b" and Operator_ID is that of the "/" operator sign. The other record with the same Cal_Num, but Component_Num = 2, will have Left_Operand_Type = "Comp", the Left_Component_Num = 1, and the Right_Operand_Type

= "Var" and the Right_Var_ID = that of "c".

2.4.9   *Predicate*

The Predicate table described in this section stores complex predicate statements or database conditions. Each line of specification or code i.e. position, has a type that is stored in the Dataflow table. If the Pos_Type is
either a decision, while loop, for loop, or database operation, the statement
details at this line of code are stored in this Predicate table. In case the
Pos_Type is a decision, a decision statement can have a simple or a complex
predicate. An example of a simple predicate is:

```
if a = b then ...
```

In this case, there are no calculations, or several components. There
is only one component, and the left and right operands are variables, or a
variable and a constant. An example of a complex predicate is

```
if (a+b) < (d+11) then ...
```

In this case there are several components. The components are them-
selves calculations. Records are inserted in this table starting with the low-
est level components, i.e. the ones that contain only simple predicates, with
just variables and constants, i.e. from the inside out. Finally the predicate
that represents the relationship among the already described components
is inserted, using their respective component numbers. The technique is
very similar to that used in the calculation table.

```
Prompt Creating Table 'Predicate'
Create Table Predicate
  (Pos_Num Number(7) Not Null
  ,Component_Num Number(7) Not Null
  ,Left_Operand_Type Varchar2(5) Default 'Comp'
  ,Left_Component_Num Integer
  ,Left_Var_Id Integer
  ,Left_Text_Const Varchar2(240)
  ,Left_Num_Const Float(12)
  ,Left_Table_Name Varchar2(240)
  ,Left_Table_Alias Varchar2(240)
  ,Left_Db_Field Varchar2(240)
  ,Left_Func_Name Varchar2(240)
```

```
,Left_Func_Inst_Num Integer
,In_Db_Oper_Num Integer
,Right_Operand_Type Varchar2(5) Default 'Comp'
,Right_Component_Num Integer
,Right_Var_Id Integer
,Right_Text_Const Varchar2(240)
,Right_Num_Const Float(12)
,Right_Func_Name Varchar2(240)
,Right_Func_Instance_Num Integer
,Right_Table_Name Varchar2(240)
,Right_Table_Alias Varchar2(240)
,Right_Db_Field Varchar2(240)
,Operator_Id Integer Not Null )
```

In case the Pos_Type is a loop, the beginning of a loop is a decision statement, determining whether another iteration through the loop should take place or to exit. The same discussion above for decisions is followed to describe a while loop predicate.

In case the Pos_Type is a database operation, the predicate table is where the WHERE clause of a database operation is stored. Similar to the above examples, the operands are database fields compared to one another, for example: S.ID = Emp.ID. In this example, both the left and the right operands are database fields in the S and Emp tables (or table alias) This example will be treated as a simple predicate, as discussed above. However, if the WHERE clause had many such simple predicates joined with AND or other operators, they will again be treated each as a component. The components will be stored in this predicate table as in the above examples involving only variables, and/or constants.

Finally, if the Pos_Type is a set operation, the predicate table is also used to express operations on sets. For example:

$$N \notin Known$$

In this case, the Component_Type is a DB_Table. Database tables are sets. And the records are the tuples belonging to the sets. Left and/or Right_Table_Name are the fields from the predicate table used in this case.

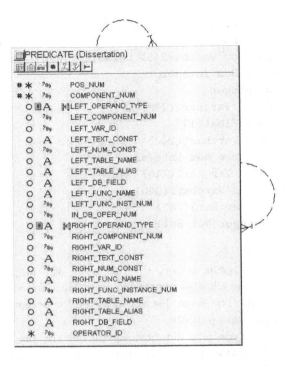

Fig. 2.12   Representing Predicates and Conditions in SpecDB

2.4.10   *Database Operations*

In   this   subsection
the DB_Operations, DB_Operation_Tables and DB_Operation_Details tables of SpecDB are discussed. The DB_Operation_Tables table contains a list of tables involved in each DB operation defined in the DB_Operations table. The DB_Operations table stores database structure query language (SQL) statements. Good references for SQL standards can be found in [73, 74, 75, 76]. Since a database select, insert, delete, update, etc, each is considered one statement; this table is dedicated to describe those operations. There is one record for each SQL statement in the software specification or code.

The WHERE clause is stored in the Predicate table, in this case the DB_Condition_Num is the same as the Pos_Num in the predicate table. This table is a representation of some of the database operations that can be programmed in a PL-SQL environment. Following the same technique all the SQL Language functions can be represented in this table. The functions listed and accounted for by fields are just a few examples to show the feasibility of the approach. The DB_Operation_Details table contains the details of the Database operations described in the DB_Operations table.

For SELECT Statements, the variables to select are inserted in the Field_Name attribute. If there are n fields to select, there will be n records in this table with the same DB_Operation_Num. If the database operation was SELECT *, then there will be only one record, and the Field_Name will be *. The INTO portion of the select statement, i.e. the variable where the selected value should be stored is indicated in the INTO_Var_ID attribute. The FROM clause is listed in the DB_Operation_Tables table, with the same DB_Operation_Num. Finally, The WHERE clause is stored in the Predicate table, with the POS_NUM = the DB_Conition_Num. If a cursor is defined, the Operation_Type is in the DB_Operations table is 'Cursor', it is treated the same as a SELECT statement, with the addition of identifying the cursor name in the Cursor_Name field in the DB_Operations table. Also if a Fetch operation on a cursor is the DB_Operation type, again the Cursor_Name field is populated, and all the details of which values to put from the cursor SELECT statement into variables are all stored in the DB_Operation_Details table.

For INSERT INTO statements, the table name will be in DB_Operation_Tables table, with the same DB_Operation_Num. If specific attributes (sequence) is specified, there will be a record in this table for each attribute. The attributes will be stored in the Field_Name field. The corresponding value that should be inserted in this attribute, will be placed in the appropriate Type_Value field, according to its type.

```
Prompt Creating Table 'DB_Operations'
Create Table Db_Operations
 (Db_Operation_Num Integer Not Null
 ,Operation_Type Varchar2(6)
 ,Db_Name Varchar2(30)
 ,Db_Condition_Num Integer
 ,Sequence_Name Varchar2(240)
 ,Sequence_Incremented Varchar2(1) Default 'T'
 ,Sequence_Start_With Number
 ,Savepoint_Rollback_Name Varchar2(30)
 ,Role_Priv_Name Varchar2(6)
 ,Role_Priv_User Varchar2(20)
```

```
,Priv_Object_Name Varchar2(240)
,Grant_Admin_Option Varchar2(1) Default 'T'
,View_Name Varchar2(240)
,Table_Name Varchar2(240)
,As_Table_Name Varchar2(240)
,As_Select Varchar2(1) Default 'T'
,Cursor_Name Varchar2(240) )
```

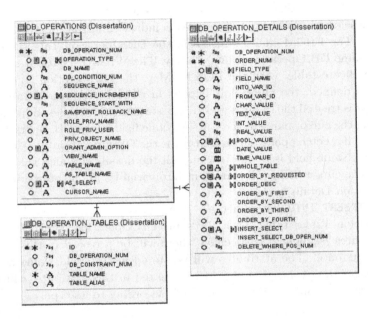

Fig. 2.13   Representing Database Operations in SpecDB

If the database statement is an INSERT statement of the following type:

```
INSERT INTO table_name SELECT ...
```

then the Insert_Select attribute will be set to true. Also the Insert_Select_DB_ Oper_Num will be filled out and the Select operation will be defined in the DB_Operations table, under its Pos_Num.

```
Prompt Creating Table 'DB_Operation_Details'
Create Table DB_Operation_Details
 (Db_Operation_Num Integer Not Null
```

```
,Order_Num Integer Not Null
,Field_Type Varchar2(10)
,Field_Name Varchar2(30)
,Into_Var_Id Number
,From_Var_Id Integer
,Char_Value Varchar2(1)
,Text_Value Varchar2(100)
,Int_Value Integer
,Real_Value Real
,Bool_Value Varchar2(1) Default 'T'
,Date_Value Date
,Time_Value Date
,Whole_Table Varchar2(1) Default 'T'
,Order_By_Requested Varchar2(1) Default 'T'
,Order_Desc Varchar2(1) Default 'T'
,Order_By_First Varchar2(15)
,Order_By_Second Varchar2(15)
,Order_By_Third Varchar2(15)
,Order_By_Fourth Varchar2(15)
,Insert_Select Varchar2(1) Default 'T'
,Insert_Select_Db_Oper_Num Number
,Delete_Where_Pos_Num Number )
```

For DELETE datbase operations, the table name will be in the DB_Operation_Tables table, with the same DB_Operation_Num. If there is no WHERE clause, then all the rows in the table will be deleted, and thus the Whole_Table field should be set to True. If there are some specific rows to be deleted, identified by a where clause, the where clause will be specified in the Predicate table, under a POS_Num equal to the position number of the current Delete Statement in the code. This number will be also stored in the Delete_Where_Pos_Num field.

For UPDATE database operations, the table name will be in the DB_Operation_Tables table, with the same DB_Operation_Num. The attributes will be stored in the Field_name field. The corresponding value that should be inserted in this attribute, will be placed in the appropriate 'type'_Value field, according to its type. (if there are n fields to update, there will be n records in this table with the same DB_Operation_Num). The WHERE clause is stored in the Predicate table, with the Pos_Num = the DB_Condition_Num. In a query, the same table can be used twice, with different aliases. Accordingly, the key to the DB_Operation_Tables table is a numeric ID.

```
Prompt Creating Table 'DB_Operation_Tables'
Create Table Db_Operation_Tables
 (Id Integer Not Null
 ,Db_Operation_Num Integer
 ,Db_Constraint_Num Number
 ,Table_Name Varchar2(30) Not Null
 ,Table_Alias Varchar2(240) )
```

2.4.11   Classes, Objects and Class Relations

SpecDB can represent object oriented constructs; classes, objects, and relationships between classes. In this section, the entities of SpecDB that hold object oriented constructs are explained.

```
Prompt Creating Table 'Classes'
Create Table Classes
 (Class_Name Varchar2(240) Not Null  )

Prompt Creating Table 'Objects'
Create Table Objects
 (Object_Name Varchar2(240) Not Null
 ,Class_Name Varchar2(240) Not Null )
```

The Classes table holds the names of the classes, the rest of the information of the class are stores in their respective locations, with an identifier pointing to their respective classes. For example, class variables, methods and encapsulations rules are stored in the Var table, in Section 2.4.4 and either the Procedure or the Function_Declaration tables respectively, explained in Section 2.4.5. Relationships between classes, like inheritance, association, aggregation, etc, are stored in the Class_Relationships table. In the Class_Relationships table, two classes from the Classes table are referenced in Class_Name1 and Class_Name2. The relationship type between those two classes is stored in the relationship attribute in the Class_Relationships table. This attribute has a list of valid values representing the types of relationships among classes, e.g. inheritance, encapsulation, etc.

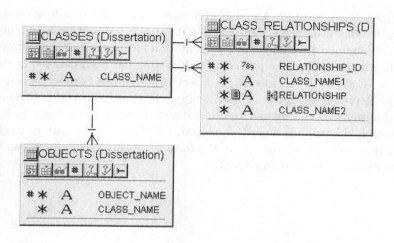

Fig. 2.14   Representing Object Oriented Classes and Objects in SpecDB

```
Prompt Creating Table 'Class_Relationships'
Create Table Class_Relationships
 (Relationship_Id Number Not Null
 ,Class_Name1 Varchar2(240) Not Null
 ,Relationship Varchar2(30) Not Null
 ,Class_Name2 Varchar2(240) Not Null)
```

2.5   Conclusion

Chapter 2 contains the design of SpecDB, a database designed to hold parts of a software's specification for the purpose of unit testing. Through the use of SpecDB, automated tools can become more intelligent in generating and testing programs, based on an understanding of what the software under construction is supposed to do, by design. The database design of SpecDB is scalable and thus, allows for any additions to accommodate more programming constructs or requirements, new forms of input or output, etc., as they evolve in the future. The motivation behind building SpecDB has been expressed in Chapter 1, mainly to aid the process of automated intelligent test case generation based on a machine-readable specifications. In order to create these machine-readable specifications to allow for automated reasoning, SpecDB was created. A scalable representation of the specification was chosen for this task, namely the use of a database specification representation. The standard technique of representation for portions of the software specifications to aid in the process of unit testing has been detailed in this chapter. Each of the entities of the SpecDB design has been explained in detail. Representing software specifications in this manner aids the process of automated testing and automated code generation. In Chapter 3, a translation algorithm is described, to translate formal specifications to the SpecDB representation, showing how a formal software specification can be automatically represented in SpecDB.

2.5.1 *Other SpecDB Entities*

The following SQL Scripts are for creating other SpecDB Entities:

```
Prompt Creating Table 'Actual_Input'
Create Table Actual_Input
 (Ai_Ip# Number(7) Not Null
 ,Ai_Test_Case# Number(7) Not Null
 ,Ai_Category Varchar2(4)
 ,Ai_Db_Table Varchar2(20)
 ,Ai_Db_Field Varchar2(25)
 ,Ai_Db_Row# Integer
 ,Ai_Value Varchar2(240)
 ,Ai_Into_Var Varchar2(240)
 ,Ai_Gui_Window Varchar2(240)
 ,Ai_Gui_Component Varchar2(240)
 ,Ai_Type Varchar2(240)
 ,Ai_File_Name Varchar2(240))

Prompt Creating Table 'Pre_Post_Conditions'
Create Table Pre_Post_Conditions
 (Db_Operation_Id Number Not Null
 ,Condition_Num Number Not Null
 ,Pre_Post Varchar2(4))

Prompt Creating Table 'Actual_Output'
Create Table Actual_Output
 (Ao_Op# Integer Not Null
 ,Ao_Test_Case# Integer Not Null
 ,Ao_Category Varchar2(240)
 ,Ao_Db_Table Varchar2(240)
 ,Ao_Change Varchar2(240)
 ,Ao_Field Varchar2(240)
 ,Ao_Row# Varchar2(240)
 ,Ao_Value Varchar2(240)
 ,Ao_From_Var Varchar2(240)
 ,Ao_Window Varchar2(240)
 ,Ao_Component Varchar2(240)
 ,Ao_Type Varchar2(240)
 ,Ao_Name Varchar2(240)
```

```
,Ao_Text Varchar2(240)
,Ao_Destination Varchar2(240)
,Ao_User Varchar2(240)
,Ao_Role_Id Varchar2(240)
,Ao_Privilege_Id Varchar2(240)
,Ao_Database Varchar2(240)
,Ao_Savepoint Varchar2(240))

Prompt Creating Table 'Db_Test_Operation'
Create Table Db_Test_Operation
 (Operation_Id Number Not Null)

Prompt Creating Table 'Test_Templates'
Create Table Test_Templates
 (Db_Operation_Id Number Not Null
 ,Test_Template_Name Varchar2(30) Not Null)

Prompt Creating Table 'Expected_Output'
Create Table Expected_Output
 (Eo_Op# Integer Not Null
 ,Eo_Test_Case# Integer Not Null
 ,Eo_Category Varchar2(240)
 ,Eo_Db_Table Varchar2(240)
 ,Eo_Change Varchar2(240)
 ,Eo_Field Varchar2(240)
 ,Eo_Row# Varchar2(240)
 ,Eo_Value Varchar2(240)
 ,Eo_From_Var Varchar2(240)
 ,Eo_Window Varchar2(240)
 ,Eo_Component Varchar2(240)
 ,Eo_Type Varchar2(240)
 ,Eo_Name Varchar2(240)
 ,Eo_Text Varchar2(240)
 ,Eo_Destination Varchar2(240)
 ,Eo_User Varchar2(240)
 ,Eo_Role_Id Varchar2(240)
 ,Eo_Privilege_Id Varchar2(240)
 ,Eo_Database Varchar2(240)
 ,Eo_Savepoint Varchar2(240))
```

```
Prompt Creating Table 'Attributes_Changed'
Create Table Attributes_Changed
 (Db_Operation_Id Number Not Null
 ,Attribute Varchar2(30)
 ,Of_Table Varchar2(40))

Prompt Creating Table 'Host_Var'
Create Table Host_Var
 (Db_Operation_Id Number Not Null
 ,Host_Var_Name Varchar2(30) Not Null
 ,Input_Output Varchar2(1)
 ,Direct_Indirect Varchar2(1)
 ,On_Attribute Varchar2(30)
 ,Of_Table Varchar2(40))

Prompt Creating Table 'Constants'
Create Table Constants
 (Constant_Name Varchar2(240)
 ,Type Varchar2(240)
 ,Int_Value Integer
 ,Real_Value Real
 ,Char_Value Char
 ,String_Value Varchar2(240)
 ,Date_Value Date
 ,Time_Value Date)

Prompt Creating Table 'Expected_Input'
Create Table Expected_Input
 (Ei_Ip# Number(7) Not Null
 ,Ei_Test_Case# Number(7) Not Null
 ,Ei_Category Varchar2(4)
 ,Ei_Db_Table Varchar2(20)
 ,Ei_Db_Field Varchar2(25)
 ,Ei_Db_Row# Integer
 ,Ei_Value Varchar2(240)
 ,Ei_Into_Var Varchar2(240)
 ,Ei_Gui_Window Varchar2(240)
 ,Ei_Gui_Component Varchar2(240)
 ,Ei_Type Varchar2(240)
,Ei_File_Name Varchar2(240))
```

```
Prompt Creating Table 'Char_Equiv_Class'
Create Table Char_Equiv_Class
 (Var_Name Varchar2(15) Not Null
 ,Range_From Varchar2(1) Not Null
 ,Range_To Varchar2(1) Not Null
 ,Equivalence_Class Integer Not Null)

Prompt Creating Table 'Equivalence_Class'
Create Table Equivalence_Class
 (Db_Operation_Id Number Not Null
 ,Data_Group_Name Varchar2(240) Not Null
 ,Attribute_Name Varchar2(30)
 ,Int_Value Integer
 ,Real_Value Real
 ,Char_Value Varchar2(1)
 ,Text_Value Varchar2(100)
 ,Date_Value Date)

Prompt Creating Table 'Real_Equiv_Class'
Create Table Real_Equiv_Class
 (Var_Name Varchar2(15) Not Null
 ,Range_From Real Not Null
 ,Range_To Real Not Null
 ,Equivalence_Class Integer Not Null)

Prompt Creating Table 'Data_Group_Values'
Create Table Data_Group_Values
 (Db_Operation_Id Number Not Null
 ,Data_Group_Name Varchar2(240) Not Null
 ,Int_Value Integer
 ,Real_Value Real
 ,Char_Value Varchar2(1)
 ,Text_Value Varchar2(100)
 ,Date_Value Date)

Prompt Creating Table 'Int_Equiv_Class'
Create Table Int_Equiv_Class
 (Var_Name Varchar2(15) Not Null
 ,Range_From Integer Not Null
 ,Range_To Integer Not Null
```

```
 ,Equivalence_Class Integer Not Null)

Prompt Creating Table 'Constraints'
Create Table Constraints
 (Constraint_Name Number Not Null
 ,Constraint_Num Integer Not Null
 ,Constraint_Or_Assertion Varchar2(1)
 ,On_Field Varchar2(1) Default 'T'
 ,Field_Name Varchar2(240)
 ,Table_Name Varchar2(240)
 ,Db_Name Varchar2(240)
 ,Condition_Predicate_Pos_Num Integer
 ,Constraint_Clause_Pred_Num Number
 ,Db_Op_Tables_Id Integer
 ,Db_Operation_Num Integer
 ,Db_Operator_Id Number(2))

Prompt Creating Table 'Data_Group'
Create Table Data_Group
 (Db_Operation_Id Number Not Null
 ,Data_Group_Name Varchar2(240) Not Null
 ,Attribute Varchar2(30) Not Null
 ,Of_Table Varchar2(40)
 ,Probability Number)

Prompt Creating Table 'Table_Field_Description'
Create Table Table_Field_Description
 (Has_Constraint_Restriction Varchar2(1) Default 'T'
 ,Table_Name Varchar2(240) Not Null
 ,Field_Name Varchar2(240) Not Null
 ,Field_Id Varchar2(240)
 ,Field_Type Varchar2(240)
 ,Is_Primary_Key Varchar2(1) Default 'T'
 ,Is_Unique Varchar2(1) Default 'T'
 ,Mandatory Varchar2(1) Default 'T'
 ,Length Integer
 ,Has_Lov_Restriction Varchar2(1) Default 'T'
 ,Has_Table_Restriction Varchar2(1) Default 'T'
 ,Has_Range_Restriction Varchar2(1) Default 'T'
 ,Num_Of_Constraint_Restriction Integer
```

```
,Which_Table Varchar2(240)
,Which_Field  Varchar2(240))
```

Chapter 3

Representing Formal Specifications in SpecDB: A Translation Algorithm

In this chapter, an algorithm that translates formal software specifications to the SpecDB representation is described. The design of the underlying database that stores the formal specifications, SpecDB was detailed in Chapter 2. SpecDB is a database created to represent and store parts of the formal specifications of software, in a queryable / machine-readable format. The Z formal specification language has been chosen as an example for the translation algorithm. Similar algorithms to that shown in this chapter can be created to translate other formal specification languages such as VDM to SpecDB.

The outcome of the translation algorithm is a machine readable representation of the original formal specifications. It is worth noting that the users of SpecDB do not have to learn formal languages. The specifications can be entered directly in SpecDB as easily as populating any database with the correct data. However, this chapter gives an algorithm to transform formal specifications to the SpecDB representation, if formal specifications were used in the specification and design phases of the software development. The representation of SpecDB for formal specifications, aids in the processes of both automated software code generation and automated software unit testing, based on the actual software specifications.

3.1 A Translation Algorithm from Formal Specifications to SpecDB

The entities of the SpecDB database store the (formal) specifications of a program. SpecDB is designed to efficiently and accurately represent portions of the software formal specifications required to carry out the testing techniques. In Section 2.4, the methodology of how SpecDB covers and represents different forms of formal specifications is shown. In this section, formal specification fragments will be given together with their corresponding algorithm fragments to store these specifications in SpecDB. SpecDB was also designed to be generic, upgradeable and modifiable as described in Chapter 1.

The examples given from the Z formal specification language in Section 3.4 are from [18, 19, 23]. In each subsection in Section 3.4 the Z specification syntax is shown together with the algorithm steps. The algorithm is numbered to identify it from the explanation text. Not all the syntax of the Z specification language is discussed in this chapter. Examples are given for some of the algorithms. Numbers corresponding to the step in the translation algorithm in that specific case are shown to show the reader how each step is executed. Examples of the most commonly used syntax to document specifications are discussed. These examples are demonstrated to prove the feasibility and show the reader how an automated translation tool can execute this algorithm and transform formal specifications into the machine-readable SpecDB representation. The main idea of the algorithm can be understood from the steps of the algorithm in the following subsections. Similar algorithms can be created to translate other formal specification syntax to the SpecDB representation as needed. The understanding of the SpecDB design detailed in Chapter 2 is needed to generate extensions to the algorithm explained in this chapter.

3.2   Assumptions and Restrictions

All operators are prefixed with the "&" sign to distinguish them from variables, user-defined types and schemas. User-defined variables and types should not have dashes, spaces or Z symbols.

Var_Name? is an input.

VAR_Name! is an output.

No user-defined variable can end with 1. Variables ending with 1 are temporary variable names given by the algorithm, until the variable is given a name in the specifications to follow its type definition.

3.3 Preparing for Translation

Before translating the formal specifications into the representation of SpecDB, the following preparation phase tasks have to be performed:

(1) Going through the formal specifications, expand complex operations to simple operations.

(2) Number the specifications, giving the main control (main menu) the number 1, and increments of 10 to the rest of the specifications in order to allow room for future enhancements, and add on specifications where needed.

(3) Put either the OR symbol before a schema name that is the first line after another schema definition, according to the meaning of the schema.

(4) Add necessary brackets to complex calculations and predicates to accurately describe the required output. A couple of brackets should enclose one operator and one LHS operand or component within brackets and one RHS operand or component within brackets.

3.4 Storing Formal Specifications

In the following subsections of Section 3.4, a number of formal specification statement syntax is provided together with its corresponding translation into the SpecDB representation. The most commonly used statements are discussed. Other formal specification statements can also be translated into the SpecDB representation by expanding on the given algorithm with the same scheme after understanding the entities of the SpecDB design detailed in Chapter 2.

3.4.1 *Defining Schema Names and Operations*

Program units in the Z formal specifications language are given the name schemas. In Figure 3.1, the unit or schema name is AddBirthday.

To represent Z schemas in SpecDB, the following steps in the algorithm have to be executed.

(1) Set Current_SR to the Schema Name. (SR stands for subroutine.)

(2) Insert the name of the Schema in the Procedure table, setting the Num_of_Inputs field to 0 and Changes_System_State to F for False. The operation remains a procedure until proven to be a function, then

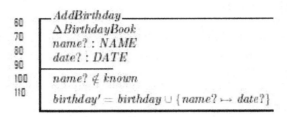

60	┌─ *AddBirthday* ─────────────
70	│ $\Delta BirthdayBook$
80	│ *name?* : $NAME$
90	│ *date?* : $DATE$
100	│ *name?* $\notin$ *known*
110	│ *birthday'* = *birthday* $\cup$ { *name?* $\mapsto$ *date?* }

Fig. 3.1   AddBirthday Schema

it is removed from this table and placed in the Function_Declaration
table instead.

Example: in the "AddBirthday" schema in Figure 3.1, executing the
above algorithm, will result in the following state:

(1) Current_SR := AddBirthday
(2) Insert record in the Procedure table (AddBirthday, 0, F)

3.4.2   *Types of Schemas*

After specifying the name of a schema, as in Section 3.4.1, the nature of
the schema is specified. For example, in Figure 3.1, in line 70 of the AddBirthday schema:

$\Delta Schema1$ or $\Delta BirthdayBook$

This means that AddBirthday is part of another schema (program in
this case) i.e. AddBirthday is a subroutine definition. It indicates that
after the execution of the AddBirthday schema, a change will occur in the
state of the system, defined by Schema1, or BirthdayBook in this case. In
this case, i.e. a or change in the state, the following should be executed:

(1) Update the Function_Declaration or Procedure record for the Cur-
     rent_SR, set Changes_System_State to T for True.

Example from Figure 3.1: Update the record of the Currecnt_SR (AddBirthday) in the Procedure table from (AddBirthday, 0, F) to (AddBirthday, 0, T) Another type of schema exists, one that does not change the

state of the system, identified in Z by:

$\Theta Schema1$

This symbol indicates that after the execution of the schema, No change should occur in the state of the system, defined by Schema1. In this case:

(1) Update the Function_Declaration or Procedure record for the Current_SR, set Changes_System_State to F for False, if not already F.

3.4.3 *User-Defined Types*

User-defined types are built from the basic system-defined types, possibly with some constraints. In the Z syntax below, the basic system-defined types for these user-defined types are not specified:

[Type1, Type2, ...]

(1) Insert a record in the Types table for every type until the closing bracket is reached.

For the example [NAME, DATE], two records should be inserted in the Types table as follows: NAME, , , N and DATE, , , N.

3.4.4 *Specifying Values for a User-Defined Type*

A variable type can be a pre-defined system type, or as part of the specification, other higher level types can be defined. The Z syntax for values of user-defined types is as follows:

FreeType1::= val1 | val2 | ...

In this case all the values after the ' ::= ' sign form a list of valid values that FreeType1 can take, and implicitly specifies, that FreeType1 cannot be of any other value. In order to represent this list of values for a user-defined type in SpecDB, the following algorithm should be executed:

(1) Insert a record in the Types table (FreeType1,,,L) where L is the value of the LRN field that specifies that the type has a list of valid values.
(2) Set LOV_Count = 0

(3) Increment LOV_Count. Also each time a "—" symbol is observed LOV_Count is incremented indicating that another valid value is specified next, then repeat step number 4.
(4) Insert a record in the Types_LOV table (FreeType1,LOV_Count, , ,Val, ...) Val is placed in the Text_Value field since the type is not specified. Repeat steps 3 and 4 if the next symbol read is a "—"

For example, line 290 below in Figure 3.2, the REPORT free type definition, has three possible values: ok, already_known, not_known.

290 *REPORT ::= ok | already_known | not_known.*

Fig. 3.2   Report, User-defined Type Specification

3.4.5 *Defining Sets or Tables*

Sets can be defined to describe entities in the software design. An example of a set is the set of known people in the birthday book. Sets are defined in formal languages as follows:

A: ρ B

Where A is a set, and B is a type. In order to represent the set definition above in SpecDB the following algorithm should be executed:

(1) If not already inserted, insert B in Types: Type_Name (leave all the rest of the attributes empty).
(2) Create a new table, give it the name A, with one attribute (key) of type B and name this attribute: A.
(3) Add A to the List_of_Tables table.
(4) Add a record in the Table_Descriptions table (A, A, B).

As an example, assume **known:** ρ **NAME**

(1) Output: record added to Types table: (known, , ,).
(2) Table Known created, with one attribute, known, of type NAME.
(3) Record added to List_of_Tables table: (Known).
(4) Record added to Table_Descriptions table: (Known, known, NAME).

3.4.6 *Function Declaration*

System functions are higher level processes or subroutines that collectively define the system operations. A function returns a value, unlike a procedure. Functions are defined as follows in the Z formal specification language syntax:

$$\text{Func1: } B \rightharpoonup C$$

Where Func1 is a function or set, B is an input type and C is an output type. In order to declare a function in SpecDB the following algorithm should be executed:

(1) Put all types (B and C) in the Types table (Type_Name), if not already inserted.
(2) Consider variables B1 and C1 of types B and C respectively. Insert in the Var table two records: (#, B1, B, Func1, , T, ...), where T is true for the Is_Input attribute and # is the next sequence number in the Var table, (#, C1, C, Func1, , , T, ...), where T is true for the Is_Output attribute.
(3) Insert in the Function_Declaration table (Func1, 1, C, #, C1, F, F, F). Where # is the VarID for C1. All restrictions are false by default until restrictions are declared.
(4) Insert in the Subroutine_IP_Description table (Func1, 1, B, B1, F, F , F)
(5) Add Func1 to the List_of_Tables table.
(6) Create a new table, give it the name Func1, with two fields of type B and C, and give them the name B1 and C1.
(7) Add a record in the Table_Description table (Func1, B1, B).
(8) Add a record in the Table_Description table (Func1, C1, C).

3.4.7 *Subroutine Input Restriction*

The input of a subroutine (procedure or function) can have restrictions, based on the business rules. Subroutine Input restrictions are defines as follows in Z:

A = dom Func1

In order to represent that input restriction above in SpecDB the following algorithm should be executed:

(1) Current_Var_ID := the input Var_ID for Func1.
(2) In the Subroutine_IP_Description table, update the fields of the Func1 function to reflect this restriction: (Func1, 1, B, , F, T, F), i.e. the input B has a table restriction.
(3) Insert a record in the Var_Table_Restrictions table (#, Current_Var_ID, Func1, , I, BelongsTo, A, ,) where # is the current serial number for this table.

Example: known = dom birthday

In this case the specifications state that the set of people already in the birthday book are the set of known individuals. New people added to the birthday book should not thus be already in it, to avoid multiple entries for the same person. Given the above specification, the algorithm trace should look as follows:

(1) Current_Var_ID := 1.
(2) Update the Birthday record in the Subroutine_IP_Description table: (Birthday, 1, NAME, Name1, F, T, F).
(3) Add record to the Var_Table_Restrictions table: (1, 1, Birthday, , 1, BelongsTo, Known, Known,).

3.4.8 *Variable Definition and Subroutine Inputs*

Variables can be defined as inputs to a subroutine, or in the main operation, they are defined in the Z formal specification language as follows:

Var1? : Type1

In order to introduce the subroutine inputs above in SpecDB the following algorithm should be executed:

(1) Insert in table Var (#, var1, Type1, Current_SR, , T, , ...). Current_SR is inserted in that record if the variable is defined in the context of an operation definition (Schema). Set the value of the Is_Input field to true.
(2) Set Current_Var_ID = serial number given above (#).
(3) If the variable is defined in the context of an operation, update the Procedure table, increment num_of_inputs, where proc_name = Current_SR.

For example, line 80 in Figure 3.1:

`name?: NAME`

In this case a variable named: name is of the type NAME. This is a general variable definition statement. However in line 80 in Figure 3.1, it is defined within the scope of a subroutine definition, accordingly, it is an input variable to that subroutine. The trace of the algorithm for the above example should be as follows:

(1) Add record in the Var table: (3, name, NAME, AddBirthday, , T, . . .).
(2) Current_Var_ID := 3.
(3) AddBirthday record updated in the Procedure table (AddBirthday, 1, F).

3.4.9 *Output Definition*

Variables can also be defined as an output of some subroutine. In Z output variables are defined by appending an exclamation mark to the right of the variable name as follows:

`Var3! : Type2`

In order to define output variables in SpecDB the following algorithm should be executed:

(1) Insert in table Var (#, var3, Type2, Current_SR, , T, , . . .). Current_SR is inserted in that record if the variable is defined in the context of an operation definition (Schema). Set the value of the Is_Output field to true (T).
(2) Set Current_Var_ID = serial number given above (#).
(3) If the variable is defined in the context of an operation, then this operation is a function, not a procedure. Add Current_SR to the Function_Declaration table, with the same data in its record in the procedure table. Set Op_type = Type2, and Op_Var_ID = Current_Var_ID. Set all restrictions to False in this record.
(4) Remove Current_SR from procedure table.

For example in line 180 in Figure 3.3, below: `date!: DATE`
In this case, it is understood from the specifications that FindBirthday is a function that returns an output of type DATE. The algorithm trace should result in the following:

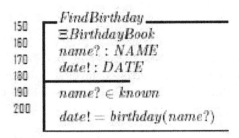

Fig. 3.3   FindBirthday Schema

(1) Add a record to the Var table: (6, date, DATE, FindBirthday, , , T, ...).
(2) Current_Var_ID := 6.
(3) Record added to the Function_Declaration table (FindBirthday, 1, DATE, 6, date, F, F, F, F, ...).
(4) Delete the FindBirthday record from the Procedure table.

3.4.10   *Setting Variable Restrictions*

A variable can be restricted based on the business rules or software specifications. In Z, variables restrictions are defined as follows:

$$\text{Var?} \notin \text{Set1}$$

The symbol $\notin$ can be substituted with any other operator. Any set operators can be expressed using the same method. Also, the type of the variable, it can be an input with the symbol ? or an output variable with the symbol ! appended to the right of the variable name. In order to represent the variable restrictions defined above in SpecDB the following algorithm should be executed:

(1) Set Current_Var_ID = Var_ID of Var1 in the Var table, for the Current_SR if any.
(2) Update the Var table for Current Var_ID, Set has_table_restrictions to true.
(3) Insert record in the Var_Table_Restrictions table (#, Current_Var_ID, Current_SR, Var1, , NotBelongTo, Set1,). Current_SR is inserted in that record if the variable is defined in the context of an operation

definition (Schema) (Substitute NotBelongTo with any other operator used.)

For example in line 100 in Figure 3.1:

Name? ∉ Known

The trace of the above algorithm should yield the following results:

(1) Current_Var_ID := 3.
(2) The record for name in the Var table is modified: (3, name, NAME, AddBirthday, , T, ..., T, ...).
(3) Add a record to the Var_Table_Restrictions table: (2, 3, AddBirthday, Name, , NotBelongTo, Known, ...).

3.4.11 *Assignment Operation*

The syntax for assignment operations in Z is simple, and defined as follows:

A = B

Where A is a variables and B can be a variable or a constant. However, the same assignment syntax can be used for more complex constructs than simple variables. The following algorithm should be executed to represent an assignment operation in SpecDB:

(1) Insert a record in the DateFlow table (Pos#, Suc#, Assign, ...) If this statement is the end of a schema definition (operation) then set Suc# to 1, i.e. main menu control or the Suc# from the calling subroutine.
(2) Set Current_Var_ID = Var_ID for A in its subroutine context, if this statement is defined in a subroutine.
(3) If B is a variable, set Current_Var_ID2 = Var_ID for B in its subroutine context, if this statement is defined in a subroutine.
(4) Insert record in the Assignment table (Pos#, Current_Var_ID, Var, ..., Current_Var_ID2, ...). If B is a constant then replace Var for Const in the RHS_Type field, and put the value in the correct location according to its type.

For example, in line 320 in Figure 3.4:

result! = ok:

Below is the trace of the algorithm for the above example:

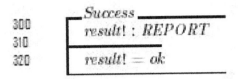

Fig. 3.4   Success Schema

(1) Add record to the DataFlow table (320, 1, Assign, ...).
(2) Current_Var_ID := 9.
(3) Add record to the Assignment table (320, 9, Const, Text, ... , ok, ...).

3.4.12   *Assignment Operation and Function Call*

The following formal specification syntax is more complex than those discussed above. In formal specifications, we can define both a function call and an assignment at the same time, where the output of the function call is directly assigned to a variable, as in the below example:

```
Var2! = func1(var1?)
```

In order to define function calls and assign the output to a variable in SpecDB the following algorithm should be executed:

(1) Insert a record in the DataFlow table (Pos#, Suc#, Assign, ...) If this statement is the end of a schema definition (operation) then set Suc# to 1, i.e. main menu control or the Suc# from the calling position.
(2) Set Current_Var_ID = Var_ID for Var1? in its subroutine context, if this statement is defined in a subroutine.
(3) Insert record in the Subroutine_IP_Instance table (#, 1, func1, Var, Current_Var_ID, ...). Repeat this for all inputs to the function in order, incrementing the IP_Order field every time.
(4) Inst_Num
= # used in the above record inserted in Sub_Routine_IP_Instance. Set Current_Var_ID = Var_ID for Var2! in its subroutine context, if this statement is defined in a subroutine.
(5) Insert record in the Assignment table (Pos#, Current_Var_ID, Func, , , func1, Inst_Num, ...), where the RHS is set to Func, and FuncName

to func1.

For example, in line 200 in Figure 3.3:

```
date! = birthday(name?)
```

The function `birthday` is called, given the variable `name?` as an input, the output of the function is stored in the output variable `date!`. Below is the trace of the algorithm for this example:

(1) Add record to the DataFlow table (200, 1 , Assign, . . .).
(2) Current_Var_ID = 5; where 5 is the Var_ID for date in its subroutine.
(3) Add record to the Sub_Routine_IP_Instance table: (1, 1, Birthday, Var, 5).
(4) Inst_Num := 1.
(5) Current_Var_ID := 6.
(6) Add record to the Assignment table(200, 6, Func, . . . , Birthday, 1, . . .).

3.4.13 *Assignment Operation for Sets*

As mentioned in Subsection 3.4.11, assignment operations can be more complex, in this section the syntax of assigning a set to the value of another set is discussed.

```
Set1 = Set2
```

This is a complex operation that will be broken down at the preparation phase. The resulting operations will be as follows:

```
Pos#: Drop Set1
Pos#+1: Set1 = Set2 (Create new table Set1 = Set2)
```

(1) If Set1 already exists, drop table, and continue. Add record in the DataFlow table (Pos#, Pos#+1, DB, , , , , ,).
(2) Add a record in the DB_Operations table (Pos#, Drop_Table, , , . . . , Set1) where Set1 is the Table_Name field.
(3) Create a new table, give it the name Set1, and with fields equal to Set2
(4) Add record in the DataFlow table (Pos#, Suc#, DB, , , , , ,). If step 1 was executed, then Pos#+1 should replace Pos#. (also in the next step). If this statement is the end of a schema definition (operation) then set Suc# to 1, i.e. main menu control or the Suc# from the calling subroutine.

(5) Add a record in the DB_Operations table (Pos#(+1), Create_Table, , , ..., Set1, Set2, ...) where Set1 and Set2 are the Table_Name and As_Table_Name fields.

3.4.14 *Assignment Operation for a Set*

The formal specification syntax allows for resetting set values, as follows:

$$\text{Set1} = \Phi$$

In this case, the assignment operation on the set deletes all the elements in the set, i.e. all the table records and changes the state of the system as follows:

(1) Add record in the DataFlow table (Pos#, Suc#, DB, , , , ,). If this statement is the end of a schema definition (operation) then set Suc# to 1, i.e. main menu control or the Suc# from the calling subroutine.
(2) Add a record in the DB_Operations table (Pos#, Delete, ...).
(3) Add a record in the DB_Operation_Tables table (Pos#, Set1, ...).
(4) Add a record in the DB_Operation_Details table (Pos#, 1, ..., T, ...) where the Whole_Table field is set to T for True.

An example is shown in line 280 in the InitBirthdayBook schema in Figure 3.5 below.

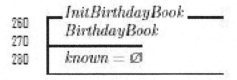

260	*InitBirthdayBook*
270	*BirthdayBook*
280	*known = ∅*

Fig. 3.5   InitBirthdayBook Schema

3.4.15 *Inserting Records in the Database*

Elements can be added to sets, in other words, records can be added to tables, the syntax in Z is as follows:

$$Set1 = Set1 \cup Var1 \rightarrow Var2$$

In order to represent this complex operation in SpecDB, the following algorithm should be executed:

(1) Add a record in the DataFlow table (Pos#+2, Suc#, DB, ...). If this statement is the end of a schema definition (operation) then set Suc# to 1, i.e. main control or the Suc# from the calling subroutine.
(2) Add a record in the DB_Operations table (Pos#+2, Insert, ...).
(3) Add a record in the DB_Operation_Tables Table (Pos#+2, Set3, ...).
(4) Set Current_Var_ID = Var_ID for Var1 in its specific subroutine, if any.
(5) Add a record in the DB_Operation_Details Table (Pos#+2, 1, ..., Current_Var_ID).
(6) Set Current_Var_ID = Var_ID for Var2 for its specific subroutine, if any.
(7) Add a record in the DB_Operation_Details Table (Pos#+2, 2, ..., Current_Var_ID). Current_Var_ID is inserted in the From_Var_ID field for steps 5 and 7).

For example, in line 110 in Figure 3.1.

$$birthday = birthday \cup name? \rightarrow date?$$

This is a complex operation that is broken down in the preparation phase. The resulting operations will be as follows:

(1) Line 110: $birthday_- = birthday$ (creating a new table)
(2) Line 112: $birthday_- = birthday_- \cup name? \rightarrow date?$
(3) For line 112, add record in the Data flow table: (112, 1, DB, ...). Add record to DB_Operations (112, Insert, ...). Add record to DB_Operation_Tables (112, Birthday_, ...). Set Current_Var_ID := Var_ID of name?, e.g. 3, add a record to DB_Operation_Details (112, 1, ..., 3, ...) and set Current_Var_ID := Var_ID of date?, e.g. 4. Add record to DB_Operation_Details (112, 2, ..., 4, ...).

3.4.16 *Output Table Definition*

A system may give various forms of outputs, sets can be one of those output types. In Z, output sets are defined as follows:

$$Var1! : \rho \, Type1$$

Var1 is an output table or set in this case, this is known from its type. In order to represent this definition in SpecDB, the following algorithm should be executed:

(1) If not already inserted, insert Type1 in types: Type_Name (all the rest of the attributes will be empty).

(2) If table Var1 exists, drop it, add record in dataflow table (Pos#, Suc#, DB ...) and add record in DB_Operations table (Pos#, Drop_Table, ..., Var1, ...) where Var1 is the table name. if not, add Var1 to the List_of_Tables table.

(3) Create a new table, give it the name Var1, with one attribute (key) of type Type1 and name this attribute: Var1.

(4) Add a record in the Table_Description table (Var1, Var1, Type1, ...).

(5) Insert a record in the Var table (#, Var1, DB_Table, Current_SR, T, ...). Current_SR is inserted in that record if the variable is defined in the context of an operation definition (Schema). The true (T) is the value of the Is_Output field.

If Var1 is the only output for a subroutine, then it cannot be considered as a function, since tables cannot be passed as parameters. The output table will be available for access outside the scope of the current subroutine, and when this subroutine or any other utilizing the same table is invoked, this table will be dropped, and recreated, and the data redefined. For example, see line 240 in Figure 3.6.

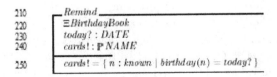

210	_Remind_
220	$\Xi BirthdayBook$
230	$today? : DATE$
240	$cards! : \mathbb{P}\, NAME$
250	$cards! = \{\, n : known \mid birthday(n) = today? \,\}$

Fig. 3.6   Remind Schema

In this case the algorithm will add cards to the list of tables. Create table Cards, with one attribute (key), named Cards of type NAME. Add record in the Table_Description table (Cards, Cards, NAME). Add record in the Var table (8, Cards, , Remind, ..., T, ...) Where the Is_Output attribute is set to T.

3.4.17 *Assigning a Variable to a Calculation*

The Z syntax of a calculation assigned to a variable is given below.

$$Var1 = ((Func2 \wedge Func3) \vee Func4)$$

Any other right hand side calculation can be represented using the same method. The operators can be of any type that match the results of the operands. In order to represent this complex calculation and assignment operation in SpecDB the following algorithm should be executed:

(1) Add record in the DataFlow table (Pos#, Suc#, Assign, ...). If this statement is the end of a schema definition (operation) then set Suc# to 1, i.e. main menu control or the Suc# from the calling subroutine.

(2) Set Current_Var_ID = Var_ID for Var1 in its subroutine context, if this statement is defined in a subroutine.

(3) Insert record in the Assignment table (Pos#, Current_Var_ID, Calc, , Pos#, ...), where Cal_Num = Pos#.

(4) Reset Component_Num = 0.

(5) Increment Component_Num.

(6) Follow the rule for inserting components in the calculation or predicate table: Insert record in the Calculation table (Pos#, Component_Num, func, ..., func2, , func, ..., func3, , AND).

(7) If the Calculation is a complex one like the one shown, then repeat steps 5 and 6 until all the components of the calculation are specified.

390	┌ *R AddBirthday* ────────────────
400	│ Δ *BirthdayBook*
410	│ *name?* : *NAME*
420	│ *date?* : *DATE*
430	│ *result!* : *REPORT*
	└
440	result! = (AddBirthday AND Success) OR AlreadyKnown

Fig. 3.7   RAddBirthday Schema

Line 440 in Figure 3.7 is an example of assigning the variable result to either the value "Success" or "AlreadyKnown" depending on whether the input name and date already exist in the Birthday Book or not.

3.4.18   *Constraining the Values Populating a Table or Set*

The syntax of the formal specification language Z allows for specifying how a set or a table can be populated. In the below example, a table is populated from specific results of a query on another table:

$$Var1! = n : Set1 | func1(n) = var2$$

Var1 is an output table set, func1 is also a table or a database operation on a table. The first part of this statement: "Var1!=" populates a table (set). Assigning a variable to a set means that this variable is really a set of values, which may include 0 or more values.

The second part, "n:Set1 —" introduces a new variable, of type Set1, also it indicates that n is in itself a set (table) that is a subset of Set1. The third part of the statement "—" indicates that there is a restriction on n, it does not include all the values of set1, just those that satisfy the condition on the right hand side. The final part of this statement is the restriction "func1(n)=var2".

These four steps will be defined in SpecDB as in this translation of this SQL statement:

```
INSERT INTO var1
  (SELECT Field_Name1 FROM Set1
  WHERE Field_Name1 IN
      (SELECT Field_Name2 FROM func1
      WHERE Field_Name3 = Var2))
```

The statement will be specified as above, using Pos# and Pos#+1. The field names and the quantities will be known by inquiring the already specified design.

(1) Insert into DataFlow table (Pos#, Suc#, DB, ...) If this statement is the end of a schema definition (operation) then set Suc# to 1, i.e. main menu control or the Suc# from the calling subroutine. In this case Suc# should not = Pos#+1, since this value will be used to define this statement.

(2) Insert record in DB_Operations table (Pos#, Insert, ,Pos#, ..., Var1, , T) where the value of DB_Condition_Num = Pos#, Table_Name = Var1 and AS_Select = T for true.

(3) Set Field_Name = Field_Name from the Table_Description table where Table_Name = Set1.

(4) Insert record in DB_Operation_Details (Pos#, 1, , Field_Name, ...).

(5) Insert record in DB_Operation_Tables (Pos#, Set1, ...).

(6) Reset Component_Num = 1 for the predicate table, with each insert in the Predicate table with the same Pos#, increment the Component #.

(7) Insert record in Predicate (Pos#, Component_Num, DB_Field, ..., Field_Name, ..., Pos#+1, ...) where Left_DB_Field = Field_Name and In_DB_Oper_Num = Pos#+1.

(8) Insert record in DB_Operations table (Pos#+1, Select, , Pos#+1, ...).

(9) Set Field_Name = SR_Var_name from the SubRoutine_IP_Description table where Subroutine_Name = func1 and type = that of Set1.

(10) Insert record in DB_Operation_Details (Pos#+1, 1, , Field_Name, ...).

(11) Insert record in DB_Operation_Tables (Pos#+1, func1, ...).

(12) Set Field_Name = OP_Var_name from the Function_Declaration table where Func_Name = func1.

(13) Set Current_Var_ID to Var_ID of var2.

(14) Reset Component_Num = 1 for the predicate table, with each insert in the Predicate table with the same Pos#, increment the Component #.

(15) Insert record in Predicate (Pos#, Component_Num, DB_Field, ..., Field_Name, VAR,, Current_Var_ID, ..., Equal) where Left_DB_Field = Field_Name and Right_Component_Type = Var and Right_Var_ID = Current_Var_ID and Operator_ID = Equal.

Figure 3.6, line 250 is an example of such an operation. The steps for this complex operation will be defined in SpecDB as in the following translation of the statement at Line 250:

```
cards! = {n: known | birthday(n) = today?}:

INSERT INTO cards
  (SELECT known from Known
   WHERE known in
       (SELECT Name1 FROM Birthday
        WHERE Date1 = today))
```

The statement should be specified as above, using Pos# 250 and Pos# 251.

3.5 Conclusion

In Chapter 3, some examples were given to show how software specifications written in a formal specification language like Z, can be represented in

SpecDB. The formal specification language Z was chosen because it is one of the more common formal specification languages. Representing formal specifications in SpecDB aids the process of automated testing and automated code generation. Through the use of SpecDB, automated tools can become more intelligent in generating and testing programs, based on an understanding of what the software under construction is supposed to do, by design. The scalable database design of SpecDB allows for any additions to accommodate more programming constructs, new forms of input and output, etc., as they evolve in the future, as well as formal specification syntax. In this chapter,a method to translate some examples of formal specifications written in Z into the unified SpecDB representation was illustrated. Automated tools can easily be developed based on the algorithm in this chapter to carry out the translation from well formulated and complete specifications written in Z, or other formal methods, into the SpecDB representation. Once the specifications have been stored in SpecDB, automated tools can reason about the specifications and thus can automatically and intelligently create test cases to test the software.

In Chapter 4 and Chapter 5 two examples of such intelligent testing applications are discussed, namely an automated constraint generator to enforce business rules at the database level of applications and a reverse engineering automated testing tool.

Chapter 4

An Automated Constraint Generator

In this chapter, one of the tools that uses SpecDB as a foundation, is introduced; namely, an automated database constraints generator that creates constraints from represented business rules in SpecDB. The main advantage of database level constraints creation is the enforcement of some of the business rules at the database level, and not only at the application level. If business rules are enforced only within the code of the application running on a database, with no database level constraints on the tables, the possibility exists for inconsistent, unreliable data. Errors in the data may result from not enforcing and coding the business rules every time a change is made to the state of the database, or when the data is altered at the database level, and not the application level.

The algorithm used to automatically generate database-level constraints is explained. The tables from SpecDB used in this process are also highlighted, together with an explanation of how the individual entities collectively describe business rules represented by the specifications expressed in SpecDB.

The use of SpecDB illustrated in this chapter focuses only on the specifications required to generate database-level constraints for the purpose of testing the validity of the data in a database. In some production databases, database administrators prefer to drop constraints in order to offer more speed during system use. This strategy may result in faster system response, but can threaten data consistency, completeness, correctness and integrity. However, if the technique explained in this chapter is used to generate constraints, data errors will be more easily managed, providing an opportunity to clean the production data. The tool described in this chapter can also be used in the data cleaning phase of building a data warehouse. The algorithm in this chapter automatically generates a script for all the database constraints representing data-level business rules. When the script is run, all the constraints are created in the database. Once the constraints are created and the database integrity is checked, all the data that does not satisfy the constraints will be identified.

The algorithm also creates another script for dropping all the constraints, if the system administrators would like to drop them again. As business rules are added to or changed in SpecDB, the algorithm in this chapter will generate the new or modified constraints. A new table is introduced into SpecDB, in adding to the original design. In addition, an extension to one of the tables in SpecDB is made. The addition and extension of these tables was necessary to store some business rules to aid the process of automatic constraint generation. The new table is Constraints. Table_Description was extended to include some required business rules for the fields. From SpecDB, Type_Restrictions, Type_LOV, Predicate and DB_Operation_Tables are also used.

The extensions of the above mentioned tables highlight the scalability of the SpecDB implementation. Other tools can be created, like the one presented in this chapter to test other sections of the software applications. If an extension of SpecDB is required to represent the specifications necessary for the completion of the new automated tool's function, a similar technique as the one used in this chapter can be adopted to extend the SpecDB representation.

4.1   The Design of Additional Tables in SpecDB

In this section, each of the new entities of the SpecDB design is defined and explained in more detail.

SpecDB is a database. The entities of the SpecDB database store the specifications of a software application. SpecDB is designed to efficiently and accurately represent the specifications necessary for unit testing database applications. In the representation described in Chapter 2, it was stated that the same concept used to create SpecDB can be used to add more tables to represent other aspects of a software application, as needed. In this chapter, a new table is added to SpecDB, in addition to the original design. Also Table_Description was extended to become Table_Fields_Description. The addition and extension of these tables was necessary to store some business rules. The storage of business rules in SpecDB provided the basis for automatic constraint generation for the purpose of data validity testing. The new table added is Constraints. From SpecDB the List_Of_Tables, Type_Restrictions, Type_LOV, Predicate and DB_Operation_Tables are also used.

Throughout the subsections of Section 4.1 are figures showing parts of the relational database schema of the design of SpecDB. The relations between the entities are shown. Also the types of the fields in each entity are shown, numeric types are identified by the symbol "789", text or character types are identified by "A", and date or time types are identified by a calendar card. Table keys are identified by a "#" symbol, and whether each field is optional or mandatory is identified by "O" or "*" respectively. There are some fields that are constrained to a list of valid values, for example Is_Unique is a Boolean value that can be only T for true or F for false. On the left of the field name, if the value of the field is constrained to a list of valid values, a relation symbol is viewed. All the tables specifications and field types are given in an SQL script in Chapter 2.

4.1.1 *Using List_of_Tables and Table_Description*

The List_of_Tables entity stores all the software/user-defined sets and tables. The Table_Description entity stores the names of the fields of the user-defined tables and sets, and their respective types.

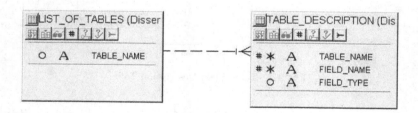

Fig. 4.1   SpecDB Original Table_Description

The SpecDB Table_Description entity is extended to become Table_Fields_ Description as shown in Figure 4.2. This table stores the fields of the user defined tables and the features of each field. The Field_ID is the Table_Name concatenated with the Field_Name, separated by an underscore symbol. Accordingly, Field_ID is a unique value in this table for every record (since a table cannot have two fields with the same name). The Field_ID is used to identify a specific field in the rest of the tables joined to Table_Fields_Description. Is_Primary_Key is a Boolean value indicating whether the field is a primary key to its entity. If Is_Primary_Key is true, then the algorithm in Section 4.2 will create a Primary Key constraint for this field on its table. Is_Unique is also a Boolean value indicating whether the field is unique for each record. If Is_Unique is true, then the algorithm in Section 4.2 will create a UNIQUE constraint for this field on its table. Optional_Mandatory can be either O for optional or M for mandatory; indicating whether the field can be left blank or not for each record. If the value of Optional_Mandatory is "M", then the algorithm in Section 4.2 will create a NOT NULL constraint for this field on its table. Has_LOV_Restrictions is also a Boolean value used to identify whether there is a list of valid values that the values of this field have to take.

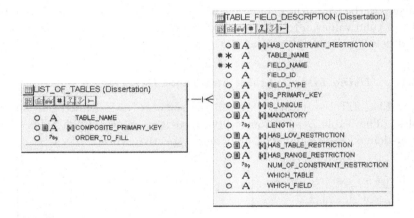

Fig. 4.2   Table_Field_Description Table

Any other value not in the valid list of values is considered invalid. If the value of Has_LOV_Restrictions is true, the algorithm in Section 4.2 creates a table for the valid values referencing the Types_LOV table. It also creates a REFERENCES constraint on this field. Has_Table_Restrictions is similar to Has_LOV_Restrictions. If the value of this field is true then Which_Table will store the table that this field references, and Which_Field will store the specific field referenced in the specified table. The algorithm in Section 4.2 will create a REFERENCES constraint if the value of Has_Table_Restrictions is true. Finally, Has_Range_Restrictions indicates whether there are other restrictions on the range of valid values for the specified field. For example, the value of "Age" for employees should be any value between 18 and 74. A list of values can be created, however, for some situations as the latter, it is best to specify a valid range of values. In this case, the algorithm in Section 4.2 creates a check constraint. For the latter example, the range is specified in the Types table, and a check constraint is created to enforce that range of valid values.

Other restrictions like specifying a range excluding one or more values are stored in the Type_Excluded_Ranges table. Also the Constraints table stores some other more complex restrictions as shown in Section 4.1.2. If there are other types of constraints, the Has_Constraint_Restrictions attribute is set to true. Multiple constraints can exist for the same field, accordingly, the number of the other constraints is specified in the Num_of_Constraint_Restrictions attribute.

4.1.2 Adding the Constraints Table to SpecDB and Using the Predicate Table

The constraints table was added to specify certain business rules that are not represented in the Table_Fields_Description table and dynamic business rules. The constraints specified in Section 4.1.1 in table Table_Fields _Description are static, meaning that the requirement that a SSN is unique is a fact that needs to be enforced on the tables, regardless of the data in the tables. However, there are some business rules that are dynamic. For example, the salary of an employee should not exceed that of his/her manager. The salary of the manager is not pre-set, i.e. not static. Also different employees have different managers. In order to enforce such business rules on the database level, the Constraints table is added to SpecDB and use other tables like DB_Operation_Tables and Predicate to store those business rules. The Constraints table is defined in Figure 4.3. The tables used by the constraint are stored in DB_Operation_Tables. The conditions and rules are stored in the Predicate table. On_Field is a Boolean value indicating whether the constraint is on a specific field. If On_Field is True, then there is a constraint for the value of the specified field. For example the employee age has to be greater than 18. In this case the Field_Name, DB_Name and Table_Name will store the name of the field (age in this example), name of the database where the table is, and the name of the table containing the specified field, respectively. Field_Name and Table_Name are required to link between the Table_Field_Description table and the Constraint table. A snapshot of the data in the SpecDB tables to represent constraint 1 is shown in Tables 4.1, 4.2 and 4.3.

Table Name	Field Name	Is Unique	Has Range Restrictions	Has Constraint Restrictions	...
Employee	Age	F	F	T	...

Constraint Name	Constraint Num	On Field	Field Name	Table Name	Constraint Clause Pred Num	...
EAgeGT18	27	T	Age	Employee	79	...

Pos Num	Component Num	Left Operand Type	Left Table Name	Left DB Field	Right Operand Type	Right Num Const	Operator ID	...
79	1	DB_Field	Employee	Age	Const	18	>=	..

$$Constraint1 : employee\_age \geq 18$$

If On_Field is false, this means that the constraint is not defined on only one field, there is a number of tables involved, or a number of fields involved in the constraint, i.e. a complex business rule or constraint. In this case the relation between the values of some fields in specified tables has to be governed by the constraint.

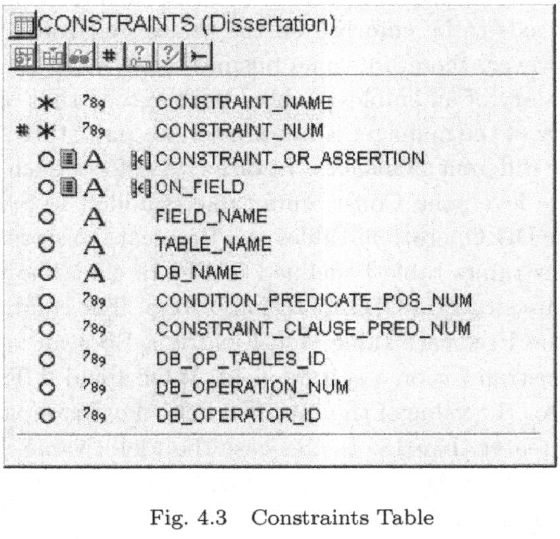

Fig. 4.3   Constraints Table

If for example, the salary of an employee should not exceed that of his/her manager, the employee entity is involved twice in this constraint, once as a manager and once as an employee. There are also other fields

Table Name	Field Name	Is Unique	Has Range Restrictions	Has Constraint Restrictions	...
Employee	Salary	F	F	T	...

Constraint Name	Constraint Num	On Field	Field Name	Table Name	Constraint Clause Pred Num	Condition Pred Pos Num	...
ESalLTMSal	28	T	Salary	Employee	80	81	...

Pos Num	Component Num	Left Oper Type	Left Tab Name	Left Tab Alias	Left DB Field	Right Oper Type	Right Tab Name	Right Tab Alias	Right DB Field	Oper ID
80	1	DB_Field	Employee	E	Salary	DB_Field	Employee	M	Salary	LE
81	1	DB_Field	Employee	E	Mgr_ID	DB_Field	Employee	M	Emp_ID	EQ

involved, the salary, manager's ID, and the employee ID. The employee table will be used twice in this case, once to reference the employee and once to reference the manager. For each case, the alias is different, for example, alias 'e' is used for employee and alias 'm' is used for manager. The rule itself is placed in the Predicate table. To illustrate how this constraint is represented in SpecDB, snapshots of the tables of SpecDB needed to represent this constraint number are shown:

$$Constraint2 : employee\_salary \leq manager\_salary$$

The constraint in Table 4.5 states that there exists a restriction on the employee salary. The constraint itself (*employeesalary* < *managersalary*) is placed in the Predicate table at Pos_Num 80. However, this is not the complete constraint. The relation between the two employees is also stated in the Predicate table at Pos_Num 81. Using the Predicate table in combination with the Constraint table allows for a wide variety of constraints and business rules that can be represented in SpecDB.

Among the complex business rules for example, the age of an employee should not exceed 125% of his manager, or the increase in an employee's salary should not exceed 125% of his previous salary. The latter example is a comparison between the value of a field before and after a modification. To identify old and new values, :old.field_name and :new.field_name are used respectively in the Left or Right_DB_Field field in the Predicate table. This complex business rule requires a constraint clause in the Predicate table with two components. The first component is the :old.salary *1.25, in this case the left operand is a database field, the right operand is a constant and the operator is the multiplication operator. The second

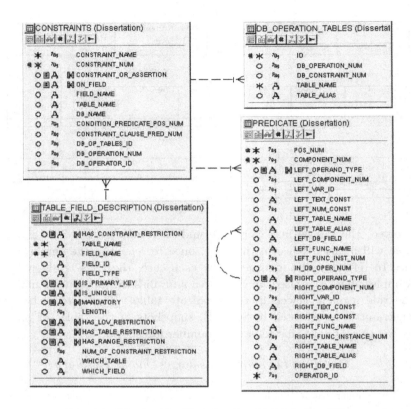

Fig. 4.4   Representing Complex Constraints in SpecDB

component is constraining :new.salary ¡= component 1, i.e. :new.salary ¡= :old.salary*1.25.

4.1.3   *Other Tables from SpecDB, Type_Excluded_Ranges and Types_LOV*

SpecDB allows for user-defined restrictions on the fields of the user-defined entities. Restrictions can be one of the following outlined below.

4.1.3.1   *Representing List of Values (LOV) Restrictions*

A list Of valid values, i.e. any variables of the specified type can only take on one of the values in the LOV. In this case, the valid values for a specified field are placed in the Types_LOV entity, each in a separate record.

4.1.3.2 *Representing Other Field Restrictions*

If a type is restricted to a set of valid values that are better specified by ranges, the ranges are specified in the Types table. If there are some values to exclude from the range of valid values, or even a range to exclude, the rules are stored in the Type_Excluded_Ranges entity to define the allowable range. For example, for a Month_Number type, valid values are between the range of 1 to 12. Hence the value of the Has_Range field in the Types table is set to True, Range_From_Int is set to 1 and Range_To_Int is set to 12. A range can also be specified, with exclusions, e.g. integers from -150 to 150 excluding 0. In this case the range -150 to 150 is specified in the Types table, and the exclusion of the value 0 is specified in the int_value in a record in the Types_Excluded_Ranges table.

4.2 Automatically Generating Database-Level Constraints: The Algorithm

After describing the tables used by the automatic constraint generator tool in Section 4.1, in this section the algorithm required to generate the different types of constraints, outlined in the previous section, is described. There are two main modules for the automated constraint generator. The first is a straight forward constraint generator with little computational intelligence, that generates static constraints on the values of the fields. This first module creates database level constraints. The second is a more complex constraint generator that handles complex business rules that cannot be expressed by simple database constraint operations, but need more complex stored procedures for triggers to enforce the business rules on the database level. Accordingly, the second module automatically generates stored procedure code for triggers, as opposed to simple database constraints as in the first module. In Section 4.2.1 simple constraints are generated. In Section 4.2.2, the algorithm for the stored procedure code generator is described.

4.2.1 *Generating Simple Static Constraints: The Algorithm*

In this section, the simple algorithm for the automatic constraint creator is described. The actual code is provided in Section 4.2. The constraints that are automatically created by the algorithm in this section handle primary key constraints, uniqueness, not-null (mandatory) constraints, simple check

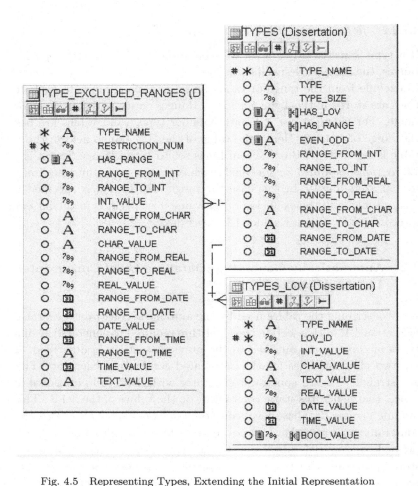

Fig. 4.5   Representing Types, Extending the Initial Representation

constraints for static field values, list of values and relation constraints.

(1) Open a file to create the SQL script to create the constraints , and another to drop all the constraints.

(2) For every constraint created and appended to the SQL constraint script file, append a drop constraint command to the drop constraints script, so as to create two SQL script files, one for creating all the constraints, and one for dropping them.

(3) For every record in the Table_Field_Description table do:

    (a) If the field is a primary key (Primary_Key = T) then append to the script file the ALTER TABLE command to add a primary key constraint on the current table (Table_Name) for the current field (Field_Name).

(b) If the field value is unique (Is_Unique = T) then append to the script file the ALTER TABLE command to add a unique constraint on the current table for the current field.

(c) If the field is mandatory (Mandatory = T) then append to the script file the ALTER TABLE command to add a Not Null constraint on the current table for the current field.

(d) If the field should have a value from a list of valid values then append to the script file the following commands:

    i. CREATE TABLE command, give it the name of the table appended to the field, for example Employee_MonthHired.

    ii. INSERT INTO the newly created table all the values from Types_LOV where the Type_Name is the current field ID.

    iii. ALTER the current table, add a foreign key constraint referencing the newly created table.

(e) If the field has table restrictions, i.e. it references another field in the same or another table, this is a referential integrity constraint that is enforced by the RDBMS once the constraint is created. Accordingly, there is no need to create triggers to check whether the constraint is satisfied upon the deletion of a record in the referenced table, since this is enforced by the RDBMS. However, we need to specify the referential integrity constraint as done below: then append to the script file, a foreign key constraint for the current field and table, referencing the specified table and field in the Which_Table and Which_Field attributes.

(f) If the field has a range of valid values then query the Types table for the range boundaries (Range_From, Range_To values).

    i. Append a check constraint to the script file to check that the value of the field is between the two boundary values obtained.

    ii. Since there can be one or more exclusions from the range of valid values, query the Type_Restrictions table for exclusions on the valid range specified.

    iii. For each exclusion, determine whether it is a single value excluded from the range, or a range to exclude.

    iv. Append to the script another check constraint to check that the value of the field is not between the specified range in the exclusion, or is not the value of the specific value to exclude.

(g) If the field has constraint restrictions then query the Constraint table for the current table and field. If a constraint exists in the Constraint table where the On_Field attribute is set to true then using the Constraint_Clause_Pre_Num attribute for that constraint, query the Predicate table for the record(s) where Pos_Num = Constraint_Clause_Pre_Num. If there is only one record (i.e. on component) in the Predicate table then append to the script file a check constraint on the current field and table to satisfy the predicate, i.e. left operand followed by the operator ID, followed by the right operand. Else if there are multiple components:

    i. Use a temporary empty text variable or database text fields to build the constraints from the predicate components.

    ii. Starting with component 1, append to the text variable, or DB field the left operand, operator ID, right operand.

    iii. Repeat with the following components until all components are built.

    iv. Append all components to one another with the correct operators into one predicate.

    v. Append to the script file the check constraint for the current field and table with the built predicate.

(4) Commit the changes to the open files and close the two script files. Case 3.g is for representing a constraint similar to constraint 1 discussed in Section 4.1.2. In this case, a constraint can have a simple one component predicate e.g. Age $>$=18, or it can have a complex multiple component predicate, e.g. Annual Salary + Annual Bonus $<=$ 125% Annual Salary. For the latter example, the constraint can be on the Annual Bonus field. Three components in the Predicate table represent this constraint.

    (a) Annual Salary * 1.25 ... Component 1.

    (b) Annual Salary + Annual Bonus ... Component 2.

    (c) Component 2 $<=$ Component 1.

The algorithm specifies the use of a temporary database table containing one text field or a text variable to form the complete constraint predicate statement before appending the constraint to the script file. The PL/SQL code for the constraints generator can be found in Appendix 7.4.

4.2.2 Enforcing Complex Dynamic Business Rules

More complex business rules involving dynamic checking of the data cannot be enforced using simple database constraints as those described in Section 4.2.1. Complex operations similar to constraint 2 discussed in Section 4.1.2 can only be created using database assertions or stored procedure code for a database level trigger. Accordingly, in this section we look at how the automated constraint generator automatically generates procedure code for triggers to enforce those complex dynamic business rules.

The algorithm is better understood following an example. Constraint 2 is used for this purpose in this section:

$$Constraint2 : employee\_salary \leq manager\_salary$$

There are two main criteria for specifying such a constraint. First, the constraint that needs to be satisfied, in this example the constraint that needs to be satisfied is:

$$e.salary \leq m.salary$$

The second criterion is the relationship(s) between the entities that are involved in this constraint. For Constraint 2, the relationship is the specification: 'Employee's Manager', translated into PL/SQL code renders a WHERE clause: e.manager_id = m.employee_id. The first criterion, the constraint that needs to be satisfied, can be a simple one component constraint, or it can be more complex, and more than one component is needed in the Predicate table to represent it. The same applies for the second criterion, the relationship between the entities. Again there can be none, one or more components in the predicate to specify the relationship between the entities involved in the constraint. The data needed for the automatic creation of procedure code to enforce complex dynamic constraints is represented in the design of the SpecDB tables discussed in Section 4.1. From Tables 4.4, 4.5 and 4.6, the algorithm in Section 4.2.2.1 automatically generates the procedure code for a trigger to enforce Constraint 2.

The full algorithm to include all cases is decision intensive, one of the cases is given here to illustrate the idea. However, there are other paths in the algorithm where the operands are not database fields, but constants for example. Handling components in the predicate table with constant operands is simpler than handling database field operands, as shown in Section 4.2.2.1. An example of handling constants was shown in Section 4.2.1.

There are also paths through the algorithm where the key to the table being modified is a composite key, meaning that there are multiple fields that
collectively form the database table key. Before the below algorithm is executed, the database table key is identified from the Table_Field_Description
table. In the algorithm shown in Section 4.2.2, a simple database table
key is handled. The same technique can be used to handle composite keys.
Finally, there are cases in the algorithm involving multiple components for
a predicate. The technique used in the algorithm in Subsection 4.2.1 to
handle complex multiple component predicates is also used here for the
same purpose.

4.2.2.1 *Complex Business Rules Generation Algorithm*

The algorithm described in this section is a part of the whole decision
intensive algorithm to automatically create procedure code for database
level triggers, as discussed in the previous section. A file is created for the
procedure code of the trigger, every statement in the algorithm is appended
to the file. Also the value of the fields involved in the update operation that
triggers the execution of the procedure code created by the algorithm below,
are referred to as 'New' values, in the 'Current' record.

(1) Open a file to create the procedure code to create the constraint, and
    another to drop all the triggers created.
(2) Determine the field name that is being updated from the Constraint
    table.
(3) Create   a
    trigger to be executed before update on Constraint.Table_Name and
    Constraint.Field_Name.
(4) Determine the type of the field stored in right_DB_field in the Predicate
    table.
(5) Create a variable of the type identified in step 4, append the alias to
    the right_DB_field and use the result as the name of the variable (e.g.
    m_salary).
(6) Determine the key to the table being modified.
(7) Query the Constraint table to identify the pos_num for the predicate
    representing the constraint clause (a), and that representing the condi-
    tion (b).
(8) Append a SELECT statement as follows:

    (a) SELECT right_table_alias.right_db_field INTO variable (created in step 5) (the data used here is from (a) in step 7).

(b) FROM   right_table_name   right_table_alias, left_table_name left_table_alias (the data used here is from (a) in step 7).

(c) WHERE
left_table_alias.Key = CurrentKey AND left_table_alias.left_db_field operator_id right_table_alias.right_db_field (the data used here is from (b) in step 7).

(9) From step 2, the field being updated has been determined, now determine the new value of that field (Current_Field and New_Field_Value).

(10) Append to the file the decision statement: If not (New_Field_Value operator_id right_table_alias.right_db_field) (the data used here is from (a) in step 7) then raise and handle exception.

From Constraint 2 the following SELECT statement is created from the above algorithm:

```
SELECT m.salary INTO m_salary
FROM employee m, employee e
WHERE e.employee_id = Current_Emp_ID AND
      m.employee_id = e.manager_id;

if not(Current_Salary <= m_salary) then raise_exception
```

4.3 Conclusion

In this chapter, an automated tool to generate business rules is described. It is also shown how to enforce them on the database level by automatically creating constrains and assertions using the data stored in SpecDB. The design of a new table, Constraints is explained, and an extension on the design of the Table_Description table of SpecDB, renamed Table_Fields_Description is shown. Those modifications were made in order to allow for the representation of business rules. The automated tool introduced in this chapter creates a script to create all the constraints and another one for dropping them. System administrators can choose to run production systems with fewer constraints or no constraints at all in order to enhance the speed of production systems. They can run the tool described in this chapter to create and automatically test the validity of the data during certain times of the week to maintain the accuracy, completeness and integrity of the system. Through the use of SpecDB, automated tools can become more intelligent in generating and testing programs, based on an understanding of what the software under construction is supposed to do, by design. The design of SpecDB allows for future enhancement and additions to accommodate more programming constructs or requirements, as they evolve in the future.

Chapter 5

A Reverse Engineering Testing Tool

A lot of testers and available automated testing tools decide that a software application has passed a specific test case, if the expected outputs match the actual outputs, based on black box testing. This is, however not an absolute proof of the quality of the software for that test scenario. The fact that the software application can commit implicit errors that are not directly reflected in the output values still exists. For example, in a database application, if a social security number of an employee is given as an input, and the system outputs the correct required information about that employee, as specified for an operation; this is not enough for a tester or a testing tool to decide that the software passed for this operation. There is a possibility that some other records in the database have been mistakenly deleted, or modified, during or after the output is displayed. In another example from a system security perspective, an e-commerce tool takes credit card information, and customer confidential data, carries out the transaction, sends a sales order to the correct department, and accomplishes all the tasks to complete the sale transaction. The system could also send the credit card information to an external non-secure site. Such a security leak cannot be identified by black box testing techniques.

The motivation to think of a testing tool solution that would spot such implicit errors of commission was aroused from the above discussion where the software application carries out operations that are not part of the specifications. White box testing techniques, where the actual implementation code is examined is a good basis for such a solution. If a testing tool can trace an implemented program and generate the specifications of the actual implementation, and present it in a non-technical form that can be understood by testers and end users, it will then be very easy to spot such errors of commission. In this chapter the operation of this tool is explained. it is called a reverse engineering testing tool, or RE tool, since it generates the specification of the implemented software from its implementation. Such specifications can then be fed into SpecDB and automatically compared to another set of specifications, the required specifications of that software, i.e. compare actual to specified behavior. The determination whether the implemented software matches the required specifications can then be easily made, with the guarantee that no errors of commission have been unnoticed.

In this chapter, a reverse engineering approach is used to log the actual execution of a program, due to some input, using white box testing techniques. Once the program actions have been logged, they are presented in a way that non-technical testers and other members of the software development team can understand. The actual outcome can then be easily matched and compared to the expected output based on the program specifications and oracles that determine what the expected program output should be, based on the specified inputs. The decision-to-decision path (also known as DD-paths) white box testing technique, is modified and used to implement this tool, but for a different purpose; namely to determine the actual program behavior given specific input, and not as a coverage matrix to determine which parts of the software are covered by some test cases, or as a guide to generate test cases to cover all DD-paths in a program. Instead, the technique for DD-paths testing is used to determine actual program behavior.

5.1   Technique

The main idea of the reverse engineering (RE) testing tool is to follow the program execution and report the inputs it receives, the actual outputs it generates, and the process it follows. These three perspectives, input, actual output and the process are all represented in a simple format for non-programmers to be able to read and interpret how the program executed and whether there are specifications that it did not execute. This outcome will aid in the process of identifying and isolating faults. Faults are identified when the tester (or automated tool, using SpecDB) compares the expected output to the actual output. Faults are isolated by following the process execution behavior provided as an outcome of the approach explained in this chapter and comparing it to the specifications. For the purpose of the RE tool described in this chapter, as an outcome of the tool, for a specific test case on specified inputs, the tool will show how the executed portion of the software got its inputs, and how it generated its outputs and finally it presents all the outputs of the executed software. Those three outcomes of the tool (inputs, computation, and output) are presented in an easily understandable format described in the following sub-sections.

Triggers are programmed units that execute if a pre-specified event occurs. For example, a database trigger can be implemented to check the constraint on a table when a certain field is modified. GUI triggers execute when a certain event happens in the GUI, for example, a button click, a menu item click, a modification to the value of a field, etc. In order to make the RE tool more effective, it is designed not only to test the specified unit without noting changes that occur due to some triggers that execute as a result of the state changes, or output results of the unit being tested, but also to consider all the modifications that happen to the system, including those of triggers. A domino effect might result from following triggers, since a state change or an output of one trigger may trigger another to execute. The specifications of the RE testing tool include that it should query all triggers, and note the possible domino effect, displaying as part of its own output all the system and state changes as well as all the outputs that result from the execution of the unit being tested.

5.1.1 *Input and Pre-Condition Classification and Categorization*

The first outcome of the RE testing tool is identifying all the inputs used by the software in its execution of the test case. Inputs can be both explicit and implicit. Explicit inputs are those given to the software using a GUI component or a command line. Implicit inputs are values that the software implicitly acquires during execution, for example when the application queries a database table for the value of a specific field in a specific record. Such values are used in the computations and the outcome of the software is affected by those implicitly acquired inputs too. The format of how the RE tool identifies what kinds of input the program received, are described in this section.

Database applications are dynamic by nature depending on the input state of the database. This means that when a test case is executed on a database application with the same input values several times, the output can be different for each test case, if the data stored in the database changed, and yet, it can be decided that the application passed that test.

Database applications are thus unlike a function, which depends soley on explicit parameters. For example, a calculator application is a function, each time you enter 4 * 7 to a calculator, it should output the same value, 28. However, each time you query the social security database for the number of living citizens, it might give you a different number each time. It is thus important to testers or automated testing tools on such data-centric applications to know the inputs, both explicit and implicit, to be able to judge whether the software being tested has passed a test case.

Inputs can be presented to a program in many forms. Among the forms are GUI objects, data from a database or files. It is assumed that other forms of input like barcode readers can be considered as data from GUI components, since a barcode reader replaces manual keyboard entry of data in a text field for example. For inputs from the database, a specific database table, field and row identifier are required to determine a single value, as an input. For example the Employee table, has a salary field. There are as many values for this field as there are records, i.e. employees in this table. An employee identifier is also required to specify a single input salary value. Once the value is extracted from the database, it can be stored in a variable to be used in the computation process. For queries that return multiple values of the same field, i.e. no record identifier is chosen, during run time the results of those queries are placed in what Oracle and other databases call a cursor. Databases use cursors to handle computations on input values, one record at a time. The end result is that only one value per field is handled for computation at a time in a loop, i.e. one record at a time. Accordingly, the above assumption that one field and one row identifier provide a single input is also valid in the situation of using a cursor in a loop for computation involving a query that returns multiple rows. In this case, in each iteration through the loop, the value of the input variable will change because the row identifier changes.

Inputs to a program can also be presented through a GUI component, like a text box, choice from a drop down menu or a value from a combo box containing a list of possible values, etc. If input is taken from a GUI component, several identifiers are needed to specify which GUI component the input is taken from, unless every component has a unique identification number or name. For example, if an employee first name is entered in a text field, the window identifier where that text field is located should be given as part of the component definition, together with an identifier of that specific text field, since there can be many text fields in the same window, a text field for last name, for example. The type of the component should

Category	Classification				
Database(DB)	Table	Field	Row #	Value	Into Variable
GUI	Window	Component	Type	Value	Into Variable
File	Name/Location	Type	Into Variable		

Input #	Category	Table	Field	Row #	Value	Into Variable
5	DB	Employee	Salary	14	56000	Salary
6	DB	Employee	Emp_LName	14	Gibson	Last_Name

also be specified for computation purposes; in this example the component is a text field.

If input is taken from a file, the file name and location are specified. The type of file is necessary for computation purposes, for example text or binary file, etc. In Table 5.1, the specifications for the different forms of input are summarized, i.e. how each input to the software being tested is identified as an outcome of the RE testing tool. The RE testing tool, determines all the kinds of inputs given or acquired by a software application during execution of a scenario (test case). Each input is given a number, and described depending on its type, i.e. obtained from a database (DB), GUI or file. Types of GUI components can include "Button", "Text", "Radio" for a Radio Button and "Check" for a check box, etc.

In the next subsections examples of the first set of outputs provided by the RE testing tool are given, namely identified inputs of a software execution. The examples will show how the tool communicates the implicit and explicit inputs to a software in a non-technical form, following the categorization in Table 5.1.

5.1.1.1   Examples of Inputs from the Database Displayed by the RE Testing Tool

Table 5.1 summarizes how the RE testing tool represents a database input to the program. In Table 5.2 are some examples of database inputs. Table 5.2 specifies that the fifth input given to/acquired by the program is from the database. Specifically from the salary field in the employee table, at row 14. The value of Salary at row 14 when read was 56000 and this value was assigned to the internal variable "Salary".

Input #	Category	Window	Component	Type	Value	Into Variable
1	GUI	Employee_Entry	Save	Button	Click	
2	GUI	Employee_Entry	SSN	TextBox	123456789	E_SSN
3	GUI	Employee_Entry	Female	Radio	Y	Female
4	GUI	Employee_Entry	Speak_Spanish	Check	N	Spanish

5.1.1.2   Examples of Inputs from the GUI Displayed by the RE Testing Tool

Table 5.1 also summarizes how the RE testing tool represents a GUI input to the program. In table 5.3 show some examples of how the testing tool identifies GUI inputs.

Table 5.3 indicates that the first input to the program was a click on the Save button in the Employee_Entry window. This might have triggered the operation or computation that is taking place. The Second input was the social security number, take from the SSN field, the value taken was 123456789 and it was saved in the E_SSN variable to be used in the computation. The outcome of the RE tool is represented in this simple form to aid the process of specification comparison and isolating the sources of errors. The outcome can also be stored directly in SpecDB as will be shown in Section 5.3.

5.1.2   Output and Post-Condition Classification and Categorization

The second and most important outcome of the RE testing tool is the full set of outputs, and system state modifications that resulted from the software during execution of a scenario or test case. There is a great advantage to displaying to the development and quality assurance team all the outputs, and changes to the system state, both explicit and implicit so that hidden errors of commission and security flaws can be identified easily. Similar to the inputs, there are explicit and implicit outputs or post-conditions of the execution of the software application being tested. Explicit outputs are those that can be observed by functional black box testing techniques. Examples are the display of data in GUI components, a printed report, etc. However, there also can be some other implicit outputs like exporting information to another non-secure location. This kind of implicit output that cannot be spotted by black box testing techniques, and the post conditions of the software execution of a particular test case scenario are all displayed as an outcome of the RE testing tool.

In the remaining part of this section, an explanation is given to show how the RE tool describes different forms of system outputs and post conditions including system state modifications, and both implicit and explicit system outputs. In table 5.4, the RE testing tool description of twelve different forms of outputs or post conditions is given, namely, database (DB), GUI, file, sequence, message, report, variable, role, privilege, commit, savepoint and finally rollback . A modification in the database state could either be a deleted or updated row, or a newly inserted row. If the change in the database is a deletion or insertion, only the row number is presented. If the database change resulted from a row or field in a row that has been updated, the row number, field name, value stored and from which variable the value was taken are all presented by the RE testing tool. In case a database sequence has been modified, the new value is displayed by the RE tool. Also if a message is displayed, the message text is given. If a report is one of the outputs of the software, the RE testing tool displays the destination of the report, whether it is to a printer, file, or on the screen. If a global (persistent) variable has been modified after an operation is completed, the RE testing tool also displays its new value. This can help isolate the source of some errors depending on calculations based on global variables. If the software application created or changed a database user's role or gave or revoked database privileges, the RE tool also indicates the role or privilege, specified by the displayed identifier, has been granted or revoked to the specified user. In database applications if modifications have been made but not committed (saved), and a rollback operation has been triggered, the data in the database could become corrupt, or the operation that triggered the rollback command, would reverse the operation of another unit previously executed.

System testers can have a hard time isolating where the resulting database state got its values, after testing a previous operation, and affirming that it operates correctly. When the RE testing tool displays all the operations of the system and all the implicit outputs, the software development and testing team can immediately recognize that the execution of the scenario being tested committed the changes to the database or not, and accordingly, spot errors of omission, namely, forgetting to commit the changes to the database, so that the executed operations will not be rolled back accidentally.

The RE testing tool also displays the operation of the unit or software being tested, displaying step-by-step how the outputs got their value. In Section 5.3, an example of the execution of the RE testing tool is given and also how it displays the inputs, outputs and operation of the software being

Category	Classification					
Database(DB)	Table	Change	Field	Row #	Value	From Variable
GUI	Window	Component	Type	Value	From variable	
Sequence	Name	New Value				
Report	Name	Destination				
File	Name/Location	Type	Change			
Variable	Name	Value	From Variable			
Role	User	Role Identifier	Change			
Privilege	User	Privilege Identifier	Change			
Message	Text					
Commit	Database Name					
Savepoint	Database Name		Savepoint Name			
Rollback	Database Name		Savepoint Name			

Output #	Category	Window	Component	Type	Value	From Variable
1	GUI	Employee_Date	Emp_Picture	Image		C:/images/gib1.jpg

tested.

5.1.2.1   Examples of Outputs Displayed by the RE Testing Tool

As described in the previous section, the RE testing tool displays several types of outputs or post conditions resulting from the execution of a scenario. In table 5.5 an example of how the RE testing tool displays a GUI output of the application being tested is shown. In this case, the RE testing tool identifies that the application running the scenario, has displayed the employee picture from a file located in C:/images/gib1.jpg, in the Emp_Picture GUI component on the Employee_Data window or screen.

5.2   Case Study

After discussing the benefits and motivation for the creation of the RE testing tool, and explaining how it displays different forms of inputs and outputs of the software being tested, in this section a case study to show the functionality of the tool is described. In order to demonstrate the RE testing tool's ability to spot errors, three examples are shown in Section 5.3. The examples are all different scenarios for the same case study or program. The program is an implementation of the vacation salesman commission problem, adopted from [5]. In Section 5.2.1 the specifications, or problem statement is explained. In Section 5.2.2 the implementation of the program is shown, to allow for white box testing, and to show how the RE testing tool traces the computation of the software and how it displays its outputs. As noted previously in this chapter, a technique similar to DD-paths testing is adopted by the RE testing tool. Accordingly, the DD-paths and the DD-path graph are shown in Section 5.2.3 and 5.2.4. In Section

5.3 three different examples using the problem described in Section 5.2 are given, to show how the RE testing tool can spot help spot errors.

5.2.1 *Problem Statement for the Vacation Salesman Commission Program*

The problem of the vacation salesman commission is adopted from [5]. Since it a reasonably small problem, it serves to show the functionality of the RE testing tool. In this section the problem statement of the vacation salesman commission program is given. A vacation planner sells cruises, airline tickets and books shore excursions, to a certain destination. A one week Cruise costs $450 per person, airline tickets from his city to the cruise port of departure cost $300 per person and the shore excursions and activities package costs $250 per person. The vacation planner hires a salesmen. A salesperson has to sell at least one cruise ticket, one airline ticket and one shore activities package per month to stay in the job! The vacation planner has a maximum of 70 cruise tickets to book per month, 80 airline tickets and 90 shore excursions packages. Occasionally, the salesman should report to his employer what vacation items he sold and how many of each. The employer enters this in the database, under the salesman name, and the date of sale. At the end of the month, the employer computes the salesman's commission, by simply clicking on a button. The salesman gets a 10% on all sales up to and including $10,000, 15% on the next $8,000 and 20% on any sales in excess of $18,000. An on-screen report is issued for the salesman showing the total number of cruises, airline tickets and shore excursions sold. The report also shows the salesperson's total dollar sales and his commission.

5.2.2 *Implementation*

In order for the RE testing tool to accomplish its task of identifying faults of commission or omission, a technique similar to DD-path testing is used in the RE testing tool to determine how the software application being tested computed the resulting output, and what output values it gave, and how it modified the system state, if it did. A detailed explanation of DD-Path testing can be found in [5]. White box testing techniques require the actual implementation code. Accordingly, the implementation of the vacation salesman commission program is presented in this section. The implementation will be used in the examples in Section 5.3, and also to

generate the DD-paths for this program. Each line of the implementation
of the Vacation Salesman Commission program shown below, is numbered,
to help in the DD-paths generation in the subsequent sections and the
examples following the case study. The code is written in PL/SQL, it is
triggered when the calculate commission GUI button is clicked, to calculate
the commission of the given salesman, in the given month and year. The
salesman ID, month and year are three text fields in the GUI window, the
implementation uses these values as inputs. After querying the database
for all the sales of the month, and calculating the commission, the program
displays total cruises, air tickets and activities sold, total sales, and the
commission in five text fields on the same GUI window. Accordingly, the
system takes three GUI inputs, other information from the database, and
displays five outputs. All exception handling and error checking code has
been taken out for simplicity. The code given is executed after checking that
all input is given and in the correct format. Below is the implementation:

```
1   -- Salesman Commission Calculate Button
        WHEN-BUTTON-PRESSED Trigger
2   Declare cruise, air_ticket, activity integer;
3   Cruise_Price, Air_Price, Activity_Price,
    Total_Sales, commission real;
4   Number_Cruises_Sold, Number_Air_Tickets_Sold,
    Number_Activities_Sold integer;
5   Dollar_Cruise_sales, Dollar_Air_sales,
    Dollar_Activity_sales real;

6   cursor all_month_sales is
7       select s_cruise,
                s_air_ticket,
                s_activity from sales
8       where s_salesman = :salesman and
                s_month = :month and
                s_year = :year;

9   Begin
10      Cruise_Price = 450;
11      Air_Price = 300;
12      Activity_Price = 250;
13      Number_Cruises_Sold=0;
```

```
14      Number_Air_Tickets_Sold=0;
15      Number_Activities_Sold=0;

16      open all_month_sales;
17      loop
18          fetch all_month_sales into cruise,
                                        air_ticket,
                                        activity;
19          exit when all_month_sales %notfound;
20          Number_Cruises_Sold = Number_Cruises_Sold +
                                   cruise;
21          Number_Air_Tickets_Sold = Number_Air_Tickets_Sold +
                                        air_ticket;
22          Number_Activities_Sold = Number_Activities_Sold +
                                       activity;
23      end loop
24      close all_month_sales;

25      :Total_Cruises_Sold := Number_Cruises_Sold;
26      :Total_Air_Ticket_Sold := Number_Air_Tickets_Sold;
27      :Total_Activities_Sold := Number_Activities_Sold;
28      Dollar_Cruise_sales = Cruise_Price *
                               Number_Cruises_Sold;
29      Dollar_Air_sales = Air_Price *
                            Number_Air_Tickets_Sold;
30      Dollar_Activity_sales = Activity_Price *
                                 Number_Activities_Sold;
31      Total_Sales = Dollar_Cruise_sales +
                       Dollar_Air_sales +
                       Dollar_Activity_sales;
32      :Total_Sales = Total_Sales;

33      if  Total_Sales > 18000
34      then
35          commission = 0.10 * 10000;
36          commission = commission +
                          (0.15 * 8000);
37          commission = commission +
                          (0.20 *
```

```
                              (Total_Sales - 18000))
38             else if Total_Sales > 10000
39                then
40                    commission = 0.10 *
                                   10000;
41                    commission = commission +
                                   0.15 *
                                   (Total_Sales - 10000)
42                    else commission = 0.10 *
                                       Total_Sales;
43                 end if;
44        end if;
45        :Salesman_Commission := commission;
46 end;
```

5.2.3 *Program Graph for the Salesman Commission Problem*

From the implementation in Section 5.2.2, one could come up with the program graph shown in Figure 5.1. A detailed description of how to obtain this program graph from the implementation is available in [5]. The program graph is stage needed to be created to complete extract the program's DD-paths. The DD-paths are used to trace program behavior on specific input.

5.2.4 *DD-Path Graph for the Salesman Commission Problem*

From the program graph in Figure 5.1 in Section 5.2.3, the DD-Paths graph in Figure 5.2 can be automatically derived, from the DD-paths in table 5.6. Jorgensen in [5] explains how to derive the DD-Paths for an implementation. The DD-paths graph and the DD-paths are used in the RE tool. Figure 5.3 shows a higher level module diagram for the RE tool. The unit implementation and test case scenario are taken as inputs. The different process of the RE tool are shown, followed by the RE tool output, representing the inputs, and outputs to and from the system, both implicit and explicit, and also the process that the software executed to reach the results shown.

In the following Section 5.3 three examples to test the vacation salesman

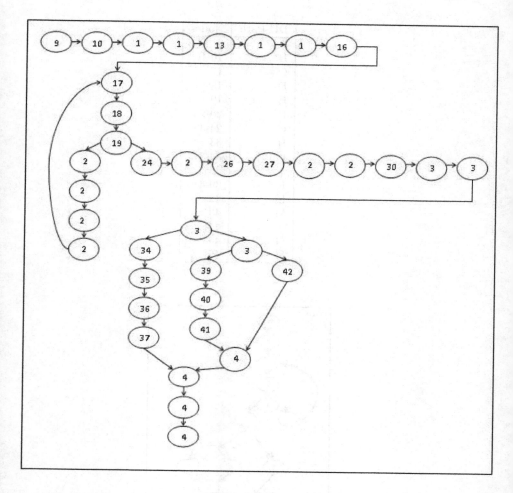

Fig. 5.1   Program Graph for the Vacation Salesman Commission Problem

commission program is demonstrated, based on the implementation, DD-paths, and DD-paths graph generated in this section.

5.3   Examples

From the case study described in Section 5.2, three examples are discussed in the following subsections. Each of the examples demonstrates a different state showing how the RE tool can spot errors of commission that are not

DD-Path	Nodes
A	9
B	10-16
C	17
D	18
E	19
F	20-23
G	24-32
H	33
I	34-37
J	38
K	39-41
L	42
M	43
N	44
O	45
P	46

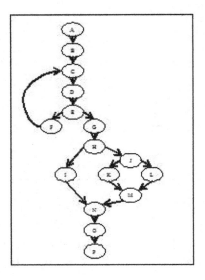

Fig. 5.2   DD-Path Graph for the Vacation Salesman Commission Problem

identified by other tools, discussed in the literature review in Chapter 1.

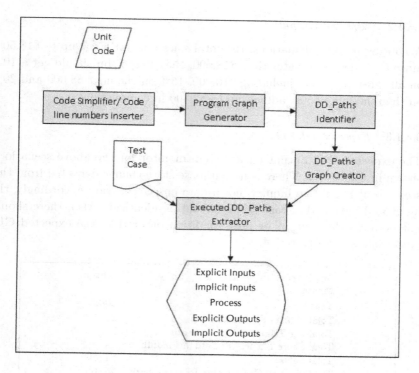

Fig. 5.3 RE Tool Model Diagram

5.3.1 *Test Case 1*

5.3.1.1 *Scenario*

This example demonstrates a simple scenario for the purpose of showing how the RE tool technique works. The salesman reports twice to his employer during a certain month. The first time he reports that he sold 8 cruise vacations, 12 airline tickets and 10 shore excursion packages. The second time he reports that he sold 12 cruise vacations, 8 airline tickets and 4 shore excursion packages. The employer enters the data as sent each time. Therefore the database will have two records in the sales table for that specific salesman for that specific month. At the end of the month the employer enters the employee name, and the required month and year, then hits the Calculate Commission Button to view the report on the screen.

5.3.1.2   *Expected Output*

According to the calculations, the total sales amount sums up to $18,500. Since the sales are greater than $18,000, the salesperson should get a 10% on all sales up to and including $10,000, 15% on the next $8,000 and 20% on the remaining $500, adding up to $2,300 in commission.

5.3.1.3   *Expected GUI Output*

The expected GUI output on a system monitor for the above scenario is shown in Figure 5.4. There is no database state change expected from this scenario, it is only an inquiry, not a data entry scenario. Accordingly, the input and output database states should be identical. Also, there should not be any other output or post condition other than the expected GUI output shown in Figure 5.4.

Employee	Gibson
Month	9
Year	2003
Total Cruises Sold this month	20
Total Air Tickets Sold this month	20
Total Shore Excursions Sold this month	14
Total Sales	$18500
The Salesman Commission for this month	$2300

Fig. 5.4   The Expected GUI Output for Test Case 1

5.3.1.4   *The RE Tool Display of the Inputs for Test Case 1*

When the RE testing tool runs on the vacation salesman commission program implementation and for the specified scenario in test case 1, it will produce three main outcomes. First, it will give all the inputs, implicit and explicit, those inputs are shown in this section. It will also show which DD-paths have been traversed and the operation of the software, and how it computed each value of the outputs. The third outcome of the RE testing tool is a list of all the outputs produced by the software during execution of the test case. The operation and outputs are shown in the next section.

Tables 5.7 and 5.8 show how the RE testing tool displays the inputs of the vacation salesman commission program, given from the GUI and

Input #	Category	Table	Field	Row #	Value	Into Variable
1	DB	Sales	s_cruise	15	8	cruise
2	DB	Sales	s_air_ticket	15	12	air_ticket
3	DB	Sales	s_activity	15	10	activity
4	DB	Sales	s_cruise	17	12	cruise
5	DB	Sales	s_air_ticket	17	8	air_ticket
6	DB	Sales	s_activity	17	4	activity

Input #	Category	Table	Field	Row #	Value	Into Variable
7	GUI	Commission_Calc	Salesman	TextBox	Gibson	NULL
8	GUI	Commission_Calc	Month	TextBox	9	NULL
9	GUI	Commission_Calc	Year	TextBox	2003	NULL

acquired from the database. These two tables are the first outcomes of the RE testing tool for test case 1.

5.3.1.5 Process and Actual Output

The RE testing tool also displays two other outcomes, the outputs of the program after executing the test case, and how the values of the output where obtained. The RE testing tool displays the following for test case 1:

At execution time the program traversed the following nodes in the DD-path graph: ABCDEF CDEF CDE GHI NOP. The following statements where executed: 9, 10-16, 17, 18, 19, 20-23, 17, 18, 19, 20-23 ,17 ,18 ,19, 24-32, 33, 34-37,44, 45, 46.

The RE testing tool also displays the outputs and the calculations. The development or testing team can refer to the calculations if errors where found in the output. The calculations or the operation of the program unit resulting in the values of the outputs will thus aid in fault isolation. The following is what the RE testing tool displays for test case 1, outputs are tabulated.

```
Cruise_Price = 450
Air_Price = 300
Activity_Price = 250
Number_Cruises_Sold=0
Number_Air_Tickets_Sold=0
Number_Activities_Sold=0
Number_Cruises_Sold = 0 + 8
Number_Air_Tickets_Sold = 0 + 12
```

Output #	Category	Window	Component	Type	Value	From Variable
1	GUI	CommissionCalc	Tot_Cruises_Sold	Textbox	20	Num_Cruises_Sold
2	GUI	CommissionCalc	Tot_Air_Tkts_Sold	Textbox	20	Num_Air_Tkts_Sold
3	GUI	CommissionCalc	Tot_Activities_Sold	Textbox	14	Num_Activities_Sol

Output #	Category	Window	Component	Type	Value	From Variable
4	GUI	CommissionCalc	Total_Sales	TextBox	18500	Total_Sales

Output #	Category	Window	Component	Type	Value	From Variable
5	GUI	CommissionCalc	Salesman_Commission	TextBox	2300	Commission

```
Number_Activities_Sold = 0 + 10
Number_Cruises_Sold = 8 + 12
Number_Air_Tickets_Sold = 12 + 8
Number_Activities_Sold = 10 + 4


Dollar_Cruise_sales = 450 * 20
Dollar_Air_sales = 300 * 20
Dollar_Activity_sales = 250 * 14
Total_Sales = 9000 + 6000 + 3500


Since Total_Sales > 18000
  Then commission = 0.10 * 10000;
       commission = 1000 + (0.15 * 8000);
       commission = 2200 + (0.20 * (18500 - 18000))
```

The RE testing tool thus displays and confirms that there were only five outputs to the program for test case 1. No other database outputs, or state changes have been executed by the unit.

5.3.1.6   *Comparing Expected to Actual Output*

In this example it is clear that the expected output in Figure 5.4 matches the actual output in tables 5.9, 5.10 and 5.11, and therefore, there is little or no need to look at the execution process. The specifications of the implemented program can be fed in SpecDB, together with the actual outputs and the calculation formulas. Those implementation specifications can then be compared to the required specifications and an automated tool can make the correct decision whether the implementation matches the required specifications or not. Also if there are discrepancies between the actual and the required specifications, the outcome of the RE tool helps in isolating the

Output #	Category	Window	Component	Type	Value	From Variable
1	GUI	Sale_Report	Total_Cruises_Sold	Textbox	20	Num_Cruises_Sold
2	GUI	Sale_Report	Tot_Air_Ticket_Sold	Textbox	20	Num_Air_Tickets_Sold
3	GUI	Sale_Report	Tot_Activities_Sold	Textbox	14	Num_Activities_Sold
4	GUI	Sale_Report	Tot_Sales	Textbox	18500	Tot_Sales
5	GUI	Sale_Report	Salesman_Comm	Textbox	2300	Commission

source of the errors. The RE tool numbers the implementation statements, accordingly, it indicates which statement caused the difference between the required specifications and actual the actual code.

SpecDB is used to host the specifications resulting from the RE testing tool. An example is given in this section to clarify how an automatic comparison between the required specifications and those resulting from the RE testing tool, i.e. the specifications of the implemented code, can be made using SpecDB. A snapshot of a portion of the specifications will be used to demonstrate the principal technique. The specifications of the output for the vacation salesman commission problem are then represented in SpecDB. In test case 1, both the required and actual specifications match. However, to show how a discrepancy can be automatically identified, a software error is injected. Software fault injection was researched in [102]. The software error injected is not in the code, but in the GUI, assuming that in the vacation salesman commission program, after the employer entered the correct salesman identification, month and date, and pressed the calculate commission button, the program did not display anything in the five fields. The same program unit implementation in test case 1 is used again. By viewing the result of the RE testing tool in the previous sections, it is noticed that the program unit executed correctly. For the purpose of comparing the actual specifications to the faulty ones, the scenario is changed, so that the output is not displayed in the correct window, it is displayed in another screen. From the user's point of view, no output is generated when he clicks the button, although, the output is computed correctly, is it displayed in the wrong location in the GUI. Given the required specifications and the observed specification of the fault injected test case 1, resulting from the RE tool, it is a trivial task to note where the error occurred. After injecting this fault the results in tables 5.9, 5.10 and 5.11 should be slightly different, the window name should change and/or the component name should change. The RE testing tool outputs in this case, display the values in table 5.12.

For the purpose of this example, only a snapshot of the original required specifications represented in SpecDB is shown in the Var, Dataflow and the Input_Output_Definition tables.

Pos_Num	Suc_Num	Pos_Type	...
16	17	Output	...
17	18	Output	...
18	19	Output	...
23	24	Output	...
34	35	Output	...
...	...	...	...

Var_ID	Var_Name	Type	Is_Input	Is_Output	Is_GUI_Object	GUI_Window
7	Total_Sales	Real	F	F	F	
8	Commission	Real	F	F	F	
9	Number_Cruises_Sold	Int	T	F	F	
10	Number_Air_Tickets_Sold	Int	T	F	F	
11	Number_Activities_Sold	Int	T	F	F	
18	Total_Cruises_Sold	Int	F	T	T	CommissionCalc
19	Total_Air_Sold	Int	F	T	T	CommissionCalc
20	Total_Activities_Sold	Int	F	T	T	CommissionCalc
21	Total_Sales	Real	F	T	T	CommissionCalc
22	Salesman_Commission	Real	F	T	T	CommissionCalc

Tables 5.13, 5.14 and 5.15 contain a snapshot of the data in Var, Dataflow and the Input_Output_Definition tables in SpecDB, reflecting the original required specifications. The snapshots of only the fields in the SpecDB tables that are used are shown here, for simplification. Note that the program line numbers in Section 5.2.2 are not the same as the POS_Num in the specifications. Line numbers used in the implementation in Section 5.2.2 was for the purpose of extracting the DD-paths graph. However, when the specifications are entered in SpecDB directly or through the RE tool, different POS_Num numbers are given, to reflect actual program actions. For example, all variable or type declarations are inserted in the Var or Type tables prior to inserting the dataflow data in the Dataflow table.

The above specifications in the Dataflow table in SpecDB, specify that the sixteenth operation that should take place in the implemented vacation salesman program should be an output. The output is defined in the Input_Output_Definition table, shown in Table 5.15 below. But before we examine the snapshot of the specifications in the Input_Output_Definition table, we need to look at the snapshot of the SpecDB Var table, in Table 5.14.

The Var table snapshot shown in Table 5.14 above, indicates that the variable Total_cruise_sold is an output, specifically, a GUI numeric output, that should be displayed in the CommissionCalc window or screen. Finally, from the above two SpecDB tables and the snapshot of the SpecDB Input_Output_Definition table in Table 5.15, it is specified that the output in the sixteenth operation is a display of the value of the Number_Cruises_Sold

Pos_Num	Var_ID	Input_Output	Source_Destination_Var_ID	...
16	18	Output	9	...
17	19	Output	10	...
18	20	Output	11	...
23	21	Output	7	...
34	22	Output	8	...
...	...	...	...	...

variable (Var_ID = 9) in the Total_cruise_sold GUI output (Var_ID = 18) in the CommissionCalc window or screen.

After examining the snapshot of the required specifications stored in SpecDB above, we will now examine a snapshot of the specifications of the program after fault injection. After examining how the specifications of the faulty program are represented in SpecDB as a result of the RE testing tool, the errors can then be automatically isolated and reported; as shown below.

In Table 5.12, the RE testing tool display of the GUI output for test case 1 after fault injection was demonstrated. The outcome of the RE tool is then stored in SpecDB, in the same three tables, Var, Dataflow and Input_Output_Definition. The RE testing tool specifications stored in the Dataflow, and Input_Output_Definition tables can be identical to that of Tables 5.13 and 5.15 for test case1 after fault injection. Table 5.16, however, shows that the five output variables (text boxes) are in another GUI window.

Comparing the two instances of the SpecDB tables to one another, those of the required specifications and the specifications of the implemented code resulting from the RE tool, is a trivial task that is easy to automate. However, much attention should be dedicated to identifying variable names and matching the implemented variable names to those of the specified design. If the implementation uses the exact variable names as those in the specifications and design, then comparing the RE testing tool specifications of the implementations to those of the required specifications is an easy task. The same consideration should be given to the operation position numbers in the Dataflow table, to facilitate the comparison of the two sets of specifications and the process of fault isolation. The next example illustrates how RE Testing tool helps spot faults of commission, when either the program generates an unspecified output, or generates the wrong output, due to an incorrect computation.

Var_ID	Var_Name	Type	Is_Input	Is_Output	Is_GUI_Object	GUI_Window
7	Total_Sales	Real	F	F	F	
8	Commission	Real	F	F	F	
9	Number_Cruises_Sold	Int	T	F	F	
10	Number_Air_Tckts_Sold	Int	T	F	F	
11	Number_Activities_Sold	Int	T	F	F	
18	Total_Cruises_Sold	Int	F	T	T	Sale_Report
19	Total_Air_Sold	Int	F	T	T	Sale_Report
20	Total_Activities_Sold	Int	F	T	T	Sale_Report
21	Total_Sales	Real	F	T	T	Sale_Report
22	Salesman_Commission	Real	F	T	T	Sale_Report

5.3.2   Test Case 2

In test case 1, the correct implementation in Section 5.2.2 was used. In order to demonstrate how the RE testing tools spots errors of commission, a slight modification of the code will be used. If the program was generated using the wrong implementation, it will result in errors. Accordingly, in test case 2, only one line of code will be modified to show how the RE testing tool identifies errors of commission. Per the specifications, the data entered by the employer in salesman identification, month and year fields should not change, when the user clicks on the calculate commission button. The five other fields should display the required information. In test case 2, we will assume that the program in line 25 has the following statement:

```
25 :month := Number_Cruises_Sold;
```

Instead of

```
25 :Total_Cruises_Sold := Number_Cruises_Sold;
```

In this case, the programmer made an error while writing the statement at line 25. Consequently, the actual output displayed on the GUI screen, will appear as in Figure 5.5. The GUI output clearly indicates an error, since the total sales and commission amounts are based on more than just 20 air tickets and 14 shore excursions. The value in the month field has also changed to 20. Since an error is obvious in the GUI output, the report displayed by the RE testing tool helps to isolate where the error occurred. In test case 2, the actual input is identical to that in test case 1, detailed in Tables 5.7 and 5.8.

5.3.2.1   Process and Actual Output

For test case 2, the DD-paths and all calculations will be identical to the process shown in test case 1. Hence, the calculation for the total number of cruises sold is correct. The program just does not display it in its position.

Employee	Gibson
Month	20
Year	2003

Calculate Commission

Total Cruises Sold this month	
Total Air Tickets Sold his month	20
Total Shore Excursions Sold this month	14
Total Sales	$18500
The Salesman Commission for this month	$2300

Fig. 5.5   GUI Output for Test Case 2

Out #	Category	Window	Component	Type	Value	From Variable
1	GUI	CommmissCalc	Month	Textbox	20	Num_Cruises_Sold
2	GUI	CommmissCalc	Tot_Air_Tckts_Sold	Textbox	20	Num_Air_Tckts_Sold
3	GUI	CommmissCalc	Tot_Activities_Sold	Textbox	14	Num_Activities_Sold

The only difference between the correct program implementation in test case 1 and that of test case 2 is in the first set of outputs shown in Table 5.17. Output number 1 in Table 5.17 highlights the error.

From the GUI output, the tester can see that the text box Total_Cruises_Sold was not modified by the program. Whereas the text box Month was modified. This is a fault of commission since the program should not have modified that Month field. In this example it was given the value 20. If there was a restriction on the integer values that are allowed to be put in the Month text box from 1 to 12, this might have cause a run time error as well. There is also a fault of omission since the program did not modify the Total_Cruises_Sold text box, when it should have displayed in it the value of the variable Number_Cruises_Sold. These faults could be easily spotted by the testers, once the RE testing tool displays the program output shown in Table 5.17 and the process. The RE testing tool technique thus far helped to identify the faults by showing the source of the

Var_ID	Var_Name	Type	Is_Input	Is_Output	Is_GUI_Object	GUI_Window
18	Tot_Cruises_Sold	Int	F	T	T	CommissCalc

Pos_Num	Var_ID	Input_Output	Source_Destination_Var_ID	...
16	16	Output	9	...

implementation's actual output. It also helped isolate where the problem occurred.

As in test case 1, the results of the RE testing tool in SpecDB are shown. The required specifications represented in SpecDB remain unchanged for all the test cases. In test case2, the specifications from the RE testing tool help to automatically isolate the source of the error and show that the specifications of the implementation are different, in only one location. Table 5.18 shows that the information in the Var table resulting from the RE tool specifications are correct, and match the original specifications.

Table 5.19 shows the specifications of the implemented code, resulting from the RE testing tool. Compared to table 5.15, the error lies in the Var_ID, it should have been 18 (Total_Cruise_Sold) not 16 (Month). By a trivial automated comparison of the two sets of specifications, the error can be easily isolated.

5.3.3   *Test Case 3*

The third example shows errors that most probably are missed by the testing team. These involve errors that happen in the database and are not immediately visible to the tester. This happens when all the items in the GUI return the expected output, however, more happens in the background that the user/tester cannot immediately see, but that also leaves the database in a possibly unstable or wrong state, i.e. containing erroneous data. Such implicit errors, can be easily identified when the tester examines the outcome of the RE tool containing the list of program outputs given in the actual output table. The following example demonstrates those invisible errors of commission and/or omission. The advantage of this technique lies in its ability to recognize errors as early as on the first run of a specific unit. To demonstrate this, the following scenario is used.

The expected output for the calculate commission unit is only to fill out five fields in the GUI and not to change the underlying database in any way. In this example the code in line 46 is changed to show how the

RE testing tool identifies hidden errors of commission. The original code in line 46 of the implementation was:

```
46    END;
```

Instead, we will assume that the programmer put the following lines of code:

```
46        UPDATE sales
47            SET cruise = Total_Sales
48            WHERE s_salesman = :salesman AND
                  s_month = :month AND
                  s_year = :year;
49    END;
```

This updates the entire salesman's cruise sales in that specific month and year (taken from the GUI text fields) to the total dollar sales amount for the cruises, air tickets and activities. This is of course an error. However, it is not visible to the tester if she uses ordinary testing procedures to compare the actual output to the expected output. The outputs are correct for the first time, but then the data in the database changes, and when the unit is executed again with the same inputs, it will generate different values each time! Such an error may not be visible during the testing phase when the database is not fully populated, but after a year of usage, if the employer wants a report of each salesman's sales and commissions earned, the report will give him totally different numbers than those the salesmen actually received during the year.

In order to spot such an error early one needs to either compare the database state before and after the execution of the unit code or use the proposed technique in this chapter. The output process will reveal the following, making it very easy for the tester to identify that there is an error of commission where the unit was programmed to execute more than it was intended to execute. The actual input will be identical to that in test case 1. The process will have statements 46-48 extra. The output will be identical to that in Example 1; five GUI outputs will be displayed by the RE testing tool in addition to the database modification in Table 5.20, which is displayed by the RE testing tool at the end of the program operation.

Similar errors can be spotted when executing the COMMIT database command after an error is introduced to the database. Also other errors

Output #	Category	Table	Change	Field	Row #	Value	From Var
6	DB	Sales	Updated	S_Cruise	15	18500	Tot_Sales
7	DB	Sales	Updated	S_Cruise	17	18500	Tot_Sales

that can be spotted with this technique are those resulting from deleting correct information that was stored in the database earlier but was not yet committed and a ROLLBACK command is executed. When the results of the RE tool are fed into SpecDB, it will be shown that there are an extra set of specifications resulting from the implementation that do not exist in the original specifications, thus isolating an error of commission.

5.4   Conclusion

In this chapter one of the applications of SpecDB is introduced, namely a reverse engineering approach is used to log the actual execution of the program from the code due to some input. Once the program actions have been logged, the specifications of the actual executed code are fed into SpecDB to allow for automatically matching them with the expected specifications. In this chapter, the actual program outputs and post conditions are presented as an output of the RE testing tool, in a way that non-technical testers and other members of the software development team can understand. As a result of the RE testing tool, the development team will have two documents, the required specifications, and the actual implemented specifications. The observed output can then be easily compared to the expected output based on the program specifications and oracles that determine what the expected program output should be. If the specifications of the software were already logged in SpecDB, then both specifications can be automatically compared, and errors automatically reported.

Specifications can be stored in SpecDB from the RE tool without a specific input for a test case. The RE testing tool as shown in the case study in this chapter automates the process of program DD-paths graph creation. After traversing the code with the white box testing technique shown using DD-paths testing, the specifications of the implemented code can be fed directly into SpecDB following the dataflow of the implementation. Accordingly, the RE tool can feed the specifications of the implemented units into SpecDB without the input data from a test case.

Chapter 6

Enhancing Other Testing Tools Using SpecDB

A survey of related literature, and the current available tools in the market was presented in Chapter 1. In Chapter 2, the motivation for the creation of SpecDB was discussed together with its detailed design. In Chapters 4 and 5 two testing tools using SpecDB were introduced. The focus of this chapter is on using the machine-readable specifications represented in SpecDB to enhance other tools developed or designed by other researchers. As an example, AGENDA, developed by Chays et. al. in [27,28,54,55,56,57] is chosen. The research involving AGENDA was summarized in Chapter 1. After introducing the machine interpretable specification of SpecDB, The research on AGENDA and other testing tools or techniques, can be enhanced. One of the motivations for the creation of SpecDB was to enhance already existing automated testing systems and complete current recommendations in some of the published literature. This chapter focuses on improvements that can be made to those testing tools using the SpecDB representation and tools described in the previous chapters of this book. In each section a limitation of other tools is emphasized, a solution using SpecDB is proposed and examples of the solution are provided.

6.1 Using the SpecDB Database to Expand the Testing Domain

Some testing tools, including AGENDA use linked lists and files to store the information needed about rules, constraints and the schema of the application being testing. Tools implemented using linked lists are not as easy to modify as database queries, and database schema design. Accordingly, scalability is an advantage of using a database design over a linked list representation. Using the database of SpecDB is a more effective approach, since much more software specifications can be stored in the SpecDB database. Storing more specifications result in an ability to automatically create better test cases and automate the process of test case result verification. Also enforcing the constraints on the SpecDB database is a means of testing the specifications for completeness and accuracy to a great extent. The authors of AGENDA investigated memory capacity and had concerns about size limits for the linked lists used. However, the use of tables in SpecDB to store and access the information as opposed to linked lists, eliminates the concern about memory overflow.

To demonstrate how storing more specifications result in an ability to automatically create better test cases, examples are given to show what test cases can be generated by AGENDA, compared to those created by a tool using SpecDB. In the vacation salesman program described in Chapter 5, the only portion of the unit implementation in Section 5.2.2 that can be tested using AGENDA, is line 7 and 8 (the Select statement) of the 46 lines of code. Tools built on SpecDB, however, can test imperative code as well. The RE testing tool demonstrated in Chapter 5 traversed the entire unit implementation, as shown by the three examples in Chapter 5. Accordingly, better test coverage is possible using the SpecDB.

However, in order to benefit from the results of other research, it is shown how it is an uncomplicated task to expand SpecDB, to represent the information used by other tools. Again this is demonstrated with an example from AGENDA. Translating the information needed and used by AGENDA in linked lists to the SpecDB representation is shown in Figure 6.1. In this figure it is shown how the data used to generate test cases and database states in AGENDA can be stored in SpecDB. Each database operation to test in AGENDA is given a unique query ID. This ID is stored in SpecDB in the DB_Test_Operation table (operation ID). The host variables used or updated in the query or database operation are stored in the Host_Var table. The descriptions of those variables are also stored in this table. In AGENDA only one attribute is observed for modifications per operation. The SpecDB representation offers an expansion to this simple query testing, to include multiple attributes per operation. This is achieved by storing those attributes in the Attributes_Changed table. Also multiple pre and post-conditions can be observed and tested using the SpecDB representation.

The Pre_Post_Conditions table in SpecDB stores the operation number together with the corresponding condition number. The conditions are stored in the Predicate table, using the Condition_Num to reference the

Pos_Num. The authors of AGENDA explained that only simple mathematical numeric calculations for post-conditions can be carried out using their tool. In the SpecDB representation, very complex operations can be specified using the generality of the Predicate table and the complicated operations that it can represent. The Predicate table can also represent operations on other types besides numbers, text, date, etc.

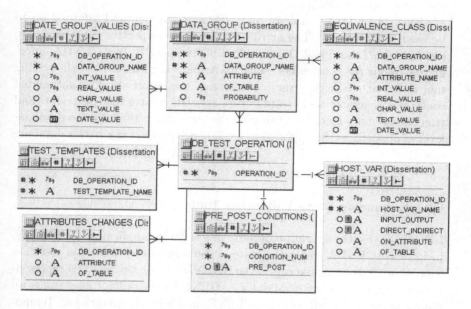

Fig. 6.1   Representing AGENDA Data in SpecDB

Data groups used by AGENDA to populate the tables with valid data and generate test cases, are stored in the Data_Group table in SpecDB, together with the probability of each group. The probability was used by AGENDA as a heuristic to guide the processes of data and test case generation. Boundary values are stored in the Type_Restrictions table in SpecDB. In SpecDB every attribute in every table can have a distinct type with specified allowable values. Using this representation in SpecDB allows for more complex type restrictions. For example specifying a range (from -5 to 5) excluding one or more values (like 0). If the attribute is better described by a list of values (e.g. month name), the values are represented in the Types_LOV table.

Using AGENDA, the tester specifies some valid values for each of the data groups to guide the database state and test case generation processes. The list of valid values for each data group is stored in the Equivalence_Class table in SpecDB. The attribute can be of any basic type. Other more complex types can also be used by expanding this table in SpecDB, giving the tester more flexibility.

The List_of_Tables table in SpecDB hosts the required information to be used by AGENDA. The attributes and their respective constraints are stored in the Table_Field_Description table in SpecDB. In AGENDA, information about the constraints on each attribute is gathered, but the testing tool does not check that these constraints are enforced. The automated constraint generator tool created using SpecDB creates the constraints and triggers for more complex business rules, to enforce them at the database level as opposed to the application level, hence producing a better database state.

6.2   Enhancing State Validation Tools

In order to test the operation of parts of a software application, some tools, including AGENDA, check the state of the resulting system after test case execution. Another enhancement to AGENDA using the SpecDB representation is that of the state validator. AGENDA creates log tables to log all modifications after UPDATE, DELETE, or INSERT operations. Temporary constraints are created on the log tables to check the database state changes after test execution. AGENDA also checks whether the number of affected rows by the test execution equals that expected. If not, AGENDA reports an error, indicating that either too many or too few records have been affected. It is then the tester's task to go through the database and the log tables to identify the error.

There are some errors that can occur and not be identified by AGENDA, in its current technique. AGENDA checks that the post-conditions have been fulfilled using the temporary constraints on the log tables. However, if the specified post-conditions have been fulfilled but also other modifications have been made to the database too, i.e. errors of commission, that would be a software error that would not be spotted by AGENDA. For example, if the operation being tested, gives an annual raise to all employees. AGENDA will make sure that the salaries of all the employees have been raised by the specified corresponding rates. It would not however, spot an error of a change in the department made to all employees. Also AGENDA would not spot other operations in the database that are not linked to the database table being updated, for example, the operation grants privileges to unauthorized users to access the database. Using SpecDB the latter problem is solved by using the reverse engineering approach to testing the database system, where all of the modifications made to a system are logged for each test operation. Those errors of commission can be spotted very easily using this technique, as shown in the examples of Chapter 5.

If the number of rows modified equals that expected, and the post-conditions are satisfied by the temporary constraints on the log tables, but other attributes have been modified too, AGENDA does not report an error. In order to report such software errors, using SpecDB more constraints can be automatically generated on the log tables to check if all the other attributes, (other than the ones in the Attributes_Changed table for the specified operation) have not changed. If the post-conditions have not been satisfied or satisfied but other attributes have also been modified, the SpecDB tool will report an error.

6.3 Expanding the Testing Scope Beyond Variables and Database States

Testing tools that are not based on the specifications are limited in the areas of the software that can be tested. AGENDA for example, tackles only variables and database states. It does not tackle operations where the variables or database states are not changed, but files, printed reports, magnetic card programming, etc. Many Applications use numerous types of inputs and outputs, besides database state changes. SpecDB's reverse engineering testing tool handles all these different types of inputs and outputs, as seen in Chapter 5.

6.4 Testing Different Operation Types

Another limitation of testing tools not based on software specifications, is the types of operations tested. AGENDA tests four types of database operations, namely, INSERT, UPDATE, DELETE and SELECT. Using the reverse engineering testing tool of SpecDB, other database operations can also be tested, and their results logged, together with imperative code, i.e. programming constructs that are not related to database operations. Using SpecDB the following database operations can be tested: GRANT, REVOKE, CREATE and DROP (view, table, constraint, sequence, etc.), role, access sequence, COMMIT, ROLLBACK and SAVEPOINT. Other operations can be easily added to SpecDB. The reverse engineering SpecDB testing tool observes all modifications that happen in the database state, the application variables, and the input/output domains. Each of the above mentioned database operations are logged with all its details, as described in the reverse engineering SpecDB testing tool in Chapter 5.

6.5   Conclusion

In this chapter four areas of enhancements offered by the SpecDB representation of the software specifications over other tools, were proposed. As an example from the literature, Chays et. al. AGENDA research [27,28,54,55,56,57] using SpecDB was chosen. Additional enhancements can be made using the SpecDB representation to other tools in the market and the literature using the machine interpretable specifications stored in SpecDB. Many researchers confirmed that having such a queryable formal specification that can be used in their tools will automate and enhance the process of test case generation and result verification [31, 34, 38]. As a future work, many of those proposals can be implemented using the SpecDB technique, to test systems using the proposed automated ideas published, based on the software's machine-readable specifications. In Chapter 7 a comparison between different testing tools discussed in Chapter 1, and SpecDB, is given to show some of the limitations and strengths of each tool. Also a conclusion of all the concepts presented in this book and recommend future work is summarized in Chapter 7.

Chapter 7

Conclusion and Future Work

In this book the design of SpecDB, a database created to represent and host software specifications in a machine-readable format is introduced. This scalable database representation was shown to aid in the processes of both automated software code generation and automated software testing, based on the actual software specifications. One of the main objectives in this book is the creation of SpecDB; its design is detailed in Chapter 2. SpecDB is the database that can hold the specifications required for unit testing database software. An algorithm was created to show how to translate formal specifications into the SpecDB representation, in Chapter 3. Some examples of automated testing applications benefiting from the machine-retable specifications represented in SpecDB were created and discussed in Chapters 4 and 5.

This chapter concludes with a summary of the books objectives. In Section 7.1 the new ideas introduced in this book are summarized in points. Some proposals for future work and research continuation ideas are given in Section 7.2.

7.1 A Comparison Between Testing Tools

In this section six test case scenarios are described. With these scenarios, the ability of different tools to identify errors in the software being tested, is highlighted. The testing tools used in this comparison are AGENDA, SpecDB, RE testing tool, Rational, Mercury and Segue set of tools. In the next section the test cases and scenarios are describe first. Then a comparison follows denoting the strength and limitations of each of the tools mentioned above.

7.1.1   *Test Cases and Scenarios*

7.1.1.1   *Test Case 1*

The case study and implementation in Chapter 5 is also use for this test case. The salesman reports twice to his employer during a certain month. The first time he reports that he sold 8 cruise vacations, 12 airline tickets and 10 shore excursion packages. The second time he reports that he sold 12 cruise vacations, 8 airline tickets and 4 shore excursion packages. The employer enters the data as sent each time. Therefore the database will have two records in the sales table for that specific salesman for that specific month. At the end of the month the employer enters the employee name, and the required month and year, then hits the Calculate Commission Button to view the report on the screen. In this scenario, it is assumed that the code is all error free and matches the specifications. No errors should be identified.

7.1.1.2   *Test Case 2*

In test case 2 the same implementation as in test case 1 is used, changing only one line of code. Per the specifications, the data entered by the employer in salesman identification, month and year fields should not change, when the user clicks on the calculate commission button. The five other fields should display the required information. In test case 2, we will assume that the program in line 25 has the following statement:

```
25    :month := Number_Cruises_Sold;
```

Instead of:

```
25    :Total_Cruises_Sold := Number_Cruises_Sold;
```

Accordingly, the testing tool should identify 2 errors, changing the month value, and not displaying the total cruises sold.

7.1.1.3   *Test Case 3*

In this test case, the same code used in test case 1 is adopted, however, the code in line 46 is changed. The original code in line 46 of the implementation was:

```
46    END;
```

Instead, we will assume that the programmer put the following lines of code:

```
46      UPDATE sales
47         SET cruise = Total_Sales
48         WHERE s_salesman = :salesman AND
               s_month = :month AND
               s_year = :year;
49   END;
```

As discussed in Chapter 5, this results in modifying the data in the database, and will generate different values for the same salesman for the same month, each time the test case is run. One error of commission, resulting in two database modifications should be identified as an error. If this case is run only once, the error will not be identified by any functional testing tool. If run several times on the same inputs, functional testing techniques might spot the difference in the output data each time.

7.1.1.4 *Test Case 4*

Again in test case 4, the implementation is the same as that used in test case 1, with the exception of displaying the output in another window. From the user's point of view, no output is generated when he clicks the button, although, the output is computed correctly, it is displayed in the wrong location in the GUI. Ten errors should be identified, since the five outputs are misplaced.

7.1.1.5 *Test Case 5*

In this test case, the same code used in test case 1 is adopted here too, changing only the code in line 46. The original code in line 46 of the implementation was:

```
46      END;
```

Instead, we will assume that the programmer put code that grants privileges to all users to enter sales information, a privilege that should only be given to the employer. The same test case can also include other database operations including, Revoke, Create and Drop(view, table, constraint, sequence, etc.), Roles, Access Sequence, Commit, Rollback and Savepoint. All functional testing tools will not identify this error, no matter how many times the test is run. Although this test case involves a database operation, yet AGENDA cannot test it, because it is a Grant operation.

7.1.1.6   Test Case 6

Test case 6 is similar to test case 5, but assuming that code is written before the end to export employee names, social security number, addresses and dates of birth to another remote location. No functional testing tools will identify this error even if the test is run many times. Also high volume testing techniques [77,79] cannot spot this kind of error. High volume testing is when the test case is executed many times, even if the result is correct, to identify whether the system will react differently over time, for example from a slight system clock difference that accumulates over time.

7.1.2   Comparing the Ability of Testing Tools to Identify Errors

In this section, a comparison is given between different testing tools to show how each reacts to the errors in each scenario of the test cases in Section 7.1.1. A tabulation of the results is presented in Table 7.1. The test cases selected in Section 7.1.1, were chosen to show only the differences between the tools. There are a lot of test cases that show that all or most of the tools react similarly. In Table 7.1 the instances where the tools react similarly are removed, and only the areas where they differ in capability are included. The functionality of each testing tool is observed and compared based on a variety of factors. Among the comparison criteria is whether the tool can test imperative code, i.e. performs structural testing; also coverage of the test cases that can be handled by the tool and the percentage of code in the implementation that can be tested by the tool.

As seen from Table 7.2, there is some functionality that can be performed by other tools, namely memory management, stress testing, etc, that cannot be performed SpecDB tools using the current design of SpecDB. However, as proved previously, the SpecDB representation is readily scalable. The current design of SpecDB detailed in Chapter 2, is for the purpose of unit testing. However, SpecDB can be expanded easily, as previously shown in Chapter 4, to accommodate more specifications to test other parts of the software including security, memory usage, etc.

There are unique capabilities associated with SpecDB based tools that other testing tools cannot perform, as shown in Figure 7.1; namely the ability of the tools to automatically identify errors in execution if the inputs to a test case changed, and the ability to isolate the source of errors in the code. Functional testing tools like regression testing, stress testing, scenario testing, record and play back tools, need the testers to verify the result of the test case. Since the tools have no specifications, the testers need to identify the expected outputs of each test case. The latter tools can run the same test scenarios with different input and automatically verify the output, unless for each of those test cases the expected output is given to the tool.

For SpecDB based tools however, since the specifications are available, tools based on SpecDB can automatically generate the expected output and verify whether the test case results match the specifications. The book is concluded with a comparison between some of the most widely used testing tools and the capabilities of SpecDB based tools. In the next sections the main concepts introduced in this book are summarized, and recommendations for future work on SpecDB and SpecDB based tools are suggested.

7.2 Book Concepts at a Glance

In this section the main concepts presented in this book are outlined. In Section 7.3 the book's concepts are discussed in more detail.

In this book, the following was introduced:

(1) Software specifications are represented in a machine-readable format. This was detailed in Chapter 2, the design of SpecDB.

(2) A translation algorithm to transform formal specifications written in a formal language like Z, to the machine-readable SpecDB representation is created. This was shown in Chapter 3, storing specifications in SpecDB.

(3) Useful applications that make use of the data stored in SpecDB are developed.

(a) In Chapter 4 an automated tool that generates constraints and business rules to enforce them at the database level as opposed to the application level, for better reliability, was introduced.

(b) Also in Chapter 5 a reverse engineering approach to testing database applications was introduced. It uses specifications to intelligently test software.

(4) Enhancements on published work in the field of testing database ap-

plications, were suggested using the techniques described in this book in Chapter 6; some enhancements on AGENDA, created by Chays et. al. were proposed.

7.3   Conclusion and Recommendations

In this book, the design of SpecDB, a database to hold parts of a software's specification is being investigated. Representing software specifications in this manner aids the process of automated testing and automated code generation. Through the use of machine-readable specifications represented in SpecDB, automated tools can become more intelligent in generating and testing programs, based on an understanding of what the software under construction is supposed to do, by design. The design of SpecDB, discussed in Chapter 2, is scalable, allowing for any enhancement and additions to accommodate more forms of specifications or business requirements, as they evolve in the future, or other parts of the software application, e.g. security or networking, etc. In this book, the idea of a unified standard to represent portions of the software specifications necessary to aid in the process of unit testing was explained.

In Chapter 3, an algorithm is implemented to translate examples of formal specification statements to the SpecDB representation. In this chapter, it was shown how a formal software specification can be automatically represented in SpecDB. Additional future work is to implement this algorithm, and also create an automated intelligent software testing tool to test software represented by SpecDB.

As an application of SpecDB and to illustrate its purpose, an automated tool was developed to generate business rules and enforce them on the database level by automatically creating constrains and assertions using the data stored in SpecDB, in Chapter 4. The design of a new table in SpecDB was discussed, Constraints, and the design of the Table_Description table of SpecDB was extended, and renamed Table_Fields_Description. Those modifications were made in order to allow for the representation of business rules, and to show how the main idea of SpecDB is scalable and easily modifiable to accommodate the representation of different specifications, proving that the design of SpecDB allows for future enhancement and additions to accommodate more types of requirements, as they evolve in the future. The automated tool introduced in Chapter 4 creates a script to create all the constraints and another to drop them. System administrators can choose to run the production system with fewer constraints or no constraints at all in order to enhance the speed of production systems. They can run the automatic constraint generator demonstrated in Chapter 4 to create and automatically test the validity of the data during certain times of the week to maintain the accuracy, completeness and integrity of the system. Through the use of SpecDB, automated tools can become more intelligent in generating and testing programs, based on an understanding of what the software under construction is supposed to do, by design.

The use of SpecDB in a reverse engineering testing tool is discussed in Chapter 5. The reverse engineering approach was used to log the actual execution of the program due to some input, extracting the specifications of the implemented code. Once the program actions have been logged, they are presented in simple explanations which non-technical testers and other members of the software development team can understand. The actual outcome can also be easily matched and automatically compared to the expected output based on the program specifications in SpecDB that determine what the expected program output should be based on the specified inputs. Finally, in Chapter 6, some enhancements to published literature, using SpecDB was suggested; as an example AGENDA [27,28,54] was used to show the suggested enhancements.

The representation of specifications in SpecDB can be used as a completion of AGENDA, the work done by Chays et. al. in [27,28,54,55,56,57]. AGENDA uses the database schema as a specification of the relations between the different entities in the software, and to extract the constraints on the relations or the values of the attributes. AGENDA depends on the tester's user scenarios to test a database system, as well as the testers input files describing attribute valid values and their respective distribution probabilities. It generates tests by parsing application code not the specifications and design of the software. AGENDA then generates test data to populate the database and use the populated database as the input state for the testing operations. The data generated by AGENDA are only valid data that satisfy the constraints imposed by the implemented schema.

The tester suggests inputs for test cases and finally AGENDA provides a log file of the output database state after each test operation has executed, providing the modifications that happened in the database and output variables as a result of the test execution. The tester then uses those log files to determine whether the database output state is correct given the database input state before the test was executed with the given input variable values. The authors repeatedly highlight the fact that the process of verifying the test results cannot be automated since there is no formal specification available (i.e. available for automated to tool to interpret and reason with).

SpecDB is the machine-readable specifications that can be used with tools like AGENDA to automate the process of verifying the output state of a database after a test has been executed. In production level databases when there are thousands and maybe millions of records, it will be difficult and time consuming for the testers to verify the output and determine whether or not a test failed. SpecDB can also be used as an input to systems like AGENDA, instead of testers log files. When relying on testers to supply values for attributes or values for test inputs, are leaving the tester with the task of attaining good system coverage with the values he/she provides. Automating this process too, will provide the coverage needed without any coverage gaps.

Given SpecDB as one of the inputs to AGENDA, better data can be generated to populate the database. Also better test cases can be generated, with better system coverage. As another future work suggestion, the specifications stored in SpecDB can be used to automate the process of mathematically inducing rules to determine the validity of the output state after a test has been performed, and thus fully automate the process of output validation.

7.4   Future Work at a Glance

From the above description of the concepts introduced in this book, and the ideas to continue on the work summarized in Section 7.2, the following future work ideas are proposed:

(1) Expand on the design of SpecDB to allow for testing other areas of the software, including system security, networking, etc. An example of this was presented in Chapter 4, to show the scalability of the SpecDB representation.

(2) Expand on the translation algorithm using the same idea in Chapter 3 to include other forms of the formal specifications syntax, as needed for the testing process.

(3) Implement the algorithm in Chapter 3, to automatically read formal specifications and translate them to the SpecDB representation. This is a simple coding task that requires no little effort.

(4) Create a code generating tool that automatically implements software

from the specifications in SpecDB. This is mostly a coding task, and requires a good understanding of the design of SpecDB and where and how specifications are represented and the interactions between the different entities of SpecDB.

(5) Create other intelligent automated testing tools that benefit from the machine-readable software specifications in SpecDB and automatically reason with the specifications and create test cases based on an understanding of the specifications. This task requires the deployment of some artificial intelligence techniques to automatically generate intelligent and useful test cases to test software based on its specifications detailed in SpecDB. The test cases generated should attain good system coverage based on a white box system coverage matrix, maybe as part of the reverse engineering tool explained in Chapter 5. The result is a very useful intelligent automated testing tool that crowns the work established in this book. All the requirements to build this tool are detailed in this book. A lot of simple test cases can be directly generated from the basic data in SpecDB, including boundary analysis test cases, and type restrictions, as seen in Chapter 4. However, since the SpecDB representation handles more complex specifications and business rules, the proposed automated testing tool for future work can create much more intelligent, important and useful test cases, reasoning with the specifications in SpecDB, as well as generic requirements test cases including memory and exception handling, etc.

(6) Implement the automatic comparison tool, which compares actual system behavior reported by the reverse engineering tool in Chapter 5, to that specified in SpecDB and report system errors or any discrepancies.

(7) Automate the process of output state validation by mathematically inducing rules to determine the validity of the output state after a test has been performed. For database applications with thousands of records in production databases, it is very hard to manually validate the output state of the database after an operation has been executed. Accordingly, the proposed automated database output state validation tool can make the decision whether the output state matches the post conditions specified.

(8) Create a GUI - SpecDB populating tool to populate SpecDB with the software specifications. Instead of writing the formal specifications, or manually populating SpecDB with the specifications, the human language specifications can be fed into SpecDB using this GUI tool. The GUI tool can ask questions and give choices for answers. Based on the

input, the specifications are stored in the correct location in SpecDB. The GUI tool can also make sure that the specifications are complete, and non-contracting to one another, i.e. test the specifications.

All the proposed future work tasks outlined in this section can be completed following the concepts demonstrated in this book. The introduction of the machine-readable specifications of SpecDB, is certainly the most important step and the basis for creating the intelligent automated testing tools that can aid the testing task immensely, if not promise to fully automate the process of testing software applications. After the completion of the recommendations for future work in this book, there is no doubt that, a fully automated intelligent test case generator and output validator tool, based on the ideas presented of this book, will be available in the near future.

Test Case #	Tool	Can run test	Can spot error	Spot errors (different inputs for same scenario)	Can automatically verify results	Isolates error sources
1	SpecDB Based Tools	✓	✓	✓	✓	✓
2		✓	✓	✓	✓	✓
3		✓	✓	✓	✓	✓
4		✓	✓	✓	✓	✓
5		✓	✓	✓	✓	✓
6		✓	✓	✓	✓	✓
1	AGENDA	x	-	-	-	-
2		x	-	-	-	-
3		✓	User	x	To an extent	x
4		x	-	-	-	-
5		x	-	-	-	-
6		x	-	-	-	-
1	WinRunner, Robot, SilkTest	✓	✓	x	✓	x
2		✓	✓	x	✓	x
3		✓	x	x	x	x
4		✓	✓	x	✓	x
5		✓	x	x	x	x
6		✓	x	x	x	x
1	QuickTest, Functional SilkTest5	✓	✓	x	✓	x
2		✓	✓	x	✓	x
3		✓	✓	x	✓	x
4		✓	✓	x	✓	x
5		✓	x	x	x	x
6		✓	x	x	x	x
1	LoadRunner, Performance Tester, Silk Performer	✓	x	x	x	x
2		✓	x	x	x	x
3		✓	x	x	x	x
4		✓	x	x	x	x
5		✓	x	x	x	x
6		✓	x	x	x	x

Fig. 7.1   A Comparison Between How Testing Tools/Techniques Handle Test Cases 1-6

Capability / Tool	Can automatically generate tests	Readily scalable	Can automatically verify results	Unit Coverage	Memory / performance testing	Stress testing	Tests DB operations	Structural testing	Functional testing	Handles complex operations
SpecDB Based Tools	✓	✓	✓	100%	✗	✗	✓	✓	✓	✓
AGENDA	✓	✗	To an extent	Only DB operations	✗	✗	✓	✓ (DB operations only)	✓	✗
WinRunner Robot, SilkTest	✓ (only test scripts from user scenarios)	✗	✓	Based on scenarios	✗	✗	considered functions	✗	✓	observed only
QuickTest, Functional SilkTest5	✓ (only user scenario test scripts)	✗	✓	Based on scenarios	✗	✗	considered functions	✗	✓	observed only
LoadRunner Performance Tester, Silk Performer	✓	✗	✓	✗	✓	✓	✗	✗	✓	observed only
Purify	✓	✗	✓	✗	✓	✗	✗	✗	✓	observed only

Fig. 7.2   A Comparison Between the Capabilities of Testing Tools and Techniques

Appendix A

SQL Scripts for SpecDB Constraints

```
Prompt Creating Primary Key On 'Des_Sr_Post_Conds'
Alter Table Designed_Sr_Post_Conditions
 Add (Constraint Post_Cond_Pk
      Primary Key (Post_Cond_Num))/

Prompt Creating Primary Key On 'Actual_Input'
Alter Table Actual_Input
 Add (Constraint Ip_Pk
      Primary Key (Ai_Ip# ,Ai_Test_Case#))/

Prompt Creating Primary Key On 'Var'
Alter Table Var
 Add (Constraint Var_Pk Primary Key (Var_Id))/

Prompt Creating Primary Key On 'Assignment'
Alter Table Assignment
 Add (Constraint Fun_Pk Primary Key (Pos_Num))/

Prompt Creating Primary Key On 'Sr_Ip_Desc'
Alter Table Sub_Routine_Ip_Description
 Add (Constraint Sr_Id_Pk
      Primary Key (Subroutine_Name ,Ip_Order))/

Prompt Creating Primary Key On 'Predicate'
Alter Table Predicate
 Add (Constraint Pre_Pk
      Primary Key (Pos_Num ,Component_Num))/
```

```
Prompt Creating Primary Key On 'Actual_Output'
Alter Table Actual_Output
Add (Constraint Actual_Op_Pk
    Primary Key (Ao_Op# ,Ao_Test_Case#))/

Prompt Creating Primary Key On 'Class_Relationships'
Alter Table Class_Relationships
Add (Constraint Cls_Rel_Pk
    Primary Key (Relationship_Id))/

Prompt Creating Primary Key On 'Input_Output_Def'
Alter Table Input_Output_Definition
Add (Constraint Odn_Pk
    Primary Key (Var_Id, Pos_Num ,Var_Sequence))/

Prompt Creating Primary Key On 'Triggers'
Alter Table Triggers
Add (Constraint Trg_Pk
    Primary Key (Trigger_Name))/

Prompt Creating Primary Key On 'Db_Oper_Details'
Alter Table Db_Operation_Details
Add (Constraint Dit_Pk
    Primary Key (Db_Operation_Num ,Order_Num))/

Prompt Creating Primary Key On 'Types'
Alter Table Types
Add (Constraint Type_Pk Primary Key (Type_Name))/

Prompt Creating Primary Key On 'Expected_Output'
Alter Table Expected_Output
Add (Constraint A_Op_Pk
    Primary Key (Eo_Op# ,Eo_Test_Case#))/

Prompt Creating Primary Key On 'Db_Operations'
Alter Table Db_Operations
Add (Constraint Db_Oper_Pk
    Primary Key (Db_Operation_Num))/
```

```
Prompt Creating Primary Key On 'Db_Test_Operation'
Alter Table Db_Test_Operation
 Add (Constraint Dbto_Pk
      Primary Key (Operation_Id))/

Prompt Creating Primary Key On 'Test_Templates'
Alter Table Test_Templates
 Add (Constraint Tte_Pk
      Primary Key (Db_Operation_Id ,Test_Template_Name))/

Prompt Creating Primary Key On 'Type_Restrictions'
Alter Table Type_Restrictions
 Add (Constraint Type_Res_Pk Primary Key (Restriction#))/

Prompt Creating Primary Key On 'Type_Excluded_Ranges'
Alter Table Type_Excluded_Ranges
 Add (Constraint Type_Exran_Pk
      Primary Key (Restriction_Num))/

Prompt Creating Primary Key On 'Sr_Ip_Instance'
Alter Table Sub_Routine_Ip_Instance
 Add (Constraint Sr_Ipin_Pk
      Primary Key (Instance_Num ,Ip_Order))/

Prompt Creating Primary Key On 'Func_Decl'
Alter Table Function_Declaration
 Add (Constraint Func_Dcl_Pk
      Primary Key (Func_Name))/

Prompt Creating Primary Key On 'Var_Restricts'
Alter Table Var_Restrictions
 Add (Constraint Tres_Pk
      Primary Key (Restriction_Num))/

Prompt Creating Primary Key On 'Host_Var'
Alter Table Host_Var
 Add (Constraint Hvr_Pk
      Primary Key (Db_Operation_Id ,Host_Var_Name))/
```

```
Prompt Creating Primary Key On 'Objects'
Alter Table Objects
 Add (Constraint Obj_Pk Primary Key (Object_Name))/

Prompt Creating Primary Key On 'Data_Group'
Alter Table Data_Group
 Add (Constraint Dgp_Pk
     Primary Key (Data_Group_Name ,Db_Operation_Id))/

Prompt Creating Primary Key On 'Var_Table_Restrictions'
Alter Table Var_Table_Restrictions
 Add (Constraint Fvtr_Pk Primary Key (Var_Tab_Rst_Id))/

Prompt Creating Primary Key On 'Operator'
Alter Table Operator
 Add (Constraint Opr_Pk Primary Key (Operator_Id))/

Prompt Creating Primary Key On 'Expected_Input'
Alter Table Expected_Input
Add (Constraint Exop_Pk
     Primary Key (Ei_Ip# ,Ei_Test_Case#))/

Prompt Creating Primary Key On 'Char_Equiv_Class'
Alter Table Char_Equiv_Class
 Add (Constraint Ces_Pk
     Primary Key (Var_Name ,Range_From ,Range_To))/

Prompt Creating Primary Key On 'Runtime_Sr_Post_Conds'
Alter Table Runtime_Sr_Post_Conditions
 Add (Constraint Rt_Post_Cond_Pk
     Primary Key (Post_Cond_Num ,Instance_Num))/

Prompt Creating Primary Key On 'Dataflow'
Alter Table Dataflow
 Add (Constraint Daw_Pk
     Primary Key (Pos_Num ,Suc_Num))/

Prompt Creating Primary Key On 'Real_Equiv_Class'
```

```
Alter Table Real_Equiv_Class
 Add (Constraint Rec_Pk
      Primary Key (Var_Name ,Range_From ,Range_To))/

Prompt Creating Primary Key On 'Procedure'
Alter Table Procedure
 Add (Constraint Proced_Pk Primary Key (Proc_Name))/

Prompt Creating Primary Key On 'Db_Operation_Tables'
Alter Table Db_Operation_Tables
 Add (Constraint Otab_Pk Primary Key (Id))/

Prompt Creating Primary Key On 'Int_Equiv_Class'
Alter Table Int_Equiv_Class
 Add (Constraint Iec_Pk
      Primary Key (Var_Name ,Range_From ,Range_To))/

Prompt Creating Primary Key On 'Var_Lov'
Alter Table Var_Lov
 Add (Constraint Fvlov_Pk Primary Key (Id))/

Prompt Creating Primary Key On 'Constraints'
Alter Table Constraints
 Add (Constraint Con_Pk Primary Key (Constraint_Num))/

Prompt Creating Primary Key On 'Calculation'
Alter Table Calculation
 Add (Constraint Calc_Pk
      Primary Key (Cal_Num ,Component_Num))/

Prompt Creating Primary Key On 'Classes'
Alter Table Classes
 Add (Constraint Cls_Pk Primary Key (Class_Name))/

Prompt Creating Primary Key On 'Table_Field_Descn'
Alter Table Table_Field_Description
 Add (Constraint Tdesc_Pk
      Primary Key (Table_Name ,Field_Name))/
```

```
Prompt Creating Unique Key On 'Sub_Routine_Ip_Desc'
Alter Table Sub_Routine_Ip_Description
 Add (Constraint Sr_Ipdesc_Uk
     Unique (Subroutine_Name ,Sr_Var_Name))/

Prompt Creating Unique Key On 'Sr_Ip_Instance'
Alter Table Sub_Routine_Ip_Instance
 Add (Constraint Sr_Ipinst_Uk Unique (Subroutine_Name))/

Prompt Creating Unique Key On 'Table_Field_Desc'
Alter Table Table_Field_Description
 Add (Constraint Field_Id_Uk Unique (Field_Id))/

Prompt Creating Check Constraint On 'Valid_Operands'
Alter Table Valid_Operands
 Add (Constraint Avcon_1144240102_Opera_000
 Check (Operand1_Type
 In ('Char', 'Text', 'Int', 'Real', 'Bool', 'Date',
     'Time', 'Db_Table')))/

Prompt Creating Check Constraint On 'Valid_Operands'
Alter Table Valid_Operands
 Add (Constraint Avcon_1144240102_Opera_001
 Check (Operand2_Type
 In ('Char', 'Text', 'Int', 'Real', 'Bool', 'Date',
     'Time', 'Db_Table')))/

Prompt Creating Check Constraint On 'Valid_Operands'
Alter Table Valid_Operands
 Add (Constraint Avcon_1144240102_Opera_002
 Check (Operand3_Type
 In ('Char', 'Text', 'Int', 'Real', 'Bool', 'Date',
     'Time', 'Db_Table')))/

Prompt Creating Check Constraint On 'Valid_Operands'
Alter Table Valid_Operands
 Add (Constraint Avcon_1144240102_Over__000
 Check (Over_Loaded In ('T', 'F')))/
```

```
Prompt Creating Check Constraint On 'Des_Sr_Post_Conds'
Alter Table Designed_Sr_Post_Conditions
 Add (Constraint Avcon_1144240102_Sourc_000
 Check (Source In ('Var', 'Const', 'Gui', 'Seq')))/

Prompt Creating Check Constraint On 'Des_Sr_Post_Conds'
Alter Table Designed_Sr_Post_Conditions
 Add (Constraint Avcon_1144240102_Post__000
      Check (Post_Cond_Type
 In ('Gui', 'Db', 'File', 'Report',
     'Seq', 'Message', 'Priv_Role',
     'Global_Var', 'Commit', 'Savepoint', 'Rollback',
     'Run_Exe', 'Open_Win', 'Terminate')))/

Prompt Creating Check Constraint On 'Des_Sr_Post_Conds'
Alter Table Designed_Sr_Post_Conditions
 Add (Constraint Avcon_1144240102_From__000
 Check (From_Bool_Const_Value In ('T', 'F')))/

Prompt Creating Check Constraint On 'Des_Sr_Post_Conds'
Alter Table Designed_Sr_Post_Conditions
 Add (Constraint Avcon_1144240102_To_Fi_000
 Check (To_File_Change In ('Create', 'Delete',
                           'Moidify', 'Relocate')))/

Prompt Creating Check Constraint On 'Designed_Sr_Post_Conds'
Alter Table Designed_Sr_Post_Conditions
 Add (Constraint Avcon_1144240102_To_Re_000
 Check (To_Report_Destination In ('Printer',
                            'Screen', 'File')))/

Prompt Creating Check Constraint On 'Actual_Input'
Alter Table Actual_Input
 Add (Constraint Avcon_1144240102_Ai_Ca_000
 Check (Ai_Category In ('Db', 'Gui', 'File')))/

Prompt Creating Check Constraint On 'Var'
Alter Table Var
 Add (Constraint Avcon_1144240102_Is_In_000
```

```
Check (Is_Input In ('T', 'F')))/

Prompt Creating Check Constraint On 'Var'
Alter Table Var
 Add (Constraint Avcon_1144240102_Is_Ou_000
 Check (Is_Output In ('T', 'F')))/

Prompt Creating Check Constraint On 'Var'
Alter Table Var
 Add (Constraint Avcon_1144240102_Is_Co_000
 Check (Is_Constant In ('T', 'F')))/

Prompt Creating Check Constraint On 'Var'
Alter Table Var
 Add (Constraint Avcon_1144240102_Is_Gu_000
 Check (Is_Gui_Object In ('T', 'F')))/

Prompt Creating Check Constraint On 'Var'
Alter Table Var
 Add (Constraint Avcon_1144240102_Encap_000
 Check (Encapsulation In ('Private',
                          'Protected', 'Public')))/

Prompt Creating Check Constraint On 'Var'
Alter Table Var
 Add (Constraint Avcon_1144240102_Has_L_000
 Check (Has_Lov_Restrictions In ('T', 'F')))/

Prompt Creating Check Constraint On 'Var'
Alter Table Var
 Add (Constraint Avcon_1144240102_Has_T_000
 Check (Has_Table_Restrictions In ('T', 'F')))/

Prompt Creating Check Constraint On 'Var'
Alter Table Var
 Add (Constraint Avcon_1144240102_Has_O_000
 Check (Has_Other_Restrictions In ('T', 'F')))/

Prompt Creating Check Constraint On 'Var'
```

```
Alter Table Var
 Add (Constraint Avcon_1144240102_Manda_000
 Check (Mandatory In ('T', 'F')))/

Prompt Creating Check Constraint On 'Var'
Alter Table Var
 Add (Constraint Avcon_1144240102_Defau_000
 Check (Default_Bool_Value In ('T', 'F')))/

Prompt Creating Check Constraint On 'Var'
Alter Table Var
 Add (Constraint Avcon_1144240102_Curre_000
 Check (Current_Bool_Value In ('T', 'F'))) /

Prompt Creating Check Constraint On 'List_Of_Tables'
Alter Table List_Of_Tables
 Add (Constraint Avcon_1144240102_Compo_000
 Check (Composite_Primary_Key In ('T', 'F')))/

Prompt Creating Check Constraint On 'Types_Lov'
Alter Table Types_Lov
 Add (Constraint Avcon_1144240102_Bool__000
 Check (Bool_Value In (0, 1)))/

Prompt Creating Check Constraint On 'Assignment'
Alter Table Assignment
 Add (Constraint Avcon_1144240102_Rhs_T_000
 Check (Rhs_Type
 In ('Calc', 'Func', 'Var',
    'Const', 'Seq', 'Gui', 'Db')))/

Prompt Creating Check Constraint On 'Assignment'
Alter Table Assignment
 Add (Constraint Avcon_1144240102_Const_000
 Check (Const_Type
 In ('Char', 'Text', 'Int', 'Real', 'Bool', 'Date',
    'Time', 'Db_Table')))/

Prompt Creating Check Constraint On 'Assignment'
```

```
Alter Table Assignment
 Add (Constraint Avcon_1144240102_Const_001
 Check (Const_Boolean_Value In ('T', 'F')))/


Prompt Creating Check Constraint On 'Sr_Ip_Desc'
Alter Table Sub_Routine_Ip_Description
 Add (Constraint Avcon_1144240102_Has_L_001
 Check (Has_Lov_Restrictions In ('T', 'F')))/


Prompt Creating Check Constraint On 'Sr_Ip_Desc'
Alter Table Sub_Routine_Ip_Description
 Add (Constraint Avcon_1144240102_Has_T_001
 Check (Has_Table_Restrictions In ('T', 'F')))/


Prompt Creating Check Constraint On 'Sr_Ip_Desc'
Alter Table Sub_Routine_Ip_Description
 Add (Constraint Avcon_1144240102_Has_O_001
 Check (Has_Other_Restrictions In ('T', 'F')))/


Prompt Creating Check Constraint On 'Predicate'
Alter Table Predicate
 Add (Constraint Avcon_1144240102_Left__000
 Check (Left_Operand_Type
 In ('Comp', 'Var', 'Const',
     'Func', 'Db_Table', 'Db_Field')))/


Prompt Creating Check Constraint On 'Predicate'
Alter Table Predicate
 Add (Constraint Avcon_1144240102_Right_000
 Check (Right_Operand_Type
 In ('Comp', 'Var', 'Const', 'Func',
     'Db_Table', 'Db_Field')))/


Prompt Creating Check Constraint On 'Pre_Post_Conds'
Alter Table Pre_Post_Conditions
 Add (Constraint Avcon_1144240102_Pre_P_000
 Check (Pre_Post In ('Pre', 'Post')))/


Prompt Creating Check Constraint On 'Actual_Output'
```

```
Alter Table Actual_Output
Add (Constraint Avcon_1144240102_Ao_Ca_000
Check (Ao_Category
In ('Db', 'Gui', 'File', 'Sequence',
    'Message', 'Report', 'Variable',
    'Role', 'Privilege', 'Commit',
    'Savepoint', 'Rollback')))/

Prompt Creating Check Constraint On 'Class_Relations'
Alter Table Class_Relationships
Add (Constraint Avcon_1144240102_Relat_000
Check (Relationship
In ('Inherits_From', 'Aggregate_Of',
    'Associated_With', 'Uses',
    'Instantiates', 'Metaclass_Of')))/

Prompt Creating Check Constraint On 'Triggers'
Alter Table Triggers
Add (Constraint Avcon_1144240102_On_Gu_000
Check (On_Gui_Object In ('T', 'F')))/

Prompt Creating Check Constraint On 'Triggers'
Alter Table Triggers
Add (Constraint Avcon_1144240102_On_Ta_000
Check (On_Table In ('T', 'F')))/

Prompt Creating Check Constraint On 'Triggers'
Alter Table Triggers
Add (Constraint Avcon_1144240102_Befor_000
Check (Before_After In ('B', 'A')))/

Prompt Creating Check Constraint On 'Triggers'
Alter Table Triggers
Add (Constraint Avcon_1144240102_Table_000
Check (Table_Insert In ('T', 'F')))/

Prompt Creating Check Constraint On 'Triggers'
Alter Table Triggers
Add (Constraint Avcon_1144240102_Table_001
```

```
Check (Table_Delete In ('T', 'F')))/

Prompt Creating Check Constraint On 'Triggers'
Alter Table Triggers
Add (Constraint Avcon_1144240102_Table_002
Check (Table_Update In ('T', 'F')))/

Prompt Creating Check Constraint On 'Triggers'
Alter Table Triggers
Add (Constraint Avcon_1144240102_For_E_000
Check (For_Each_Row In ('T', 'F')))/


Prompt Creating Check Constraint On 'Db_Oper_Details'
Alter Table Db_Operation_Details
Add (Constraint Avcon_1144240102_Field_000
Check (Field_Type
In ('Char', 'Text', 'Int', 'Real', 'Bool', 'Date',
    'Time', 'Db_Table')))/

Prompt Creating Check Constraint On 'Db_Oper_Details'
Alter Table Db_Operation_Details
Add (Constraint Avcon_1144240102_Bool__001
Check (Bool_Value In ('T', 'F')))/

Prompt Creating Check Constraint On 'Db_Oper_Details'
Alter Table Db_Operation_Details
Add (Constraint Avcon_1144240102_Whole_000
Check (Whole_Table In ('T', 'F')))/

Prompt Creating Check Constraint On 'Db_Oper_Details'
Alter Table Db_Operation_Details
Add (Constraint Avcon_1144240102_Order_000
Check (Order_By_Requested In ('T', 'F')))/

Prompt Creating Check Constraint On 'Db_Oper_Details'
Alter Table Db_Operation_Details
Add (Constraint Avcon_1144240102_Order_001
Check (Order_Desc In ('T', 'F')))/
```

```
Prompt Creating Check Constraint On 'Db_Oper_Details'
Alter Table Db_Operation_Details
 Add (Constraint Avcon_1144240102_Inser_000
 Check (Insert_Select In ('T', 'F')))/

Prompt Creating Check Constraint On 'Types'
Alter Table Types
 Add (Constraint Avcon_1144240102_Has_L_002
 Check (Has_Lov In ('T', 'F')))/

Prompt Creating Check Constraint On 'Types'
Alter Table Types
 Add (Constraint Avcon_1144240102_Has_R_000
 Check (Has_Range In ('T', 'F')))/

Prompt Creating Check Constraint On 'Types'
Alter Table Types
 Add (Constraint Avcon_1144240102_Even__000
 Check (Even_Odd In ('E', 'O')))/

Prompt Creating Check Constraint On 'Expected_Output'
Alter Table Expected_Output
 Add (Constraint Avcon_1144240102_Eo_Ca_000
 Check (Eo_Category
 In ('Db', 'Gui', 'File', 'Sequence',
     'Message', 'Report', 'Variable',
     'Role', 'Privilege', 'Commit',
     'Savepoint', 'Rollback')))/

Prompt Creating Check Constraint On 'Db_Operations'
Alter Table Db_Operations
 Add (Constraint Avcon_1144240102_Opera_003
 Check (Operation_Type
 In ('Select *', 'Select', 'Insert', 'Update',
     'Delete', 'Create View',
     'Create_Table', 'Drop_Table', 'Create_Seq',
     'Access_Seq', 'Create Role', 'Select Count',
     'Commit', 'Rollback', 'Savepoint',
```

```
                    'Grant', 'Revoke', 'Cursor', 'Fetch')))/

Prompt Creating Check Constraint On 'Db_Operations'
Alter Table Db_Operations
 Add (Constraint Avcon_1144240102_Seque_000
 Check (Sequence_Incremented In ('T', 'F')))/

Prompt Creating Check Constraint On 'Db_Operations'
Alter Table Db_Operations
 Add (Constraint Avcon_1144240102_Grant_000
 Check (Grant_Admin_Option In ('Grant', 'Admin')))/

Prompt Creating Check Constraint On 'Db_Operations'
Alter Table Db_Operations
 Add (Constraint Avcon_1144240102_As_Se_000
 Check (As_Select In ('T', 'F')))/

Prompt Creating Check Constraint On 'Type_Restricts'
Alter Table Type_Restrictions
 Add (Constraint Avcon_1144240102_Inclu_000
 Check (Include_Exclude In ('I', 'E')))/

Prompt Creating Check Constraint On 'Type_Restricts'
Alter Table Type_Restrictions
 Add (Constraint Avcon_1144240102_Has_R_001
 Check (Has_Range In ('T', 'F')))/

Prompt Creating Check Constraint On 'Type_Restricts'
Alter Table Type_Restrictions
 Add (Constraint Avcon_1144240102_Even__001
 Check (Even_Or_Odd In ('E', 'O')))/

Prompt Creating Check Constraint On 'Type_Excl_Ranges'
Alter Table Type_Excluded_Ranges
 Add (Constraint Avcon_1144240102_Has_R_002
 Check (Has_Range In ('T', 'F')))/

Prompt Creating Check Constraint On 'Sr_Ip_Instance'
Alter Table Sub_Routine_Ip_Instance
```

```
Add (Constraint Avcon_1144240102_Sourc_001
Check (Source_Type In ('Var', 'Const', 'Gui', 'Seq')))/

Prompt Creating Check Constraint On 'Sr_Ip_Instance'
Alter Table Sub_Routine_Ip_Instance
 Add (Constraint Avcon_1144240102_From__001
 Check (From_Bool_Const_Value In ('T', 'F')))/

Prompt Creating Check Constraint On 'Func_Decl'
Alter Table Function_Declaration
 Add (Constraint Avcon_1144240102_Op_Ha_000
 Check (Op_Has_Lov_Restrictions In ('T', 'F')))/

Prompt Creating Check Constraint On 'Func_Decl'
Alter Table Function_Declaration
 Add (Constraint Avcon_1144240102_Op_Ha_001
 Check (Op_Has_Table_Restictions In ('T', 'F')))/

Prompt Creating Check Constraint On 'Func_Decl'
Alter Table Function_Declaration
 Add (Constraint Avcon_1144240102_Op_Ha_002
 Check (Op_Has_Other_Restrictions In ('T', 'F')))/

Prompt Creating Check Constraint On 'Func_Decl'
Alter Table Function_Declaration
 Add (Constraint Avcon_1144240102_Chang_000
 Check (Changes_System_State In ('T', 'F')))/

Prompt Creating Check Constraint On 'Func_Decl'
Alter Table Function_Declaration
 Add (Constraint Avcon_1144240102_Encap_001
 Check (Encapsulation In ('Private',
                          'Protected', 'Public')))/

Prompt Creating Check Constraint On 'Var_Restricts'
Alter Table Var_Restrictions
 Add (Constraint Avcon_1144240102_Input_000
 Check (Input_Or_Output In ('I', 'O')))/
```

```
Prompt Creating Check Constraint On 'Var_Restricts'
Alter Table Var_Restrictions
 Add (Constraint Avcon_1144240102_Inclu_001
 Check (Include_Exclude In ('I', 'E')))/

Prompt Creating Check Constraint On 'Var_Restricts'
Alter Table Var_Restrictions
 Add (Constraint Avcon_1144240102_Has_R_003
 Check (Has_Range In ('T', 'F')))/

Prompt Creating Check Constraint On 'Var_Restricts'
Alter Table Var_Restrictions
 Add (Constraint Avcon_1144240102_Even__002
 Check (Even_Or_Odd In ('E', 'O')))/

Prompt Creating Check Constraint On 'Host_Var'
Alter Table Host_Var
 Add (Constraint Avcon_1144240102_Input_001
 Check (Input_Output In ('I', 'O')))/

Prompt Creating Check Constraint On 'Host_Var'
Alter Table Host_Var
 Add (Constraint Avcon_1144240102_Direc_000
 Check (Direct_Indirect In ('D', 'I')))/

Prompt Creating Check Constraint On 'Var_Tab_Restricts'
Alter Table Var_Table_Restrictions
 Add (Constraint Avcon_1144240102_Input_002
 Check (Input_Or_Output In ('I', 'O')))/

Prompt Creating Check Constraint On 'Operator'
Alter Table Operator
 Add (Constraint Avcon_1144240102_Over__001
 Check (Over_Loaded In ('T', 'F')))/

Prompt Creating Check Constraint On 'Expected_Input'
Alter Table Expected_Input
 Add (Constraint Avcon_1144240102_Ei_Ca_000
 Check (Ei_Category In ('Db', 'Gui', 'File')))/
```

```
Prompt Creating Check Constraint On 'Run_Sr_Post_Conds'
Alter Table Runtime_Sr_Post_Conditions
 Add (Constraint Avcon_1144240102_Sourc_002
 Check (Source In ('Var', 'Const', 'Gui', 'Seq')))/

Prompt Creating Check Constraint On 'Run_Sr_Post_Conds'
Alter Table Runtime_Sr_Post_Conditions
 Add (Constraint Avcon_1144240102_Post__001
 Check (Post_Cond_Type
 In ('Gui', 'Db', 'File', 'Report',
     'Seq', 'Message', 'Priv_Role',
     'Global_Var', 'Commit', 'Savepoint',
     'Rollback', 'Run_Exe', 'Open_Win',
     'Terminate')))/

Prompt Creating Check Constraint On 'Run_Sr_Post_Conds'
Alter Table Runtime_Sr_Post_Conditions
 Add (Constraint Avcon_1144240102_From__002
 Check (From_Bool_Const_Value In ('T', 'F')))/

Prompt Creating Check Constraint On 'Run_Sr_Post_Conds'
Alter Table Runtime_Sr_Post_Conditions
 Add (Constraint Avcon_1144240102_To_Fi_001
 Check (To_File_Change In ('Create', 'Delete',
                            'Moidify', 'Relocate')))/

Prompt Creating Check Constraint On 'Run_Sr_Post_Conds'
Alter Table Runtime_Sr_Post_Conditions
 Add (Constraint Avcon_1144240102_To_Re_001
 Check (To_Report_Destination In ('Printer',
                            'Screen', 'File')))/

Prompt Creating Check Constraint On 'Dataflow'
Alter Table Dataflow
 Add (Constraint Avcon_1144240102_Pos_T_000
 Check (Pos_Type
 In ('Initial', 'Begin', 'Input', 'Assign',
     'Decision', 'Whileloop',
```

```
      'Forloop', 'Run_Exe', 'Output', 'Call',
      'Db', 'Open_Window',
      'Construct_Object', 'Destruct_Object',
      'Call_Object_Method', 'End',
      'Endfor', 'End Loop', 'Terminal')))/

Prompt Creating Check Constraint On 'Dataflow'
Alter Table Dataflow
 Add (Constraint Avcon_1144240102_Resul_000
 Check (Result In ('T', 'F')))/

Prompt Creating Check Constraint On 'Procedure'
Alter Table Procedure
 Add (Constraint Avcon_1144240102_Chang_001
 Check (Changes_System_State In ('T', 'F')))/

Prompt Creating Check Constraint On 'Procedure'
Alter Table Procedure
 Add (Constraint Avcon_1144240102_Encap_002
 Check (Encapsulation In ('Private',
                          'Protected', 'Public')))/

Prompt Creating Check Constraint On 'Var_Lov'
Alter Table Var_Lov
 Add (Constraint Avcon_1144240102_Input_003
 Check (Input_Or_Output In ('I', 'O')))/

Prompt Creating Check Constraint On 'Constraints'
Alter Table Constraints
 Add (Constraint Avcon_1144240102_Const_002
 Check (Constraint_Or_Assertion In ('C', 'A')))/

Prompt Creating Check Constraint On 'Constraints'
Alter Table Constraints
 Add (Constraint Avcon_1144240102_On_Fi_000
 Check (On_Field In ('T', 'F')))/

Prompt Creating Check Constraint On 'Calculation'
Alter Table Calculation
```

```
Add (Constraint Avcon_1144240102_Left__001
Check (Left_Operand_Type
In ('Comp', 'Var', 'Const', 'Func',
    'Db_Table', 'Db_Field')))/
```

```
Prompt Creating Check Constraint On 'Calculation'
Alter Table Calculation
Add (Constraint Avcon_1144240102_Right_001
Check (Right_Operand_Type
In ('Comp', 'Var', 'Const', 'Func',
    'Db_Table', 'Db_Field')))/
```

```
Prompt Creating Check Constraint On 'Tab_Field_Desc'
Alter Table Table_Field_Description
 Add (Constraint Avcon_1144240102_Has_C_000
 Check (Has_Constraint_Restriction In ('T', 'F')))/
```

```
Prompt Creating Check Constraint On 'Tab_Field_Desc'
Alter Table Table_Field_Description
 Add (Constraint Avcon_1144240102_Is_Pr_000
 Check (Is_Primary_Key In ('T', 'F')))/
```

```
Prompt Creating Check Constraint On 'Tab_Field_Desc'
Alter Table Table_Field_Description
 Add (Constraint Avcon_1144240102_Is_Un_000
 Check (Is_Unique In ('T', 'F')))/
```

```
Prompt Creating Check Constraint On 'Tab_Field_Desc'
Alter Table Table_Field_Description
 Add (Constraint Avcon_1144240102_Manda_001
 Check (Mandatory In ('T', 'F')))/
```

```
Prompt Creating Check Constraint On 'Tab_Field_Desc'
Alter Table Table_Field_Description
 Add (Constraint Avcon_1144240102_Has_L_003
 Check (Has_Lov_Restriction In ('T', 'F')))/
```

```
Prompt Creating Check Constraint On 'Tab_Field_Desc'
Alter Table Table_Field_Description
```

```
    Add (Constraint Avcon_1144240102_Has_T_002
    Check (Has_Table_Restriction In ('T', 'F')))/

Prompt Creating Check Constraint On 'Tab_Field_Desc'
Alter Table Table_Field_Description
 Add (Constraint Avcon_1144240102_Has_R_004
 Check (Has_Range_Restriction In ('T', 'F')))/

Prompt Creating Foreign Key On 'Valid_Operands'
Alter Table Valid_Operands
 Add (Constraint Oprnd_Opr_Fk Foreign Key
 (Operator_Id) References Operator (Operator_Id))/

Prompt Creating Foreign Key On 'Des_Sr_Post_Conds'
Alter Table Designed_Sr_Post_Conditions
 Add (Constraint Post_Cond_Proc_Fk Foreign Key
 (Sub_Routine_Name) References
 Procedure (Proc_Name))/

Prompt Creating Foreign Key On 'Des_Sr_Post_Conds'
Alter Table Designed_Sr_Post_Conditions
 Add (Constraint Post_Cond_Fdcl_Fk Foreign Key
 (Sub_Routine_Name) References
 Function_Declaration (Func_Name))/

Prompt Creating Foreign Key On 'Var'
Alter Table Var
 Add (Constraint Var_Cls_Fk Foreign Key
 (Class_Name) References Classes (Class_Name))/

Prompt Creating Foreign Key On 'Var'
Alter Table Var
 Add (Constraint Var_Odn_Fk Foreign Key
 (Var_Id) References
 Input_Output_Definition (Var_Id))/

Prompt Creating Foreign Key On 'Var'
Alter Table Var
 Add (Constraint Var_Obj_Fk Foreign Key
```

```
(Object_Name) References Objects (Object_Name))/

Prompt Creating Foreign Key On 'Types_Lov'
Alter Table Types_Lov
 Add (Constraint Tlov_Tdesc_Fk Foreign Key
 (Type_Name) References
 Table_Field_Description (Field_Id))/

Prompt Creating Foreign Key On 'Types_Lov'
Alter Table Types_Lov
 Add (Constraint Tlov_Tps_3_Fk Foreign Key
 (Type_Name) References Types (Type_Name))/

Prompt Creating Foreign Key On 'Assignment'
Alter Table Assignment
 Add (Constraint Fun_Calc_Fk Foreign Key
 (Cal_Num) References Calculation (Cal_Num))/

Prompt Creating Foreign Key On 'Assignment'
Alter Table Assignment
 Add (Constraint Fun_Daw_Fk Foreign Key
 (Pos_Num) References Dataflow (Pos_Num))/

Prompt Creating Foreign Key On 'Assignment'
Alter Table Assignment
 Add (Constraint Fun_Fdcl Foreign Key
 (Func_Name) References
 Function_Declaration (Func_Name))/

Prompt Creating Foreign Key On 'Sr_Ip_Desc'
Alter Table Sub_Routine_Ip_Description
 Add (Constraint Srp_Proc_Fk Foreign Key
 (Subroutine_Name) References
 Procedure (Proc_Name))/

Prompt Creating Foreign Key On 'Sr_Ip_Desc'
Alter Table Sub_Routine_Ip_Description
 Add (Constraint Finrest1_Fdcl_Fk Foreign Key
 (Subroutine_Name) References
```

Function_Declaration (Func_Name))/

Prompt Creating Foreign Key On 'Sr_Ip_Desc'
Alter Table Sub_Routine_Ip_Description
 Add (Constraint Srp_Tps_3_Fk Foreign Key
 (Type) References Types (Type_Name))/

Prompt Creating Foreign Key On 'Predicate'
Alter Table Predicate
 Add (Constraint Pre_Opr Foreign Key
 (Operator_Id) References Operator (Operator_Id))/

Prompt Creating Foreign Key On 'Predicate'
Alter Table Predicate
 Add (Constraint Right_Var_Fk Foreign Key
 (Right_Var_Id) References Var (Var_Id))/

Prompt Creating Foreign Key On 'Predicate'
Alter Table Predicate
 Add (Constraint Pre_Trg_Fk2 Foreign Key
 (Pos_Num) References Triggers (Trigger_Name))/

Prompt Creating Foreign Key On 'Predicate'
Alter Table Predicate
 Add (Constraint Pre_Daw_Fk Foreign Key
 (Pos_Num) References Dataflow (Pos_Num))/

Prompt Creating Foreign Key On 'Predicate'
Alter Table Predicate
 Add (Constraint Left_Comp_Fk Foreign Key
 (Left_Component_Num ,Pos_Num) References
 Predicate (Component_Num ,Pos_Num))/

Prompt Creating Foreign Key On 'Predicate'
Alter Table Predicate
 Add (Constraint Left_Var_Fk Foreign Key
 (Left_Var_Id) References Var (Var_Id))/

Prompt Creating Foreign Key On 'Predicate'

```
Alter Table Predicate
Add (Constraint Right_Comp_Fk Foreign Key
(Right_Component_Num  ,Pos_Num) References
Predicate  (Component_Num ,Pos_Num))/


Prompt Creating Foreign Key On 'Pre_Post_Conds'
Alter Table Pre_Post_Conditions
Add (Constraint Ppn_Dbto_Fk Foreign Key
(Db_Operation_Id) References
Db_Test_Operation (Operation_Id))/


Prompt Creating Foreign Key On 'Class_Relations'
Alter Table Class_Relationships
Add (Constraint Crfk2 Foreign Key
(Class_Name2) References Classes (Class_Name))/


Prompt Creating Foreign Key On 'Class_Relations'
Alter Table Class_Relationships
Add (Constraint Crfk1 Foreign Key
(Class_Name1) References Classes (Class_Name))/


Prompt Creating Foreign Key On 'Input_Output_Def'
Alter Table Input_Output_Definition
Add (Constraint Odn_Daw_Fk Foreign Key
(Pos_Num) References Dataflow (Pos_Num))/


Prompt Creating Foreign Key On 'Db_Oper_Details'
Alter Table Db_Operation_Details
Add (Constraint Op_Detail_Fk Foreign Key
(Db_Operation_Num) References
Db_Operations (Db_Operation_Num))/


Prompt Creating Foreign Key On 'Test_Templates'
Alter Table Test_Templates
Add (Constraint Tte_Dbto_Fk Foreign Key
(Db_Operation_Id) References
Db_Test_Operation (Operation_Id))/


Prompt Creating Foreign Key On 'Type_Restricts'
```

```
Alter Table Type_Restrictions
Add (Constraint Type_Res_Tdesc_Fk Foreign Key
(Type_Name) References
Table_Field_Description (Field_Id))/


Prompt Creating Foreign Key On 'Type_Restricts'
Alter Table Type_Restrictions
Add (Constraint Type_Res_Tps_Fk Foreign Key
(Type_Name) References
Types-Old (Type_Name))/


Prompt Creating Foreign Key On 'Attrs_Changed'
Alter Table Attributes_Changed
Add (Constraint Ace_Dbto_Fk Foreign Key
(Db_Operation_Id) References
Db_Test_Operation (Operation_Id))/


Prompt Creating Foreign Key On 'Type_Excl_Ranges'
Alter Table Type_Excluded_Ranges
Add (Constraint Tres_1_Tps_3_Fk Foreign Key
(Type_Name) References Types (Type_Name))/


Prompt Creating Foreign Key On 'Type_Excl_Ranges'
Alter Table Type_Excluded_Ranges
Add (Constraint Type_Exran_Tdesc_Fk Foreign Key
(Type_Name) References
Table_Field_Description (Field_Id))/


Prompt Creating Foreign Key On 'Sr_Ip_Instance'
Alter Table Sub_Routine_Ip_Instance
Add (Constraint Srp_1_Obj_Fk Foreign Key
(Object_Name) References
Objects (Object_Name))/


Prompt Creating Foreign Key On 'Sr_Ip_Instance'
Alter Table Sub_Routine_Ip_Instance
Add (Constraint Sp_Fk Foreign Key
(Subroutine_Name  ,Ip_Order) References
Sub_Routine_Ip_Description (Subroutine_Name,
```

                    Ip_Order))/

Prompt Creating Foreign Key On 'Sr_Ip_Instance'
Alter Table Sub_Routine_Ip_Instance
 Add (Constraint Srp_Proc1_Fk Foreign Key
 (Subroutine_Name) References
 Procedure_1 (Proc_Name))/

Prompt Creating Foreign Key On 'Sr_Ip_Instance'
Alter Table Sub_Routine_Ip_Instance
 Add (Constraint Finrest_Fdcl_Fk Foreign Key
 (Subroutine_Name) References
 Function_Declaration_1 (Func_Name))/

Prompt Creating Foreign Key On 'Func_Decl'
Alter Table Function_Declaration
 Add (Constraint Fdcl_Cls_Fk Foreign Key
 (Class_Name) References Classes (Class_Name))/

Prompt Creating Foreign Key On 'Func_Decl'
Alter Table Function_Declaration
 Add (Constraint Fdcl_Tps_3_Fk Foreign Key
 (Op_Type) References Types (Type_Name))/

Prompt Creating Foreign Key On 'Var_Restricts'
Alter Table Var_Restrictions
 Add (Constraint Varres_Fdcl_Fk Foreign Key
 (Sub_Routine_Name) References
 Function_Declaration (Func_Name))/

Prompt Creating Foreign Key On 'Var_Restricts'
Alter Table Var_Restrictions
 Add (Constraint Varres_Var_Fk Foreign Key
 (Var_Id) References Var (Var_Id))/

Prompt Creating Foreign Key On 'Var_Restricts'
Alter Table Var_Restrictions
 Add (Constraint Var_Fip_Fk Foreign Key
 (Sub_Routine_Name ,Var_Name) References

```
Sub_Routine_Ip_Description (Subroutine_Name,
                            Sr_Var_Name))/

Prompt Creating Foreign Key On 'Host_Var'
Alter Table Host_Var
Add (Constraint Hvr_Dbto_Fk Foreign Key
(Db_Operation_Id) References
Db_Test_Operation (Operation_Id))/

Prompt Creating Foreign Key On 'Objects'
Alter Table Objects
Add (Constraint Obj_Cls_Fk Foreign Key
(Class_Name) References Classes (Class_Name))/

Prompt Creating Foreign Key On 'Data_Group'
Alter Table Data_Group
Add (Constraint Dgp_Dbto_Fk Foreign Key
(Db_Operation_Id) References
Db_Test_Operation (Operation_Id))/

Prompt Creating Foreign Key On 'Var_Table_Restricts'
Alter Table Var_Table_Restrictions
Add (Constraint Vtr_Fip_Fk Foreign Key
(Sub_Routine_Name ,Var_Name) References
Sub_Routine_Ip_Description (Subroutine_Name,
                            Sr_Var_Name))/

Prompt Creating Foreign Key On 'Var_Table_Restricts'
Alter Table Var_Table_Restrictions
Add (Constraint Vtr_Fdcl Foreign Key
(Sub_Routine_Name) References
Function_Declaration (Func_Name))/

Prompt Creating Foreign Key On 'Var_Table_Restricts'
Alter Table Var_Table_Restrictions
Add (Constraint Vtr_Var_Fk Foreign Key
(Var_Id) References Var (Var_Id))/

Prompt Creating Foreign Key On 'Runtime_Sr_Post_Conds'
```

```
Alter Table Runtime_Sr_Post_Conditions
 Add (Constraint Post_Con_1_Fdcl_Fk Foreign Key
 (Sub_Routine_Name) References
 Function_Declaration (Func_Name))/

Prompt Creating Foreign Key On 'Runtime_Sr_Post_Conds'
Alter Table Runtime_Sr_Post_Conditions
 Add (Constraint Post_Con_1_Proc_Fk Foreign Key
 (Sub_Routine_Name) References Procedure (Proc_Name))/

Prompt Creating Foreign Key On 'Equivalence_Class'
Alter Table Equivalence_Class
 Add (Constraint Ecs_Dgp_Fk Foreign Key
 (Db_Operation_Id  ,Data_Group_Name) References
 Data_Group (Db_Operation_Id  ,Data_Group_Name))/

Prompt Creating Foreign Key On 'Dataflow'
Alter Table Dataflow
 Add (Constraint Daw_Obj_Fk Foreign Key
 (Object_Name) References Objects (Object_Name))/

Prompt Creating Foreign Key On 'Dataflow'
Alter Table Dataflow
 Add (Constraint Daw_Cls_Fk Foreign Key
 (Class_Name) References Classes (Class_Name))/

Prompt Creating Foreign Key On 'Procedure'
Alter Table Procedure
 Add (Constraint Proc_Cls_Fk Foreign Key
 (Class_Name) References Classes (Class_Name))/

Prompt Creating Foreign Key On 'Db_Operation_Tables'
Alter Table Db_Operation_Tables
 Add (Constraint Otab_Con_Fk Foreign Key
 (Db_Constraint_Num) References
 Constraints (Constraint_Num))/

Prompt Creating Foreign Key On 'Db_Operation_Tables'
Alter Table Db_Operation_Tables
```

```
Add (Constraint Otab_Db_Oper_Fk Foreign Key
(Db_Operation_Num) References
Db_Operations (Db_Operation_Num))/

Prompt Creating Foreign Key On 'Data_Group_Values'
Alter Table Data_Group_Values
Add (Constraint Dgv_Dgp_Fk Foreign Key
(Db_Operation_Id ,Data_Group_Name) References
Data_Group (Db_Operation_Id ,Data_Group_Name))/

Prompt Creating Foreign Key On 'Var_Lov'
Alter Table Var_Lov
Add (Constraint Vlov_Fdcl Foreign Key
(Var_Id ,Sub_Routine_Name) References
Function_Declaration (Op_Var_Id ,Func_Name))/

Prompt Creating Foreign Key On 'Var_Lov'
Alter Table Var_Lov
Add (Constraint Vlov_Var_Fk Foreign Key
(Var_Id) References Var (Var_Id))/

Prompt Creating Foreign Key On 'Var_Lov'
Alter Table Var_Lov
Add (Constraint Vlov_Srp_Fk Foreign Key
(Sub_Routine_Name ,Var_Name) References
Sub_Routine_Ip_Description (Subroutine_Name,
                            Sr_Var_Name))/

Prompt Creating Foreign Key On 'Constraints'
Alter Table Constraints
Add (Constraint Con_Db_Oper_Fk Foreign Key
(Db_Operation_Num) References
Db_Operations (Db_Operation_Num))/

Prompt Creating Foreign Key On 'Constraints'
Alter Table Constraints
Add (Constraint Con_Tdesc_Fk Foreign Key
(Field_Name ,Table_Name) References
Table_Field_Description (Field_Name ,Table_Name))/
```

```
Prompt Creating Foreign Key On 'Constraints'
Alter Table Constraints
 Add (Constraint Con_Opr_Fk Foreign Key
 (Db_Operator_Id) References
 Operator (Operator_Id))/


Prompt Creating Foreign Key On 'Calculation'
Alter Table Calculation
 Add (Constraint Right_Comp_Var_Fk2 Foreign Key
 (Right_Var_Id) References Var (Var_Id))/


Prompt Creating Foreign Key On 'Calculation'
Alter Table Calculation
 Add (Constraint Left_Comp_Var_Fk2 Foreign Key
 (Left_Var_Id) References Var (Var_Id))/


Prompt Creating Foreign Key On 'Calculation'
Alter Table Calculation
 Add (Constraint Left_Comp_Calc_Fk Foreign Key
 (Left_Component_Num ,Cal_Num) References
 Calculation (Component_Num ,Cal_Num))/


Prompt Creating Foreign Key On 'Calculation'
Alter Table Calculation
 Add (Constraint Right_Comp_Calc_Fk Foreign Key
 (Right_Component_Num ,Cal_Num) References
 Calculation (Component_Num ,Cal_Num))/


Prompt Creating Foreign Key On 'Calculation'
Alter Table Calculation
 Add (Constraint Calc_Opr_Fk Foreign Key
 (Operator_Id) References Operator (Operator_Id))/


Prompt Creating Foreign Key On 'Calculation'
Alter Table Calculation
 Add (Constraint Calc_Fdcl_Fk2 Foreign Key
 (Right_Func_Name) References
 Function_Declaration (Func_Name))/
```

```
Prompt Creating Foreign Key On 'Calculation'
Alter Table Calculation
 Add (Constraint Calc_Fdcl_Fk Foreign Key
 (Left_Func_Name) References
 Function_Declaration (Func_Name))/
```

Appendix B

PL/SQL Constraint Generator Code

```
PROCEDURE "CONSTRAINT_LOADER"
   (Directory IN VARCHAR2,
    File_Name IN VARCHAR2) IS

T_Table_Name                        varchar2(240);
T_Field_ID                          varchar2(240);
T_Field_Name                        varchar2(240);
T_Field_Type                        varchar2(240);
T_Primary_Key                       varchar2(1);
T_Is_Unique                         varchar2(1);
T_Mandatory                         varchar2(1);
T_LENGTH                            INT;
T_Has_LOV_Restrictions              varchar2(1);
T_Has_Table_Restrictions            varchar2(1);
T_Has_Range_Restrictions            varchar2(1);
T_Has_Constraint_Restrictions       varchar2(1);
T_Num_of_Constraint_Restrictions    INT;
T_Which_Table                       varchar2(240);
T_Which_Field                       varchar2(240);

F_Type_Name           VARCHAR2(20);
F_Restriction_Num     int;
F_Include_Exclude     VARCHAR2(1);
F_Has_Range           char(1);
F_Even_Or_Odd         VARCHAR2(1);
F_Range_From_Int      int;
F_Range_To_Int        int;
```

```
F_Int_Value              int;
F_Range_From_Char        char;
F_Range_To_Char          char;
F_Char_Value             char;
F_Range_From_Real        real;
F_Range_To_Real          real;
F_Real_Value             real;
F_Range_From_Date        date;
F_Range_To_Date          date;
F_Date_Value             date;
F_Range_From_Time        date;
F_Range_To_Time          date;
F_Time_Value             date;
F_Text_Value             VARCHAR2(240);

ScriptFile UTL_FILE.FILE_TYPE;

LOV_Tab_Name varchar2(20);

included_ranges int;
excluded int;
int1 int;
int2 int;
char1 char;
char2 char;
real1 real;
real2 real;
date1 date;
date2 date;

CURSOR  Field_Desc IS
  Select *
From Table_Field_Description;

CURSOR  field_restrictions IS
  Select *
From TYPE_EXCLUDED_RANGES
Where Type_Name = T_Field_ID;
```

```
BEGIN
  -- A  file should already be created,
  -- or else, an exception handler should be created
ScriptFile := UTL_FILE.FOPEN (Directory, File_Name, 'w');


OPEN Field_Desc;
LOOP
  Fetch Field_Desc Into T_Table_Name,
  T_Field_ID,
T_Field_Name,
T_Field_Type,
T_Primary_Key,
T_Is_Unique,
T_Mandatory,
T_LENGTH,
T_Has_LOV_Restrictions,
T_Has_Table_Restrictions,
T_Has_Range_Restrictions,
T_Has_Constraint_Restrictions,
T_Num_of_Constraint_Restrictions,
T_Which_Table,
T_Which_Field;

EXIT WHEN Field_Desc%NOTFOUND;

-- Primary key constraint.
IF (T_Primary_Key = 'T') OR
   (T_Primary_Key = 't') THEN
  UTL_FILE.PUTF(ScriptFile,
  'ALTER TABLE %s ' ||
  'ADD CONSTRAINT %s_PK PRIMARY KEY (%s); \n',
  T_Table_Name,
  T_Field_Name,
  T_Field_Name);
END IF;
```

```
-- Unique field constraint.
IF (T_Is_Unique = 'T') OR
   (T_Is_Unique = 't') THEN
  UTL_FILE.PUTF(ScriptFile,
    'ALTER TABLE %s ' ||
    'ADD CONSTRAINT %s_UK UNIQUE (%s); \n',
    T_Table_Name,
    T_Field_Name,
    T_Field_Name);
END IF;

-- Not NULL constraint.
IF (T_Mandatory = 'M') OR
   (T_Mandatory = 'm') THEN
  UTL_FILE.PUTF(ScriptFile,
  'ALTER TABLE %s ' ||
  'MODIFY %s NOT NULL; \n',
    T_Table_Name,
    T_Field_Name);
END IF;

-- Has a List of Valid Values.
IF (T_Has_LOV_Restrictions = 'T') OR
   (T_Has_LOV_Restrictions = 't') THEN
  LOV_Tab_Name := T_Table_Name || '_' ||
                  T_Field_Name || '_LOV';
  UTL_FILE.PUTF(ScriptFile,
    'CREATE TABLE %s (%s Varchar2, valid_value %s);\n',
    LOV_Tab_Name, T_Field_Name,T_Field_Type);

  IF T_Field_Type IN ('INTEGER','SMALLINT') THEN
    UTL_FILE.PUTF(ScriptFile,
      'INSERT INTO %s ' ||
      'SELECT INT_Value ' ||
      'FROM Types_LOV ' ||
      'WHERE Type_Name = %s ;\n',
      LOV_Tab_Name,
      T_Field_ID);
```

```
ELSIF T_Field_Type IN ('CHAR'
                       'VARCHAR2',
                       'LONG') THEN
  UTL_FILE.PUTF(ScriptFile,
  'INSERT INTO %s ' ||
  'SELECT CHAR_Value ' ||
  'FROM Types_LOV ' ||
     'WHERE Type_Name = %s ;\n',
     LOV_Tab_Name,
     T_Field_ID);

ELSIF T_Field_Type IN ('NUMBER','REAL') THEN
  UTL_FILE.PUTF(ScriptFile,
     'INSERT INTO %s ' ||
     'SELECT REAL_Value ' ||
     'FROM Types_LOV ' ||
     'WHERE Type_Name = %s ;\n',
     LOV_Tab_Name,
     T_Field_ID);

ELSIF T_Field_Type IN ('BOOLEAN') THEN
  UTL_FILE.PUTF(ScriptFile,
     'INSERT INTO %s ' ||
     'SELECT BOOL_Value ' ||
     'FROM Types_LOV ' ||
     'WHERE Type_Name = %s ;\n',
     LOV_Tab_Name,
     T_Field_ID);

-- Other types can also be added here,
-- if in the Types_LOV table,
-- just change the string value of the
-- T_Field_Type and the name of the field
-- after the Select statement.
END IF;

UTL_FILE.PUTF(ScriptFile,
  'ALTER TABLE %s ' ||
  'ADD CONSTRAINT %s_FK FOREIGN KEY(%s) ' ||
```

```
   'REFERENCES %s(valid_value); \n',
   T_Table_Name,
   T_Field_Name,
   T_Field_Name,
   LOV_Tab_Name);
END IF;


-- References another field in the same
-- or a different table.
IF (T_Has_Table_Restrictions = 'T') OR
   (T_Has_Table_Restrictions = 't') THEN
  UTL_FILE.PUTF(ScriptFile,
    'ALTER TABLE %s ' ||
    'ADD CONSTRAINT %s_FK FOREIGN KEY(%s) ' ||
    'REFERENCES %s(%s); \n',
    T_Table_Name,
    T_Field_Name,
    T_Field_Name,
    T_Which_Table,
    T_Which_Field);
END IF;


-- If there are other restrictions in the range,
-- including certain numbers, or excluding others, etc.
IF (T_Has_Range_Restrictions = 'T') OR
   (T_Has_Range_Restrictions = 't') THEN

   -- Look in TYPE, to define the range
   -- of valid values.
   IF T_Field_Type IN ('INTEGER','SMALLINT') THEN
     select min(range_from_int),
            max(range_to_int) into int1, int2
     from TYPES
     where type_name = T_Field_ID and
           upper (has_range) = 'T';

     UTL_FILE.PUTF(ScriptFile,
       'ALTER TABLE %s ' ||
       'ADD CONSTRAINT %s_CK ' ||
```

```
        'CHECK(%s BETWEEN %s AND %s); \n',
        T_Table_Name,
        T_Field_Name,
        T_Field_Name,
        int1,
        int2);

ELSIF T_Field_Type ='CHAR' THEN
   select min(range_from_char),
          max(range_to_char)
          into char1, char2 from TypeS
   where type_name = T_Field_ID and
          upper (has_range) = 'T';

   UTL_FILE.PUTF(ScriptFile,
     'ALTER TABLE %s ' ||
     'ADD CONSTRAINT %s_CK ' ||
     'CHECK(%s BETWEEN %s AND %s); \n',
     T_Table_Name,
     T_Field_Name,
     T_Field_Name,
     char1,
     char2);

ELSIF T_Field_Type IN ('NUMBER','REAL') THEN
   select min(range_from_REAL),
          max(range_to_REAL)
          into REAL1, REAL2
   from TypeS
   where type_name = T_Field_ID and
          upper (has_range) = 'T';

   UTL_FILE.PUTF(ScriptFile,
     'ALTER TABLE %s ' ||
     'ADD CONSTRAINT %s_CK ' ||
     'CHECK(%s BETWEEN %s AND %s); \n',
     T_Table_Name,
     T_Field_Name,
     T_Field_Name,
```

```
        REAL1,
        REAL2);

  ELSIF T_Field_Type ='DATE' THEN
    select min(range_from_DATE),
           max(range_to_DATE)
           into DATE1, DATE2
    from TypeS
    where type_name = T_Field_ID and
          upper (has_range) = 'T';

    UTL_FILE.PUTF(ScriptFile,
      'ALTER TABLE %s ' ||
      'ADD CONSTRAINT %s_CK ' ||
      'CHECK(%s BETWEEN %s AND %s); \n',
      T_Table_Name,
      T_Field_Name,
      T_Field_Name,
      DATE1,
      DATE2);

--ELSIF T_Field_Type = 'TIME' THEN
--   select range_from_TIME, range_to_TIME
--           into TIME1, TIME2
--   from Type_Restrictions
--   where type_name = T_Field_ID;

--   UTL_FILE.PUTF(ScriptFile,
--     'ALTER TABLE %s ' ||
--     'ADD CONSTRAINT %s_CK ' ||
--     'CHECK(%s BETWEEN %s AND %s); \n',
--     T_Table_Name,
--     T_Field_Name,
--     T_Field_Name,
--     TIME1,
--     TIME2);
  END IF;

  Select count(*)
```

```
          into excluded
From TYPE_EXCLUDED_RANGES
Where Type_Name = T_Field_ID ;

-- There is at least one
-- excluded range or value.
IF excluded > 0 THEN
  OPEN field_restrictions;
  LOOP
    Fetch field_restrictions Into
      F_Type_Name,
      F_Restriction_Num,
      F_Has_Range,
      F_Range_From_Int,
      F_Range_To_Int,
      F_Int_Value,
      F_Range_From_Char,
      F_Range_To_Char,
      F_Char_Value,
      F_Range_From_Real,
      F_Range_To_Real,
      F_Real_Value,
      F_Range_From_Date,
      F_Range_To_Date,
      F_Date_Value,
      F_Range_From_Time,
      F_Range_To_Time,
      F_Time_Value,
      F_Text_Value;
  EXIT WHEN field_restrictions%NOTFOUND;

IF T_Field_Type IN ('INTEGER','SMALLINT') THEN
  IF upper(F_Has_Range) = 'T' THEN
    UTL_FILE.PUTF(ScriptFile,
      'ALTER TABLE %s ' ||
      'ADD CONSTRAINT %s_CK ' ||
      'CHECK(%s NOT BETWEEN %s AND %s); \n',
      T_Table_Name,
      T_Field_Name,
```

```
            T_Field_Name,
            F_Range_From_Int,
            F_Range_To_Int);

        ELSE
          UTL_FILE.PUTF(ScriptFile,
            'ALTER TABLE %s ' ||
            'ADD CONSTRAINT %s_CK ' ||
            'CHECK(%s <> %s); \n',
            T_Table_Name,
            T_Field_Name,
            T_Field_Name,
            F_Int_Value);

        END IF; -- Has range or single value.

      ELSIF T_Field_Type = 'CHAR'  THEN
        IF upper(F_Has_Range) = 'T' THEN
          UTL_FILE.PUTF(ScriptFile,
            'ALTER TABLE %s ' ||
            'ADD CONSTRAINT %s_CK ' ||
            'CHECK(%s NOT BETWEEN %s AND %s); \n',
            T_Table_Name,
            T_Field_Name,
            T_Field_Name,
            F_Range_From_Char,
            F_Range_To_Char);

        ELSE
          UTL_FILE.PUTF(ScriptFile,
            'ALTER TABLE %s ' ||
            'ADD CONSTRAINT %s_CK ' ||
            'CHECK(%s <> %s); \n',
            T_Table_Name,
            T_Field_Name,
            T_Field_Name,
            F_Char_Value);
        END IF; -- Has range or single value.
```

```
ELSIF T_Field_Type IN ('NUMBER','REAL') THEN
  IF upper(F_Has_Range) = 'T' THEN
    UTL_FILE.PUTF(ScriptFile,
      'ALTER TABLE %s ' ||
      'ADD CONSTRAINT %s_CK ' ||
      'CHECK(%s NOT BETWEEN %s AND %s); \n',
      T_Table_Name,
      T_Field_Name,
      T_Field_Name,
      F_Range_From_Real,
      F_Range_To_Real);

  ELSE
    UTL_FILE.PUTF(ScriptFile,
      'ALTER TABLE %s ADD CONSTRAINT %s_CK
      CHECK(%s <> %s); \n', T_Table_Name,
      T_Field_Name,
      T_Field_Name,
      F_Real_Value);
  END IF; -- Has range or single value.

ELSE IF T_Field_Type ='DATE' THEN
  IF upper(F_Has_Range) = 'T' THEN
    UTL_FILE.PUTF(ScriptFile,
      'ALTER TABLE %s ADD CONSTRAINT %s_CK
      CHECK(%s NOT BETWEEN %s AND %s); \n',
      T_Table_Name,
      T_Field_Name,
      T_Field_Name,
      F_Range_From_Date,
      F_Range_To_Date);
  ELSE
    UTL_FILE.PUTF(ScriptFile,
      'ALTER TABLE %s ADD CONSTRAINT %s_CK
      CHECK(%s <> %s); \n',
      T_Table_Name,
      T_Field_Name,
      T_Field_Name,
      F_Date_Value);
```

```
    END IF;  -- Has range or single value.

    END LOOP;
  END IF;  -- There is an excluded range or value.

END IF;
END LOOP;

  UTL_FILE.FFLUSH(ScriptFile);
  CLOSE Field_Desc;

  COMMIT;
  UTL_FILE.FCLOSE (ScriptFile);
END;
```

Bibliography

1. Brown, Norm, Industrial - Strength Management Strategies. IEEE Software, July 1996.

2. Mann, Charles, Why Software Is So Bad, Technology Review, July / August 2002.

3. Kaner, Cem., The Impossibility of complete testing, Software QA, Volume 4, Number 4, p. 28, 1997.

4. Whittaker, James, What is Software Testing? And Why is it So Hard?, IEEE Software, January/February 2000.

5. Jorgensen, Paul, Software Testing: A Craftsman's Approach. CRC Press, 2nd Ed. 2002.

6. Kaner, Cem, Jack Falk and Hung Q. Nguyen, Testing Computer Software, John Wiley & Sons, Inc., 1999.

7. Sommerville, Ian, Software Engineering. 4th ed., Addison-Wesley, 1993.

8. Langer, Arthur M., Analysis and Design of Information Systems. Springer-Verlang New York, Inc. Second ed., 1997.

9. Jackson, Michael, Software Requirements & Specifications: A lexion of Practice, Principles, and Prejudices. ACM Press Books 1995.

10. Kaner, Cem and James Bach, Paradigms of black box software testing, 16th International Conference and Exposition on Testing Computer Software, Washington, D.C., June, 1999.

11. Eriksson, Hans-Erik and Magnus Penker, Business Modeling with UML: business patterns at work. John Wiley & Sons, Inc., 2000.

12. Naiburg, Eric and Robert Maksimchuk, UML for Database Design. Addison Wesley, 2001.

13. Fowler, Martin, UML Distilled: Applying the standard object modeling language. Addison Wesley Longman, Inc. 1997.

14. Eriksson, Hans-Erik and Magnus Penker, Business Modeling with UML: Business Patterns at work. John Wiley & Sons, Inc. 2000.

15. Bowen, Jonathan, Formal Specification & Documentation using Z: A case study approach. International Thomson Publishing Inc., 1996.

16. Pagan, Frank, Formal Specifications of Programming Languages: A Panoramic Primer. Prentice-Hall, Inc. 1981.

17. Nissanke, Nimal, Formal Specifications: techniques and applications. Springer-Verlag, 1999.

18. Spivey, J.M. The Z Notation: A Reference Manual, Second Ed. Prentice Hall, Int. Series in Computer Science, 1989.

19. Spivey, J.M. An Introduction to Z and formal specifications. ACM Software Engineering Journal, Vol. 4 Issue 1, January 1989.

20. Potter, Ben and Jane Sinclair, An Introduction to Formal Specification and Z. Second Edition, Prentice Hall 1996.

21. Jacky, Jonathan, The way of Z: Practical programming with formal methods. Cambridge University Press,1997.

22. Feijs, L.M and H. Jonkers, Formal Specifiications and Design. Cambridge University press, 1992.

23. Spivey, J.M, Understanding Z: A specification language and its formal semantics. Cambridge University Press 1988.

24. Harry, Andres, Formal Methods Fact File: VDM and Z. John Wiley & Sons 1996.

25. ApTest: Software Testing Resources and Tools.
www.aptest.com/resources.html

26. Oracle Developer Suite: Product Editions.
http://www.oracle.com/tools/index.html?tools_editions.html

27. David Chays , Yuetang Deng, et al., Demonstration of AGENDA tool

set for testing relational database applications, Proceedings of the 25th international conference on Software engineering, p.802, May 03-10, 2003, Portland, Oregon.

28. David Chays, Saikat Dan, et al., A framework for testing database applications, Proceedings of the International Symposium on Software Testing and Analysis, Portland, Oregon, Pages: 147 - 157, 2000.

29. Zhang, Jian and S.C. Cheung, Automated test case generation for the stress testing of multimedia systems. ACM Software Practice & Experience, Volume 32, Issue 15, December 2002.

30. Dustin, Elfriede, Jeff Rashka and John Paul, Automated Software Testing: introduction, management and performance. Addison Wesley, 1999.

31. Kuhn, Richard et. al., Cost effective use of formal methods in verification and validation. Technical report, National Institute of Standards and Technology and NASA Langley Research Center, 2000.

32. Atif M. Memon, Martha E. Pollack, and Mary Lou Soffa, Hierarchical GUI Test Case Generation Using Automated Planning, IEEE transactions on software engineering, vol. 27, no. 2, February 2001.

33. Hung, Edward, Testing of Database Applications, Technical Report, Dec 2001.
http://www.cs.umd.edu/~ehung/research_brief.htm

34. Antoy, Sergio and Dick Hamlet, Automatically checking an implementation against its formal specification. IEEE Transactions on Software engineering, Vol. 26, No.1, January 2000.

35. Bernot, Gilles et al, Software testing based on formal specifications: a theory and a tool. Software engineering Journal, November 1991.

36. Chan, Man and Shing Cheung, Applying White box Testing to Database Applications. Technical report, The Hong Kong University of Science and Technlogy, January 1999.

37. Guida, Giovanni and Giancarlo Mauri, Evaluating performance and quality of knowledge-based systems: Foundation and Methodology. IEEE Transactions on Knowledge and Data Engineering, Vol. 5, No.2, April 1993.

38. Nota, Giancarlo and Giuliano Pacini Querying of Executable software specifications. IEEE Transactions on Software Engineering Vol 18 No 8 Au-

gust 1992.

39. Stocks, Phil and David Carrington, A framework for specification-based testing. IEEE Transactions on Software Engineering Vol 22 No 11 November 1996.

40. Tsai, W., et. al., Automated Test Case Generation for Programs specified by Relational Algebra Queries. IEEE Transactions on Software Engineering, Vol 16, No.3, March 1990.

41. Rational Software: Rational Robot.
http://www-306.ibm.com/software/awdtools/
tester/robot/index.html

42. Rational Software: Rational Functional Tester.
http://www-306.ibm.com/software/awdtools/tester/
functional/index.html

43. Rational Software: Rational Performance Tester.
http://www-306.ibm.com/software/awdtools/tester/
performance/index.html

44. Rational Software: Rational Purify.
http://www-306.ibm.com/software/awdtools/purify/unix/

45. Mercury Software Products: www.merc-int.com.

46. Mercury Software Products: Mercury WinRunner.
http://www.mercury.com/us/products/quality-center/
functional-testing/winrunner

47. Mercury Software Products: Mercury QuickTest.
http://www.mercury.com/us/products/quality-center/
functional-testing/quicktest-professional

48. Mercury Software Products: Mercury LoadRunner.
http://www.mercury.com/us/products/
performance-center/loadrunner

49. Segue, SilkCentral Test Manager, www.segue.com.

50. Segue, Silk Test.
http://www.segue.com/products/
functional-regressional-testing/silktest.asp

51. Segue, Silk Performer.

http://www.segue.com/products/
load-stress-performance-testing/silkperformer.asp

52. Compuware: QA-Run: www.compuware.com.

53. Kit, Edward, Software Testing in the real world: improving the process. ACM Press and Addison-Wesley, 1995.

54. Chays, David, et al., Testing Database Transactions with AGENDA, Proceedings of the International Conference on Software Engineering ICSE, St. Louis Missouri, May 15-21 2005.

55. Chays, David, et al., AGENDA: A test Generator for Relational Database Applications - A Technical Report. Polytechnic University TR-CIS-2002-04.

56. Chays, David, et al., An Agenda to test relational database applications. Journal of software testing, verification and reliability, March 2004.

57. Chays, David, Test Data Generation for Relational Database Applications. Ph.D. thesis, Polytechnic University, January 2004.

58. M. Y. Chan and S.C. Cheung. Testing database applications with SQL semantics. Proceedings of the 2nd International Symposium on Cooperative Database Systems for Advanced Applications, pages 363-374, March 1999.

59. Davies, R., R. Beynon and B. Jones, Automating the testing of databases. Proceedings of the first international workshop on automated program analysis, testing and verification, June 2000.

60. Lyons, N.R., An automatic data generating system for database simulation and testing. Database, 8(4):10-13, 1977.

61. Zhang, Jian, C. Xu and S.C. Cheung, Automatic generation of database instances for white-box testing. Proceedings of the 25th Annual International Computer Software and Applications Conference, October 2001.

62. Deng, Y, P. Frankl and J. Wang, Testing web database applications. Workshop on Testing, Analysis and Verification of Web Services, 2004.

63. Kapfhammer, G and M. Soffa, A family of test adequacy criteria for database-driven applications. ESEC/FSE, September 2003.

64. Cabal, M and J. Tuya, Using an SQL coverage measurement for testing database applications. FSE, October 2004.

65. Butler, Michael, Introductory notes on Specifications with Z, Technical Report, University of Southampton, 2001.

66. Robertson, Suzanne, An early start to testing: How to test requirements, Euro Star, Amsterdam 1996.

67. El Masry, Ramez and Shamkant Navathe, Fundamentals of Database Systems. Addison-Wesley Publishing Company, 2nd Ed., 1994.

68. Daou, B., R. Haraty and B. Mansour, Regression testing of database applications. Proceedings of the 16th ACM SAC 2001 symposium on applied computing, 2001, pages 285-289.

69. Gotlieb, A., B. Botella, M. Rueher, Automatic test data generation using constraint solving techniques. Proceedings of the 1998 International symposium on software testing and analysis. Pages 53-62. ACM Press, March 1998.

70. Gupta, N., A Mathur and M. Soffa, Automatic test data generation using an interactive relaxation method. Proceedings of Foundations of Software Engineering, ACM Press, November 1998.

71. Gray, J., et. al. Quickly generating billion record synthetic databases. SIGMOD Record, 23(2):243-252, June 1994.

72. Lyons, N., An automatic data generating system for database simulation and testing. Database, 8(4):10-13, 1977.

73. Date, C., H. Darwen, A Guide to the SQL Standard. Addison-Wesley, New York, 1977.

74. Earp, Richard and Sikha Bagui, Learning SQL: A step by step guide using oracle. Addison-Wesley 2002.

75. Urman, Scott, Oracle 8i: Advanced PL/SQL Programming, Osborne / McGraw-Hill, 2000.

76. Urman, Scott, Oracle 8i: PL/SQL Programming, Osborne / McGraw-Hill, 1997.

77. Berndt, D. and A Watkins, High Volume Software Testing using Genetic Algorithms, Proceedings of the 38th Annual Hawaii International Conference on System Sciences (HICSS'05) - Track 9 p. 318b, 2005.

78. Kaner, C., J. Bach, and B. Pettichord, Lessons Learned in Software Testing, John Wiley & Sons, 2002.

79. Kaner, C., W. Bond, and P. McGee, High Volume Test Automation, Keynote Address, International Conference on Software Testing Analysis and Review (STAR), Orlando, Florida, May 20, 2004.

80. Pargas, R, M. Harold and R. Peck, Test Data Generation Using Genetic Algorithms, Software Testing, Verification And Reliability, 9: 263-282, 1999.

81. Watkins, A, The Automatic Generation of Software Test Data using Genetic Algorithms, Proceedings of the Fourth Software Quality Conference, 2: 300-309, Dundee, Scotland, July, 1995.

82. Borgelt, K, Software Test Data Generation from a Genetic Algorithm, Industrial Applications of Genetic Algorithms, CRC Press 1998.

83. Lin, J-C. and P-U. Yeh, Automatic Test Data Generation for Path Testing using GAs, Information Sciences, 131: 47-64, 2001.

84. Michael, C., G. McGraw and M. Schatz, Generating Software Test Data by Evolution, IEEE Transactions On Software Engineering, 27(12), December 2001.

85. Watkins, A, Test Data Generation Using Genetic Algorithms, Proceeding of Software Engineering and Applications SEA (397) - 2003.

86. Beizer, B. Software Testing Techniques. Van Nostrand Reinhold, 2nd edition, 1990.

87. Chang, K, W. Carlisle, J. Cross , and D. Brown. A heuristic approach for test case generation. Proceedings of the 1991 ACM Computer Science Conference, pages 174-180. ACM, 1991.

88. Deason, W., D. Brown, K. Chang, and J. H. Cross. A rule-based software test data generator. IEEE Transactions on Knowledge and Data Engineering, 3(1):108-117, March 1991.

89. DeMillo, R. A. and A. J. Offutt. Constraint-based automatic test data generation. IEEE Transactions on Software Engineering, 17(9):900-910, September 1991.

90. Ferguson, R.and B. Korel. The chaining approach for software test data generation. IEEE Transactions on Software Engineering, 5(1):63-86, January 1996.

91. Ferguson, R. and B. Korel. Generating test data for distributed software using the chaining approach. Information and Software Technology,

38(1):343-353, January 1996.

92. Korel, B. Automated software test data generation. IEEE Transactions on Software Engineering, 16(8):870-879, August 1990.

93. Korel, B. and A. M. Al-Yami. Assertion-oriented automated test data generation. In Proceedings of the 18th International Conference on Software Engineering (ICSE), pages 71-80. IEEE, 1996.

94. Michael, C. and G. McGraw. Automated software test data generation for complex programs. In 13th IEEE International Conference on Automated Software Engineering, pages 136-146, October 1998.

95. Ramamoorthy, C, S. F. Ho, and W. T. Chen. On the automated generation of program test data. IEEE Transactions on Software Engineering, SE-2(4):293-300, December 1976.

96. Corning, Michael, Building Executable Software Test Specifications, Software Testing Analysis and Review, STAREAST 2003.

97. Hayes, I. and C. Jones. Specifications are not (necessarily) executable. High-Integrity System Specification and Design [12], pages 563-581, 1990.

98. Fuchs, N. E. Specifications are (preferably) executable. Software Engineering Journal, 7(5):323-334, September 1992.

99. Abdallah, A., et. al., A Provably Correct Functional Programming Approach to the Prototyping of Formal Z Specifications. Proceedings of the ACS/IEEE International Conference on Computer Systems and Applications (AICCSA'03), Tunis, Tunisia, 14-18 July 2003.

100. Abdallah, A., Filter promotion transformation strategies for deriving efficient programs from Z specification. Third IEEE International Conference on Formal Engineering Methods ICFEM, pages 157-167, 2000.

101. Yang, Sarah, Google, Microsoft and Sun Fund New UC Berkeley Internet Research Center. Press Release, University of California, Berkeley, Dec 2005.

102. Voas, Jeffrey and Gary McGraw, Software Fault Injection: Inoculating Programs Against Errors, John Wiley and Sons, inc, 1998.

Index